THE ECONOMIC GEOGRAPHY READER

THE ECONOMIC GEOGRAPHY READER

Producing and Consuming Global Capitalism

Edited by

John Bryson, Nick Henry, David Keeble and Ron Martin

John Wiley & Sons, Ltd

Chichester • New York • Weinheim • Brisbane • Singapore • Toronto

Published 1999 by John Wiley & Sons Ltd,
Baffins Lane, Chichester,
West Sussex PO19 1UD, England

National 01243 779777
International (+44) 1243 779777
e-mail (for orders and customer service enquiries): cs-books@wiley.co.uk
Visit our Home Page on http://www.wiley.co.uk
or http://www.wiley.com

Other Wiley Editorial Offices

John Wiley & Sons, Inc., 605 Third Avenue,
New York, NY 10158-0012, USA

WILEY-VCH Verlag GmbH, Pappelallee 3,
D-69469 Weinheim, Germany

Jacaranda Wiley Ltd, 33 Park Road, Milton,
Queensland 4064, Australia

John Wiley & Sons (Asia) Pte Ltd, 2 Clementi Loop #02-01,
Jin Xing Distripark, Singapore 129809

John Wiley & Sons (Canada) Ltd, 22 Worcester Road,
Rexdale, Ontario M9W 1L1, Canada

British Library Cataloguing in Publication Data

A catalogue record for this book is available from the British Library

ISBN 0-471-98527-9 (hardback)
ISBN 0-471-98528-5 (paperback)

Typeset in 10/12pt Times by Mayhew Typesetting, Rhayader, Powys
Printed and bound in Great Britain by Bookcraft (Bath) Ltd, Midsomer Norton
This book is printed on acid-free paper responsibly manufactured from sustainable forestry, in which at least two trees are planted for each one used for paper production.

CONTENTS

ACKNOWLEDGEMENTS

Producing a reader may appear at first sight to be a comparatively simple task. All one has to do is identify the papers, write a short introduction and send the manuscript to the publishers. As we have found, it is not nearly as simple as this. Producing this Reader involved numerous meetings, constant emailing, consultation and contributions from over 60 individuals. We owe a particular debt of gratitude to the authors and publishers who have granted permission for us to edit and use the selections in this Reader. Specific acknowledgements are made in a separate section at the back of this book.

We are particularly indebted to those authors who have made it possible to incorporate revisions, corrections and additions: to Doreen Massey for allowing us to publish, for the first time, one of the Occasional Papers from the Open University's South East Programme and to Alison Stenning and Mike Bradshaw for rising to the challenge of writing, at short notice, a special article for this collection on the transition economies. We found it difficult to identify a published paper that took into consideration the most recent developments in the Post-Socialist world.

Our initial conception for this Reader consisted of 12 papers with a commentary. We are deeply indebted to Tristan Palmer, formerly of Wiley, and the anonymous reviewers for challenging us to be more ambitious in our choice of papers and by encouraging us to produce a Reader containing over 50 edited papers placed within a structured framework. The final shape of this Reader is as much Tristan's as it is ours. Tristan left Wiley after we had signed the contract for this book and his role has been ably undertaken by Louise Portsmouth. Louise solved innumerable problems and supported the editorial team by her enthusiasm, patience and encouragement for the project. The front cover was 'designed' by the editors, but the realization of our 'vision' was made possible by the enthusiasm of Jim Wilkie, the cover designer.

When we set out to write this anthology we surveyed existing second-year courses in economic geography, not only in our respective departments in Birmingham and Cambridge, but also elsewhere. We are grateful to Peter Wood, Doug Watts, Jane Wills and Colin Mason for supplying us with copies of their second-year reading lists. The choice of papers was influenced by this exercise as well as by the valuable suggestions made by the eight reviewers of the original book proposal.

At Birmingham, we should like to thank Ann Ankcorn and Kevin Burkhill for redrawing the diagrams and maps and Peter Brealey for scanning some of the papers. Editing published papers would have been impossible without access to a good quality scanner. We invaded the office of Sue Lane and Lynne Pearce and appropriated their scanner for over a week. Thank you for putting up with us and for supplying coffee and the occasional cake. At Cambridge, Jane Robinson provided invaluable assistance in helping to transfer our editing to the scanned versions of the papers.

This project grew out of years of teaching and we would be seriously remiss if we did not thank our former and current students. It is student readers who will be the ultimate judges of how successful we have been in our choice and editing of the readings.

No academic or group of academics are 'islands unto themselves'. Our 'reading' of economic geography comes from our research activity as well as discussions and debates with numerous colleagues. We should like to thank John Allen for encouraging us to be more explicit about our 'take' on economic geography, and Mike Bradshaw, Peter Daniels, Cheryl McEwan, Jane Pollard, Terry Slater and Alison Stenning for their encouragement, support and suggestions. The Reader's inadequacies are of course our own.

PART ONE

Introduction

Section 1.1 Constructing the Reader

Selecting the readings for this volume has not been easy. One possible model would have been to focus on just a limited number of 'landmark' contributions, and to provide a detailed interpretative commentary. The problem is that the economic geography literature is now so vast that the choice of which readings to include and which to exclude is by no means straightforward. A small set of papers, reproduced with interpretative notes, may work in the case of a specific theme or subfield, but could not possibly capture the full diversity and development of the subject. This task is better accomplished, we believe, by bringing together a large number of closely edited contributions which cover the bulk of the field, and letting those writings 'speak for themselves'. This is the model followed here.

Our purpose in this Reader, then, has been to bring together, in edited form, those papers and other writings that, in our joint view, provide a representative map of the current vitality and breadth of economic geography. In particular, our aim was to produce a comparative international Reader which focuses on the evolving economic geography of Europe, North America and the Pacific Rim, within the framework of the global economy (or, at the very least, the processes of economic globalization currently underway). No doubt our choice can be criticized on various grounds, both for what we have included and what we have excluded. Certainly, we are aware that there are many different types of economy, and that this Reader represents a 'Western' reading of economic processes and their maps of transformation. To some extent, however, the readings included here were self-selecting, in the sense that they seemed to all of us to contain readily accessible – and highly readable – discussions of the key issues, concepts and arguments that have played a formative role in the development of the subject over the past decade or so. Numerous important statements and authors have inevitably been omitted, but we hope we have paid due acknowledgement to most of these both in this part of the Reader and in the guides to further reading.

Whilst we do consider that the papers 'speak for themselves' (sometimes in numerous ways), the Introductions to each Part provide a guide to the organization of the material. The Introductions are used to contextualize the papers and in so doing reveal the rationale ('the approach') to paper selection, their insertion into sections and their ordering within sections. In essence, then, the Introductions illustrate our particular readings of the papers. Furthermore, the aim is that when the Introductions are combined with Part One, including the introductory chapter on the field of economic geography, our particular 'approach' to economic geography may be seen.

Most especially, whilst the empirical aim of our approach was outlined above, we have been keen to produce a Reader which provides a foundation in many of the key ideas influential in contemporary economic geography. The Introductions and Readings combined try to capture economic geography's evolution in content, theory and methodology and provide, ultimately, an overview and illustration of recent developments which still draws on the sub-discipline's rich, if young, traditions.

Similarly, in delivering this overview, we have attempted to be sensitive to our audience. Whilst the ambition of the Reader grew over the course of its construction, it has always been aimed, first and foremost, at second-year/third-year (UK) students reading economic geography. Whilst we hope the book will reach beyond this market, it is these students' requirements that have been uppermost. Further selected readings have been provided as well as a comprehensive set of references. Papers have been carefully selected for 'readability'; we have attempted to remove many of the signs of editing and when papers included in the Reader are referred to elsewhere they have been highlighted in bold type. Whilst the Reader is not a 'written-through' textbook, we do view it as a companion volume for a lecture course, if not a course text. The aim has been to produce high quality resource material presented and interpreted for student use in a teacher- and student-friendly way. Thus, the Reader does have a clear organizational structure of five parts, which includes a chapter to orientate students within the sub-discipline of economic geography itself (Part One) and the 'signposts' of the Introductions and section headings. Indeed, it is these five Parts which are reflected in the design of the front cover.

Considering the sub-title of the Reader, 'Producing and Consuming Global Capitalism', the use of a globe speaks for itself, although its subdivision as a series of cogs is meant to highlight the 'interdependence' of spaces and places in the global economy today. The five pictures inserted in the cogs are each 'replicated' from the 'front' pages of each Part of the book. The series of 'twisted' globes moving on a trajectory into the distance are used to depict the changing hegemonic 'world views' (paradigm shifts) that have occurred through the intellectual history of economic geography, and which are the subject of the Reading Economic Geography section of Part One. The top cog and picture returns to the theme of 'global' interdependence as the motif for Part Two: The Economy in Transition: Globalization and Beyond? Part Three: Spaces of Production is signified by the picture of 'Carhenge' (Alliance, Nebraska) which forms part of a 10 acre sculpture park devoted to cars. This has become a tourist attraction and highlights the blurring of the manufacturing/services divide which is a key organizational theme of Part Three. The picture of the child 'devouring' a burger is a striking, if obvious, representation of Part Four: Spaces of Consumption. Finally, Part Five: Work, Employment and Society, which takes as its theme the wider economic and social implications of economic change, is represented by the picture of the 'vagrant' or 'down-and-out'.

The front cover, like the rest of this Reader, evolved from numerous joint ideas and discussions between the editors (and with others; see Acknowledgements). Similarly, whilst individuals took 'charge' of particular parts and pieces of the book, the choice of papers and words throughout the book was a collective effort. In regard to this, we have decided not to attribute individual elements of the Reader and would rather it be recognized in its entirety as a joint effort. We believe it is all the better for it, and highlights the construction of knowledge as always a joint effort (whether or not this is recognized).

Section 1.2 Reading the Reader

How then should these readings be read? The readings have been grouped under four core themes: the economy in transition: globalization and beyond; new spaces of production; new spaces of consumption; and work, employment and society. In part this division is for pedagogic reasons, to assist the reader in charting a course through the diversity of contemporary economic geography. The theoretical and empirical efforts of economic geographers are not, of course, neatly compartmentalized or hermetically isolated one from another: research – let alone reality – is not like that. However, the grouping we have used does capture what have been the dominant spheres of substantive and theoretical work in the discipline: the key 'research programmes' of economic geography. How these programmes emerge and evolve is not just determined by external events, although as we argue in Reading Economic Geography, these do play an influential role in this respect. The division of academic labour within economic geography, as in any other subject, is also a sociological process, in which researchers necessarily seek to divide the field into intellectually manageable and meaningful areas of knowledge and expertise. In this way not only do specialisms and foci emerge, each dealing with a specific set of issues or perspectives, but each of the respective communities of scholars also tends to develop its own set of concepts and discourses, which in turn serve to further distinguish the subfield from others. But, as the papers collected here clearly demonstrate, there are numerous recurring themes and issues that intersect the subject as a whole, and which link developments in the different subfields. This is not to argue, however, that the ultimate goal is or should be the construction of some sort of overarching, integrating synthesis of economic geography research. Rather, it is to acknowledge that our 'carving of the disciplinary joints' (our approach) is always somewhat arbitrary, and thus that intersections and linkages between major research themes need to be made and examined.

This issue arises in the case of the groupings defined here. Thus, the fact that the first group of readings deals with globalization and the transition of the global economy does not imply that these developments should be taken to be separate from, or logically or structurally prior to, issues of production, consumption and work. Each of these spheres of economic activity is partially dependent on the others. None completely controls the others; yet none is fully independent of the others either. Understanding globalization and the changing global economy is impossible without understanding the shifting landscapes of production, consumption and work. But equally, the latter can only be understood in an increasingly global and globalized context. Thus while the book can be read – and (given the structure of most undergraduate courses on economic geography), for heuristic reasons is likely to be read – in a linear way, the story the readings depict is in fact a much more complex and multidirectional one: the plot, so to speak, moves constantly back and forth. Indeed, we would argue that you could start from any of the Parts and work 'outwards'.

Not only do the readings collected here map out the changing economic geographies of contemporary capitalism, they also demonstrate the pluralism of method and perspective that now characterizes the discipline. As we argue in Section 1.3, Reading Economic Geography, modern economic geography is multiperspectival, open to a broad range of theories, methods, approaches and discourses on the domains of spatial economic reality and how they are constituted and interact. Any given perspective or theoretical vantage point is, of course, selective, and unavoidably mediated by one's own pregiven 'world view', assumptions, values and interests. This relativistic and hermeneutic character of modern economic geography is clearly evident in the readings that follow. To stress this feature of the discipline is not to imply that we have entered a phase in which 'anything goes', nor is it to sanction a postmodernist multiplicity of fragmented microperspectives and interpretations. While economic geographers are now rightly suspicious of the possibility and validity of 'grand theory', they still recognize the need for theories that endeavour to conceptualize, describe and interpret the structural and systemic forces at work in uneven capitalist development (such as technological change, tertiarization, informationization, cultural commodification, state regulation, globalization, and so on). Without such macrotheories, we lose sense of the systemic features and forces that shape the economic geographies of capitalism. The readings that follow do not, therefore, abandon the macro for the micro, or the general for the particular, but, in their different ways, endeavour to integrate and articulate the two.

All this should not be read as suggesting that the development of economic geography over the past decade or so has progressed evenly on all fronts. While theoretical and empirical advance has been impressive, much less so has been the subject's engagement with public policy. At a time when states throughout the capitalist world have been restructuring, reforming and abandoning policies; when, as a result of these policy shifts on the one hand, and the direction of economic, technical and global developments on the other, many countries have experienced a growth in socio-economic and spatial inequalities; and when the search for new modes of intervention and regulation – from the global scale down to the local – continues unresolved, economic geographers should be playing a much more prominent role in public policy debate and discourse. No doubt the excitement of theoretical and substantive work has meant that policy analysis has tended to be pushed towards the back of the stage. Contributing to the theoretical development of economic geography is certainly seen as more intellectually demanding and rewarding than the evaluation and discussion of policy. Thus, in their eagerness to theorize, economic geographers have tended to pay much less attention to the need to translate their theories and concepts into practical policy terms. But in addition, with the shift away from the radical political economy approaches of the 1980s has gone a corresponding decline in the sense of progressive politics that had permeated much economic geographical writing during that period. Perhaps this decline has itself been compounded by the 'policy vacuum' that has opened up between the outmoded Keynesian–welfarism of the post-war period and the discredited neoliberalism that replaced it in the 1980s and early-1990s (typified by the current Anglo-American search for a 'Third Way'). And to this still further might be added the widespread view that the power of nation-states to intervene in their economies is rapidly waning anyway.

Yet, the scope for an economic geography of policy has expanded rather than waned. If, as many economic geographers argue, regional and local economies are assuming increasing salience in the new global capitalism, then it is also at this scale that new opportunities for public policy experimentation and debate are likely to expand. Economic geographers thus have a potentially important role to play in shaping the direction of this experimentation and debate. Responding to the policy issue should not be interpreted as somehow less exciting or less important than theoretical work. To the contrary, good policy research is demanding, requiring not only firm theoretical and empirical foundations, but an ability to use policy analysis to feed back into the revision of theory. Moreover, as social scientists, economic geographers have an obligation to use their understanding of the world to try to help improve it. This means they should interrogate existing policies, and seek to influence the policy-making process. In this context, policy studies provide an opportunity to integrate the multifarious strands of theory, method and empirical enquiry that make up contemporary economic geography. Hopefully, over the coming years, the development of a greater political commitment and a greater involvement in policy analysis by economic geographers will go hand in hand with the continued expansion of the subject that we are certain will take place. As you read on, we hope you remember the preceding paragraph, for you are part of that continued expansion of the subject both as students and more widely within your daily lives.

Section 1.3 Reading Economic Geography

THE EVOLVING PROJECT OF ECONOMIC GEOGRAPHY

It has long been recognized that academic disciplines do not evolve in a steady, incremental and cumulative manner, or indeed according to any mechanistic pattern. To be sure, there is a high degree of intellectual path dependence, of intellectual sunk capital as it were, that imparts a strong continuity of subject-matter and conceptual approach. But the historical evolution of a discipline also involves periodic shifts and changes, bursts of new activity, and perhaps even changes in trajectory, as new empirical events, theoretical movements, methodologies and indeed new generations of academics, promote the emergence of what philosophers of science have variously labelled as a new paradigm, a new research programme, or new episteme. In economics, the 'Keynesian revolution' of the 1930s and 1940s is often cited as an example of how disciplines evolve; in geography, the 'quantitative revolution' of the 1950s and 1960s is frequently viewed in similar terms. At such junctures, new avenues of research are opened up, new substantive issues are explored and new theoretical frameworks are developed. A general sense of intellectual excitement and challenge pervades the subject and its practitioners. Since the late 1980s, economic geography has been in the throes of just one such wave of renewal and expansion.

Although economic geography has existed as an identifiable sub-discipline for over a century, and has occupied a central position with human geography as a whole throughout this time, its modern development only really dates from the 1950s. Its development over the past half century has not only been dramatic, but has also involved a complex series of twists and turns, not just of substantive content but also of theoretical orientation and discursive style (Barnes, 1996). Up until the 1950s, economic geography, like human geography more generally, had been largely idiographic, with a dual focus on regional synthesis – the construction of integrated descriptive accounts of specific regions and areas – and, by extension, regional differentiation – explaining why and in what ways regions differ from one another in terms of, for example, their economic structures, demographic patterns and social characteristics. During the 1950s, the discovery and translation of the earlier location-theoretic models developed by German spatial economists (von Thünen, 1826; Weber, 1929, Christaller, 1933; Lösch, 1939), in combination with a growing desire to make geography a spatial science, provided the basis of a 'new economic geography' in which the focus was firmly on the search for, and formal analysis of, systematic locational patterns and dynamics in the economic landscape. For more than two decades, up until the mid-1970s, this new, positivistic economic geography, based on a fusion of quantitative procedures, neoclassical economic principles, and the *esprit géometrique* bequeathed by the earlier German location theorists, provided the dominant paradigm for the study of the spatial organization of the modern economy (see, for example, Smith, 1971; Amedeo and Golledge, 1975, Haggett *et al*, 1977). During this period, economic geography was, to all intents and purposes, *industrial* geography, overwhelmingly concerned with the industrial aspects of regional economic development, the trajectories of which were assumed to be well-behaved and predictable. Interest in the geographies of services, labour and money, for example, was minimal. Nevertheless, a feature of this period in economic geography's development was that at least much of the work in the subject was empirically orientated. This was in stark contrast to regional science, the other field that evolved out of the German spatial economics tradition, in which the focus was firmly on the mathematical modelling of abstract, hypothetical economic landscapes.

By the early-1970s, location-theory-based economic geography had already passed its zenith. For one thing, diminishing returns had set in: dissatisfaction with the limited empirical reach and narrowly positivistic underpinnings of locational analysis had stimulated the search for alternative perspectives. Thus Massey (1976), for example, had begun to argue for a refigured industrial geography in which a central role was assigned for history and the social relations of production. At the same time, the cessation of post-war economic growth – the end of the so-called 'golden age of capitalism' (Webber and Rigby, 1996) – and the spread during the 1970s of world-wide stagnation and systemic regional and urban problems, presented economic geographers with a new set of issues and questions which existing methods and models seemed ill-equipped to address. Stimulated by a growing Marxist critique of capitalism within the social sciences more generally, economic geographers turned from neoclassical economics towards political economy, and especially Marxian political economy, for theoretical inspiration. The most influential example of this abrupt change in theoretical direction was David Harvey, who in the space of just four years refocused his theoretical lens away from neoclassical economic geography to a Marxist-based geographical political economy (compare Harvey, 1969, with Harvey, 1973). This 'Marxist turn' shifted the research agenda in economic geography away from a preoccupation with the statistical laws of spatial distribution and industrial location to the historical and material processes of uneven urban and regional development inherent in the 'laws of motion' and 'crisis tendencies' of capitalism.

Although other methods and approaches continued to have their followers (including behavioural models and studies in the location-theoretic tradition), and although criticism was levied from certain quarters of the subject against the structural and economic determinism of orthodox Marxism, it was Marxian economic geography that dominated the research frontier from the early-1970s to the mid-1980s. Geographers working in this transformed analytical framework produced an impressive collection of studies over the 1970s and early-1980s, dealing with a wide range of regional and urban problems and processes, including the logic and dynamics of urban space, regional development and crisis, the geographies of the labour process, regional industrial restructuring, and internationally uneven development. Several path-breaking economic geography books in the Marxist mould appeared during this period, including David Harvey's *The Limits of Capital* (1982), Doreen Massey's *Spatial Divisons of Labour* (1984), and Neil Smith's *Uneven Development* (1984). While these works were primarily theoretical, they stimulated a new genre of economic–geographical research and writing which was distinctively more critical and politically–orientated, even radically prescriptive. Not only was the inescapable political content of geographical research given explicit recognition, the political critique of capitalism was accepted as a valid goal of the research effort (see Massey and Meegan, 1985).

Yet, despite the stimulus given by the Marxist 'turn', by the mid-1980s economic geography no longer seemed quite so charged by the excitement and progress it had displayed in the late-1970s and early-1980s. In part this loss of momentum reflected a growing disillusionment throughout the social sciences with the conceptual and empirical relevance of Marxist social and economic theory. As the 1980s progressed, so the critique of

Marxist theory intensified, reinforced both by dramatic new developments in the nature and trajectory of capitalism, and by the emergence of new, post-structualist and postmodern philosophical and epistemological movements which threw into question the validity – or certainly the hegemony – of deep-structure theories and metanarratives of the Marxist kind (with its appeal to the ineluctable laws of capitalist development). Added to this, despite the instabilities that continued to afflict the capitalist world during the 1980s (and which persisted into the 1990s), it was in the socialist bloc that an historic process of economic and political collapse was unleashed, throwing substantive doubt on the credibility of Marxist social theory as a result (see Fukuyama, 1989). In combination, the advent of new realities, the rise of new epistemological debates, and the problems faced by the ex-socialist economies undergoing the transition to capitalism, have played a key role in promoting another wave of conceptual, methodological and empirical renewal of economic geography. By the early-1990s, not only had economic geography recovered its dynamism, but the theoretical and empirical scope of the subject had expanded dramatically. Economic geography has witnessed a renaissance.

A NEW PHASE OF CAPITALISM AND NEW ECONOMIC LANDSCAPES

The emergence of what Drucker (1989) calls 'new realities' has played a leading role in this renaissance. One of capitalism's constant features is change. Capitalism cannot stand still: its central imperative – the search for profit – drives a perpetual process of economic flux (Storper and Walker, 1989), what Joseph Schumpeter graphically described as a process of 'creative destruction' which 'incessantly revolutionises the economic structure from within, incessantly destroying the old one, incessantly creating a new one' (Schumpeter, 1943). However, as Schumpeter himself also recognized, economic change is not simply smooth and incremental: neither the pace nor the direction of change remain constant over time. Standing on the pinnacle of the present and looking back, we know that, historically, economic evolution has also been periodically disrupted by phases of particularly rapid and extensive upheaval and reorganization – what Schumpeter called 'gales' of creative destruction. During these periodic gales, economic development undergoes a sort of climacteric or 'bodily change', a transition from one form to another which, for a while at least, provides the basis for renewed growth. There have been two such historic transformations over the past hundred years. The first occurred during the 1920s to 1940s, when the economic structures, technologies, state–economy relations, financial arrangements and patterns of international trade and competition that had characterized nineteenth-century capitalism were finally superseded by new structures and systems of economic accumulation and regulation. As we pass from the twentieth into the twenty-first century, this mode of economic development in its turn is being reconfigured, as world capitalism once again undergoes a major transformation.

This reconfiguration has provided economic geographers with a rich vein for theoretical and empirical enquiry. New realities provide a major stimulus to new ideas, new ways of thinking. For the one thing that all social theories have in common is that they all, in time, become obsolete as historical events unfold along lines that pre-existing theories could not possibly have anticipated. Then comes the process of theoretical exploration, and the search for new concepts and theoretical frameworks more appropriate to the changed realities. This is precisely what has taken place in economic geography and other social sciences over the past two decades. Of course, making sense of major change whilst the process is underway, and distinguishing what might eventually prove to be ephemeral movements from the more fundamental transformations, is far from straightforward. It is not surprising, therefore, that there have been numerous interpretations of the current transition. While there is a widespread consensus that we are moving into a new phase or wave of capitalist development, there is much less agreement about what, precisely, this new form is.

Within geography, the conceptual architecture of French regulation theory has proved highly influential, with its thesis that capitalism is in transition from its twentieth-century Fordist regime of accumulation and regulation to a new, post-Fordist regime. Advocates of this interpretative framework highlight the distinctively large-scale, organized nature of the Fordist phase of capitalist development, in which the emphasis was on mass production of standardized consumer goods, supported by mass consumer demand, in turn promoted by mass advertising and state policies aimed at maintaining national aggregate demand. Allied to these features, Fordism encouraged collectivistic social relations and structures (for example, unions) and was supported by various forms of mass or universal welfare (i.e. a social wage) which not only ensured social harmony but also helped to maintain consumer spending. Of course, different nations developed their own socio-cultural-institutional variants of this economic model (Lipietz, 1987), but all shared certain common elements. The more so because these national Fordisms were linked together and stabilized by the Bretton Woods system of international financial regulation. Fordism generated specific economic geographies: such as regions specializing in mass production, a consumption landscape of surburbanization, and a public space economy of social infrastructure and social services. In most industrialized countries, Fordism was associated with a general process of convergence of regional incomes and employment.

Over the past two decades or so this regime of economic accumulation, social regulation and regional development has been in retreat in the face of a shift towards a new regime, labelled post-Fordism (for a survey of this interpretation, see Amin, 1994). Fuelled by new technologies, the search for new forms of economic organization to revive the stagnation in profits and productivity growth that had afflicted Fordism, the rise of new centres of competition in the global economy (especially the newly industrializing nations of South East Asia), the breakdown of the Bretton Woods system, and by the shift in state regulation away from extensive welfare and intervention programmes to neo-liberal policies of deregulation, privatization and monetary austerity, an intense upheaval and reorganization of the socio-economy has taken place. Production has become more customized and flexible, services have expanded, consumption has become more individualized and differentiated, the public sector has been slimmed down and welfare systems made increasingly selective, and social collectivism has given way to individualism. Again these changes have proceeded at different rates and taken different specific forms in different countries, and there has been debate over the precise contours and extent of this new post-Fordist configuration. Nevertheless, most economic geographers talk of the emergence of a new post-Fordist landscape, in which new industrial spaces of small-firm-based specialized flexible production and areas of high-technology development have risen to the fore (Scott, 1988; Markusen, 1996). Old Fordist regions are faced with a painful restructuring, of having to undergo a Schumpeterian process of localized 'creative destruction', if they are to compete in the new post-Fordist era.

At the same time, the mass consumption culture of the post-war period has exploded into a new culture of consumption that is simultaneously more individualized, internationalized and multidimensional. The introduction of the 'instant credit economy', the increasing assertiveness and acquisitiveness of a new middle class, changes in tastes and lifestyles, the revolution in the media and advertising, the increasing differentiation of products, and the rise of a whole new 'culture industry' based on the commodification of the visual, aesthetic and symbolic, all of these have stimulated new patterns and landscapes of consumption (especially the phenomenon of malls and other out-of-town shopping centres) in which instant gratification, positionality and image are now as important as use value. These developments have stimulated not only the growth of a new geography of retailing and consumption (Wrigley and Lowe, 1996), but an awareness of the centrality of the cultural in the formation and reformation of the economic landscape (Zukin, 1991; Lash and Urry, 1994; Thrift and Olds, 1996).

Arguably, new information technologies have been central to this post-Fordist reshaping of the economy and its geographies. According to many authors, we have truly entered the information or post-industrial age. Although there are different theories of post-industrialization, the unifying idea is that we are at last leaving behind the era of industrial capitalism and moving to a form of economic organization and development in which service activities, knowledge and information are the key elements (see Webster, 1995, for an excellent review of the literature). Some take this interpretation further to argue that we are witnessing the emergence of a new 'network society' in which information networks of firms, social groups, individuals and institutions are constantly being formed and reformed at a variety of interlinked spatial scales, from the local to the global. These new 'spaces of flows' are creating geographies that transcend our conventional notion of geography as a 'space of places' (Castells, 1989, 1996). In a not unrelated vein, other authors see the rapidly growing importance of information, services and consumption as marking a break with the phase of 'organized' industrial capitalism, and a shift to a 'disorganized' consumer capitalism characterized by intensely rapid circulation, not only of material goods and commodities but also of 'dematerialized' commodities such as information, ideas, knowledge, symbols and signs. In this new configuration, space – geography – is simultaneously stretched and compressed, itself dematerialized (Lash and Urry, 1987, 1994).

Another recurring theme in the debates and discussions of the new capitalism is that one of its prominent imperatives is the 'flexibilization' of economic structures, processes and institutions (Harvey, 1989; Knudsen, 1996). Thus firms are using new technologies, new sourcing and supply strategies, and new forms of working practices, to flexibilize their production, cut costs and increase their ability to respond to the pressures and instabilities of intense global competition. For their part, states have also been swept up in the pursuit of flexibility. Since the early-1980s, state after state has embarked on a programme of deregulation and privatization in an effort to 'free up the market' and give the maximum room for manoeuvre for capital. Once unleashed (particularly by the rise of neo-liberalism in the US and UK in the early-1980s), this process spread across the globe as nations raced to ensure their economies remained competitive in the new deregulated marketplace. In its wake, flexible accumulation has increased the uncertainty of economic life. The search for flexibility, and the hypermobility of capital it engenders, has exposed workers, firms and localities to considerable risk and insecurity (Beck, 1992; Elliott and Atkinson, 1998). While some local labour markets have prospered, where high technology and related employment growth has been concentrated, others have experienced increasing casualization of their employment structures, as part-time work and forms of contingent working have grown as full-time employment has stagnated or even declined. In the new global context, the flexibility of capital means that firms readily disinvest in one locality to seek out cost, market and knowledge advantages in another. One result is that the local geographies of unemployment have become much more differentiated. Another is that localities are increasingly pitched one against another in a desperate attempt to hold onto or attract mobile firms and jobs. The 'new flexibility' is thus a two-edged sword: it may bring gains to particular firms, groups of workers and localities, but it subjects other firms, workers and localities to increasing uncertainty, insecurity and vulnerability. Indeed, the question arises as to whether the rapid change that permeates the contemporary economy is not simply the turmoil that might be expected to occur in the midst of transition from one pattern and form of economic development to another, but rather is itself the defining hallmark of the new (dis)order.

Meanwhile, at the global scale the geographies of economic development and underdevelopment are changing in fundamental ways. Until the past few years, economic geographers were accustomed to thinking about the world economy in terms of three major groups of countries: the First World (the advanced capitalist nations, or 'West'), the Third World (of developing and underdeveloped nations, referred to as the 'South'), and the less talked about Second World (the 'Socialist World' of state-communist economies). This division, always problematic, has become outmoded. The past two decades have seen the dramatic emergence of the newly industrialized economies of South East Asia, where growth rates have far outstripped those of the advanced countries. These countries have become major global players shaping the geographies of trade, money and capital flows, and now exert a significant influence on the path of global economic development (Thurow, 1992, 1997). Equally profound has been the collapse of many of the former socialist countries, and their difficult transition to capitalism. In both cases, these new spaces of capitalism have proved problematic. In the late-1990s, the phenomenal growth of Japan and South East Asia suddenly turned into financial and economic crisis, sending shock waves across the global system. Equally, the transition to capitalism in Russia has failed to take off, sending the country spiralling into debt and internal economic and social turmoil. Given the globalization of economic relations, involving the increasing integration and interdependence of economic flows and relationships across nations, crises in one area of the world now quickly spill over into other areas far removed from the original sources of instability. Yet, these instabilities notwithstanding, there can be no doubt that we are entering a new age in which the trajectory of the global economy will be much more centred on the development of Asian capitalism (especially China and India). Not only are these various shifts in the global spread of capitalism fostering all sorts of regional-bloc trading and integration agreements (Anderson and Blackhurst, 1993) across the developed and developing world, their impacts at the local levels are likely to be profound.

Economic geographers have thus been confronted by an expanding kaleidoscope of 'new economic geographies'. Mapping these new geographies would itself be an important task. But what has made the role of economic geography all the more significant is that the contemporary shifts and transformations of capitalism appear not just to be geographically uneven, they also seem to involve the re-emergence and reassertion of regional and local economies. As national boundaries come under greater strain in the face of globalization, widespread deregulation and new information technologies, so the salience of local and regional economies as locii of economic accumulation and social regulation has increased. As never before, we seem to be moving towards a global system comprised of networks of local and regional economies (Amin and Thrift, 1992; Ohmae, 1995; Storper, 1997; Scott, 1998). Understanding these new economic landscapes, and their shifting and interweaving of scales, is one of the key research challenges in economic geography.

THE RETHINKING OF THE ECONOMIC IN ECONOMIC GEOGRAPHY

This research agenda is all the more challenging because, hand-in-hand with the 'transformations' of the new economic realities, there is an on-going 'transformation' in the way economic geographers are viewing the economy. Recent times have seen one of those periodic shifts and changes in the historical evolution of the field of economic geography mentioned earlier; what has been termed in rather crude shorthand as 'the cultural turn'. More than anything, within economic geography, this 'turn' has imbued a rethinking of just what is meant by 'the economic' and 'the economy' (Massey, 1988; but see Sayer, 1994; 1997; Thrift and Olds, 1996; Crang, 1997). As Crang (1997, p. 3) puts it:

> Content is being rethought in terms of what social and spatial portions of life count as economic, what portions (if any) are therefore non-economic, and how these designated spheres of the economic and non-economic interrelate.

Moreover, this thinking is two-fold. Part of this rethinking is because of the new, or previously neglected, realities of the transforming economy. For example, how does the concept of the 'firm' help us to understand the new business service organizations made up of ever-shifting groups of self-employed individuals (Bryson, 1997) or help us to grasp the realities of multinational contract catering companies which employ tens of thousands of people around the world yet whose 'assets' are negligible, including hardly a kitchen to their name (Allen and Henry, 1998)? The other part of this rethinking is due to an historical challenge to the ontological and epistemological base of economic geography, and the social sciences more generally. Not only are our concepts changing and developing in the light of our empirical spotlights but the actual process of *how we conceptualize* is also under scrutiny and undergoing transformation. To take a classic example from feminist (economic) geography: Why is it that certain forms of 'work' (principally paid, and especially full-time paid) come to be classified as part of the economy whilst others (unpaid, domestic, voluntary, for example) do not? (For a short but comprehensive review of this debate see WGSG, 1997; especially pp. 56–65 and 116–26). More generally, the challenge of 'postmodernism' and the postmodern critique is fundamental in questioning *how* we theorize, *how* we represent the world and *how* we attempt to explain it (Martin, 1994; Graham, 1997).

Taking the first element of this rethinking, as Crang (1997) argues, there has been no single cultural turn, and no single understanding of what culture and the cultural might be, which has been drawn on by the latest version of the 'new economic geography', but it is clear that 'the cultural' and 'the economic' and the relationship between the two has moved to centre stage in economic geography's research agendas. This engagement can be viewed from at least three contrasting positions; namely, continued non-engagement; the 'economization' of the cultural; or the 'culturalization' of the economic.

Continued non-engagement stems from a belief that the economic and the cultural have been kept analytically distinct for good reason; hence, for example, the sub-disciplinary split between economic geography and cultural geography. Particularly associated with the hard won gains of political–economic analysis, an engagement with culture is viewed as holding inherent dangers of muddying the analytical waters. An emphasis on beliefs, (life)styles, ideas, languages, representations, imaginations and the like is a distraction from the continued need to analyse the (often grinding) material realities of the world: work, poverty, famine, decent housing, environmental degradation and so forth.

An alternative position (on what is best viewed as a continuum) focuses more on the relationship between the two realms of economy and culture but tends to do so in a uni-directional manner which ascribes greater importance, or determination, to the economic. As Thrift (1989) has written, in its extreme characterization this position views culture as 'a subordinate outgrowth of economic and social change, with no independent life of its own, and so requiring no independent analysis' (p. 13). Whilst that view may be extreme, if, as Marx suggested, 'production is also immediately consumption' (Marx, 1973, p. 90), then 'lifestyle' consumption, for example, is about the deliberate and continued (re)fashioning of individual consumer tastes by producers in order to increase the turnover time of invested capital and the rate of profit. Similarly, in this light, postmodern architecture is the latest refashioning of urban capital in the relentless cycle of capitalism. The question is then, as Harvey (1989) asks, 'what kinds of social practice, what sets of social relations, are being reflected in different aesthetic movements?' (p. 114). Harvey provides the answer of capitalist social relations, 'precisely because capitalism is expansionary and imperialistic, cultural life in more and more areas gets brought within the grasp of the cash nexus and the logic of capital circulation' (p. 344). Harvey highlights the argument for the 'economization of culture' and sets the scene for his reading of the new economic realities of the cultural turn. One reality, for example, is the growth and expansion of the *business* of 'the production of culture'; the growth of what have been termed by some as the 'cultural industries' (e.g. advertising, museums, entertainment, theme parks, multimedia, etc.) in the contemporary economy. 'Doing culture' is now very big business as the global corporations of Sony, Disney and Viacom Inc. (owners of MTV) will tell you.

Yet, interestingly, the empirical fact of the cultural industries can be read in a different way. Above, it is used to highlight the continued economization of culture. For others, however, it is one manifestation of the third position on the relationship between economy and culture; it highlights not the economization of culture but, rather, the culturalization of the economy. One can highlight this third position by drawing on the work of Lash and Urry (1994, p. 15):

> What is increasingly being produced are not material objects, but signs. These signs are primarily of two types. Either they have a primarily cognitive element and thus are post-industrial or informational goods. Or they primarily have an aesthetic, in the broadest sense of aesthetic, content and they are primarily postmodern goods (Eagleton, 1989). This is occurring, not just in the proliferation of non-material objects which comprise a substantial aesthetic component (such as pop music, cinema, magazines, video, etc.), but also in the increasing component of sign value or image in *material* objects.

What is implied here is not only that 'goods' are increasingly cultural, whether the 'leisure experience' of the theme park or the 'brand name' of trainers, but the production of these goods involves the production process itself becoming more cultural in a number of ways. For example, the heightened 'cultural' aspect of the production process may vary from the increased 'design intensity' of software production to recognizing that the continued agglomeration of the City of London is based on a production process involving traders and dealers deciding whether or not an economic indicator (sign) means 'buy' or 'sell' by *understanding* what all the other 'experts' around them think of this sign (Thrift, 1994). Or cultural production might mean, literally, as in the retail shop or in the restaurant, understanding how geographies of display ('have a nice day' and all that) become critical as part of the production of a good ('the shopping experience' or 'the meal'). And yet another example is the largest industry in the world, tourism, where the production process involves making the tourists themselves do the imaginative work; to make them 'imagine the moment' whilst staring at (nothing but) a sign marking where President Kennedy was shot or whilst kissing a rock historically named the Blarney Stone (Urry, 1990). So, the driving forces of the economy are increasingly cultural, and we are seeing the 'culturalization' of the economy.

In fact, as the different interpretations of 'the growth of the cultural industries' reveal, it may be less a case of predetermining how 'the economic' and 'the cultural' combine and more a case of asking how they are *mutually constitutive* of one another in different ways over time and space, to produce particular characteristic forms of economy (e.g. feudal economy, socialist economy, East Asian capitalism, etc.) and particular objects which make up economies (for example, firms, in their varying forms from multinationals and their 'corporate cultures' to the 'entrepreneurial' small firm; or labour, whether it be 'skilled', 'flexible', 'informal' or 'green'). Indeed, it is the question of how 'the economic' and 'the cultural' combine, but at the level of the process of the *production of theory*, which has shaken the historical roots of economic geography. The most fundamental aspect of 'the cultural turn' has been its encapsulation as shorthand for the challenge of postmodernism. This is the second part of the rethinking of the economy involving, as it does, thinking about how theory has been, and is, produced in economic geography (whether about longstanding objects in economic geography such as profit, labour and work, or rather newer ideas such as the cultural industries). To quote Daly (1991, p. 93):

> we have moved from the metaphysical realm to the metaphorical: that is to say, we have moved from the idea that economic 'truth' is discovered and towards the idea that it is made.

What Daly is alluding to here is the 'postmodern' argument that theories 'make reality'. That is, all theories are 'metaphors' or 'stories' (if massively thought about ones) about the world; in economic geography they are 'readings' of the economic landscape and as such in all these readings certain aspects of reality are selected while others are blocked out. This inevitably leads to the question of who makes these stories and why some stories are deemed more important than others and so to a recognition of *cultures of knowledge production*. Thus the earlier example of the feminist critique of the definition of work which has become dominant in economic geography, and society more widely; namely, a definition centred on paid, full-time work in the public realm the majority of which is done by men, as against, for example, the unpaid work of the home, the majority of which is done by women. For feminists, given that men have historically dominated the factories of knowledge production, such as the academy and government, it is of little surprise that work has come to be defined in masculinistic terms.

Yet this is not to say that myriad other sites of knowledge production, other stories and readings of the world do not exist; rather, it is to say that their ability to be heard has been constrained: muffled, for example, by the 'objective' metanarratives of neoclassicism, Marxism and the like. For economic geography, as across the social sciences, the message of this element of the cultural turn is to let these readings, their voices, be heard. Moreover, this means not only that there is a need 'to open up the floor' but to recognize the *validity* of these multiple voices: for they may be no less 'selective' of reality in their theories than the rather grander theories that have come to dominate the discipline. However, recognizing the challenge of postmodernism – its argument for the multiperspectival, multidimensional and multivocal – is not to sanction a postmodern multiplicity of fragmented micro perspectives and interpretations. It is not a licence for 'anything goes'. Whilst the postmodern turn encourages multiple voices and accounts, it also compels us to examine the positionality and authority of knowledge claims, whatever the voice. This is no less true of the 'selective reading' (or approach) of this Reader. Not only does the book contain particular readings of the economic landscape, powerful stories penned by some of the most influential 'knowledge producers' in the discipline, and further reinforced by the backing of a major international publisher, the very selection we have made reflects our own particular 'take' on both the substantive and literary landscape as editors. The positionality and authority claims of our reading – and the papers included here – should *not* be accepted uncritically.

The postmodern challenge as to how we theorize and represent the world is still young in terms of histories of knowledge production and responses to it. Nevertheless, within economic geography, in one arena in particular, *methodology*, a response can be discerned. For whilst academic economic geographers are still very much the 'organizers' of stories, readings and world views, there has been a clear attempt to expand the range of methods applied in order to better identify the array of stories of the economy 'out there'. The cultural turn has introduced economic geography to a new set of qualitative methodologies, for example textual analysis, iconography, semiotics, ethnography, participant observation and action research, to name but a few. The first flushes of such methodologies often tended to be couched in terms of rejecting the quantitative methods associated with the previous paradigm of positivism, and have equally been received with a certain amount of suspicion within the discipline (representative as they are of different world views and different ways of producing world views). A more considered response, however, highlights the impressive range of methodologies open to economic geographers, from multiple regression to postal questionnaire to participant observation.

Thus, our reading of economic geography today is framed in a very different way to our historical counterparts, highlighting a plurality of theories, a plurality of methods and, ultimately, a plurality of economic geographies. The readings that follow give some flavour of that plurality but, as we have suggested, they can but only be partial.

REFERENCES

Allen J and Henry N (1998) Ulrich Beck's risk society at work: Labour and employment in the contract service industries, *Transactions of the Institute of British Geographers* **22**: 180–96.

Amedeo D and Golledge R G (1975) *An Introduction to Scientific Reasoning in Geography*, New York: Wiley.

Amin A (ed) (1994) *Post-Fordism: A Reader*, Oxford: Blackwell.

Amin A and Thrift N (1992) Neo-Marshallian nodes in global networks, *International Journal of Urban and Regional Research* **16**: 571–87.

Anderson K and Blackhurst R (1993) *Regional Integration and the Global Trading System*, London: Harvester Wheatsheaf.

Barnes T (1996) *Logics of Dislocations: Models, Metaphors, and Meanings of Economic Space*, New York: Guilford.

Beck U (1992) *Risk Society*, London: Sage.

Bryson J R (1997) Business service firms, service space and the management of change, *Entrepreneurship and Regional Development* **9**: 93–111.

Castells M (1989) *The Informational City: Information Technology, Economic Restructuring, and the Urban-Regional Process*, Oxford: Blackwell.

Castells M (1996) *The Rise of the Network Society*, Oxford: Blackwell.

Christaller W (1966 [1933]) *Central Places in Southern Germany*, trans. by C W Baskin, Englewood Cliffs: Prentice-Hall.

Crang P (1997) Cultural turns and the (re)constitution of economic geography, in R Lee and J Wills (eds) *Geographies of Economies*, London: Arnold, pp. 3–15.

Daly G (1991) The discursive construction of economic space, *Economy and Society* **20**: 79–102.

Drucker P (1989) *The New Realities: In Government and Politics, in Economy and Business, in Society and in World View*, Oxford: Heinemann.

Eagleton T (1989) *The Ideology of the Aesthetic*, Oxford: Blackwell.

Elliott L and Atkinson D (1998) *The Age of Insecurity*, London: Verso.

Fukuyama F (1989) The end of history, *The National Interest*, Summer: 3–18.

Graham E (1997) Philosophies underlying human geography research, in R Flowerdew and D Martin (eds) *Methods in Human Geography*, Harlow: Longman, pp. 6–30.
Haggett P, Cliff A D and Frey A (1977) *Locational Analysis in Human Geography*, London: Edward Arnold.
Harvey D (1969) *Explanation in Geography*, London: Edward Arnold.
Harvey D (1973) *Social Justice and the City*, London: Edward Arnold.
Harvey D (1982) *The Limits of Capital*, Oxford: Blackwell.
Harvey D (1989) *The Condition of Postmodernity: An Enquiry into the Origins of Cultural Change*, Oxford: Blackwell.
Knudsen, D C (ed) (1996) *The Transition to Flexibility*, Dordrecht: Kluwer.
Lash S and Urry J (1987) *The End of Organized Capitalism*, Cambridge: Polity.
Lash S and Urry J (1994) *Economies of Signs and Space: After Organised Capitalism*, London: Sage.
Lipietz A (1987) *Mirages and Miracles: the Crises of Global Fordism*, London: Verso.
Lösch A (1939) *The Economics of Location*, New Haven: Yale University Press.
Markusen A (1996) Sticky places in slippery slopes: A typology of industrial districts, *Economic Geography* **72**, 3: 293–313.
Martin R (1994) Economic theory and human geography, in D Gregory, R Martin and G E Smith (eds) *Human Geography: Society, Space and Social Science*, London: Macmillan.
Marx K (1973) *Grundrisse*, Harmondsworth: Penguin.
Massey D (1976) *Industrial Location Theory Reconsidered*, Milton Keynes: Open University.
Massey D (1984) *Spatial Divisions of Labour*, London: Macmillan.
Massey D and Meegan R (eds) (1985) *Politics and Method: Contrasting Studies in Industrial Geography*, London: Methuen.
Massey D (1988) Uneven development: Social change and spatial divisions of labour, in D Massey and J Allen (eds) *Uneven Redevelopment: Cities and Regions in Transition*, London: Hodder & Stoughton, pp. 250–76.
Ohmae K (1995) *The End of the Nation State: The Rise of Regional Economies*, London: HarperCollins.
Sayer A (1994) Cultural studies and 'the economy, stupid', *Environment and Planning D: Society and Space* **7**: 253–76.
Sayer A (1997) The dialectic of culture and economy, in R Lee and J Wills (eds) *Geographies of Economies*, London: Edward Arnold, pp. 16–26.
Schumpeter J A (1943) *Capitalism, Socialism and Democracy*, London: Allen & Unwin.
Scott, A J (1988) *New Industrial Spaces*, London: Pion.
Scott A J (1998) *Regions and the World Economy: The Coming Shape of Global Production, Competition and Political Order*, Oxford: Oxford University Press.
Scott A J (2000) Economic geography: The great half century, in G L Clark, M Gertler and A Feldman (eds) *Handbook of Economic Geography*, Oxford: Oxford University Press.
Smith D M (1971) *Industrial Location: An Economic Geographical Analysis*, New York: Wiley.
Smith N (1984) *Uneven Development*, Oxford: Blackwell.
Storper M (1997) *The Regional World: Territorial Development in a Global Economy*, New York: Guilford.
Storper M and Walker R (1989) *The Capitalist Imperative: Territory, Technology and Industrial Growth*, Oxford: Blackwell.
Thrift N (1989) Images of social change, in C Hamnett, L McDowell and P Sarre (eds) *The Changing Social Structure*, London: Sage, pp. 12–42.
Thrift N (1994) On the social and cultural determinants of international financial centres: The case of the City of London, in S Corbridge, R Martin and N Thrift (eds) *Money, Space and Power*, Oxford: Blackwell, pp. 327–55.
Thrift N and Olds K (1996) Refiguring the economic in economic geography, *Progress in Human Geography*, **20**: 311–17.
Thurow L (1992) *Head to Head: The Coming Economic Battle Among Japan, Europe and America*, London: Nicolas Brealey.
Thurow L (1997) *The Future of Capitalism*, London: Nicolas Brealey.
Urry, J (1990) *The Tourist Gaze: Leisure and Travel in Contemporary Societies*, London: Sage.
von Thünen J (1966 [1826]) *The Isolated State*, P Hall (ed), London: Pergamon.
Weber A (1929) *Theory of the Location of Industries*, Chicago: University of Chicago Press.
Webber M and Rigby D (1996) *The Golden Age Illusion: Rethinking Post-War Capitalism*, New York: Guilford.

Webster F (1995) *Theories of the Information Society*, London: Routledge.
Women and Geography Study Group (1997) *Feminist Geographies: Explorations in Diversity and Difference*, Harlow: Longman.
Wrigley N and Lowe M (1996) *Retailing, Consumption and Capital: Towards the New Retail Geography*, Harlow: Longman.
Zukin S (1991) *Landscapes of Power: From Detroit to Disney World*, Berkeley: University of California Press.

PART TWO

The Economy in Transition: Globalization and Beyond?

PART TWO THE ECONOMY IN TRANSITION: GLOBALIZATION AND BEYOND?

Section 2.1 Introduction

THE GLOBALIZATION DEBATE AND THE TRANSFORMATION OF CAPITALISM

In the space of barely a decade, the concept of 'globalization' has become part of the standard vocabulary of the social sciences, including economic geography. A widespread view has emerged that we are witnessing an accelerating and deepening process of globalization which marks the advent of a new and qualitatively different phase of capitalist development. But, despite the common currency the notion of 'globalization' already enjoys, it has also attracted considerable debate. This is no doubt due, in part at least, to the fact that the term has gratecrashed the academic literature without paying the entrance fee of a clear and agreed definition.

Different authors use the concepts of 'globalization' and the 'global economy' in different ways, with varying degrees of precision. Our interest here is in 'globalization' as an economic process first and foremost. Even in this sense, the term has attracted multiple meanings, with different implications for how we should think about the economic process, economic relations and the economic landscape. Much of this diversity of interpretation revolves round five key questions. First, precisely what is meant by 'globalization', and what stage in the process are we at? Second, how does 'globalization' relate to the other fundamental transformations and changes that are currently reshaping the nature of capitalist development? Third, what does globalization imply for our notions of geography, of space, location and place? Fourth, does the globalization of capitalism mean the emergence of a universal model of capitalist development thoughout the world, or, rather, does it allow the emergence of novel variants and transmutations? And, fifth, does globalization spell the end of nation-states and national regulation and governance as these have been conventionally construed? These are the intersecting issues that weave through the readings in this part of the book.

What, then, is meant by such terms as economic 'globalization' and 'global capitalism'? How do we distinguish the rhetoric and realities that surround these notions (Barry Jones, 1995)? For some, 'globalization' means the worldwide spread of modern technologies of industrial production and communication of all kinds – trade, capital, production, information and money – regardless of frontiers: in effect the advent of a 'borderless world' (Ohmae, 1990, 1996). For others, globalization implies that nearly all economies are networked with other economies throughout the world (see Comor, 1994; Castells, 1996). In another sense, some use globalization as a shorthand for the linking and intermingling of cultural forms and practices that follow when societies become integrated into and dependent on world markets (Featherstone, 1990; Wallerstein, 1991). Still others equate globalization with the convergence and homogenization of capitalist economic forms, markets and relations across nations (Kerr, 1983; Fukuyama, 1992; for a review see Boyer and Drache, 1996). These various definitions have not gone uncontested, however, and have been criticized for exaggerating and misinterpreting the nature and extent of the process.

One approach to gauging the meaning and extent of globalization is to define what an ideal-typical 'globalized economy' would look like, and then to use this as a yardstick against which to compare reality. This is the strategy deployed by **Hirst and Thompson**, two sceptics of the globalization thesis. Globalization, they argue, is not just conjunctural change towards greater trade and investment flows between an existing set of national economic spaces. Rather, globalization should be taken to mean the development of a truly supranational economic system. In a truly globalized system, distinct national economies would be subsumed and rearticulated into the system by transnational processes and transactions that are socially disembedded from specific national contexts. Given this conception, they argue that while there has been increasing interdependence within the international economy, as measured for example by flows of trade and capital, we are still some way off from being a *globalized* economy. Truly global companies are still in a minority, and most still make the bulk of their profits in their 'home markets'. Furthermore, economic integration appears to be increasingly concentrated in a 'triad' of three supranational blocs (Europe, North America and South East Asia) rather than globally (on the 'triadization' of the world economy, see Ohmae, 1985; Gibb and Michalak, 1994). While this approach is to be welcomed for its attempt to impose conceptual and empirical rigour on the debate, Hirst and Thompson nevertheless still depend heavily on material flows (of trade and capital) to measure the degree of global integration (for a similar approach, see Gordon, 1988). Furthermore, an ideal-typical characterization, of the sort they deploy, can all too easily be taken as if it were the eventual end-state form to which the world system is (or should be) moving. But, as Held *et al* (1997, 1999) argue, globalization is not a singular condition, a linear process with a similar end condition for every country. Nor is it some state of complete world integration to which all economies and societies are equally converging (Gray, 1998). Rather, globalization, like previous phases of capitalism, is almost certain to generate geographically uneven development (Yeung, 1998).

Yet, one of the popular myths about globalization perpetrated by economists is that by its very nature it signals the 'end of geography' (O'Brien, 1992) and the 'death of distance' (Cairncross, 1997). In the new world of instantaneous and borderless information, knowledge flows and 'e-commerce' (trading through the Internet), so the argument goes, space and location are rendered irrelevant and no longer important. Companies can locate any screen-based activity anywhere, and serve their customers wherever they may be. Now, it is certainly true that one of the key features of globalization is what can be called 'de-localization': the uprooting of economic and social relationships from local origins and cultures. The increasing 'time–space compression' (Harvey, 1989) or 'time–space shrinkage' (Allen and Hamnett, 1995) that is at the heart of globalization means the displacement of activities that were until recently local into networks of relationships whose reach is distant or worldwide. As Giddens (1990, p. 64) puts it: 'Globalization can . . . be defined as the intensification of worldwide social relations which link distant realities in such a way that local happenings are shaped by events occurring many miles away, and vice versa.'

As a consequence, as **Amin** argues, places are becoming 'hybridized' socio-economic composites through the interactive intermingling in individual places of local and 'non-local' modes of consumption, and social, cultural and informational practices, such that the distinction between 'local' and 'external' is becoming difficult to define. However, this hybridization of the 'local' through 'globalization' does not mean that places are losing their distinctiveness and becoming similar: globalization does not mean the homogenization of social and economic relations across space. The free movement of capital, production, information, money and cultures across frontiers occurs precisely *because* of differences between localities, regions and nations. Globalization simultaneously reinforces the pluralism *of* places and pluralism *within* places. Amin's argument is that globalization is thus not something 'out there', some exogenous process that can simply be measured by the intensity of flows of material goods, capital investment, money or even technology between places, but is a complex, two-way, socio-economic relationship between the local and the global, a process that some geographers have labelled 'glocalization' (Swyngedouw, 1997). The explication of these 'glocal' economic geographies is central to our understanding of the globalization process.

What is also clear is that globalization is inextricably bound up with a host of other major shifts and changes that are central to the contemporary transformation of capitalist development. As we have already noted in Part One, economic geographers have been particularly concerned to map and conceptualize this transformation. The second group of readings in Part Two provide different vantage points on the nature and dimensions of this new landscape. For **Harvey** and many other theorists, the key defining feature of the new capitalism is 'flexible accumulation' (see also Knudsen, 1996). According to this perspective, the pursuit of flexibility in production, in labour utilization, in consumption and in state–economy relations, can be interpreted as capitalism's search for a new technological, social and spatial 'fix' to the Fordist overaccumulation crisis of the 1970s and early-1980s. Historically, capitalism has sought out such fixes before in response to previous overaccumulation crises. But what makes the current 'flexibility fix' distinctive is its unprecedently global dimensions. The globalization of capitalism increases the local instabilities of the accumulation process as well as transmitting those instabilities far beyond their areas of origin. This in turn fuels the search for greater flexibility and circulation of capital, which in turn fuels the globalization process.

The flexibilization of capitalist economic and social relations is thus associated with a vastly faster circulation of goods, objects and cultural artefacts (see **Harvey**'s paper in Part Four). Indeed, what are increasingly produced and circulated at increasingly greater velocities are not just material goods but informational and aesthetic goods, which by their nature are rapidly consumed, disposed of and depleted of meaning. In this 'weightless' informational global economic system (see Coyle, 1998), social and economic spaces are redefined, fragmented and pluralized by a new consumerism of cultural images, symbols and signs. According to **Lash and Urry** the 'informationalization' of the economy and the aestheticization of consumption are promoting the individualization and detraditionalization of society, whilst simultaneously disembedding socio-economic relations from their local contexts. In both senses, they suggest, the new flexible, globalized capitalism is a 'disorganized' capitalism, a decentred set of economies of signs in space. However, as Harvey argues, the use of the term 'disorganized' to describe this new socio-economic system is perhaps somewhat unfortunate, since it tends to overemphasize the disintegrative effects of globalization and flexibilization. As he stresses, not only are flexibilization and globalization highly coherent capital accumulation strategies, they fragment the economic landscape through the very process of integrating it.

THE GEOGRAPHIES OF GLOBAL INTEGRATION

Understanding the geographies of global integration is thus a key task. Without question, the revolution in new information technologies has itself been a primary – perhaps *the* primary – globalizing force. This revolution is a very recent phenomenon. All three of today's key communications technologies have existed for several decades: the telephone was invented in 1876, the first television in transmission was in 1926, and the electronic computer was invented in the mid-1940s. For much of that time change has been slow, but in each case a revolution has occurred since the mid-1980s. The advent of fibre-optic cables, of digital technologies, the Internet, the mobile phone and a host of related computer, communications and transmission technologies has begun radically to transform the structure and spatiality of economic and social interactions. Not until the middle of the next century will the full effects of the new communications revolution become clear, but already the contours of its impact are beginning to be apparent.

As **Warf** argues in his paper, above all the new information technologies permit the generation and expansion of ever-wider networks of rapid (indeed, near-instantaneous) information flow and exchange. As the density, capacity and speed of transmission of these networks has rapidly expanded, and the unit cost of information transmission has fallen, so the volume of information flow has escalated. A vast increase in the diffusion of knowledge and information, the basic building blocks of economic development and growth, is taking place. The result is that markets operate faster, become more integrated and extend wider geographically. But the effects of the new communications are not neutral in their geographical impacts. Thus, US companies dominate the software technologies and distribution networks along which the new information economy moves. These companies control the networks, and access to them. Moreover, as measured by the number of access nodes in each country, it is evident that the greatest access is concentrated in the most economically developed parts of the world, notably North America, Europe and Japan. By comparison, most of Africa, the Middle East and Asia (with the exceptions of India, Thailand and Malaysia) have little or no access. The information-based process of global integation seems, therefore, to be reproducing, if not accentuating, the established patterns of uneven world development, and the economic leadership of a select group of world cities.

Nowhere have the new information technologies been more influential than in the realm of financial markets and money flows. In combination with the deregulatory policies of the major national economies, the telecommunications revolution has finally set money afloat from national banking systems and financial centres. Money has gone electronic and global, even 'virtual' (Solomon, 1997). As money has gone global and become stateless, so banks and other financial institutions have also gone global. Some writers, such as O'Brien (1992), regard money as having overcome all spatial and locational constraints, as we move towards integrated, seamless global money markets. Yet, as **Martin** stresses, even in the realm of money, global integration is not annihilating the significance of location and place. Rather, it is giving new salience to financial centres, whether global, national or provincial. Moreover, in an era of rapid information flow, money is more able than ever to take advantage of even small undulations in the global financial regulatory landscape, as the offshore financial centres, or OFCs (of which there are more than 90), testify. In fact, globalization is creating a whole new set of geographies of money (see Corbridge *et al*, 1994; Laulajainen, 1998; Cohen, 1998; Martin, 1999). To be sure, as money markets become fully screen-based and automated (like NASDAQ and EASDAQ), they are released from their territorial embeddedness in established financial centres. But whether they will become truly 'placeless' is questionable. Globalization has not meant the end of national financial spaces. What is clear, however, is that the process of global financial integration has created new geographies of financial instability, whether this be the restructuring of regional banking systems through waves of mergers and rationalizations (as in the US and Europe, for example); or, as in the case of the South East Asian crash in 1997–1998, the global diffusion of market uncertainty provoked by a localized financial crisis.

This new inter-penetration of the 'global', the 'national' and the 'local' means we need to consider explicitly how these different scales relate one to another. **Fagan and Le**

Heron suggest that one way of doing this is to return to the Marxist notion of circuits of capital. But rather than focusing on the familiar divisions of 'industrial', 'finance' or 'commercial' capital, their approach (building on the work of Bryan, 1987) is to think in terms of 'national' and 'global' fractions and their variants. Under this viewpoint, it is not 'nationality', ownership or sector that identify the fractions of capital, but the ways in which production, sales and investments are linked to the global capital accumulation process. Spatial reorganization and even *in situ* restructuring of national firms can thus become part of a wider 'global shift'.

Local interactions are not subordinate to the global accumulation process, nor is is the latter distinct from local processes of investment and disinvestment. It is the interlinking of the spatial scales of capital accumulation and disaccumulation that is one of the key features of the globalization process. The dichotomy between 'global' and 'domestic' capital can thus be abandoned without removing the national and local scales from their cucial levels of importance.

THE RISE OF NEW CAPITALISMS

No economic culture anywhere in the world can resist the changes forced on it by the existence of global markets, and globally mobile capital and information. But this does not mean the convergence of national capitalisms, even less the 'Americanization' – or 'Japanization' – of capitalist forms and cultures across the globe. The national diversity of capitalisms is not about to be obliterated by globalization (Berger and Dore, 1996). Rather, globalization will engender various transmutations and novel types. The case of the Asian newly industrializing countries (NICs) illustrates this. The spectacular success of the East Asian new industrial economies (the so-called 'tiger economies') – at least until the recent financial crisis hit the subcontinent – demonstrates that there are alternative forms of capitalism to the 'Anglo-American' form. In fact, there have long been different national variants even in the industrialised West: the US, the UK, Germany and Scandinavia, for example, all have different specific forms of economic institutions, regulation and governance; that is to say, different forms of socio-institutional embeddedness (Rogers *et al*, 1997), as was the case under Fordism (**Tickell and Peck**). Similarly, East Asian capitalisms are based on yet other forms of socio-cultural and institutional frameworks. This fact has often been overlooked in those accounts that attribute the meteoric rise of the Asian NICs to their supposedly liberal-market economies. **Brohman** shows how this interpretation is quite misleading. Not only have states in these countries played an instrumental development role in their economic success (for example by taking a strategic lead in industrial investment, technological innovation, or education), pre-exisiting systems of social, cultural and religious norms have also been of crucial importance. Likewise, **Daly** emphasizes the way in which the specific nature of corporate structures in Asian nations (such as the *kigyo keiretsu* and *sogo shosha* of Japan and the *chaebols* of Korea) have operated in conjunction with governments and distinctive financial stuctures to out-compete their Western rivals. He also highlights the key role Japan has played as a supplier of investment, finance and technology, and as an exemplar of Asian style capitalism, to the other, more recently industrialized nations in the region. However, as events in 1997 and 1998 revealed, East Asian capitalism is no less prone to crisis than is Western capitalism. As Japan and the other East Asian industrialized countries struggle to resolve their financial and economic crises, whether the distinctive socio-institutional form of Asian capitalism will remain intact, or whether it will be forced to change under the impress of wider global forces and imperatives (emanating, for example, from pressures exerted by the international financial community), is an intriguing question.

Meanwhile, a different form of capitalism is struggling to emerge in the post-communist states of eastern and central Europe (**Stenning and Bradshaw**). Whereas orthodox Marxist theory sought to conceptualize the 'transition from capitalism to socialism', recent history has instead posed a rather different problem, namely the 'transition from socialism to capitalism'. Our theorization of this form of transition is not, it has to be said, well advanced. Notwithstanding this lacuna, Western states and the world organizations they dominate, such as the IMF, seem to have assumed that the marketization, liberalization and privatization which the ex-communist states have embarked upon should follow the Western capitalist model. Yet, as in the case of East Asia, it is far from clear that such a model is appropriate. Far from reversing their previous economic stagnation and decline, the transition of the former Soviet countries to modern capitalism has to date proved distrastrous, as the gap between the raised expectations of their populations on the one hand, and the harsh realities of their collapsed economies on the other, has continued to widen. Moreover, as Stenning and Bradshaw stress, the pressures and paths of transition vary considerably from region to region in the post-communist states, reflecting among other things the legacies of communist industrial location programmes. Two lessons are evident from the problems of the transition economies of eastern and central Europe. First, these economies have sought to insert themselves far too rapidly into the new global economy, requiring a level of support and accommodation from the key players in that economy – the leading capitalist nations and the major international financial institutions – which they have been unwilling to give. Second, the former communist states of central and eastern European should not be expected – and certainly not compelled – to replicate the Western models of capitalism, but to evolve their own mutations that reflect the specific, inherited socio-institutional forms of those countries. The contrast between the failed 'rush to capitalism' by these nations and the highly successful piece-wise, selected-area-based approach adopted by China is striking.

But globalization does not simply permit the co-existence of different national capitalisms. By enabling different cultural groups who are geographically scattered to interact through communications media and the rapid circulation and transfer of cultural and other information, globalization also permits the development and co-existence of 'diasporic' business communities and practices within indigenous national business cultures. The creation of local socio-institutional networks of Japanese business people and their families in areas of Japanese inward investment is one example. The development of highly flexible, culturally-based entrepreneurial 'bamboo networks' of family businesses created by the overseas Chinese is another. For example, **Mitchell** describes the case of flexible capital circulation (including modes of raising finance capital) within the highly successful and influential Chinese business community in Vancouver, and the importance of the links that key business people in that community have with business leaders and financiers back in Hong Kong. This case study highlights the cultural stength and geographical reach of local–global contacts, and how they operate not merely to transform local places, but also simultaneously to rework the global systems of contacts and networks of which such 'diasporic capitalisms' are a part.

REGULATING THE NEW CAPITALISM

A recurring theme in the debates over globalization is that of regulation. In an era of border-less flows, hypermobile capital and information, and integrated and interdependent markets spanning the globe, is economic regulation – at all levels, the global, national and local – becoming increasingly difficult and ineffective? There are those who argue that globalization signals the 'death of the nation state' as both an arena and agent of economic regulation and governance (Ohmae, 1995). In a globalized world, the role of the state, it is argued, is at best limited to the provision of infrastructure and enhancing the skills of the nation's workforce (Reich, 1991). Some – the global neoclassicists – adopt a celebratory position on this issue, and welcome the erosion of the state's powers by globalization as both inevitable and desirable, as opening up a truly global market-place and liberating free market forces. Others adopt a much more defensive position, and argue for the need to resist the erosion of national economic autonomy and sovereignty. Still others argue that the new global economy requires a new global polity which cannot simply be state-centred but which will have to build upon and extend the range of international political–economic and legal institutions that have arisen over the past few decades, including not just the IMF, World Bank, United Nations, World Trade Organization and similar bodies, but also the increasing number of international non-governmental organizations (INGOs).

In his paper, **Dicken** examines one particularly prominent element in the argument that the nation-state is being undermined by globalization, namely that economic power has been usurped by transnational corporations (TNCs) and multinational enterprises (MNEs). Some of these are now so big that they command assets and resources greater than those of many small states. As a result of their power, TNCs are seen as undermining nation-states in at least two senses: by forcing nation-states to grant concessions under threat of the TNC locating elsewhere, and by not embedding themselves in the local economy, so that local spin-offs of technology, suppliers and other linkages are often very limited. However, as Dicken argues, although TNCs undoubtedly represent a serious challenge to national autonomy and hence national sovereignty over economic matters, and pose pressing regulatory problems, it is an exaggeration to suggest they dominate individual nations. In many cases, transnational companies remain dependent upon nation-states for a range of resources, including skilled labour, investment in infrastructural items such as telecommunications, marketing structures, trade policies, and a host of other factors. Thus, national economic policies and regulatory regimes play an important role in shaping the global locational dynamics and investment, marketing and innovation strategies of TNCs. Moreover, as Porter (1990) and Reich (1991) point out, TNCs typically retain a 'national identity', an allegiance to their home base and home market (see also **Hirst and Thompson**). While asymmetries in relative power between states and TNCs exist (and are typically less favourable for developing countries), the relationship between TNCs and nation-states is best seen in terms of what Dicken calls a 'dynamic bargaining' process (see also Holton, 1998), a process that is constantly shifting with changes in the global strategies of the TNCs, in the global economy, and in national and international regulatory environments. As Dicken remarks, there is a real need for geographers to investigate the shifting contours of this bargaining process.

The shifting nature of national regulatory regimes has, of course, become one of the key features of the contemporary transition of capitalism (and this links us back to where this Introduction began). The 1980s saw the rise and global spread of a new neo-liberal (that is neo-conservative) approach to socio-economic regulation by nation-states, led by the Thatcher governments in the UK and the Reagan administration in the US. This neo-liberal project, based on deregulation, privatization, promoting socio-economic flexibility and reigning back the state, itself added to the globalization process by removing national barriers to the flows of capital and finance, and by setting off a process of 'competitive deregulation' amongst countries. During the 1980s, the neo-liberal model had spread to numerous other advanced and developing countries alike, as well as becoming firmly entrenched in the policies of the IMF. Some observers seemed to interpret this neo-liberal state project as an attempt to construct a new mode of social regulation that was accommodative of the transition of global capitalism from Fordism to post-Fordism (for example, Jessop, 1989, 1994). But, according to **Tickell and Peck**, the neo-liberal response can equally be seen as part of an on-going and unfolding *after*-Fordism crisis rather than as a new regulatory logic, a new institutional fix, for that crisis. Indeed, they argue that the market ideology prosecuted by neo-liberalism represents the absence of a new fix, and as such has contributed to the ongoing crisis and instabilities of the *after*-Fordist world. Rather than provide the basis for a new global order, neo-liberalism has contributed to the widening of spatial inequalities, and has even justified those inequalities as necessary for the flexibility that the new global economy requires. For Tickell and Peck, the geographies of after-Fordism are geographies of crisis, not a new stable post-Fordism. In this context they are pessimistic about the likely success of the increased localized regulation that the 'hollowing-out' of the neo-liberal after-Fordist state is supposedly engendering.

CONCLUSION

If a common theme emerges from the readings in this part of the book, it is that globalization poses major challenges for economic geography, both of an analytical and normative kind. The analytical challenges centre on the problem of conceptualizing a complex and multi-dimensional process: globalization is not a single all-embracing force, driven by some ineluctable systemic dynamic of capitalism or Western economic and cultural imperialism. It is not overwhelming local communities and nation-states (Boyer and Drache, 1996). It is not something that is eroding local difference, but is simultaneously 'delocalizing' and 'relocalizing'. The normative challenges have to do with evaluative concerns about whether globalization is 'good' or 'bad' (**Amin**; Cox, 1997; Swyngedouw, 1997). The heavy tones of rhetoric that tend to permeate the use of the term globalization and related concepts serve to set the normative stance well before conceptual precision and empirical evidence are presented. The very labels 'new world order' and 'new world disorder', which are often used in debates about the global economy capture this problem. But as Holton (1998, p. 204) concludes, 'there is . . . no definitive or even provisional balance sheet to be drawn on where globalization is headed or whether it may regarded as good or bad'. Claims that globalization must be resisted and replaced by a 'return to the local' (see, for example, Mander and Goldsmith, 1996) are thus far too one-sided. Globalization also offers fresh opportunities for the reassertion and revitalization of local economic geographies.

REFERENCES

Allen J and Hamnett C (eds) (1995) *A Shrinking World?*, Oxford: Oxford University Press.
Barry Jones R J (1995) *Globalization and Interdependence in the International Political Economy*, London: Pinter.
Berger S and Dore R (eds) (1996) *National Diversity and Global Capitalism*, Ithaca: Cornell University Press.
Boyer R and Drache D (1996) *States against Markets: The Limits of Globalization*, London: Routledge.
Bryan R (1987) The state and the internationalisation of capital: An approach to analysis, *Journal of Contemporary Asia* **17**: 253–75.

Caincross F (1997) *The Death of Distance: How the Communications Revolution Will Change Our Lives*, London: Orion Business Books.
Castells M (1996) *The Rise of the Network Society*, Oxford: Blackwell.
Cohen B (1998) *The Geography of Money*, Ithaca: Cornell University Press.
Comor E A (ed) (1994) *The Global Political Economy of Communication: Hegemony, Telecommunications and the Information Economy*, London: Macmillan.
Corbridge S, Martin R L and Thrift N J (eds) (1994) *Money, Space and Power*, Oxford: Blackwell.
Cox K (ed) (1997) *Spaces of Globalization*, New York: Guilford.
Coyle D (1988) *The Weightless World: Strategies for Managing the Digital Economy*, Oxford: Capstone Publishing.
Featherstone M (1990) *Global Culture*, London: Sage.
Fukuyama F (1992) *The End of History and the Last Man*, New York: The Free Press.
Gibb R and Michalak W (eds) (1994) *Continental Trading Blocs: The Growth of Regionalism in the World Economy*, London: Wiley.
Giddens A (1990) *The Consequences of Modernity*, Cambridge: Polity.
Gordon D (1988) The global economy: New edifice or crumbling foundations? *New Left Review* **68**: 24–64.
Gray J (1998) *False Dawn: The Delusions of Global Capitalism*, London: Granta Books.
Harvey D (1989) *The Condition of Postmodernity*, Oxford: Blackwell.
Held D, Goldblatt D, McGrew A and Perraton J (1997) The globalization of economic activity, *New Political Economy* **2**: 257–77.
Held D, Goldblatt D, McGrew A and Perraton J (1999) *Global Flows, Global Transformations: Concepts, Theories and Evidence*, Cambridge: Polity Press.
Holton R J (1998) *Globalization and the Nation-State*, London: Macmillan.
Jessop B (1989) Conservative regimes and the transition to postFordism: The cases of Great Britain and West Germany, in M Gottdiener and N Komninos (eds) *Capitalist Development and Crisis Theory: Accumulation, Regulation and Spatial Restructuring*, London: Macmillan.
Jessop B (1994) The transition to postFordism and the neo-Schumpeterian workfare state, in R Burrows and B Loader (eds) *Towards a Post-Welfare State?* London: Routledge.
Kerr C (1983) *The Future of Industrial Societies: Convergence or Divergence?* Cambridge: Harvard University Press.
Knudsen D C (ed) (1996) *The Transition to Flexibility*, Dordrecht: Kluwer.
Laulajainen R (1998) *Financial Geography*, Göteborg: University of Göteborg Press.
Mander J and Goldsmith E (eds) (1996) *The Case Against the Global Economy and for a Turn Toward the Local*, San Francisco: Sierra Books.
Martin R L (ed) (1999) *Money and the Space Economy*, London: Wiley.
O'Brien R (1992) *Global Financial Integration: The End of Geography*, London: Royal Institute of International Affairs–Pinter Publishers.
Ohmae K (1985) *Triad Power*, New York: The Free Press.
Ohmae K (1990) *The Borderless World*, London: Fontana.
Ohmae K (1995) *The End of the Nation State*, New York: The Free Press.
Ohmae K (1996) *The End of the Nation State: The Rise of Regional Economies*, London: HarperCollins.
Porter M (1990) *The Competitive Advantage of Nations*, London: Macmillan.
Reich R (1991) *The Work of Nations: Preparing Ourselves for 21-st Century Capitalism*, London: Simon & Schuster.
Rogers Hollingsworth J and Boyer R (eds) (1997) *Contemporary Capitalism: The Embeddedness of Institutions*, Cambridge: Cambridge University Press.
Solomon E H (1997) *Virtual Money: Understanding the Power and Risks of Money's High Speed Journey into Electronic Space*, Oxford: Oxford University Press.
Swyngedouw E (1997) Neither global nor local: 'Glocalization' and the politics of scale, in K Cox (ed) (1997) *Spaces of Globalization*, London: Guilford.
Wallerstein I (1991) *Geopolitics and Geoculture: Essays on the Changing World System*, Cambridge: Cambridge University Press.
Yeung H W-c (1998) Capital, state and space: Contesting the borderless world, *Transactions of the Institute of British Geographers*, N.S., **23**: 291–310.

SELECTED FURTHER READING

Allen J and Hamnett C (eds) (1995) *A Shrinking World?* Oxford: Oxford University Press.

Cox K (ed) (1997) *Spaces of Globalization*, New York: Guilford Press.

Daniels P W and Lever W F (eds) (1996) *The Global Economy in Transition*, London: Longman.

Dicken P (1998) *Global Shift: The Transformation of the World Economy*, third edition, London: Paul Chapman.

Harvey D (1989) *The Condition of Postmodernity*, Oxford: Blackwell.

Held D, Goldblatt D, McGrew A and Perraton J (1999) *Global Flows, Global Transformations: Concepts, Theories and Evidence*, Cambridge: Polity Press.

Section 2.2 The Globalization Debate and the Transformation of Capitalism

Paul Hirst and Graham Thompson

'Globalization – A Necessary Myth?'

from *Globalization in Question: The International Economy and the Possibilities of Governance* (1996)

Globalization has become a fashionable concept in the social sciences. It is widely asserted that we live in an era in which the greater part of social life is determined by global processes, in which national cultures, national economies and national borders are dissolving. Central to this perception is the notion of a rapid and recent process of economic globalization. A truly global economy is claimed to have emerged or to be in the process of emerging, in which distinct national economies and, therefore, domestic strategies of national economic management are increasingly irrelevant. The world economy has internationalized in its basic dynamics, it is dominated by uncontrollable market forces, and it has as its principal economic actors and major agents of change truly transnational corporations, that owe allegiance to no nation state and locate wherever in the globe market advantage dictates.

This image is so powerful that it has mesmerized analysts and captured political imaginations. But is it the case? We began this investigation with an attitude of moderate scepticism. However, the closer we looked the shallower and more unfounded became the claims of the more radical globalists. In particular we began to be disturbed by three facts: first, the absence of a commonly accepted model of the new global economy and how it differs from previous states of the international economy; second, in the absence of a clear model against which to measure trends, the tendency casually to cite examples of internationalization of sectors and processes as if they were evidence of the growth of an economy dominated by autonomous global market forces; and third, the lack of historical depth, the tendency to portray current changes as both unique and without precedent and firmly set to persist long into the future.

[G]lobalization, as conceived by the more extreme globalizers, is largely a myth. Thus we argue that:

1. The present highly internationalized economy is not unprecedented: it is one of a number of distinct conjunctures or states of the international economy that have existed since an economy based on modern industrial technology began to be generalized from the 1860s. In some respects, the current international economy is *less* open

and integrated than the regime that prevailed from 1870 to 1914.

2. Genuinely transnational companies (TNCs) appear to be relatively rare. Most companies are nationally based and trade multinationally on the strength of a major national location of production and sales, and there seems to be no major tendency towards the growth of truly international companies.
3. Capital mobility is not producing a massive shift of investment and employment from the advanced to the developing countries. Rather, foreign direct investment (FDI) is highly concentrated among the advanced industrial economies and the Third World remains marginal in both investment and trade, a small minority of newly industrializing countries apart.
4. As some of the extreme advocates of globalization recognize, the world economy is far from being genuinely 'global'. Rather, trade, investment and financial flows are concentrated in the Triad of Europe, Japan and North America and this dominance seems set to continue.
5. These major economic powers, thus have the capacity, especially if they coordinate policy, to exert powerful governance pressures over financial markets and other economic tendencies.

It is one thing to be sceptical about the concept of globalization, it is another to explain the widespread development and reception of the concept since the 1970s. It will not do to wheel out the concept of 'ideology' here, for this view is so widespread that it covers the most diverse outlooks and social interests. It covers the political spectrum from left to right, it is endorsed in diverse disciplines – economics, sociology, cultural studies and international politics – and it is advanced by both theoretical innovators and traditionalists. The literature on globalization is vast and diverse.

We are well aware that there are a wide variety of views that use the term 'globalization'. Even among those analysts that confine themselves to strictly economic processes, some make far more radical claims about changes in the international economy than others. Some less extreme and more nuanced analyses that employ the term 'globalization' are well established in the academic community and concentrate on the relative internationalization of major financial markets, of technology, and of certain important sectors of manufacturing and services, particularly since the 1970s. Emphasis is given in many of these analyses to the increasing constraints on national-level governance that prevent ambitious macroeconomic policies that diverge significantly from the norms acceptable to international financial markets.

There are, however, very real dangers in not distinguishing clearly between certain trends toward internationalization and the strong version of the globalization thesis. It is particularly unfortunate if the two become confused by using the same word, 'globalization', to describe both.

The strong version of the globalization thesis requires a new view of the international economy, one that subsumes and subordinates national-level processes. Whereas tendencies toward internationalization can be accommodated within a modified view of the world economic system, that still gives a major role to national-level policies and actors. Undoubtedly, this implies some greater or lesser degree of change: firms, governments and international agencies are being forced to behave differently, but in the main they can use existing institutions and practices to do so. In this way it makes more sense to consider the international economic system in a longer historical perspective, to recognize that current changes, while significant and distinctive, are not unprecedented and do not necessarily involve a move toward a new type of economic system.

However, the question remains to be considered of how the myth of the globalization of economic activity became established as and when it did. In answering one must begin with the end of the post-1945 era in the turbulence of 1972–3. A number of significant changes ended a period of prolonged economic growth and full employment in the advanced countries, sustained by strategies of active national state intervention and a managed multilateral regime for trade and monetary policy under US hegemony. Thus we can point to:

1. The effects of the collapse of the Bretton Woods system and the OPEC oil crisis in producing turbulence and volatility in all the major economies through the 1970s into the early 1980s.
2. The efforts by financial institutions and manufacturers, in this period of turbulence and inflationary pressure, to compensate for domestic uncertainty by seeking wider outlets for investments and additional markets. The results were widespread bank lending to the Third World during the inflationary 1970s, the growth of the Eurodollar market, and the increasing foreign trade to GDP ratios in the advanced countries.
3. The public policy acceleration of the internationalization of financial markets by the widespread abandonment of exchange controls and other market deregulation in the late 1970s and early 1980s.
4. The tendency towards 'de-industrialization' in Britain and the United States and the growth of long-term unemployment in Europe, promoting fears of foreign competition, especially from Japan.
5. The relatively rapid development of a number of newly industrializing countries (NICs) in the Third World and their penetration of First World markets.
6. The shift from standardized mass production to more flexible production methods, and the change from large nationally rooted oligopolistic corporation towards a more complex world of multinational enterprises (MNCs), less rigidly structured major firms, and the increased salience of smaller firms – summed up in the widespread and popular concept of 'post-Fordism'.

These changes were highly disturbing to those conditioned by the unprecedented success and security of the post-1945 period in the advanced industrial states. If the widespread consensus of the 1950s and 1960s was that the future belonged to a capitalism without losers, securely managed by national governments acting in concert, then the later 1980s and 1990s are dominated by a consensus based on contrary assumptions, that global markets are uncontrollable and that the only way to avoid becoming a loser – whether as nation, firm or individual – is to be as competitive as possible. The notion of an ungovernable world economy is a response to the collapse of expectations schooled by Keynesianism and sobered by the failure of monetarism to provide an alternative route to broad-based prosperity and stable growth.

MODELS OF THE INTERNATIONAL ECONOMY

We can only begin to assess the issue of globalization if we have some relatively clear and rigorous model of what a global economy would be like and how it represents both a new phase in the international economy and an entirely changed environment for national economic actors. Globalization in its radical sense should be taken to mean the development of a new economic structure, and not just conjunctural change toward greater international trade and investment within an existing set of economic relations. An extreme and one-sided ideal type of this kind enables us to differentiate *degrees* of internationalization, to eliminate some possibilities and to avoid confusion between claims. Given such a model it becomes possible to assess it against evidence of international trends and thus enables us more or less plausibly to determine whether or not this former phenomenon of the development of a new supranational economic system is occurring. In order to do this we have developed two basic contrasting ideal types of international economy: a fully globalized economy, and an open international economy that is still fundamentally characterized by exchange between relatively distinct national economies and in which many outcomes, such as the competitive performance of firms and sectors, are substantially determined by processes occurring at the national level. These ideal types are valuable in so far as they are useful in enabling us conceptually to clarify the issues, that is, in specifying the difference between a new global economy and merely extensive and intensifying international economic relations.

Type 1: An Inter-National Economy

An *inter-national economy* is one in which the principal entities are national economies. Trade and investment produce growing interconnection between these still national economies. Such a process involves the increasing integration of more and more nations and economic actors into world market relationships. Trade relations, as a result, tend to take on the form of national specializations and the international division of labour. The importance of trade is however progressively replaced by the centrality of investment relations between nations, which increasingly act as the organizing principle of the system. The form of interdependence between nations remains, however, of the 'strategic' kind. Interactions are of the 'billiard-ball' type: international events do not directly or necessarily penetrate or permeate the domestic economy but are refracted through national policies and processes. The international and the domestic policy fields either remain relatively separate as distinct levels of governance or work 'automatically'.

Perhaps the classic case of such an 'automatic' adjustment mechanism remains the Gold Standard, which operated at the height of the *pax Britannica* system from mid-century to 1914. Great Britain acted as the political and economic hegemon and the guarantor of this system. But it is important to recognize that the system of the Gold Standard and the *pax Britannica* was merely one of the several structures of the international economy in this century. The lifetime of a prevailing system of international economic relations in this century has been no more than thirty to forty years. Indeed, given that most European currencies did not become fully convertible until the late 1950s, the full Bretton Woods system after World War II only lasted upwards of thirteen to fourteen years.

The period of this world-wide international economic system is also typified by the rise and maturity of the multinational corporation (MNC), as a transformation of the large merchant trading companies of a past era. [T]he important aspect of these MNCs is that they retain a clear national home base; they are subject to the national regulation of the mother country, and by and large they are effectively policed by that home country.

The point of this ideal type drawing on the institutions of the *belle époque* is not, however, a historical analogy: for a simple and automatically governed international economic system like that before 1914 is unlikely to reproduce itself now. The current international economy is relatively open, but it has real differences from that prevailing before World War I. The pre-1914 system was, nevertheless, genuinely international, tied by efficient long-distance communications and industrialized means of transport.

The late twentieth century communications and information technology revolution has further developed, rather than created, a trading system that could make day-to-day world prices. In the second half of the nineteenth century the submarine inter-continental telegraph cables enabled the integration of world markets. Modern systems dramatically increase the possible volume and complexity of transactions, but we have had information media capable of sustaining a genuine international trading system for over a century. If the theorists of globalization mean that we have an economy in which each part of the world is linked by markets sharing close to real-time information, then that began not in the 1970s but in the 1870s.

Type 2: A Globalized Economy

A *globalized economy* is an ideal type distinct from that of the international economy and can be developed by contrast with it. In such a global system distinct national economics are subsumed and rearticulated into the system by international processes and transactions. The international economy, on the contrary, is one in which processes that are determined at the level of national economics still dominate and international phenomena are outcomes that emerge from the distinct and differential performance of the national economies.

The global economy raises these nationally based interactions to a new power. The international economic system becomes autonomized and socially disembedded, as markets and production become truly global. Domestic policies, whether of private corporations or public regulators, now have routinely to take account of the predominantly international determinants of their sphere of operations. As systemic interdependence grows, the national level is permeated by and transformed by the international.

The first major consequence of a globalized economy would thus be the fundamental problematicity of its governance. The principal difficulty is to construct both effective and integrated patterns of national and international public policy to cope with global market forces. The systematic economic interdependence of countries and markets would by no means necessarily result in a harmonious integration in which world consumers benefit from truly independent allocatively efficient market mechanisms. On the contrary, it is more than plausible that [i]nterdependence would then readily promote *disintegration* – i.e. competition and conflict – between regulatory agencies at different levels. Enthusiasts for the efficiency of free markets and the superiority of corporate control over that of public agencies would see this as a rational world order freed from the shackles of obsolete and ineffective national public interventions. Others, less sanguine but convinced globalization is occurring, may see it as a world system in which there can be no generalized or sustained public reinsurance against the costs imposed on particular localities by unfavourable competitive outcomes or market failures.

It is also clear that this ideal type highlights the problem of weak public governance for the major corporations. Even if such companies were truly global, they will not be able to operate in all markets equally effectively and, like governments, will lack the capacity for reinsurance against unexpected shocks from their own resources alone.

A second major consequence of the notion of a globalizing international economy would be the transformation of MNCs into transnational corporations (TNCs) as the major players in the world economy. The TNC would be genuine footloose capital, without specific national identification and with an internationalized management, and at least potentially willing to locate and relocate anywhere in the globe to obtain either the most secure or the highest returns. In the financial sector this could be achieved at the touch of a button and in a truly globalized economy would be wholly dictated by market forces, without reference to national monetary policies. In the case of primarily manufacturing companies the TNC would source, produce and market at the global level as strategy and opportunities dictated.

Julius (1990) and Ohmae (1990, 1993, 1995a), for example, both consider this trend toward true TNCs to be well established. Ohmae argues that such 'stateless' corporations are now the prime movers in an Inter-linked Economy (ILE) centred on North America, Europe and Japan. He contends that macroeconomic and industrial policy intervention by national governments can only distort and impede the rational process of resource allocation by corporate decisions and consumer choices on a global scale. Ohmae argues that such corporations will pursue strategies of 'global localization' in responding on a worldwide scale to specific regionalized markets and locating effectively to meet the varying demands of distinct localized groups of consumers.

A third consequence of globalization would be the further decline in the political influence and economic bargaining power of organized labour. Globalized markets and TNCs would tend to be mirrored by an open world labour market. This market would operate not primarily by *actual* labour mobility from country to country but by mobile capital selecting locations with the best deal in terms of labour costs and supply. Thus while companies requiring highly skilled and productive labour might well continue to locate in the advanced countries, with all their advantages, rather than merely seek low wages, the trend of the global mobility of capital and the relative national fixity of labour would favour those advanced countries with the most tractable labour forces and the lowest social overheads relative to the benefits of labour competence and motivation. 'Social democratic' strategies of enhancement of working conditions would thus only be viable if they assured the competitive advantage of the labour force, without constraining management prerogatives, and at no more overall cost in taxation than the average for the advanced world.

A final and inevitable consequence of globalization is the growth in fundamental multipolarity in the international political system. The hitherto hegemonic national power could no longer impose its own distinct regulatory objectives in either its own territories or elsewhere, and lesser agencies (whether public or private) would thus enjoy enhanced powers of denial and evasion vis-à-vis any aspirant 'hegemon'. A variety of bodies from international voluntary agencies to TNCs would thus gain in relative power at the expense of national governments and, using global markets and media, could appeal to and obtain legitimacy from consumers/citizens across national boundaries. Thus the distinct disciplinary powers of national states would decline, even though the bulk of their citizens, especially in the advanced countries, remained nationally bound. In such a world, national military power would become less effective as the rationality of the objectives of 'national' state control in respect of the economy evaporated. As economics and nationhood pulled apart the international economy would become even more 'industrial' and less 'militant' than it is today. War would be increasingly localized, and wherever it threatened powerful global economic interests it would be subject to devastating economic sanction.

THE ARGUMENT IN OUTLINE

The strong concept of a globalized economy outlined above acts as an ideal type which we can measure against the actual trends within the international economy. This globalized economy has been contrasted with the notion of an inter-national economy in order to distinguish its particular and novel features. The opposition of these two types for conceptual clarity conceals the possibly messy combination of the two in fact. This would make it difficult to determine major trends on the basis of the available evidence. These two types of economy are not inherently mutually exclusive: rather, in certain conditions the globalized economy would *encompass and subsume* the international economy. The globalized economy would rearticulate many of the features of the inter-national economy, transforming them as it reinforced them. If this phenomenon occurred there would thus be a complex combination of features of both types of economy present within the present conjuncture. The problem in determining what is happening is to identify the dominant trends: either the growth of globalization or the continuation of the existing inter-national patterns.

It is our view that such a process of hybridization is not taking place, but it would be cavalier not to consider and raise the possibility. Central in this respect is the weak development of TNCs and the continued salience of MNCs and also the ongoing dominance of the advanced countries in both trade and FDI. Such evidence is consistent with a continuing inter-national economy, but much less so with a rapidly globalizing hybrid system. Moreover, we should remember that an international economy is one in which the major nationally based manufacturers and the major financial trading and service centres are strongly externally oriented, emphasizing international trading performance. The opposite of a globalized economy is thus not a nationally inward-looking one, but an open world market based on trading nations and regulated to a greater or lesser degree both by the public policies of nation states and by supra-national agencies. Such an economy has existed in some form or another since the 1870s, and has continued to re-emerge despite major setbacks, the most serious being the crisis of the 1930s. The point is that it should not be confused with a global economy.

Ash Amin

'Placing Globalization'

Theory, Culture and Society (1997)

The more we read about globalization from the mounting volume of literature on the topic, the less clear we seem to be about what it means and what it implies. We are assailed by opposing interpretations. Globalization is the triumph of capitalism on a world scale over national and local autonomy and identity; not to be stopped if you side with neo-liberalism, or to be resisted through transnational anti-capitalist or social-democratic forces if you take an opposite view. Less dramatically, it is nothing more than the intensification of exchange between distinct national social formations and, as such, still governable through the inter-state system. Somewhere in between, it symbolizes the blurring of traditional territorial and social boundaries through the interpenetration of local and distant influences, therefore requiring hybrid and multi-polar solutions.

The article begins with a critique of an increasingly trendy position among commentators on the economic aspects of globalization, that nations and states remain at the core of the international economic order. It then draws on sociological and cultural analyses of globalization to offer a broader and less polarized interpretation along the lines of the third perspective suggested above. Drawing on the idea of globalization as multi-layering, hybridity and interdependence, and basing the illustration on an account of urban trends, the article suggests that these aspects might be more salient than asking whether globalization represents the end of geography.

GLOBALIZATION AS BACK TO THE NATION

Recently, there has been a backlash against the thesis that globalization represents the triumph of world capitalism and transnational institutions over national economies and nation-states. [A]pologists such as Kenichi Ohmae (1990, 1995b) see the nation-state as an outmoded institution in the way of world markets and borderless transnational corporations or, at best, little more than a low-cost provider of infrastructure and public goods for global business.

The book *Globalization in Question* by **Paul Hirst and Grahame Thompson** (1996b) is written as a no-nonsense, once-and-for-ever rebuttal of this thesis. It is set to become a cult classic among those who wish to be reassured that the world economy is still governable through national politics and policies. The line of attack is based on debunking the idea that a *new, transnational* world economic system is replacing an old, international world system.

The argument is centred around four interrelated claims. The first is that economic activity is still nationally based and therefore also under the influence of national institutions.

This claim is supported through data to show that transnational corporations (TNCs) do not dominate world trade and output, and continue to retain on average two-thirds of their assets in the home-base (consequently highly sensitive to national economic policies). Further afield, support for this kind of claim can be found in the neo-Schumpeterian literature highlighting both the persistence of national systems of innovation and learning which draw on country-specific institutions and traditions (Zysman, 1996) and the tendency of transnational corporations to continue to rely on home-base R&D facilities (Cantwell, 1995). Further support comes from the institutionalist literature on enduring national varieties of capitalism which derive competitive advantage and specificity in the global economy from a combination of state policy choices and embedded social, cultural and institutional traditions (Hutton, 1995; Palan and Abbott, 1996).

The second of Hirst and Thompson's claims, thus, is that if we wish to use the term 'globalization' we should do so to signify an 'open and international economy with large and growing flows of trade and capital investment between countries' (1996a, p. 48), not an economic system articulated at the global scale. However, conceptualized in these terms, there is nothing new about globalization.

The third claim is that contemporary trade and other flows are not ubiquitously global but confined to self-contained regional groupings and triads. There is no flight of capital and jobs to the Third World, nor are cheap imports from the latter flooding markets in the advanced economies to exert a downward pressure on wages and welfare expenditure. Instead, most investment and trade continues to be among partners within the triad markets of Europe, the Americas and South and East Asia, therefore leaving considerable scope for national economic specialization as well as regulation through triad institutions and agreements.

Fourth, and as a consequence, there is no irreversible imperative to replace interventionist national or supra-national policies by global market forces and neo-liberal solutions. National economic regulation remains possible since we continue to live in a world of nation-states and national economies. However, Hirst and Thompson agree that national-level regulation alone is not sufficient because of the growth of regional groupings above and below the nation-state. They propose a division of labour between national policies dedicated to securing social consensus, distributional justice and macroeconomic stability; international Keynesian agreements designed to expand world demand, control short-term capital flights and reduce Third World debt; and sub-national interventions to fine tune supply-side support for increasingly flexibly specialized firms and industries.

Hirst and Thompson's critique of globalization, all told, is a denial based on a caricature of the term and a superficial quantitative evaluation of the phenomenon. In their zeal to allay alarmist fears associated with the 'end of geography' thesis, they leave us at the opposite pole with a comforting sense of business as usual. The argument fails to reassure, however, largely because it does not provide any sense of trends and changes in the world economic system which might be genuinely challenging the balance between national and global influences.

These points can be illustrated through five observations. First, the substitution of quantitative measures for qualitative argument leads to false parallels between now and the past.

Second, the counter-intuitive claim that contemporary globalization does not consist of the rise and power of transnational economic circuits – from finance to production and exchange – is open to challenge. According to UNCTAD (1994), TNCs now account for up to a third of world output, 80% of global investment and two-thirds of world trade, with around a quarter of world trade restricted to branches of the same company.

Third, it is amusing that the same enemy – neo-liberalism – unites the opposite perspectives offered by Hirst and Thompson and the socialist critiques of globalization. Both criticize the market philosophy peddled by big business, international regulatory organizations, influential conservative governments and their ideologues, that the only way forward for national and global prosperity is the application of the familiar cocktail of policies including deregulation, privatization, reduced public expenditure, wage-competition, unrestrained trade and market integration and tight international control over exchange and interest rates. [T]he question, as aptly put by Nederveen Pieterse (1997, p. 4) in a recent review essay, is that 'if the target is neo-liberalism and the unfettered market economy, then why attack globalization?'

Fourth, part of the obfuscation lies in the failure to take seriously the amalgam of social processes which we would have to acknowledge if we adopted a broader understanding of globalization. Even a cursory awareness of the globalization literature cannot fail to cast doubt over the idea of globalization as neo-liberal conspiracy or simply a system of trade and investment exchange between nations.

Fifth, and finally, Hirst and Thompson's refusal to engage with any of the above aspects of globalization leaves us with no sense of how local and national possibilities are being reshaped in light of them. Let us take the vexed issue of state national economic policy as an example. Contrary to Hirst and Thompson's caricature, the debate on national economic sovereignty has moved on beyond first warnings of the end of the nation-state, towards a general consensus that the rise of transnational economic relations and plural authority structures represents a reworking, rather than abandonment, of national economic policies.

At the same time, the increasingly free and contested nature of markets has made it imperative for nation-states to develop economic policies aimed at securing global competitiveness, at the expense of growth models oriented towards the domestic economy or national social cohesion (Reich, 1991). There appear to be a number of policy shifts associated with the emerging 'competition' or 'entrepreneurial' state (Cerny, 1995; Jessop, 1995).

GLOBALIZATION AS OUT THERE–IN HERE CONNECTIVITY

Globalization needs to be taken seriously, not reduced to the international nor swept into the past. Equally, it should not be misconstrued or demonized as an 'out there' phenomenon standing above, and set to destroy, the geography of territorial states, economies and identities. Implicit in my critique of Hirst and Thompson is an understanding of globalization which consists of two phenomena. [F]irst, a growing number of chains of economic, social, cultural and political activity that are world-wide in scope, and, second, the intensification of levels of interaction and interconnectedness between states and societies. Thus, Held (1995, p. 20), building on Giddens's concept of time–space shrinkage, proposes that:

> globalisation can be taken to denote the stretching, and deepening of social relations and institutions across space and time such that, on the one hand, day-to-day activities are increasingly influenced by events happening on the other side of the globe and, on the other hand, the practices and decisions of local groups can have significant global reverberations.

This definition replaces a territorial idea of the local, national and the global as separate spheres of social organization and action, by a relational understanding of each as a nexus of multiple and asymmetric interdependencies involving local and wider fields of influence. It is the resulting interconnectedness, multiplexity and hybridization of social life at every level – spatial and organizational – that strikes me as perhaps the most distinctive aspect of contemporary globalization. Viewed in this way, to think of the global as flows of dominance and transformation and the local as fixities of tradition and continuity is to miss the point, because it denies the interaction between the two as well as the evolutionary logics of both. Thus finding a place in a world of 'stretched and deepened' time–space connectivity might now be more a matter of how this connectivity is negotiated or worked to your advantage, rather than resisted or kept out.

In the rest of this article, I wish to explore the implications of globalization construed in this particular way for developing what Doreen Massey (1993) has called a global or progressive sense of place. I shall take contemporary urban life as my example because it strikes me that the social life of cities is a stark and vivid expression of the complexities and problems associated with the 'in here–out there' connectivities of globalization.

In contemporary urban analysis, cities have been rediscovered both as important nodes or 'basing points' (Friedman, 1995) for the economy of global flows and as 'coordinates' of the entrepreneurial state responding directly to the situated needs of global capital (Jessop, 1995; Brenner, 1996). As regards the first role, three strands of urban rediscovery have been especially important. The first strand of work, deriving from such authors as Saskia Sassen (1991, 1994), Manuel Castells (1989b) and John Friedman (1995), stresses the importance of large metropolises as key command and control centres within the interlocking globalizing dynamics of financial markets, producer services industries, corporate headquarters and associated service industries (telecommunications, business conferences, transport, property development, etc.). The second strand of work stresses two aspects of cities as economic motors: the city as knowledge-base; and as source of vital agglomeration economies. The third strand reasserts the importance of city centres by stressing how urban culture, the media, entertainment, sport and education may, with appropriate policies, interlace positively within a framework of public space to support the emergence of 'creative cities' (Landry and Bianchini, 1995). The central assertion here is that cultural revitalization based on reclaiming public spaces and animating them with diverse cultural activities can be an important source of urban renewal through the attraction of global consumers but, above all, creative or 'expressive specialists' (Hannerz, 1996).

Clearly, this is a reading of cities based upon the experience of a small number of usually global cities, and one which tends to be extrapolated to all cities (Amin and Graham, 1997; Thrift, 1996). [I]t is a reading based on a false opposition between the local as immobile or grounded and the global as mobile and ubiquitous. Underlying all three strands of the new urbanism is an emphasis on the special effects of spatial proximity in an otherwise deterritorialized and disembedded world – cities as clusters of knowledge and knowledgeable people, agglomerations of specialized firms, a critical mass of cultural creativity and informal exchange.

But the reading of globalization as an intermingling of 'in here–out there' processes, resulting in heterogeneity, shifting identities and multipolarity, seems to me to be much more consistent with contemporary urban reality. One of the most striking features of most cities today is their character as a set of spaces of juxtaposed fragments and contrasts, where diverse relational webs might coalesce, interconnect or disconnect. The city does not possess a unitary identity or homogeneous spaces, perhaps it never has.

There are clearly sites where urban propinquity does still matter as a unique asset among global flows and connections – the financial districts, the cultural zones, the industrial districts. But there are also many more zones where fragmentation between adjacent units is more the norm, as exchange and interchange becomes disembedded from the immediate locale, through fast transport and advanced telecommunications systems (Giddens, 1990).

This *multiplex* city, importantly for our purposes, is in part the product of globalization. This is made very clear in the book *Living the Global City* (1997) recently edited by John Eade, which explores the effects of globalization on daily life among different social groups and boroughs in London. In one of the essays, Martin Albrow attempts to conceptualize the character of places like Tooting which are composed of residents of different social and ethnic origins, expectations, lifestyles, perceptions of each other and usages of the locality. For Albrow, such places belong to 'socioscapes' of shifting and fluid social relations under conditions of global cultural flow (adding to the list of terms with the suffix 'scape' coined by Arjun Appadurai, 1990). He traces their distinctiveness to four aspects of globalization consistently stressed by cultural theorists writing on the global–local nexus: first, values that draw from the world (globalist sentiment); access to and influence by events elsewhere (globality); direct interaction with other parts of the world via telematics (time–space compression); and maintenance of lifestyles and life routines in new places by migrants (disembedding). The outcome is that 'people can reside in one place and have their meaningful social relations almost entirely outside it and across the globe' and 'use the locality as site and resource for social activities in very different ways' (Albrow, 1997, p. 53).

The resulting multiplexity is a characteristic of the identity of individuals themselves. Importantly, this is not just a feature of those with the greatest global or diaspora connections, but of virtually all city inhabitants: locals and cosmopolitans, traditionalists and non-traditionalists, mobile and immobile. This is precisely the nature of globalization, in forcing 'out there–in here' stretching and deepening on everybody, regardless of their personal experiences and perceptions.

In summary, the picture of urban life that we can draw on the basis of an understanding of globalization based on social theorists such as Anthony Giddens, David Held, Stuart Hall, Doreen Massey, Roland Robertson and Ulf Hannerz is one juxtaposing hybrid places and hybrid identities. It is a picture of multiple internal and external connections that implies neither collapse nor new forms of cultural intermingling and community, but simply unity in diversity.

CONCLUDING OBSERVATIONS: NEGOTIATING GLOBALIZATION

Thus I have distanced myself from the *territorial* idea of sequestered spatial logics – local, national, continental and global – pitted against each other. Instead, I have chosen to interpret globalization in *relational* terms as the interdependence and intermingling of global, distant and local logics, resulting in the greater hybridization and perforation of social, economic and political life.

This distinction between a territorial and relational understanding of globalization has not only analytical but also practical significance, across different policy fields. In the regulation of economic life, for example, the territorial logic is prone to defend one spatial scale of organization against another or seek a neat division of duties between the local, national and supra-national state. The relational logic, in contrast, is more likely to be interested in the ways in which hybrid networks of interdependence and influence locking together economic agents and institutions may be made to work for particular sites within them. Thus, in the area of industrial policy, to take one example, regional or national effort might focus on developing the supply-base (from skills through to education, innovation and communications) and the institutional base (from development agencies to business fora and political voice), in order to make particular sites into key staging points or centres of competitive advantage within respective global industrial filières and value chains. This would replace a territorial policy stance based on protecting or keeping out certain industries. In addition, attention might be paid to identifying firm interdependencies, exchange relations and rationalities of behaviour (e.g. reciprocity, trust and interactive decision-making) that work to local advantage, and identifying those which hinder the development of local capabilities and virtuous networks of entrepreneurship. This would replace a firm-centred approach, say, promoting domestic firms and preventing inward investors or vice versa. Similarly, in institutional terms, a relational industrial policy is likely to privilege an approach based on pluralist decision-making, reliance on inter-institutional voice and negotiation across different spatial scales, and sensitivity to uneven power relations within institutional networks, instead of an approach privileging institutional autarchy and purely rationalist decision-making.

The same kind of practical logic is applicable to other areas of policy action seeking to negotiate globalization. The common response of city leaders across the developed world to 'globalization', no matter how defined or understood, has been to make them attractive to inward investors, the international business community, global decision-makers and their cronies, tourists and 'expressive specialists' (Hannerz, 1996). This has involved the construction of glittering and expensive buildings, city-centre business districts, retail centres and shopping malls, up-market housing, tourist attractions, festivals, city spectacles, and other high-consumption projects.

An alternative, however, is to work with the urban social fragments and hybridity generated by 'in here–out there' mingling. The challenge is to mobilize diversity as a source of both social cohesion and urban economic competitiveness. [A] city disposed toward tolerating difference, sharing public spaces and active citizenship is also a source of entrepreneurship and creativity based on the mobilization of confidence, social interchange and the pooling of diversity. [M]eeting the challenge of globalization need not mean the sacrifice of a progressive sense of place; a civic solidarity drawing upon local diversity and difference.

David Harvey

'Theorizing the Transition . . . Flexible Accumulation – Solid Transformation or Temporary Fix?'

from *The Condition of Postmodernity* (1989)

The transition from Fordism to flexible accumulation has, in fact, posed serious difficulties for theories of any sort. Keynesians, monetarists, neo-classical partial equilibrium theorists, appear just as befuddled as everyone else. The transition has also posed serious dilemmas for Marxists. The only general point of agreement is that something significant has changed in the way capitalism has been working since about 1970.

The first difficulty is to try to encapsulate the nature of the changes we are looking at. [T]ables 1, 2, and 3 summarize three recent accounts of the transition. The first, a rather celebratory account by Halal (1986), emphasizes the positive and liberatory elements of the new entrepreneurialism. The second, by Lash and Urry (1987), emphasizes power relations and politics in relation to economy and culture. The third, by Swyngedouw (1986), provides much more detail on transformations in technology and the labour process while appreciating how the regime of accumulation and its modes of regulation have shifted. In each case, of course, the opposition is used as a didactic tool to emphasize the differences rather than the continuities, and none of the authors argues that matters are anywhere near as cut and dried as these schemas suggest. Halal appears closer to Schumpeter's theory of entrepreneurial innovation as the driving force of capitalism, and tends to interpret Fordism and Keynesianism as an unfortunate interlude in capitalist progress. Lash and Urry see the evolution in part as the collapse of the material conditions for a powerful collective working-class politics, and attempt to probe the economic, cultural, and political roots of that collapse. By the very use of the terms 'organized' and 'disorganized' to characterize the transition, they emphasize more the disintegration than the coherence of contemporary capitalism. Swyngedouw, on the other hand, by emphasizing changes in the mode of production and of industrial organization, locates the transition in the mainstream of Marxian political economy while clearly accepting the regulation school's language.

FLEXIBLE ACCUMULATION – SOLID TRANSFORMATION OR TEMPORARY FIX?

[T]here has certainly been a sea-change in the surface appearance of capitalism since 1973, even though the underlying logic of capitalist accumulation and its crisis-tendencies remain the same. We need to consider, however, whether

Table 1 The new capitalism according to Halal

	The old capitalism (Industrial paradigm)	The new capitalism (Post-industrial paradigm)
Frontier of progress	hard growth	smart growth
Organization	mechanistic structure	market networks
Decision making	authoritarian command	participative leadership
Institutional values	financial goals	multiple goals
Management focus	operational management	strategic management
Economic macro-system	profit-centred big business	democratic free enterprise
World system	capitalism versus socialism	hybrids of capitalism and socialism

Source: Halal (1986)

Table 2 Contrast between organized and disorganized capitalism according to Lash and Urry

Organized capitalism	Disorganized capitalism
concentration and centralization of industrial banking, and commercial capital in regulated national markets	de-concentration of rapidly increasing corporate power away from national markets. Increasing internationalization of capital and in some cases separation of industrial from bank capital
increasing separation of ownership from control and emergence of complex managerial hierarchies	continued expansion of managerial strata articulating their own individual and political agendas quite distinct from class politics
growth of new sectors of managerial, scientific, technological intelligentsia and of middle-class bureaucracy	relative/absolute decline in blue-collar working class
growth of collective organizations and bargaining within regions and nation states	decline in effectiveness of national collective bargaining
close articulation of state and large monopoly capital interests and rise of class-based welfare statism	increasing independence of large monopolies from state regulation and diverse challenges to centralized state bureaucracy and power
expansion of economic empires and control of overseas production and markets	industrialization of third world and competitive de-industrialization of core countries which turn to specialization in services
incorporation of diverse class interests within a national agenda set through negotiated compromises and bureaucratic regulation	outright decline of class-based politics and institutions
hegemony of technical–scientific rationality	cultural fragmentation and pluralism coupled with undermining of traditional class or national identities
concentration of capitalist relations within relatively few industries and regions	dispersal of capitalist relations across many sectors and regions
extractive-manufacturing industries dominant sources of employment	decline of extractive-manufacturing industries and rise of organizational and service industries
strong regional concentration and specialization in extractive-manufacturing sectors	dispersal, diversification of the territorial–spatial division of labour
search for economies of scale through increasing plant (work-force) size	decline in plant size through geographical dispersal, increased sub-contracting, global production systems
growth of large industrial cities dominating regions through provision of centralized services (commercial and financial)	decline of industrial cities and deconcentration from city centres into peripheral or semi-rural areas resulting in acute inner city problems
cultural-ideological configuration of 'modernism'	cultural-ideological configurations of 'postmodernism'

Source: after Lash and Urry (1987)

Table 3 Contrast between Fordism and flexible accumulation according to Swyngedouw

Fordist production (based on economies of scale)	Just-in-time production (based on economies of scope)
A THE PRODUCTION PROCESS	
mass production of homogeneous goods	small batch production
uniformity and standardization	flexibility and small batch production of a variety of product types
large buffer stocks and inventory	no stocks
testing quality ex-post (rejects and errors detected late)	quality control part of process (immediate detection of errors)
rejects are concealed in buffer stocks	immediate reject of defective parts
loss of production time because of long set-up times, defective parts, inventory bottlenecks etc.	reduction of lost time, diminishing 'the porosity of the working day'
resource driven	demand driven
vertical and (in some cases) horizontal integration	(quasi-) vertical integration sub-contracting
cost reductions through wage control	learning-by-doing integrated into long-term planning
B LABOUR	
single task performance by worker	multiple tasks
payment per rate (based on job design criteria)	personal payment (detailed bonus system)
high degree of job specialization	elimination of job demarcation
no or little on the job training	long on the job training
vertical labour organization	more horizontal labour organization
no learning experience	on the job learning
emphasis on diminishing worker's responsibility (disciplining of labour force)	emphasis on worker's co-responsibility
no job security	high employment security for core workers (life-time employment). No job security and poor labour conditions for temporary workers
C SPACE	
functional spatial specialization (centralization/ decentralization)	spatial clustering and agglomeration
spatial division of labour	spatial integration
homogenization of regional markets (spatially segmented labour markets)	labour market diversification (in-place labour market segmentation)
world-wide sourcing of components and sub-contractors	spatial proximity of vertically quasi-integrated firms
D STATE	
regulation	deregulation/re-regulation
rigidity	flexibility
collective bargaining	division/individualization, local or firm-based negotiations
socialization of welfare (the welfare state)	privatization of collective needs and social security
international stability through multi-lateral agreements	international destabilization; increased geopolitical tensions

Table 3 (continued)

Fordist production (based on economies of scale)	Just-in-time production (based on economies of scope)
centralization	decentralization and sharpened interregional/ intercity competition
the 'subsidy' state/city	the 'entrepreneurial' state/city
indirect intervention in markets through income and price policies	direct state intervention in markets through procurement
national regional policies	'territorial' regional policies (third party form)
firm financed research and development	state financed research and development
industry-led innovation	state-led innovation
E IDEOLOGY	
mass consumption of consumer durables: the consumption society	individualized consumption: 'yuppie'-culture
modernism	postmodernism
totality/structural reform	specificity/adaptation
socialization	individualization – the 'spectacle' society

Source: after Swyngedouw (1986)

the shifts in surface appearance betoken the birth of a new regime of accumulation, capable of containing the contradictions of capitalism for the next generation, or whether they betoken a series of temporary fixes. The question of flexibility has already been the focus of some debate. Three broad positions seem now to be emerging.

The first position, primarily espoused by Piore and Sabel (1984), is that the new technologies open up the possibility for a reconstitution of labour relations and of production systems on an entirely different social, economic, and geographical basis. Not everyone shares this rosy vision of the forms of industrial organization (see, for example, Murray, 1987). There is much that is regressive and repressive about the new practices. Nevertheless, many share the sense that we are at some kind of 'second industrial divide' (to appropriate the title of Piore and Sabel's book), and that new forms of labour organization and new locational principles are radically transforming the face of late twentieth-century capitalism. The revival of interest in the role of small business (a highly dynamic sector since 1970), the rediscovery of sweatshops and of informal activities of all kinds, and the recognition that these are playing an important role in contemporary economic development even in the most advanced of industrialized countries, and the attempt to track the rapid geographical shifts in employment and economic fortunes, have produced a mass of information that seems to support this vision of a major transformation in the way late twentieth-century capitalism is working.

The second position sees the idea of flexibility as an 'extremely powerful term which legitimizes an array of political practices' (chiefly reactionary and anti-worker), but without any strong empirical or materialist grounding in the actual facts of organization of late twentieth-century capitalism. **Pollert** (1988) (see Part Five, Section 5.2), for example, factually challenges the idea of flexibility in labour markets and labour organization, and concludes that the 'discovery of the "flexible workforce" is part of an ideological offensive which celebrates pliability and casualization, and makes them seem inevitable'. Sayer (1989) likewise disputes the accounts of the new forms of accumulation in new industrial spaces as put forward by Scott (1988b) and others on the grounds that they emphasize relatively insignificant and peripheral changes. Pollert, Gordon and Sayer all argue that there is nothing new in the capitalist search for increased flexibility or locational advantage. Those who promote the idea of flexibility, they suggest, are either consciously or inadvertently contributing to a climate of opinion – an ideological condition – that renders working-class movements less rather than more powerful.

I do not accept this position. The evidence for increased flexibility (subcontracting, temporary and self-employment, etc.) throughout the capitalist world is simply too overwhelming. Nevertheless, such criticisms introduce a number of important correctives in the debate. The insistence that there is nothing essentially new in the push towards flexibility, and that capitalism has periodically taken these sorts of paths before, is certainly correct. The argument that there is an acute danger of exaggerating the significance of any trend towards increased flexibility and geographical mobility, blinding us to how strongly implanted Fordist production systems still are, deserves careful consideration. And the ideological and political consequences of overemphasizing flexibility in the narrow sense of production technique and labour relations are serious enough to make sober and careful evaluations of the degree of flexibility imperative. But I think it equally dangerous to pretend that nothing has changed, when the facts of deindustrialization and of plant relocation, of more flexible manning practices and labour markets, of automation and product innovation, stare most workers in the face.

The third position, which defines the sense in which I use the idea of a transition from Fordism to flexible accumulation here, lies somewhere in between these two extremes. The current conjuncture is characterized by a mix of highly efficient Fordist production (often nuanced by flexible technology and output) in some sectors and regions (like cars in the US, Japan, or South Korea) and more traditional production systems (such as those of Singapore, Taiwan, or Hong Kong) resting on 'artisanal', paternalistic, or patriarchal (familial) labour relations, embodying quite different mechanisms of labour control. Market coordinations (often of the subcontracting sort) have expanded at the expense of direct corporate planning within the system of surplus value production and appropriation. The nature and composition of the global working class has also changed, as have the conditions of consciousness formation and political action. Unionization and traditional 'left politics' become very hard to sustain in the face, for example, of the patriarchal (family) production systems characteristic of South-East Asia, or of immigrant groups in Los Angeles, New York, and London. Gender relations have similarly become much more complicated, at the same time as resort to a female labour force has become much more widespread. By the same token, the social basis for ideologies of entrepreneurialism, paternalism, and privatism has increased.

We can, I think, trace back many of the surface shifts in economic behaviour and political attitudes to a simple change in balance between Fordist and non-Fordist systems of labour control, coupled with a disciplining of the former either through competition with the latter (forced restructurings and rationalizations), widespread unemployment or through political repression (curbs on union power) and geographical relocations to 'peripheral' countries or regions and back into industrial heartlands in a 'see-saw' motion of uneven geographical development (Smith, 1990).

I do not see this shift to alternative systems of labour control (with all its political implications) as irreversible, but interpret it as a rather traditional response to crisis.

What does seem special about the period since 1972 is the extraordinary efflorescence and transformation in financial markets. There have been phases of capitalist history – from 1890 to 1929, for example – when 'finance capital' seemed to occupy a position of paramount importance within capitalism, only to lose that position in the speculative crashes that followed. In the present phase, however, it is not so much the concentration of power in financial institutions that matters, as the explosion in new financial instruments and markets, coupled with the rise of highly sophisticated systems of financial coordination on a global scale. It is through this financial system that much of the geographical and temporal flexibility of capital accumulation has been achieved. The nation-state, though seriously weakened as an autonomous power, nevertheless retains important powers of labour disciplining as well as of intervention in financial flows and markets, while becoming itself much more vulnerable to fiscal crisis and the discipline of international money. I am therefore tempted to see the flexibility achieved in production, labour markets, and consumption more as an outcome of the search for financial solutions to the crisis-tendencies of capitalism, rather than the other way round. This would imply that the financial system has achieved a degree of autonomy from real production unprecedented in capitalism's history, carrying capitalism into an era of equally unprecedented financial dangers.

What is surprising is the way in which indebtedness and fictitious capital formation have accelerated, at the same time as massive defaults and devaluations have been absorbed, not without trauma to be sure, within the financial apparatus of overall regulation. In the United States, for example, the banking system went into the red, for the first time since 1934, in the first half of 1987 with scarcely a murmur of panic. And we need only take the secondary market value of third world debt, and multiply it by the obligations outstanding, to get a rough estimate of the volume of devaluation current within the financial system. Compared to all of this, the extraordinary fluctuations manifest in stock and currency markets appear more as epiphenomena rather than as fundamental structural problems.

It is tempting, of course, to see this all as some prelude to a financial crash that would make 1929 look like a footnote in history.

While it would be foolish to rule that out as a very real possibility, particularly in the light of the heavy losses in world stock markets in October 1987, circumstances do indeed appear radically different this time around. Consumer, corporate, and governmental debts are much more tightly tied in with each other, permitting the simultaneous regulation of both consumption and production magnitudes through speculative and fictitious financing. It is also much easier to deploy strategies of temporal and geographical displacement together with sectoral change. Innovation within the financial systems appears to have been a necessary prerequisite to overcoming the general rigidities as well as the distinctive temporal, geographical, and even geopolitical crisis into which Fordism had fallen by the late 1960s.

Two basic (though tentative) conclusions then follow. First, that if we are to look for anything truly distinctive (as opposed to 'capitalism as usual') in the present situation, then it is upon the financial aspects of capitalist organization and on the role of credit that we should concentrate our gaze. Secondly, if there is to be any medium-term stability to the present regime of accumulation, then it is in the realms of new rounds and forms of temporal and spatial fixes that these will most likely be found.

I want to stress the tentative nature of these conclusions. Yet it does seem important to emphasize to what degree flexible accumulation has to be seen as a particular and perhaps new combination of mainly old elements within the overall logic of capital accumulation. Furthermore, if I am right that the crisis of Fordism was in large part a crisis of temporal and spatial form, then we should pay rather more attention to these dimensions of the problem than is customary in either radical or conventional modes of analysis.

Scott Lash and John Urry

'After Organized Capitalism'

from *Economies of Signs and Space* (1994)

Who now reads Marx? The 1980s have surely sealed Marx's coffin for good and confined him and his monstrous works to the dustbin of history. Even if we are not at the end of history, we are surely at the end of his history, based as it was on the analysis of the unfolding contradictions of industrial capitalism. That society and those contradictions have unequivocally gone for ever.

And yet there is another Marx, not so much the theorist of industrial capitalism, more the first analyst of 'modernity'. And there is a further Marx who may have much to contribute to the analysis of those changes in social structure that seem to be sweeping all before as we approach the turn of the twenty-first century.

These turn-of-the-century changes increasingly play themselves out in the ethereal processes of time and space. If production happens at one time and at one place, circulation allows that production to vary – as commodities are cast adrift and acquire mobility to flow through changing spaces at shifting times. It is in the first part of [*Das Capital*] that Marx addresses the circuits of capital, of how one form of capital metamorphoses into another. There are three circuits, of money-capital, of productive capital and of commodity capital. Productive capital in turn consists of the means of production or constant capital and labour power or variable capital. There are thus four types of capital involved in these processes of circulation: money-capital, commodities, the means of production, and labour power. They move through space and they work to different and changing temporalities.

In contradistinction, though, contemporary Marxists introduce temporal variation with something like the following periodization. First, in nineteenth-century, 'liberal' capitalism, the circuits of the different types of capital more or less operated on the level of the locality or region, often with relatively little intersection or overlap. Second, in twentieth-century 'organized' capitalism, money, the means of production, consumer commodities and labour power came to flow most significantly on a national scale. The advanced societies witnessed the appearance of the large bureaucratic firm, vertically and in some cases horizontally integrated nationally. There was also the replacement of locally based craft unions by industrial unions whose territorial basis was 'stretched' to cover national dimensions. Commodity markets, capital markets and even labour markets took on significance across the scope of entire national economies.

Third, in the more fragmented and flexible types of production that accompany the 'disorganization' of capitalism, this circulation takes place on an international scale. At the end of the twentieth century circuits of commodities, productive capital and money qualitatively stretch to become international in terms of increases in global trade, foreign direct investment and global movements of finance.

This transformed political economy is both 'post-Fordist', in that it succeeded the era of mass production and mass consumption, and postmodern. Three of the forms of capital – money, productive capital and commodities – that we just described as circulating through international space are objects. The fourth, variable capital or labour power, is a subject. Thus the circuits of capital which Marx described are, at the same time, circuits of objects and of subjects, which are increasingly difficult to distinguish from each other. And in the shift from organized to disorganized capitalism, the various subjects and objects of the capitalist political economy circulate not only along routes of greater and greater distance, but also – especially with the rise and increasing capacities of electronic networks – at ever greater velocity.

This faster circulation of objects is the stuff of 'consumer capitalism'. With an ever-quickening turnover time, objects as well as cultural artefacts become disposable and depleted of meaning. Some of these objects, such as computers, television sets, VCRs and hi-fi, produce many more cultural artefacts or signs ('signifiers') than people can cope with. People are bombarded with signifiers and increasingly become incapable of attaching meanings to them. In this sense, of increased profusion and speed of circulation of cultural artefacts, postmodernism is not so much a critique or radical refusal of modernism, but its radical exaggeration. It is more modern than modernism. Postmodernism hyperbolically accentuates the processes of increased turnover time, speed of circulation and the disposability of subjects and objects.

Analyses of such postmodern economies and societies have dominated debate on the left and the right for the last decade. The abstraction, meaninglessness, challenges to tradition and history issued by modernism have been driven to the extreme in postmodernism. On these counts neo-conservative analysts and many Marxists are in accord. In any event not just are the analyses surprisingly convergent, but so too are the pessimistic prognoses.

Now much of this pessimism is appropriate. But there is a way out. It is to claim that the sort of 'economies of signs and space' that became pervasive in the wake of organized capitalism do not just lead to increasing meaninglessness, homogenization, abstraction, anomic and the destruction of the subject. Another set of radically divergent processes is simultaneously taking place. These processes may open up possibilities for the recasting of meaning in work and in leisure, for the reconstitution of community and the particular, for the reconstruction of a transmogrified subjectivity, and for heterogenization and complexity of space and of everyday life.

One reason why so many analysts paint such a uniformly pessimistic scenario for the future is because of a reliance on an overly structuralist conception of social process. This conception is prevalent on the political left and right, among structuralists and post-structuralists. [W]e will endeavour to correct this through focusing upon subjectivity, and in particular on an increasingly significant reflexive human subjectivity. We shall examine the causes and consequences of a subjectivity engaged in a process of 'reflexive modernization'.

What is increasingly produced are not material objects, but signs. These signs are of two types. Either they have a primarily cognitive content and are post-industrial or informational goods. Or they have primarily an aesthetic content and are what can be termed postmodern goods. The development of the latter can be seen not only in the proliferation of objects which possess a substantial aesthetic component (such as pop music, cinema, leisure, magazines, video and so on), but also in the increasing component of sign-value or image embodied in material objects. This aestheticization of material objects takes place in the production, the circulation or the consumption of such goods.

Such aestheticization is instantiated for example, in production, in which the design component comprises an increasing component of the value of goods, while the labour process as such is less important in its contribution to value-added. This is true in the sense of increased research-and-development or 'design intensity' of even industrial production. And this increased R&D intensity is often importantly aesthetic in nature, as in clothes, shoes, furniture, car design, electronic goods and the like. Consumer durables feature as a sort of built 'microenvironment', of buildings, rooms, clothes, cars, offices and so on.

[A]fter organized capitalism, there is a distinctive 'economy of signs and space'. Contemporary global order, or disorder, is in this sense a structure of flows, a de-centred set of economies of signs in space. But alongside and against these asymmetrical networks of flows there is increasing evidence of a radically other set of developments. There is evidence that the same individuals, the same human beings who are increasingly subject to, and the subjects of, such space economics are simultaneously becoming increasingly reflexive with respect to them.

Owing partly to an increased pervasion of cultural competencies and partly to a tendential breakdown of trust in the 'expert-systems' of the new order, a growing space enables such a critical reflexivity to develop. This growing reflexivity is in the first instance part and parcel of a radical enhancement in late modernity of individualization. That is, there is an ongoing process of de-traditionalization in which social agents are increasingly 'set free' from the heteronomous control or monitoring of social structures in order to be self-monitoring or self-reflexive. This accelerating individualization process is a process in which agency is set free from structure, a process in which, further, it is structural change itself in modernization that so to speak forces agency to take on powers that heretofore lay in social structures themselves. Hence for example structural change in the economy forces individuals to be freed from the structural rigidity of the Fordist labour process. That is, it is increasingly a prerequisite for structural economic change, increasingly a precondition of capital accumulation today, that the labour force becomes increasingly self-monitoring as well as develops an even greater reflexivity with respect to the rules and resources of the workplace.

Most of the now substantial literature on reflexivity has understood the phenomenon in an almost exclusively cognitive sense. Much of this literature has its origins in the sociology of science, in which reflexivity means broadly the application of a theory's assumptions to the theory itself, or more broadly the self-monitoring of an expert system, in which the latter questions itself according to its own assumptions. Subsequently sociological theory more generally has used a still very cognitive notion of reflexivity in discussions of how social agents are able increasingly to monitor and organize their own individual life-narratives and how society itself – via social science – is even more able to be self-constituting. [W]e too devote considerable attention to this cognitive dimension of reflexivity. [W]e focus here on economic life, addressing in some detail phenomena of 'reflexive production' and 'reflexive consumption'.

But running parallel to this phenomenon is another probably just as important development in late modern societies. And this is an increasing pervasion of, not cognitive, but aesthetic reflexivity. Whereas cognitive reflexivity has its origins in the rationalist and Cartesian assumptions of the Enlightenment tradition of modernity, this other dimension of reflexivity is rooted in the assumptions and practices of aesthetic modernism. If cognitive reflexivity is a matter of 'monitoring' of self, and of social-structural roles and resources, then aesthetic reflexivity entails self-interpretation and the interpretation of social background practices. Aesthetic reflexivity is instantiated in an increasing number of spheres of everyday life. In the economy itself there is an ever growing centrality of 'design'-intensive production in many economic sectors. If knowledge intensive production of goods and services is embodied in the utility of the latter, design-intensivity is embodied in the 'expressive component' of goods and services. Consumer practices will likewise be grounded in aesthetic reflexivity, as will the 'place-myths' that tourists and travellers construct and deconstruct. Aesthetic reflexivity is embodied in the contemporary sense of time – in a widespread refusal of both clock time and any sort of utilitarian calculation of temporal organization. Aesthetic or hermeneutic reflexivity is embodied in the background assumptions, in the unarticulated practices in which meaning is routinely created in 'new' communities – in subcultures, in imagined communities and in the 'invented communities' of, for example, ecological and other late twentieth-century social movements.

As we have just suggested flows and reflexivity can be substantially contradictory and counteracting phenomena. But this is not the only possibility. The individualization thesis, presupposed in the phenomenon of reflexivity, has been registered in Western social theory of the late 1980s and early 1990s. In the United States in the pervasion of rational choice theory; in Europe through the impact of theories of reflexive modernization or reflexive modernity; and in Britain in ex-Prime Minister Margaret Thatcher's contentious assertion that there is no such thing as society, but just a set of potentially entrepreneurial individuals in the context of a strongly empowered nation-state.

We agree in a rather prosaically empirical sense that earlier existing social structures do have less power in monitoring increasingly autonomous agency than in the past. But we do not argue that this entails some sort of end to the value of structural explanation *tout court*. We propose to the contrary that there is indeed a structural basis for today's reflexive individuals. And that this is not social structures, but increasingly the pervasion of information and communication structures. We propose that there is tendentially the beginnings of the unfolding of a process in which social structures, national in scope, are being displaced by such global information and communication structures. These information and communication structures are the basis of cognitive reflexivity. Thus structured flows and accumulations of images, of expressive symbols are the condition of burgeoning aesthetic reflexivity. Thus the conditions of both cognitive and aesthetic reflexivity are economies of signs in space.

Section 2.3 Geographies of Global Integration

Barney Warf

'Telecommunications and the Changing Geographies of Knowledge Transmission in the Late 20th Century'

***Urban Studies* (1995)**

The late 20th century has witnessed an explosion of producer services on an historic scale, which forms a fundamental part of the much-heralded transition from Fordism to post-Fordism (Coffey and Bailly, 1991; Wood, 1991b). Central to this transformation has been a wave of growth in financial and business services linked at the global level by telecommunications. The emergence of a global service economy has profoundly altered markets for, and flows of, information and capital, simultaneously initiating new experiences of space and time, generating a new round of what Harvey (1989, 1990) calls time–space convergence. More epistemologically, Poster (1990) notes that electronic systems change not only what we know, but how we know it.

The rapid escalation in the supply and demand of information services has been propelled by a convergence of several factors, including dramatic cost declines in information-processing technologies induced by the microelectronics revolution, national and worldwide deregulation of many service industries, including the Uruguay Round of GATT negotiations (1992) and the persistent vertical disintegration that constitutes a fundamental part of the emergence of post-Fordist production regimes around the world. The growth of traditional financial and business services, and the emergence of new ones, has ushered in a profound – indeed, an historic – transformation of the ways in which information is collected, processed and circulated, forming what Castells (1989b) labels the 'informational mode of production'.

THE GLOBAL SERVICE ECONOMY AND TELECOMMUNICATIONS INFRASTRUCTURE

There can be little doubt that trade in services has expanded rapidly on an international basis. From the perspective of contemporary social theory, services may be viewed within the context of the enormous series of changes undergone by late 20th-century capitalism. In retrospect, the signs of this transformation are not difficult to see: the collapse of the Bretton-Woods agreement in 1971 and the subsequent shift to floating exchange rates; the oil crises of 1974 and 1979, which unleashed \$375b of petrodollars between 1974 and 1981 (Wachtel, 1987), and the resulting recession and stagflation in the West; the explosive growth of Third World debt, including a secondary debt market and debt-equity swaps (Corbridge, 1984), the growth of Japan as the world's premier centre of financial capital (Vogel, 1986); the explosion of the Euromarket (Pecchioli, 1983; Walter, 1988); the steady deterioration in the competitive position of industrial nations, particularly the US and the UK, and the concomitant rise of Japan, Germany and the newly industrialising nations, particularly in east Asia; the transformation of the US under the Reagan administration into the world's largest debtor, the emergence of flexible production technologies (e.g. just-in-time inventory systems) and computerisation of the workplace; the steady growth of multinational corporations and their ability to shift vast resources across national boundaries; the global wave of deregulation and privatisation that lay at the heart of Thatcherite and Reaganite post-Keynesian policy; and finally, the integration of national financial markets through telecommunications systems. In the 1990s, one might add the collapse of the Soviet bloc and the steady integration of those nations into the world economy. This series of changes has been variably labelled an 'accumulation crisis' in the transition from state monopoly to global capitalism (Graham *et al*, 1988), or the end of one Kondratieff long wave and the beginning of another (Marshall, 1987). What is abundantly clear from these observations is the emergence of a new global division of labour, in which services play a fundamental role.

The increasing reliance of financial and business services as well as numerous multinational manufacturing firms upon telecommunications to relay massive volumes of information through international networks has made electronic data collection and transmission capabilities a fundamental part of regional and national attempts to generate a comparative advantage (Gillespie and Williams, 1988). The rapid deployment of such technologies reflects a conjunction of factors, including: the increasingly information-intensive nature of commodity production in general; the spatial separation of production activities in different nations through globalised subcontracting networks; decreases in price and the elastic demand for communications; the birth of new electronic information services (e.g. on-line databases, teletext and electronic mail); and the high levels of uncertainty that accompany the international markets of the late 20th century, to which the analysis of large volumes of data is a strategic response (Moss, 1987b; Akwule, 1992). The computer networks that have made such systems technologically and commercially feasible offer users scale and scope economies, allowing spatially isolated establishments to share centralised information resources such as research, marketing and advertising, and management (Hepworth, 1986, 1990).

Central to the explosion of information services has been the deployment of new telecommunications systems and their merger with computerised database management (Nicol, 1985). This phenomenon can be seen in no small part as an aftershock of the microelectronics revolution and the concomitant switch from analogue to digital information formats: the digital format suffers less degradation over time and space, is much more compatible with the binary constraints of computers, and allows greater privacy (Akwule, 1992). As data have been converted from analogue to digital forms, computer services have merged with telecommunications. Numerous corporations, especially in financial services, invested in new communications technologies such as microwave and fibre optics. To meet the growing demand for high-volume telecommunications, telephone companies upgraded their copper-cable systems to include fibre-optics lines, which allow large quantities of data to be transmitted rapidly, securely and virtually error-free. By the early 1990s, the US fibre-optic network was already well in place (Figure 1). In response to the growing demand for international digital data flows beginning in the 1970s, the United Nation's International Telecommunications Union introduced Integrated Service Digital Network (ISDN) to harmonise technological constraints to data flow among its members

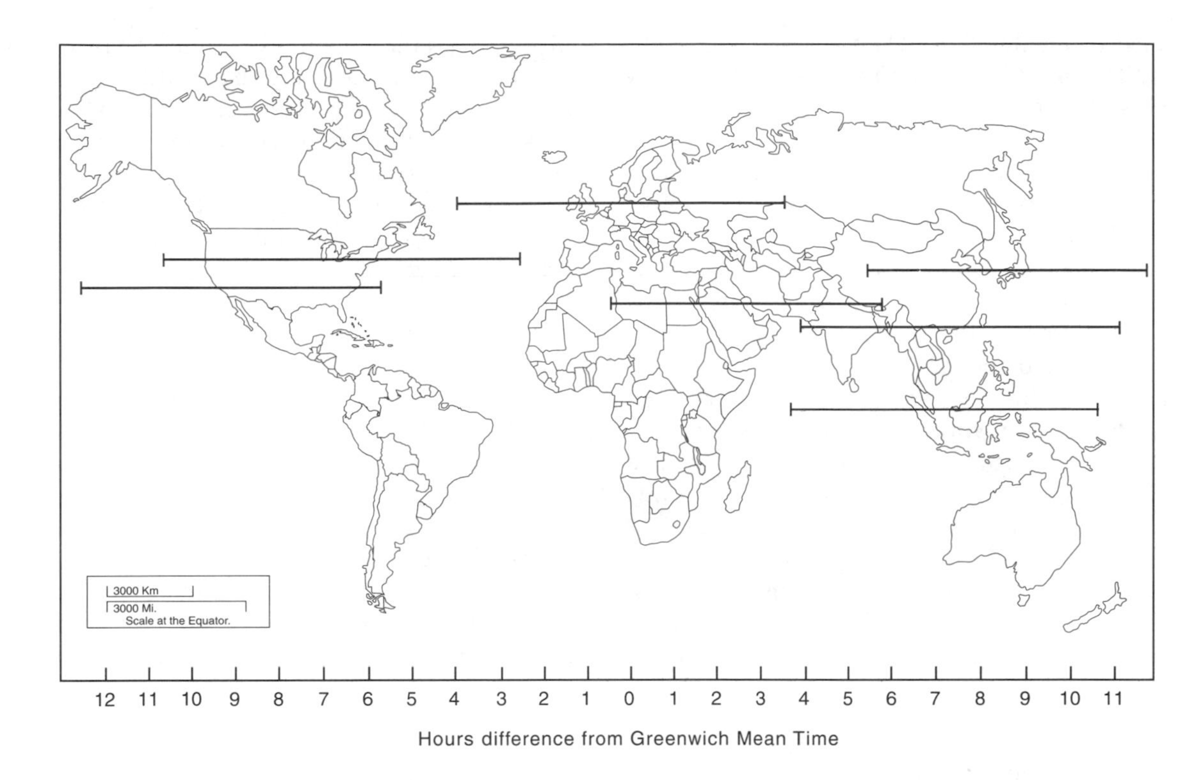

Figure 1 Trading hours of major world financial centres

(Akwule, 1992). ISDN has since become the standard model of telecommunications in Europe, North America and elsewhere.

The international expansion of telecommunications networks has raised several predicaments for state policy at the global and local levels. At the international level, issues of transborder data flow, intellectual property rights, copyright laws, etc., which have remained beyond the purview of traditional trade agreements, have become central to GATT (1985–6 and 1990–1) and its successor, the International Trade Organisation. At the national level, the lifting of state controls in telecommunications had significant impacts on the profitability, industrial organisation and spatial structure of information services.

Telecommunications allowed not only new volumes of inter-regional trade in data services, but also in capital services. Banks and securities firms have been at the forefront of the construction of extensive leased telephone networks, giving rise to electronic funds transfer systems that have come to form the nerve centre of the international financial economy, allowing banks to move capital around at a moment's notice, arbitraging interest rate differentials, taking advantage of favourable exchange rates, and avoiding political unrest (Langdale, 1985, 1989; Warf, 1989). In the securities markets, global telecommunications systems have also facilitated the emergence of the 24-hour trading day, linking stock markets through the computerised trading of stocks. As Figure 1 indicates, the world's major financial centres are easily connected even with an 8-hour trading day. The volatility of stock markets has increased markedly as hair-trigger computer trading programmes allow fortunes to be made (and lost) by staying microseconds ahead of (or behind) other markets, as exemplified by the famous crashes of 19 October 1987.

THE INTERNET: POLITICAL ECONOMY AND SPATIALITY OF THE INFORMATION HIGHWAY

Of all the telecommunications systems that have emerged since the 1970s, none has received more public adulation than the Internet. The unfortunate tendency in the popular media to engage in technocratic utopianism, including hyperbole about the birth of cyberspace and virtual reality, has obscured the very real effects of the Internet. [T]he Internet has grown at rapid rates, doubling in networks and users every year.

The Internet, which emerged upon a global scale via its integration with existing telephone, fibre-optic and satellite systems, was made possible by the technological innovation of packet switching, in which individual messages may be decomposed, the constituent parts transmitted by various channels (i.e. fibre-optics, telephone lines, satellite), and then reassembled, seamlessly and instantaneously, at the destination. In the 1990s such systems have received new scrutiny as central elements in the Clinton administration's emphasis on 'information superhighways'.

The Internet has become the world's single most important mechanism for the transmission of scientific and academic knowledge. In contrast to the relatively slow and bureaucratically monitored systems of knowledge production and transmission found in most of the world, the Internet and related systems permit a thoroughly unaltered, non-hierarchical flow of information best noted for its lack of overlords.

Despite the mythology of equal access for everyone, there are also vast discrepancies in access to the Internet at the global level (Schiller, 1993; Cooke and Lehrer, 1993). As measured by the number of access nodes in each country, it is evident that the greatest Internet access remains in the most economically developed parts of the world, notably North America, Europe and Japan. The hegemony of the US is particularly notable given that 90% of Internet traffic is destined for or originates in that nation. Most of Africa, the Middle East and Asia (with the exceptions of India, Thailand and Malaysia), in contrast, have little or no access. There is, clearly, a reflection here of the long-standing bifurcation between the First and Third Worlds. To this extent, it is apparent that the geography of the Internet reflects previous rounds of capital accumulation – i.e. it exhibits a spatiality largely preconditioned by the legacy of colonialism.

There remains a further dimension to be explored here, however; the bifurcation between the superpowers following World War II. As Buchner (1988) noted, Marxist regimes favoured investments in television rather than telephone systems: televisions, allowing only a one-way flow of information (i.e. government propaganda), are far more conducive to centralised control than are telephones, which allow multiple parties to circumvent government lines of communication. Because access to the Internet relies heavily upon existing telephone networks, this policy has hampered the emerging post-Soviet 'Glasnet'. Superimposed on top of the landscapes of colonialism, therefore, is the landscape of the Cold War.

GEOGRAPHICAL CONSEQUENCES OF THE MODE OF INFORMATION

As might be expected, the emergence of a global economy hinging upon producer services and telecommunications systems has led to new rounds of uneven development and spatial inequality. Three aspects of this phenomenon are worth noting here, including the growth of world cities, the expansion of offshore banking centres and the globalisation of back offices.

World Cities

The most readily evident geographical repercussions of this process have been the growth of 'world cities', notably London, New York and Tokyo (Moss, 1987a; Sassen, 1991), each of which seems to be more closely attuned to the rhythms of the international economy than the nation-state in which it is located. In each metropolitan area, a large agglomeration of banks and ancillary firms generates pools of well-paying administrative and white-collar professional jobs; in each, the incomes of a wealthy stratum of traders and professionals have sent real estate prices soaring, unleashing rounds of gentrification and a corresponding impoverishment for disadvantaged populations.

Long the centre of banking for the British Empire, and more recently the capital of the unregulated Euromarket, London seems to have severed its moorings to the rest of the UK and drifted off into the hyperspaces of global finance. State regulation in the City – always loose when compared to New York or Tokyo – was further diminished by the 'Big Bang' of 1986. Accordingly, the City's landscape has been reshaped by the growth of offices, most notably Canary Wharf and the Docklands. Still the premier financial centre of Europe, and one of the world's major centres of foreign banking, publishing and advertising, London finds its status challenged by the growth of Continental financial centres such as Amsterdam, Paris and Frankfurt.

Similarly, New York rebounded from the crisis of the mid 1970s with a massive influx of petrodollars and new investment funds (i.e. pension and mutual funds) that sustained a prolonged bull market on Wall Street in the 1980s (Scanlon, 1989; Mollenkopf and Castells, 1992; Shefter, 1993). Today, 20% of New York's banking employment is in foreign-owned firms, notably Japanese giants such as Dai Ichi Kangyo. Driven by the entrance of foreign firms and increasing international linkages, trade on the New York stock exchange exploded from 12m shares per day in the 1970s to 150m in the early 1990s (Warf, 1991). New York also boasts of being the communications centre of the world, including one-half million jobs that involve the collection, production, processing, transmission or consumption of information in one capacity or another (Warf, 1991).

Tokyo, the epicentre of the gargantuan Japanese financial market, is likely the world's largest centre of capital accumulation, with one-third of the world's stocks by volume and 12 of its largest banks by assets (Masai, 1989). Tokyo's growth is clearly tied to its international linkages to the world economy, particularly in finance, a reflection of Japan's growth as a major world economic power (Masai, 1989; Cybriwsky, 1991). In the 1980s, Japan's status in the global financial markets was unparalleled as the world's largest creditor nation (Vogel, 1986).

Offshore Banking

A second geographical manifestation of the new, hypermobile capital markets has been the growth of offshore banking, financial services outside the regulation of their national authorities. Traditionally, 'offshore' was synonymous with the Euromarket, which arose in the 1960s as trade in US dollars outside the US. Given the collapse of Bretton Woods and the instability of world financial markets, the Euromarket has since expanded to include other currencies as well as other parts of the world. The recent growth of offshore banking centres reflects the broader shift from traditional banking services (loans and deposits) to lucrative, fee-based non-traditional functions, including debt repackaging, foreign exchange transactions and cash management (Walter, 1989).

Today, the growth of offshore banking has occurred in response to favourable tax laws in hitherto marginal places that have attempted to take advantage of the world's uneven topography of regulation. Several distinct clusters of offshore banking may be noted, including, in the Caribbean, the Bahamas and Cayman Islands; in Europe, Switzerland, Luxembourg and Liechtenstein; in the Middle East, Cyprus and Bahrain; in southeast Asia, Singapore and Hong Kong; and in the Pacific Ocean, Vanuatu, Nauru and Western Samoa. **Roberts** (1994, p. 92) (see Part Three, Section 3.4) notes that such places 'are all part of a worldwide network of essentially marginal places which have come to assume a crucial position in the global circuits of fungible, fast-moving, furtive money and fictitious capital'.

Offshore markets have also penetrated the global stock market, where telecommunications may threaten the agglomerative advantages of world cities even as they reinforce them. For example, the National Associated Automated Dealers Quotation System (NASDAQ) has emerged as the world's fourth-largest stock market; unlike the New York, London, or Tokyo exchanges, NASDAQ lacks a trading floor, connecting half a million traders worldwide through telephone and fibre-optic lines. Similarly, Paris, Belgium, Spain, Vancouver and Toronto all recently abolished their trading floors in favour of screen-based trading.

Global Back Offices

A third manifestation of telecommunications in the world service economy concerns the globalisation of clerical services, in particular back offices. Back offices perform many routinised clerical functions. These tasks involve unskilled or semi-skilled labour, primarily women, and frequently operate on a 24-hour-per-day basis (Moss and Dunau, 1986).

Historically, back offices have located adjacent to headquarters activities in downtown areas to ensure close management supervision and rapid turnaround of information. However, under the impetus of rising central-city rents and shortages of sufficiently qualified (i.e. computer-literate) labour, many service firms began to uncouple their headquarters and back office functions, moving the latter out of the downtown to cheaper locations on the urban periphery. Most back office relocations, therefore, have been to suburbs (Moss and Dunau, 1986; Nelson, 1986). Under the impetus of new telecommunications systems, many clerical tasks have become increasingly footloose and susceptible to spatial variations in production costs.

Internationally, this trend has taken the form of the offshore office. The primary motivation for offshore relocation is low labour costs, although other considerations include worker productivity, skills, turnover and benefits. Offshore offices are established not to serve foreign markets, but to generate cost savings for US firms by tapping cheap Third World labour pools. Notably, many firms with offshore back offices are in industries facing strong competitive pressures to enhance productivity, including insurance, publishing and airlines. Several New York-based life insurance companies, for example, have erected back office facilities in Ireland, with the active encouragement of the Irish government (Lohr, 1988). Likewise, the Caribbean has become a particularly important locus for American back offices, partly due to the Caribbean Basin Initiative instituted by the Reagan administration and the guaranteed access to the US market that it provides.

The emergence of global digital networks has generated growth in a number of unanticipated places. These are definitely not the new industrial spaces celebrated in the literature on post-Fordist production complexes (Scott, 1988b), but constitute new 'information spaces' reflective of the related, yet distinct, mode of information.

CONCLUDING COMMENTS

What lessons can be drawn from these observations about the emergence of a globalised service economy and the telecommunications networks that underpin it? As part of the broad sea-change from Fordist production regimes to the globalised world of flexible accumulation, about which so much has already been said, it is clear that capital – as data or cash, electrons or investments – in the context of global services has acquired a qualitatively increased level of fluidity, a mobility enhanced by the worldwide wave of deregulation unleashed in the 1980s and the introduction of telecommunication networks. Such systems give banks, securities, insurance firms and back offices markedly greater freedom over their locational choices. In dramatically reducing the circulation time of capital, telecommunications have linked far-flung places together through networks in which billions of dollars move instantaneously across the globe, creating a geography without transport costs. There can be no doubt that this process has real consequences for places, as attested by the current status of cities such as London, New York, Tokyo and Singapore and the Cayman Islands. Generally, such processes tend to concentrate skilled, high value-added services, e.g. in global cities, while dispersing unskilled, low value-added services such as back offices to Third World locations.

In short, it is vital to note that, contrary to early, simplistic expectations that telecommunications would 'eliminate space', rendering geography meaningless through the effortless conquest of distance, such systems in fact produce new rounds of unevenness, forming new geographies that are imposed upon the relics of the past. Telecommunications simultaneously reflect and transform the topologies of capitalism, creating and rapidly recreating nested hierarchies of spaces technically articulated in the architecture of computer networks. Indeed, far from eliminating variations among places, such systems permit the exploitation of differences between areas with renewed ferocity. As Swyngedouw (1989) noted, the emergence of hyperspaces does not entail the obliteration of local uniqueness, only its reconfiguration.

Ron Martin

'Stateless Monies, Global Financial Integration and National Economic Autonomy: The End of Geography?'

from ***Money Power and Space*** **(1994)**

INTRODUCTION: THE GLOBALIZATION ISSUE

Over the past few years the concept of 'globalization' has attracted increasing attention within the social science literature. In international economics and politics, for example, the term has already become part of the standard lexicon (Holland, 1987; Gill and Law, 1988; Wallerstein, 1991; McGrew and Lewis, 1992), though not without voices of dissent (such as Gordon, 1988; **Hirst and Thompson**, 1992). Likewise, economic geographers have also begun to give prominence to the 'globalization' process and the 'global economy' (for example, Thrift, 1986; Wallace, 1990; Peet, 1991; Dicken, 1992). There is in fact a widespread acceptance that an accelerating and deepening globalization of capitalism is under way which raises several fundamental problems of analysis, theory and policy. For one thing, exactly what is meant by the concepts of 'globalization' and the 'global economy'? Different authors use these terms in different ways and with varying degrees of precision. Yet another issue concerns the relationship between the global and the local, as expressed by such notions as the 'global–local interplay' and 'glocalization'. A third and fundamental problem has to do with the impact of globalization on the nature and role of the nation state. In many ways, the question of the powers and autonomy of the state is central to the globalization issue.

There are already three divergent schools of thought on this matter. Probably the most prominent is the view that transnationalization and globalization are seriously undermining the nation considered as a geographically bounded set of political, economic and social relations. The argument is that capital is overcoming the constraints of national economic organization and regulation, subordinating nation states to global markets and forces that cross national boundaries (Radice, 1984; Gordon, 1988; MacEwan and Tabb, 1989a; Piccioto, 1991; Reich, 1991; Crook *et al*, 1992). At the other extreme there are those who argue that the claim that the nation state is being undermined by globalization is greatly exaggerated, and that individual states still exercise substantial independence and authority in the regulation and management of their domestic political economies (Porter, 1990; Pooley, 1991; **Hirst and Thompson**, 1992). Still others see the dichotomy between nation states and global capital as a false one; for them the current transnationalization of capital is a process of reorganization of the state into the 'transnational state' and, simultaneously, of the system of national regulation into one of global regulation (McMichael and Myhre, 1991).

Nowhere are these processes and debates more pertinent than in the realm of money. [I]t is within the financial sphere that globalization is arguably most developed, reflecting the much greater fungibility and convertibility of the money capital form. Indeed, according to some observers the process of financial globalization is already sufficiently well advanced to signal the 'end of geography' with respect to monetary structures and relationships (O'Brien, 1992; see also Wachtel, 1986; Ohmae, 1990).

GLOBALIZATION AND THE GROWTH OF A SUPRANATIONAL FINANCIAL SYSTEM

By globalization in this context is meant a movement beyond a system based on international financial transactions between nations. Nor is it simply synonymous with multinational banks and finance houses, or the process of internationalization associated with the increasing presence of such multinational companies in domestic finance markets. Globalization combines these elements with a strong degree of *integration* between the different national and multinational parts. It refers to the emergence of truly transnational banks and financial companies that integrate their activities and transactions across different national markets. And above all it refers to the increasing freedom of movement, transfer and tradeability of monies and finance capital across the globe, in effect integrating national markets into a new supranational system. This process of globalization stems from a number of interrelated changes: the progressive deregulation of money and financial markets, both internally and externally; the introduction of an expanding array of new financial instruments and monetary products, allowing riskier, bigger and more easily tradeable financial investments; the emergence and role of new market actors, especially institutional investors such as large pension funds; and the spread of new communications and information technologies that have extended and accelerated financial transactions.

According to Susan Strange (1971, p. 207), the Eurodollar market was the 'great technological breakthrough of international finance in the mid-twentieth century'. It had its origins in dollar deposits transferred to Europe in the immediate postwar years by Communist Bloc countries seeking to avoid their possible sequestration by the United States. [F]rom modest beginnings of about US$11 billion in 1964, the size of the Eurodollar system had grown to an estimated US$40 billion by the end of the decade. The real take-off came, however, after the 1973–4 quadrupling of oil prices by the OPEC countries. These placed a substantial part of their hugely inflated earnings of 'petrodollars' in Eurodollar accounts in the offshore branches of American banks, mainly in Europe. As the 1970s continued so the OPEC surpluses grew, and almost all of these found their way into the Eurodollar system, only to be lent out through the so-called 'petrodollar recycling' process. By the late 1970s the Eurodollar pool had grown to more than US$400 billion. Spurred on by the revolution in information processing and communications, the rapid growth of the Eurodollar system continued apace during the 1980s, reaching US$1 trillion by 1984 and an estimated US$2.8 trillion by the end of the decade.

Not only were Eurodollars the first truly supranational form of money, they fostered various other related innovations in international finance. [T]he Eurodollar market helped to breed a stateless banking system, a world of offshore trading dominated by a few dozen giant banks that operate in every corner of the globe. With modern computer technology coming on-line in the mid-1970s, it did not take long for these new offshore banking sanctuaries to become important players in the game. Similarly, in the same way that new offshore banking locations proliferated so did the range of offshore currencies: in Deutschmarks, French francs, sterling, yen and other currencies. A further related product of the growth of the Eurocurrency system has been the development of an associated market in Eurobonds, essentially securitized loans denominated in Eurocurrencies.

These developments in the stateless monies of the Euromarkets have been paralleled by the dramatic growth and increasing globalization of all the other major types of traded financial services. The switch to floating exchange rates, the growth in the need for foreign currency associated with the growth in international trade, the abolition of exchange controls and the mushrooming of the international investment flows of portfolio investors, life insurance companies and pension funds have all combined to generate a rapid expansion in the market for foreign exchange. [I]n 1979 the *daily* turnover on the world's major foreign exchange markets was reckoned to be about US$100 billion; by 1989 this was estimated to have increased to more than US$400 billion (*Euromoney*, January 1990); that is, an annual trading total probably in excess of US$100 trillion. Foreign currencies are mostly traded in six world financial centres: London, New York, Frankfurt, Bahrain, Singapore and Tokyo (*The Economist*, 1992). And within these centres the market is dominated by just a handful of global players: some 60% of the world's foreign exchange trade is accounted for by the top twenty transnational banks.

The volume of international banking, and the markets for syndicated loans, for the financing of cross-border merger and acquisition activity, and for international securities, all experienced unprecedented growth during the 1980s. For example, between 1985 and 1989 the asset value of world merger and acquisition deals, both domestic and cross-border, increased from about US$300 billion to US$1 trillion. Within this total, the value of cross-border deals rose particularly rapidly, from about US$50 billion to almost US$300 billion. [T]he global deregulation of banking activities and the opening up of world stockmarkets to foreign buyers have promoted a range of new instruments for financing merger and acquisition activity. The two processes have thus been mutually reinforcing.

While an international market in equities and bonds can be traced back to the early nineteenth century (primarily through London), the real move towards a global equities market came during the 1980s. Furthermore, the number of Third World stockmarkets has been growing, encouraged by the World Bank and the International Finance Corporation: there are now some thirty-five equity markets in developing countries, and some of these (Brazil, India, Malaysia, Korea and Taiwan) are bigger than many medium-sized European stockmarkets.

The growth of 'stateless' monies and the emergence of internationally integrated markets in an expanding array of financial services and products represent two aspects of the globalization of finance. A further dimension has been the growth of global or stateless banks and finance institutions that have offices around the world. Most financial product markets are characterized by a high degree of institutional concentration: typically, the top twenty institutions in a market account for between 40 and 60% of worldwide transactions (Smith, 1992).

The process of global financial integration is thus firmly established, and may be expected to continue, although its pace and shape will depend upon future developments in the deregulation and competition policies of individual countries and upon further innovations in financial information technologies.

THE GEOGRAPHIES OF GLOBAL FINANCIAL INTEGRATION

One of the claims of the 'end of geography' thesis is that this integration of world financial markets has undermined the significance of location, in the sense that the new communications technologies considerably widen the choice of geographical location of financial firms and allow them to serve widely dispersed markets regardless of where they happen to be. In assessing the validity of this claim it is necessary to conceptualize the impact of globalization in terms of two counterforces. In principle, modern telecommunications technologies render the need for financial centres increasingly obsolete. Market participants no longer have to be in the same centre, the same country or even the same continent for trading to take place: in terms of contact between financial firms and institutions, new information technologies allow propinquity without proximity.

Under this scenario, then, the spatial expression of global integration would be an increasing locational dispersal and specialization of markets: a relative shift of financial activity away from the established global metropolitan centres towards smaller, provincial locations, nearer to final customers.

[A]n equally plausible geographical trajectory of global integration is that the new technologies will actually reinforce the concentration of expertise and business within existing major centres, in that firms located in the latter can now easily access customers and funds wherever these are located. In other words, the new technologies allow firms even greater possibilities to develop economies of scale and scope through concentrating activities in a relatively small number of leading financial centres, selling and servicing global products from a few strategic locations. Thus, while the speed of information communication has annihilated *space* (Castells, 1989b; Harvey, 1989) it has by no means undermined the significance of location, of *place*. In theory, then, both these divergent forces, of decentralization and dispersal, and of centralization and concentration, are consistent with global financial integration, and it is the dynamic tension between them that will shape the evolving geography of the global financial system.

Technological innovation is only one of the forces at work, and its impact is dependent upon a second, that of deregulation. The wave of financial deregulation that has swept through the OECD nations over the past decade and a half has unleashed a new process of global competition. There are two interrelated aspects to this competition: between the major global financial centres themselves, as the prime financial market locations, and between financial institutions. [W]ith widespread deregulation and the move towards an increasingly 'even playing field', competition has become far more intense, and more global. The 'offshore' financial centres now face a much more open and competitive global market. And in confronting this new environment, the competitive differences of place – differences in costs, specialist skills, market opportunities, broad regional affinities, customs, any remaining restrictions, etc. – assume heightened importance as each centre seeks to maintain and improve its position within, and its share of, the global market in finance (Smith, 1992).

In short, global integration does not spell the 'end of geography' as far as the continuing overwhelming locational and trading influence of the world's financial centres is concerned; however, it does mean that market activity has become extremely sensitive to even small differences in the competitive advantage and trading performance of different financial centres. The top world centres, London, New York and Tokyo in particular, show no signs of losing their overall dominance, but as capital and trading become ever more mobile so markets shift more readily from one centre to another in response to differences and changes in transaction costs, liquidity, profits and other dimensions of relative attractiveness.

Within this context, there is the key question of whether increasing global integration is likely to promote greater convergence or divergence between the major centres in terms of their mix of different financial products and services. An 'end of geography' perspective would imply convergence, with the leading financial centres functioning as linked, largely undifferentiated 'trading nodes' within essentially uniform worldwide markets. Against this, Ricardian comparative advantage theory suggests that market deregulation and free trade should serve to accentuate existing patterns of specialization and dominance, as centres focus on the particular markets in which they already have a competitive edge. Under this scenario, different financial centres would be the primary markets for different global financial products and services. Evidence can be invoked to support both these views. One major consideration is the possibility that, geographically, global finance will evolve in the same direction that the systems of trade and foreign direct investment seem to be moving, into a 'triad' of three major regional blocs, America, Europe and the Pacific Rim (Schott, 1991; United Nations, 1991), based upon three corresponding world financial centres, New York, London and Tokyo, each having obvious competitive advantages with respect to its own regional bloc.

FINANCIAL GLOBALIZATION AND THE NATION STATE

Whatever trajectory the geographical structure of global financial integration takes, however, the question of national economic sovereignty remains a central issue. According to the 'end of geography' argument, financial market regulators are no longer able to exert control over their regulatory territories. As national financial markets have become increasingly integrated on a global scale, and the boundaries between domestic and foreign financial business have become increasingly blurred, so the 'openness' and vulnerability of the national economy to external financial shocks has increased and the ability of governments to exercise national policy autonomy has declined (Ohmae, 1990; O'Brien, 1992). In going transnational, money has outflanked the nation state by nullifying national economic policy (Drucker, 1993).

There can be no doubting the political significance of the national economic sovereignty issue. Control of the money supply, of interest rates and to some extent of the exchange rate and capital flows has traditionally been seen as quintessential to national economic sovereignty, and the development of national economic policy has always been predicated on the assumption that the state has effective jurisdiction over these monetary and financial variables. In all of the main schools of economics, the nation state is viewed as an essential unit of economy, and its role as an 'economic actor' is taken for granted. This notion of the 'national economy' is anchored in the institutions of central banking, the national currency, national income and national industry, and it underpinned the classical economics of *laissez-faire* and free trade, and even more so its twentieth-century successor, Keynesian macroeconomic demand management.

Most forms of economics, in fact, assume that capital fractures primarily along the boundaries of nation states and that the latter are sovereign actors over their domestic spaces. The rise of 'stateless monies' and the global integration of financial markets clearly challenge these assumptions. The very process of competitive financial deregulation by nation states has 'disarmed' them against the hypermobility of finance capital that deregulation has helped to unleash (Bienefeld, 1992). Individual countries seem caught in a classic prisoners' dilemma: they have moved to open up their financial systems to international competition and market forces in an attempt to attract growth, yet simultaneously have exposed themselves to financial speculation and global instability, against which national action is largely inhibited by the deregulated and globalized nature of the present system (cf. Preston and Windsor, 1992).

In some quarters this increasing national 'openness' to global financial flows and fluctuations is seen as both desirable and inevitable, as an integral part of a movement towards a new liberal economic world order, towards 'global neoclassicism' (Schor, 1990). In contrast to Keynesians and regulation-theorists, adherents of global neoclassicism by-pass the nation state and take the world as their basic unit of analysis. For them, marked differences in financial conditions and returns between nations can only be temporary, as offsetting capital flows will be rapid and powerful. Some go further and see global financial integration as leading to a single world money market and a single world capital market, with perhaps even a single world money, a single world monetary policy, and equalization of interest rates and rates of return.

Under global neoclassicism the loss of national policy autonomy is unavoidable in the face of the power of market economic forces. Any attempt by a government to follow a discretionary economic policy or to regulate its financial markets will be futile and self-defeating. The loss of national sovereignty through financial globalization is assumed to be compensated for by the increased allocative efficiency of capital markets in terms of the demand for and supply of loanable funds and investment finance. This, at least, has been the message of governments, policy-makers and the financial elite in the leading capitalist countries over the past decade and a half.

However, none of these claims is self-evident. The increasing global integration of financial markets and loss of national economic autonomy have not necessarily been because of some inexorable logic of late capitalism, or because of the driving imperatives of global competition. For it has undoubtedly been politically engineered, promoted by those same capitalist nations which, already containing the leading world financial centres, stood to gain most from encouraging a 'free market' in global finance. The crux of this change is that financial institutions have demanded and been granted less regulation and more protection at the same time that nation states themselves have prioritized the financial sphere and substituted monetary goals and policies for the employment and welfare priorities of the past. A new 'bankers' bargain' (Pringle, 1992) has replaced the former 'social bargain' with industry and labour, and in the process the state has ceded a considerable degree of its economic and political power to financiers, most of whom are transnational rather than national in orientation. The loss of national autonomy to global finance is thus not some benign outcome or necessity of world market forces, but has a political origin.

Nor are the assumed mechanisms and benefits of global neoclassicism as prominent as protagonists of the model suppose. National savings and investment rates, for example, continue to be highly correlated (Epstein and Gintis, 1992), whereas the global neoclassical model predicts they should be virtually uncorrelated. Similarly, the model suggests that the nationality mix of assets in investors' portfolios should be in rough proportion to the size of each nation in the world economy; but in practice nationals disproportionately hold assets of their own countries. Third, the equalization of short-term interest rates is not matched by the equalization of profit rates, which continue to show marked differences among the leading capitalist countries. Perhaps even more importantly, [c]apital does not always gravitate to where it is most needed or where the highest returns can be offered, but instead is typically rationed in various ways.

The global neoclassical argument that nation states should adopt a passive, accommodative stance towards financial markets is thus suspect; the more so because it is assumed that such markets are composed of myriads of small, autonomous price-taking actors, that cartels are difficult to organize and that there is no centralized rationing of funds. The global financial system is simply not like that. The institutional players in this game are very large and powerful. The importance of these large actors throws the global neoclassical view of how world financial markets work into serious doubt.

The shift towards a deregulated, globally integrated financial system, where markets are freed from and take precedence over intervention and control by nation states, does not therefore hold out the unequivocal promise of increased efficiency and wealth. Indeed, the evidence thus far suggests that the disadvantages may seriously outweigh the benefits. Compared to the postwar Keynesian era of fixed exchange rates and capital controls, the past decade and a half has seen considerable global and local financial instability. It is questionable whether this 'casino capitalism' (Strange, 1986) and the much riskier financial environment that now prevails are to be preferred to a regulated system.

CONCLUSIONS: MONEY, POWER AND SPACE

There can be no doubt that global financial integration has substantially altered and circumscribed the effective economic boundaries of the nation state. But to depict this dilution of national autonomy as the 'end of geography' is to take too narrow a view of the relationships between money, power and space. Even in the global neoclassical vision of unregulated and perfectly functioning world capital markets and complete capital mobility, financial flows and accumulation would be inherently uneven: savings and funds would continue to concentrate in the high growth countries, in much the same way that regional imbalances in the movement of capital occur between regions within the monetary union of a nation state.

Financial globalization is not 'obliterating geography'; rather it is reconfiguring the geographies of money, power and dependency. As nation states have decoupled themselves from banking capital and relinquished control over international monetary transactions, so there has been a substantial shift of power to a globally integrated hierarchy of financial centres, led by London, New York and Tokyo, and the major financial institutions located in them. These sites now shape and control the international spaces of financial flows. They constitute a system of geographical 'refuge centres' within nation states, located within and dependent upon national economic, social and political infrastructures, yet substantially shielded from national regulation.

This would not matter, perhaps, but for the increased disarray and instability that now characterizes the world financial system. If financial markets fail to fulfil their required functions, pressure for government intervention and regulation could mount. Opinions are divided over the case for and form of such re-regulation (see Adelman, 1988; Friedmann, 1991). But whatever the form, the case for re-regulation is strong: the power of global money over national economic space has already been allowed to extend too far.

Robert H. Fagan and Richard B. Le Heron

'Reinterpreting the Geography of Accumulation: The Global Shift and Local Restructuring'

***Environment and Planning D: Society and Space* (1994)[1]**

INTRODUCTION

In this paper we are concerned to take seriously the idea that 'the nation is not the 'natural' space [for] the circulation and reproduction of capital' (Bryan, 1987, p. 254). Recognition of this view is vitally important in understanding the restructuring of economic activity which has taken place since the late 1970s. Since then, there has been increased integration of production through transnational corporations (TNCs), the realisation of profit in increasingly global markets, and rapid circulation of international financial capital. Although there is now general consensus about greater integration, the concept of 'globalisation' has proven difficult to handle in both theoretical and empirical research into industrial restructuring.

Global metaphors have proliferated since the mid-1980s, with seductive ideas such as 'global markets', 'the global factory', and 'the global village' (Dicken, 1992, p. 1). Yet all of these are abstractions which can exaggerate the degree of global integration and bury the continuing importance of national and local variation (Sadler, 1992). Indeed, globalisation is now such a loose term (Robertson, 1992) that the usefulness it promised to geographers and sociologists for much of the 1980s is in danger.

Reinterpreting the geography of global accumulation has become even more urgent because of vigorous debates about the emergence of 'postmodern geographies' (Soja, 1989), and about theorising the transition from Fordism to a new regime already commonly labelled as 'flexible accumulation' (Harvey, 1989, pp. 119–97; Schoenberger, 1988; Scott, 1988a).

[C]onnections between global and local manifestations of restructuring have become a vexed question both in geography (Dicken, 1993) and in sociology (Robertson, 1992). Most of the geographical literature establishes a dichotomy between the global economy and changes occurring within specific nation-states. The categories used to express the spatially uneven impacts of structural change most commonly occur as binaries: global–local, core–periphery, centre–margin (one might add flexibility–rigidity). Reinterpreting the internationalisation of capital, however, suggests that the global–local dichotomy can be a mystification in both theoretical and empirical analysis of restructuring. A more powerful

[1] Fagan, R. and Le Heron, R. (1994) 'Reinterpreting the Geography of Accumulation: The Global Shift and Local Restructuring', *Environment and Planning D: Society and Space*, **12**: 265–285. Reprinted by permission of Pion Limited, London.

framework is needed for clarifying global and local in the 1990s in which internationalised processes of accumulation are expressed in, and reproduced through, social, economic, and political changes which remain bounded territorially within nation-states.

Adopting this view means that many path-breaking analyses of the geography of economic change in the 1980s must be assessed as partial steps towards explaining restructuring at global, national, and local scales. Collapsing the dichotomy between global and 'national' processes of accumulation strengthens both our grasp of globalisation and our capacity to guide empirical work and policy formulation on specific instances of economic restructuring, whatever the geographical scale. In rejecting the dichotomy, we argue that the notion of 'the geography of production' must be replaced with a conception of the geography of capital accumulation at a global scale. This includes, by definition, consideration of markets and international trade (realisation) and global financial flows (reproduction), as well as production. We reject the 'productionist' bias of much work on the global economy during the 1980s.

Contemporary literature on restructuring appears to contain two alternative theoretical trajectories which converge on the central question of this paper. These, loosely named, are 'internationalisation theses' and 'national restructuring themes'. In fact, each path is built up from a succession of connected theoretical developments which, largely emerging in isolation from the others, have not been brought together in a satisfactory synthesis.

INTERNATIONALISATION THESES

The New International Division of Labour (NIDL)

The NIDL thesis conceptualised capital from core industrialised countries as escaping profitability crises by finding new sources of absolute surplus value in cheap labour-pools located in the global periphery. Technological change was very important in liberating capital geographically, allowing production to shift to low-cost sites through increased automation, deskilling of labour-force requirements, and further subdivision of commodities into components, the production of some of which could be relocated to NICs (newly industrialising countries).

Further theoretical and empirical work during the 1980s exposed three serious flaws in the NIDL thesis. First, initial formulations overemphasised the role of TNCs and their control over the new geographies of production. NIDL overstressed the idea that internationalisation of production had moved away from being a struggle for control of the world's major markets towards being an 'escape' from high costs, especially labour costs, in core economies. This view severely underplayed the internal social relations of production in NICs, in particular the key roles of the state and their local capitalist classes in creating conditions for export oriented industrialisation.

Second, the importance of restructuring within developed industrialised countries themselves was buried by the over emphasis on 'spatial fixes', notably the search for low-cost labour-pools as the key strategy employed by large firms to restore accumulation. In fact, firms based in core countries have employed a 'bewildering proliferation of strategies' (Schoenberger, 1988, p. 245) in their attempts to restore profits.

Third, research into global restructuring during the 1980s demonstrated the central importance of international political factors in shaping both the long boom and the transition away from its economic structures. After 1970 the hegemonic power of the United States was challenged as Western European states moved towards greater economic and political union and, especially, by the rapid economic rise of Japan to its position from the mid-1980s as the world's major trade-surplus nation. Internationalisation of capital now strongly reflects a tripolar trading, investment, and production system linking first, the United States and Canada; second, the European Community; and third, Japan and the Asian NICs.

Table 1 Dimensions of internationalisation

Fraction	International circuit of capital[a]			Linkages with global economy[b]
	production	reproduction	realisation	
National	N	N	N	Direct: imported means of production; licences; offshore finance Indirect: local competition with firms in market-constrained and global fractions
Investment-constrained	N	I	N	Direct: exports; imported means of production; offshore finance Indirect: competition from TNCs for export markets; export franchising
Market-constrained	N (G)	N	G	Direct: TNC branch plants status; imported means of production; offshore finance Indirect: competition with firms in global fraction
Global	N (G)	G	G	Direct: part of TNC networks; global competition with other TNCs; imported means of production; offshore finance

Note. This table builds on an original schema by Bryan (1987) and includes two important modifications: two fractions are specified (national and global) with two 'hybrid' cases (investment-constrained and market-constrained); the range of empirical connections linking each fraction to the global economy are indicated.

[a] G global circuit (intracorporate); I international circuit; N nationally bound circuit.
[b] TNC transnational corporation.

Clearly, global restructuring in the 1980s has been more complex than the production relationships stressed by the NIDL thesis.

A Return to Circuits of Capital

Bryan (1987) succeeds in situating nation-states in this increasingly global capital accumulation process using the circuits of capital approach. His initial framework has been used to develop Table 1. Rather than utilising familiar divisions such as 'industrial', 'finance', or 'commercial' capital (see Andreff, 1984), Bryan focuses on 'national' and 'global' fractions and their variants. These fractions are not absolute divisions but dominant tendencies identified in relation to the internationalisation of circuits. As usual, fractions represent a higher level of abstraction than concrete categories such as individual firms or industry sectors.

A *national* fraction of capital (N) is defined in which movement of capital within the three circuits is largely confined to a particular nation-state. Firms within this fraction produce domestically for sale primarily in local markets, and allocate locally sourced finance to reproduction. The *investment-constrained* fraction (I-C) is a variant containing those firms which produce nationally and both raise and reinvest finance domestically, but which sell a significant proportion of output (realise profits) on world markets. Firms in the I-C were the major exporters during the long boom.

By contrast, a *global* fraction (G) is defined in which movement of capital within all three circuits is internationalised. Local production commonly takes place within global networks of TNC branch plants, realisation depends on world markets, and financial capital is obtained and reinvested in reproduction at a global scale. Since the rise to greater importance of this genuinely global fraction since 1980, an increasing proportion of commodity trade, especially in manufactures, has actually been international product transfer inside corporate groups.

An important variant, the *market-constrained* fraction (M-C) is also recognised in Table 1. It contains those firms which principally serve domestic markets but whose circuit of reproduction is internationalised. In other words, financial capital can be obtained outside the nation-state (foreign investment) but, in addition, surplus generated locally could be invested overseas.

The framework collapses any primary analytical distinction between domestic and 'foreign' firms or capital. This follows from the theoretical insight that it is not 'nationality', *ownership*, or sector which identify these fractions of total capital, but the ways in which production, sales, and investments are linked to the global capital accumulation process. In concrete cases, even accumulation in the national fraction is affected strongly by internationalisation of capital and global restructuring because of these links. Spatial reorganisation and even *in situ* restructuring of these national firms can thus become part of 'global shift'. Nation-states have thus encompassed a diminishing part of the geography of the capital accumulation process over time. This tendency has strengthened as the global fraction has become more important.

Yet in no way has this reduced the importance of nation-states in shaping global economic processes. Capital still *requires* nation-states to secure economic, social, and political conditions under which any accumulation can continue. Capital, labour, and the state interact in specific societies and cultures which are *territorially bounded*.

These local interactions are not subordinate to the globalised accumulation process as has been argued in various globalist theses of the diminishing economic power of nation-states. Although arguing that global financial integration in the 1980s reduced the capacity of nation-states to create specific regulatory territories, O'Brien (1992) ultimately finds the subtitle of his book – *The End of Geography* – unsustainable. Deregulation and other policy changes by the world's major economic powers have exerted a major influence on global integration, yet 'despite the apparent globalness [*sic*] of events, the motivation for these changes is primarily domestic' (O'Brien, 1992, p. 19).

NATIONAL RESTRUCTURING THEMES

Spatial Division of Labour (SDL)

Since 1980 one of the two most detailed bodies of empirical scholarship on industrial restructuring has been work on spatial division of labour (SDL) within developed countries (see, especially, Allen and Massey, 1988; Massey, 1984; Massey and Meegan, 1982; Sayer and Walker, 1992).

Despite these major contributions, [three] problems have confronted attempts to relate this work to internationalisation of capital. First, empirical work on SDL has not attempted specifically to theorise relations with the global economy. Second, empirical findings about restructuring of relations between capital, labour, and state in regions and localities would be expected to be bound by features of the society in question (Fagan, 1989); yet this has often occurred to the point of ignoring internationalisation altogether. Third, much geographical analysis in the 1980s was limited by its bias towards manufacturing (Sayer, 1985); only recently has work on agriculture, service industry, and the emerging information economy been brought into the same framework.

The Regulationist School

Some of these problems can be resolved through insights provided by the regulationist school, the second and, in geography more recent, of the two most influential bodies of scholarship on restructuring. Two interlinked categories are central to this approach, although they have often been misunderstood in recent debate (Tickell and Peck, 1992). According to Lipietz (1986, p. 22) the *regime of accumulation* and the *mode of regulation* are structural and institutional arrangements within capitalist nation-states which 'have succeeded because they ensured some regularity and permanency in social reproduction'.

The regulation school provides a basis for relating global geographies of accumulation to national spatial divisions of labour. Production, consumption, and circulation of money are central features of both accumulation and regulation. They were all central to Fordism, and analysis of the breakdown of this regime and its modes of regulation in various countries is the main project of the regulationist school. The degree of internationalisation in the circuits of capital during the Fordist long boom varied from country to country.

Regulationists need no convincing that the nation-state is the basic unit for reproducing capitalist relations of production and for establishing modes of social regulation. Hence the Fordist regime was regulated in different ways in different countries. Conversely, a particular mode of regulation is not a sufficient condition for stable accumulation (Scott and Storper, 1987; Storper and Scott, 1989). Both Lipietz (1984, 1986) and Jessop (1988a) emphasise the 'chancy' and 'improbable' nature of a regulatory order being compatible with continued accumulation.

Since the mid-1980s, the regulationist framework has been increasingly influential in economic geography (see Harvey, 1989, pp. 121–4). Partly because of its original application to Fordism in the United States, North American geographers have gone furthest in using the regulationist framework to theorise the transition to a new regime of accumulation in which Fordist mass production is replaced by a new technological paradigm (Storper and Scott, 1989) and 'flexible specialisation' (Scott, 1988a). Three major problems with this analysis are relevant. First, changes in production, new technologies, and spatial agglomerations of the new industrial forms, have been central to the flexibility thesis. Evidence from inside and outside North America, however, suggests a widespread tendency in the flexibility literature to overgeneralise from specific cases (Gertler, 1988; Hudson, 1989; **Sayer and Walker**, 1992, pp. 190–5). Second, many geographers adopting the regulationist framework have argued that its attraction lies in integrating the underlying economic processes of accumulation with the indeterminate social behaviour of agents and institutions (Cooke *et al*, 1992, p. 40). Yet, although there has been much debate about flexible *accumulation*, there has been far less research into flexible modes of *regulation* (Peck, 1992a; Tickell and Peck, 1992). Third, and arising from this, there has been very little attention paid, at least by the regulationists, to the increasingly integrated global economy and how this shapes the new modes of regulation, including new spatial divisions of labour, which could emerge within countries.

Jessop's (1988a) review of the regulationist literature highlights three matters of relevance in attempts to synthesise internationalisation and regulationist frameworks. First, Jessop distinguishes between *structural* categories (accumulation regimes, modes of regulation, hegemonic structures) and *strategic* categories (accumulation strategies, regulatory strategies, or norms of conduct) around which the exercise of state power is centred. This suggests a need to analyse the strategic restructuring behaviour of individual firms and organisations and breaks down the suggestion of a hegemony of the structural categories. Second, Jessop highlights the different ways in which interests of capital are articulated in given accumulation strategies. Third, he identifies forms of state intervention which have differential implications for particular accumulation strategies (Jessop, 1983, 1988a). These concepts place a premium on specific relations between capital, labour, and state.

TOWARDS A SYNTHESIS

The argument developed thus far indicates that important common ground can be found between the circuits of capital approach to internationalisation and the literatures from both the SDL and the regulationist perspectives. In the first approach, the state mediates conditions for the global expansion and integration of the capital accumulation process whereas in the second, national modes of regulation and spatial divisions of labour develop as accumulation is reproduced. Whereas globalist theses are severely deficient without these national modes, regulation theory is deficient in its treatment of globalised accumulation.

These complex local interactions give rise to changes in the degree of internationalisation in the circuits on the left-hand side of Figure 1. [T]he precise articulation between a nation-state and the evolving global economy cannot be read off from abstract theory but must be *determined empirically* for each regime or period of restructuring. [A] geography of accumulation is a mosaic produced at a series of interacting scales. This synthesis avoids separating the geography of the global economy and spatial divisions of labour within nation-states. It also rejects the common 'globalist' implication that the global scale dominates all others. [W]e accept the global space as the 'natural space' for capital accumulation in the 1990s but suggest that it can be comprehended only at the level of the nation-state and below. Globalised accumulation has been an uneven and not inexorable process in practice.

METHODOLOGICAL IMPLICATIONS OF THE FRAMEWORK

The process of globalisation since 1980 has not involved simple progression from predominantly national to global accumulation. For a given nation-state, the impacts of globalisation vary according to contingent circumstances and there is a two-way flow between global and local. Empirical research shows a great deal of variety at the level of individual firms and enterprises. Such variations determine the ways in which national economies are inserted into global accumulation. Table 2 suggests some of this greater variation, recognising a variety of local-scale ensembles of capital labour–state relations which shape the spatial division of labour.

The framework can be employed in sectoral or enterprise studies to relate globalisation to national and local change. This analytical task involves examining the intersection of globally organised industry with other fractions of capital at all centres of production in the commodity chain from raw material to finished product (Le Heron, 1988, 1990; Le Heron *et al*, 1989). Approaching economic activities through the entire commodity chain has the great advantage of tying together raw material, manufacturing, marketing, and consumption stages (the French concept of the *filière*). Dicken (1993) suggests more detailed research on these chains in a global context is needed urgently. The framework shown in Figure 1 enhances the commodity chain approach because it overcomes a common deficiency in its treatment of the circuit of finance, especially in relation to conglomerate enterprises which commonly cut across several commodity *filières*. This is now prevalent in food-industry commodity chains, for example where globalising corporations have made massive investments in food processing, attracted by the cash flows necessary to support their debt-raising activities in the circuit of reproduction (see Fagan, 1990; Fagan and Rich, 1991).

Commodities whose production is dominated by firms in the global fraction cannot be approached simply through strategies of the TNCs which control production. Interactions between the global fraction and other fractions determine the regional and local patterns of restructuring, as Gibson (1990) demonstrates clearly for the Australian coal industry. Nor is restructuring determined entirely by capital, state, and labour in the workplace. Gibson (1991) also shows that the relationships are strongly gendered, and processes within households have an important bearing on the construction of these 'global' coal regions. Australia's largest coal exporter, BHP Ltd, still a predominantly Australian-controlled company, moved from being a silver miner in the I-C fraction, through a long period as a nationally bound and vertically integrated monopoly steel producer at the core of Australia's national fraction, to its status since the mid-1980s as a global minerals and energy corporation with nearly one third of its productive assets offshore (Fagan, 1986).

Regimes of Accumulation	Circuits of internationalisation			Nation-state (fractions)	Social relations of production in nation-states		
	Production	Realisation	Reproduction		Capital (C) C-C	Labour (L) C-L	State (S) C-S
Extensive				N I-C			
Restructuring crisis				N I-C			
Intensive ('Fordist')				C I-C M-C (G)			
Restructuring crisis				M I-C M-C G			
Integrated				N I-C (M-C) G			

Global geography — Spatial division of labour

Geography of accumulation

G = Global
I-C = Investment Capital
M-C = Market-constrained;
N = National
( ) = Indicate emergent or relief fractions

Figure 1 A model of globalisation

Table 2 Geographical areas of accumulation

Fractions of total capital[a]	Connections of individual capitals to circuits					
	production	inputs	sales	reinvestment of surplus	ownership	competition
National (N)						
1	N	N	N	N	N	N
2	N	N,G	N,G	N	N	N,G
3	N	N,G	N	N	N/G	N,G
4	N,G	N,G	N,G	N	N	N,G
5	N	N/G	G,N	N/G	N	N
Global (G)						
6	N,G	N,G	N/G	N,G	G,N	N,G
7	N,G	G	N,G	N,G	G	N,G
8	N,G	G	G	G	G	G

Note. We are indebted to the late Steve Britton for suggesting this figure and for urging the demonstration of the empirical variety likely in concrete cases consistent with the fractions of capital identified theoretically. Britton developed his case from Fagan (1990, p. 652).

[a] *National* – Definition based on dominant circuits of capital as in Table 1. Includes family and publicly owned multiplant firms: (1, 2) without overseas operations, major exporters constituting the investment-constrained fraction of Table 1; (3) licence and franchise holders; (4) with overseas branch plants; (5) producer marketing organisations with overseas investments. Regional varieties are not shown but patterns of connection could be determined in the same way. *Global* – Subsidiaries or branch plants of TNCs: (6) serving primarily domestic markets (the market-constrained fraction of Table 1); (7) with domestic sales and exports; (8) no domestic sales (as from export processing or free-trade zones). Production systems (for example, resource-based commodity-chains involving amalgams of enterprises and sectors), and some diversified or conglomerate enterprises, will consist of *combinations* of connections.

Hence, globalisation of capital accumulation through production, trade, and financial flows since 1980 has had local outcomes which include at least the following seven:

(1) increasing centralisation of capital through large-scale mergers and acquisitions, funded increasingly by loans from transnational banks (Green, 1990);
(2) decreasing concentration of capital in some kinds of production (Allen, 1988a) as globalising firms dispose of local assets to restructure debt or rationalise lower-profit production;
(3) increased competition within corporate divisions and between firms inside nation-states leading to widespread technological changes and further rationalisation of production [Sadler (1992) details the specific expression of this process in Britain where the formerly state-owned steel industry was heavily rationalised in preparation for its privatisation];
(4) an upsurge in strategic alliances (Cooke *et al*, 1992, p. 76), formal and informal production, marketing, and technology agreements between firms operating in the major markets of the global triad, often bringing about new relationships between firms in the global and national fractions;
(5) the development by TNCs since 1980 of more or less complete production chains within each of the major triad market areas, to take account of differences in market characteristics, labour and regulatory environments [a strategy common to recent Japanese penetration of United States and Western Europe, described by Cooke *et al* (1992) as 'global localisation'];
(6) fragmentation of production, for example through increased subcontracting, and the emergence of so-called 'flexible' systems

(Sayer and Walker, 1992, pp. 162–90; Shutt and Whittington, 1987);

(7) the appearance in developed countries of locally marginalised labour processes such as sweatshops and outworking (Lipsig-Mumme, 1983; Mitter, 1986; Peck, 1992b; Soja, 1989) as an alternative to 'offshore' relocation or low-cost imports from NICs.

[T]he Australian federal system vests power over resources, utilities, regional and local issues, in the hands of individual state governments. As a result, rivalry between these governments during the long boom led directly to spatial fragmentation of [market-constrained] [M-C] branch plants [These are firms which principally serve domestic markets, but whose circuit of reproduction is internalised]. This was encouraged indirectly by the federal government's high levels of tariff protection which tightly controlled or eliminated competition from imports faced by Australia's relatively high-cost plants of national and M-C firms. Defensive mergers among larger national firms produced similarly fragmented spatial divisions of labour, usually after mergers between the leading firms in each state submarket with production plants maintained in each state capital city. Geographical fragmentation, in a relatively high-wage environment, became one of the principal cost problems for Australian manufacturers during the 1970s.

Deregulation of the circuit of realisation through the reduction of tariffs by successive Australian governments since the mid-1970s has favoured firms in the global and I-C fractions but has been opposed by firms in the national and M-C fractions (some of whom, such as Ford Australia Ltd, are in a fully global fraction elsewhere). The Australian government's floating of the currency, and deregulation of the national banking system in the early 1980s, also favoured global and M-C fractions at the expense of the other two. Accelerated penetration of the Australian economy by international financial capital after 1984 partly reflected firms' financing their globalisation strategies with overseas loans. The debt burdens often speeded up domestic rationalisation to increase cash flows and secure debt-servicing. There was also increased incentive for firms to engage in offshore production within the markets in whose currency the international loans had been raised. An example is the attempted penetration of United States and Western European markets by Australian-based TNCs, often those producing a commodity (such as beer) with a well-known brand name.

It follows that deregulation is just as much an intervention in the capital accumulation process as was the Keynesian regulatory regime which it is designed to replace. The spate of deregulation in capitalist countries during the 1980s is held by regulation theorists to be part of the breakdown of Fordism, and is seen by many geographers and sociologists as part of a struggle to create new modes of regulation to support a regime of flexible accumulation. Deregulation has, among other things, acted as positive intervention by the state on behalf of the global fraction to facilitate more rapid inflow and outflow of finance, and changing global product sourcing and marketing strategies.

Last, this framework offers much to revived debates over the 'nationality' of capital in a globally integrated economic system. Clearly, it makes little sense to ascribe national *origin* to internationalised financial capital nor even national *liability* for debt owed to transnational banks by enterprises in the global fraction. National statisticians, however, have no choice but to do this in calculating their country's balance of payments. In the case of economies such as Australia, long vulnerable to the impacts of global changes such as commodity price and currency fluctuations, chronic balance of payments deficits have become a potent political force during the 1980s (Daly, 1993). Governments have used the size of 'Australia's' international debt to legitimise keeping a tight control over domestic wage-levels and major reductions in per capita government expenditures on social welfare and infrastructure (Fagan and Bryan, 1991). Indeed, recourse to the idea of globalisation, presented as inevitable or beyond the control of government, has often been part of an ideological smokescreen designed to legitimise the introduction of economic rationalist policies. The parallels with use of the concept of 'flexibility' are striking (see Gertler, 1988; **Pollert**, 1988, p. 72).

CONCLUSION

In this paper we have shown that a fruitful convergence can be achieved between theorising the internationalisation of capital and restructuring within nation-states. We have sought to develop a framework for reinterpreting the geography of accumulation in which the stark dichotomy between global and domestic production is abandoned without removing the national and local scales from their crucial positions of importance. This development in theoretical and empirical work has become urgent as a result of increased, but widely misunderstood, global integration since 1980.

There is a need to reconstruct appropriate theoretical categories at different scales for specific contexts. Terms which are employed commonly at both macro- and meso-scales, such as internationalisation, global capital, monopoly capital, domestic firms, deindustrialisation, and even the spatial division of labour, need reinterpreting in the light of the two-way determination between global and local. Interactions between capital, labour, and state determine both the ways in which national economies are linked to global accumulation and the changing spatial divisions of labour within the countries. Yet global changes are also transmitted to national and local scales via these concrete links.

Vigorous debate will continue about the nature of new regimes of accumulation emerging from the restructuring crises of the 1970s and 1980s. Although recent empirical research has cast doubts over the generality of 'flexible specialisation' as the regime emerging to replace Fordism in developed capitalist countries (Gertler, 1988; Hudson, 1989; Lovering, 1990), there is little doubt that any new regime(s) will be shaped strongly by globalisation.

Section 2.4 The Rise of New Capitalisms

John Brohman

'Postwar Development in the Asian NICs: Does the Neoliberal Model Fit Reality?'

***Economic Geography* (1996)**

THE NEOLIBERAL VERSION OF DEVELOPMENT IN THE ASIAN NICS

Proponents of neoliberal development strategies point to the performance of the Asian NICs to validate outward-oriented, market-led development models. The development experience of the Asian NICs is seen as the result of an evolutionary process of industrially induced modernization and structural transformation, which the remainder of the South could replicate. Growth and development in the NICs are viewed as natural, inherent properties of open capitalist economies in which market forces are allowed to operate with little state interference.

Accordingly, neoliberals stress lessons derived from the supposedly laissez-faire elements of NIC policies. Third World countries are called on to drop their obsolete *dirigiste*, state-centered development strategies in favor of a neoliberal approach based on policies supposedly reflecting the successful market-led development experience of the Asian NICs. These policies include the elimination of exchange-rate controls and restrictions on international trade, deregulation of the financial sector, privatization of state enterprises, creation of an unregulated labor market, specialization according to 'comparative advantage' and market driven resource allocations, and generally defining a 'minimalist' role for the state in development (Balassa, 1981, 1991; Bhagwati, 1986; Krueger, 1986; Lal, 1983; Sachs, 1985).

Neoliberals contend that while the Asian NICs created conditions for sustained export-led growth based on enhanced international competitiveness, Latin American countries propped up an obsolete, inwardly oriented development model through expanded international borrowing and increased state intervention.

For neoliberals, this divergence in development strategies explains the contrast between high growth rates and rising per capita incomes enjoyed by the Asian NICs and the vicious circle of indebtedness, inflationary pressures, stagnant growth, and declining standards of living in Latin America (Balassa, 1991; C. Lin, 1988; J. Lin, 1989; World Bank, 1983, 1985, 1987). Therefore, neoliberals contend that Latin America (as well as much of the rest of the Third World) should abandon outmoded state-centered, inward-oriented development strategies in favor of a market-led, outward-oriented model that reflects the successful NIC experience.

Table 1 GDP growth rates in the Asian NICs and major Third World regions, 1960–1990 (percentage per year)

	1960–70	1970–80	1980–90
Asian NICs			
Hong Kong	10.0	9.2	7.1
Singapore	8.8	8.3	6.4
South Korea	8.6	9.6	9.7
Taiwan	8.8	9.8	7.7
Third World regions			
East Asia and Pacific	5.9	6.7	7.6
Latin America and Caribbean	5.3	5.4	1.7
Middle East and North Africa	n.a.	4.6	0.2
South Asia	3.9	3.5	5.6
Sub-Saharan Africa	4.2	3.6	1.7

Source: World Bank (1982, 1992, 1993); Republic of China (1975, 1991); and personal correspondence with Michael Hee, International Economics Dept., World Bank, May–June 1994.

Table 2 Share of exports in GDP (percentage)

	1960	1970	1980	1990
Asian NICs				
Hong Kong	70.9	92.2	88.0	133.9
Singapore	163.1	102.0	207.2	189.0
South Korea	3.4	14.1	34.0	31.0
Taiwan	10.5	25.2	47.8	48.4
Third World regions				
East Asia and Pacific	6.4	6.1	19.1	25.1
Latin America and Caribbean	14.8	12.6	16.0	16.8
Middle East and North Africa	n.a.	n.a.	42.2	31.5
South Asia	6.8	5.4	7.7	9.3
Sub-Saharan Africa	23.6	20.6	30.4	28.3

Source: World Bank (1982, 1992, 1993); Republic of China (1975, 1991); and personal correspondence with Michael Hee, International Economics Dept., World Bank, May–June 1994.

Table 3 Growth rate of exports in the Asian NICs and major Third World regions, 1965–1990

	Average annual growth rate of exports (%)	
	1965–80	1980–90
Asian NICs		
Hong Kong	9.1	6.2
Singapore	4.7	8.6
South Korea	27.2	12.8
Taiwan	18.9	12.1
Third World regions		
East Asia and Pacific	8.5	9.8
Latin America and Caribbean	−1.0	3.0
Middle East and North Africa	5.7	−1.1
South Asia	1.8	6.8
Sub-Saharan Africa	6.1	0.2

Source: World Bank (1992).

THE MACROECONOMIC DEVELOPMENT RECORD OF THE ASIAN NICs

The development performance of the Asian NICs has indeed been spectacular. Table 1 shows that the NICs have enjoyed strong gross domestic product (GDP) growth from 1960 to 1990, even as growth rates slowed in the South. Among the 60 countries covered in a study of per capita GDP growth from 1960 to 1985, Taiwan placed second, Hong Kong fourth, Singapore fifth, and South Korea sixth. Much of this growth resulted from increasing exports. The export share of GDP in the NICs increased rapidly between 1960 and 1990, while it remained constant or declined in most other areas of the South (Table 2). The growth rate of exports in the NICs has consistently remained well above the average in all of the South's major regions (Table 3). The Asian NICs also greatly increased their share of total world exports and Third World exports, particularly of manufactures (Table 4).

While it should be remembered that such aggregate figures may mask growing inequalities between specific groups, the NIC record of generating strong growth with relative equity, especially when compared to other Third World countries, must be deemed remarkable.

OBJECTIONS TO THE NEOLIBERAL DEPICTION OF THE ASIAN NICs AND LATIN AMERICA

Recently, a number of objections have been raised to the neoliberal 'spin' placed on the development performances of the Asian NICs and Latin America. For Banuri (1991, p. 9), 'the

Table 4 Export penetration of the Asian NICs and Third World countries, 1965–1990

	Share in world exports			Share in Third World exports		
	1965	1980	1990	1965	1980	1990
Total exports						
Asian NICs	1.5	3.8	6.7	6.0	13.3	33.9
New Southeast Asian NICs[a]	1.5	2.2	2.4	6.2	7.8	12.4
Total Asian NICs[b]	3.0	6.0	9.1	12.2	22.1	46.3
All Third World	24.2	28.7	19.8	100.0	100.0	100.0
Exports of manufactures						
Asian NICs	1.5	5.3	7.9	13.2	44.9	61.5
New Southeast Asian NICs[a]	0.1	0.4	1.5	1.1	3.8	12.0
Total Asian NICs[b]	1.6	5.7	9.4	14.3	48.7	73.5
All Third World	11.1	11.8	12.9	100.0	100.0	100.0

Source: World Bank (1993).

Notes

[a] New Southeast Asian NICs are Indonesia, Malaysia and Thailand.

[b] Total Asian NICs are the original Asian NICs (Hong Kong, Singapore, South Korea, Taiwan) and the New Southeast Asian NICs.

identification of "successful" Asia with openness, and "successless" Latin America with illiberalism is little better than a crude caricature'.

Moreover, strategies followed by the Asian NICs diverged substantially from the neoliberal ideal of laissez-faire (Appelbaum and Henderson, 1992; Bradford, 1987; Hart-Landsberg, 1993; Vogel, 1992; Wade, 1992, 1993). The NICs established strict controls over their external sectors to maximize benefits from trade and reduce vulnerabilities to fluctuations in global financial and commodity markets. Meanwhile, outward-oriented policies that increased financial openness and deepened dependency on global financial and commodity markets made Latin American economies vulnerable to fluctuations in global markets and capital market shocks, contributing to macroeconomic imbalances (e.g. Dietz, 1992; Hughes and Singh, 1991).

Latin American socioeconomic and political structures also make it difficult to replicate the East Asian model of export-oriented industrialization (EOI) based on labor-intensive manufactures. While primary-export development historically has been largely insignificant to the Asian NICs, many Latin American countries were inserted into the world economy as exporters of primary commodities.

Because most Latin American countries have had a relatively long history of labor mobilization, policies designed to support EOI by reducing real wages and standards of living would have encountered extensive opposition (Amadeo and Banuri, 1991). By contrast, wage levels during the initial stages of Asian NIC industrialization were already at levels low enough to derive a comparative advantage on world markets. Labor organizations in East Asia were too weak and fragmented to exert much political influence. Such historical variations within societal structures imply that elements of development models are only rarely directly transferable from one Third World region to another.

THE ROLE OF THE STATE IN NIC DEVELOPMENT

Neoliberals contend that NIC development is based largely on the successful implementation of a laissez-faire growth strategy. High NIC growth rates supposedly are based on an absence of state economic intervention and the ability of markets to operate smoothly, without undue regulation (Balassa, 1988, 1991; Hughes, 1988; Riedel, 1988). However, this free-market explanation of NIC development is criticized, especially from the experience of Taiwan and South Korea (Appelbaum and Henderson, 1992; Haggard, 1990; Hart-Landsberg, 1993;

Hughes, 1988; Kearney, 1990; Vogel, 1992; Wade, 1990, 1992, 1993). In their haste to fit NIC development into an ideologically driven model of free-market growth, neoliberals ignore considerable contradictory evidence.

Only Hong Kong could be said to have followed a laissez-faire type of development strategy; even there, the government's 'positive nonintervention' policies are involved in a broad range of activities (public housing, public services and social welfare, export promotion, economic diversification, and technological change). State intervention elsewhere has played a key role in stimulating growth and facilitating structural change.

Studies need to theorize the state within its broader context of social relations and structures. In particular, the rise of specific state forms and actions in the NICs is closely interrelated with patterns of capital accumulation and social reproduction which necessarily involve classes and social groups in complex processes of conflict and accommodation. These, in turn, are related to broader social structures evolved under particular historical conditions in each NIC.

The NIC experience supports the claim that an activist state may spur growth and development, particularly for 'late industrializers.' The evidence points to a 'supply-push' development model, in which the state plays a key role in stimulating capital formation and accelerating structural change (Bradford, 1987, p. 314). Rather than laissez-faire, the NICs are examples of 'guided market economies' in which state intervention is focused on 'strategic industries' based on criteria such as global demand elasticity and the potential for technological progress (Chang, 1993; Oman and Wignaraja, 1991; Onis, 1991).

Rather than conforming to free trade and laissez-faire, NIC development strategies more closely resemble classical Listian mercantilism (Burmeister, 1990; Hoogvelt, 1990; White, 1988). The nineteenth-century German economist Friedrich List ([1844]1916) claimed that comparative advantage was a doctrine of the dominant; the dominated could expect to derive little advantage from it. Instead of allowing their markets to be dominated by established industrial powers through free-trade policies, List counseled late industrializers to protect strategic infant industries to strengthen their productive forces. An examination of early European industrialization reveals that most countries pursued Listian policies of economic nationalism with striking parallels to contemporary NIC strategies, including strong state intervention, infant industry protection, and 'temporary dissociation' from international competition during early industrialization (Hoogvelt, 1990, pp. 354–5).

The Asian NICs demonstrate that selective policies supporting ISI [import-substitution industrialization] and domestically oriented sectors may be compatible with export promotion. State economic intervention promoted selective market opening in internationally competitive sectors, while protecting local markets for ISI and other noncompetitive sectors. This enabled the NICs to use the rational core of comparative advantage to enlarge their participation in international markets, while simultaneously providing conditions for more internally articulated development utilizing a range of domestic resources.

State efforts to broaden both specific technical expertise and general educational levels have been facilitated by NIC cultural traditions that revere education and achievement. As in Japan previously, the land-poor NICs stressed humans as their greatest resource to propel modernization and development. This underscored the importance of education, the spread of information, the learning of new skills, and, above all, the enhancement of human capacities to participate in the structural changes needed to create a new technologically advanced, industrially based society.

THE INFLUENCE OF EXTERNAL FACTORS ON NIC DEVELOPMENT

In addition to being shaped by internal conditions, the development performance of the

NICs was also influenced by external factors related to their geographic location and the historical period of their export-led industrialization drives. The NICs were presented with unusually favorable development opportunities both by their position within the 'new international division of labor' during a period of unparalleled global economic expansion (Browett, 1985; Douglass, 1993; Gereffi and Wyman, 1990) and by their geographic location, which enabled them to take advantage of Western geostrategic concerns and an evolving regional division of labor with Japan at the head.

[T]he location of the NICs in East and Southeast Asia has given them special development advantages that few other countries enjoy. Hong Kong occupies a pivotal position astride the trading routes between Northeast and Southeast Asia and is the main link to the outside world for southeast China. Singapore is strategically situated on the Strait of Malacca, which funnels trade flows between the Pacific and Indian oceans. It is also centrally located relative to the rest of Southeast Asia, facilitating its rise as a regional financial, commercial and administrative-managerial center (Parsonage, 1992). Moreover, all of the NICs (especially South Korea and Taiwan) are ideally located to expand trade and other ties with Japan. Important complementary factors among the economies of Japan, the NICs, and surrounding Asian countries have fostered a regional division of labor that has been profitable for all concerned (Emmerij, 1987; Kim, 1993).

The economic advance of the Asian NICs was also facilitated by broader global conditions that may be fast disappearing. Much of any country's development story typically may be attributed to external circumstances and events beyond its control. This was true of Western Europe during the Industrial Revolution, the New World during the nineteenth and twentieth centuries, and the NICs in the 1960s and 1970s (Kearney, 1990, p. 198). Rapid NIC development based on export-led industrialization was accelerated during this period by a number of fortuitous external circumstances: reduced transport costs and trade barriers for industrial products entering the United States; intensified competition within many U.S. industrial sectors; unparalleled growth in the world economy, particularly in the United States and other Organization for Economic Cooperation and Development (OECD) countries; and enhanced comparative advantages for labor-intensive products in the NICs relative to the capitalist core.

THE CHANGING GLOBAL CONDITIONS FACING ASPIRING NICS

Some new conditions may make it difficult for recent Third World industrializers to replicate the NICs' development experience (Gereffi and Wyman, 1990; Harris, 1987; Wade 1992). These conditions include the collapse of global financial circuits resulting from Third World indebtedness; crises within the market-widening strategies of many countries; the rise of new productive technologies permitting previously exported manufacturing to return to First World countries; global economic stagnation and uneven growth, both within and between countries; and the spread of protectionist sentiments, especially within the capitalist core versus Third World products. For Bello and Rosenfeld (1990a,p. 57), such conditions mean that the NIC export-led growth model may be running out of steam just as it has been enshrined as the new development orthodoxy by neoliberal theorists.

In addition to demand-side constraints, new Third World industrial exporters are also facing difficult supply-side conditions associated with increased competition from existing and aspiring NICs and from older industrialized countries trying to maintain their manufacturing base.

New production technologies and marketing techniques are also permitting the reimportation of some industries to the capitalist core that previously had been located in peripheral areas (Ariff and Hill, 1986; Jenkins, 1985; Harris, 1991). These new techniques require high quality control, increased flexibility, and rapid decision making to respond to sudden market changes – all of which enhance the locational advantages of First World rather than Third World sites.

Many new Third World exporters are thus faced with stiff competition within limited markets from both aspiring and already established NICs in a fiercely contested succession process (Athukorala, 1989). The ability of aspiring NICs to replace the original NICs depends especially on the capacity of the original NICs to shift production into higher-value, more technologically advanced sectors, leaving labor-intensive sectors at the lower end of the export market to the new arrivals.

In effect, this has subjected the NICs to a 'structural squeeze' in which they are able to graduate into only a few advanced capital-intensive sectors and are priced out of their older labor-intensive sectors by rising wage levels (Bello and Rosenfeld, 1990; Clark and Kim, 1993). The succession process by which Third World countries are supposed to gain upward mobility has largely been blocked.

SOME NEGATIVE ELEMENTS OF THE NIC MODEL

[T]he broader consequences of NIC development also deserve serious scrutiny – especially in debates over the appropriateness of the NIC model for other Third World countries. Among areas deserving more attention are the democratization process, respect for personal liberties and basic human rights, freedom of association, distribution of income and wealth, equality of opportunity among classes and social groups, working and living conditions, and environmental sustainability. Many authors (Amirahmadi, 1989; Amsden, 1989; Bello and Rosenfeld, 1990a; Douglass, 1993; Hart-Landsberg, 1993; Ogle, 1990; Petras and Hui, 1991) claim that the NICs have sacrificed progress in these areas to an all-out pursuit of rapid growth, thereby diminishing their usefulness as models of development for the rest of the South.

The role of the state in directing NIC development has not been confined to direct economic planning or exerting controls over economic institutions. Authoritarianism, repression, the exercise of strict social control, and the disciplining of the working class and other popular sectors to serve the accumulation interests of capital have also been central elements of the NICs' national development projects. None of the NICs have made much progress in creating democratic structures that would facilitate meaningful political participation by the majority.

To restrict dissent and ensure compliance to state-directed development goals, all of the NICs created large internal security apparatuses. But in South Korea and Taiwan concern for security was extreme. Both of these countries became internationally notorious for the extreme repression carried out by their internal security forces against labor, farmers, students, and popular organizations.

While the record of the NICs concerning respect for personal liberties, basic human rights, and democratization has generally been deficient, their record is more mixed in areas such as employment, poverty reduction, wage and income levels, and working conditions.

Notwithstanding the benefits that development has conferred on workers, they have also paid a high price for the NICs' export success. State policies are designed to heighten accumulation opportunities in key sectors by ensuring corporations the cheapest, most productive, and least militant workers possible. The state has disciplined the working class to accept these conditions by a number of means, including state control of labor organizations, restrictions on freedom of association and other repressive labor laws, state-directed violence against labor activists, and weak or nonenforced legislation concerning work hours and workplace conditions (Addison and Demery, 1988; Amsden, 1989; Bello and Rosenfeld, 1990a; Hart-Landsberg, 1993; Ogle, 1990).

Although other social groups have also been systematically exploited, young women have borne a disproportionate burden. Much of the labor force in export-oriented industrial sectors is composed of a youthful female 'temporary' proletariat working during the transition between school and marriage (Lin, 1989; Park, 1993). These women are concentrated in entry-level, shop-floor industrial jobs with low pay and long hours – jobs that are left vacant by older workers because the wages are too low to support a household.

The economic rationale behind the use of 'temporary' young women in export manufacturing is readily apparent – because the work force is female, transitional between generations, and generally does not have to support a family, a true 'living wage' does not have to be paid (Lin, 1989). In many cases, social reproduction rests on the mobilization of all members of extended families.

CONCLUSION

Given the weight of evidence that contradicts it, the neoliberal explanation of NIC development should be rejected. Rather than conforming to the neoliberal model of free trade and laissez-faire, the Asian NICs more closely resemble guided market economies in which an activist state has pursued policies of economic nationalism and classical Listian mercantilism.

The NIC development performance was also molded by a number of geographic and historical conditions, both internally and externally, that have been largely ignored by neoliberal explanations. Moreover, while the NICs have made great strides in some areas of development, there are also serious shortcomings to their development model, including authoritarianism and repression, pervasive environmental degradation, and the systematic exploitation of certain social groups, especially young women. These considerations call into question the transferability of the NIC development experience, via the formal, universal model of neoliberalism, for other Third World countries.

Maurice T. Daly

'The Road to the Twenty-First Century: The Myths and Miracles of Asian Manufacturing'

from ***Money Power and Space*** **(1994)**

The industrialization of Asia was, perhaps, the most spectacular economic happening of the second half of the twentieth century. Asian nations (more specifically those of the Pacific edge of Asia) compressed the process of industrialization; Japan achieved in 35 years an industrial transformation that had taken Britain and Germany over 50 years, and then Taiwan and the Republic of Korea (South Korea) reduced the span to 15 years. Moreover, this industrialization had an immediate effect on the rest of the world.

Japan had become the world's richest nation (per capita) and a clutch of other Asian countries had displayed the world's fastest rates of income growth. The future of Asia was critical to the future shape of the world economy. 'By the early twenty-first century Japan, Taiwan, South Korea and China will probably have as much weight in the world economy as North America or Europe. Taiwan and South Korea will be as rich as Great Britain and Italy' (Wade, 1990, p. 4).

The particular relationships created between finance and industry have been vital to Pacific Asia. The status of this relationship is critical to understanding the direction of change in the 1990s, especially in Japan. The other essential factor is the development of technology in the new industrial age, and Pacific Asia's place within this.

THE WORLD ECONOMIC ORDER OF THE 1990S

The growth enjoyed by so many nations in the decades after the Second World War was a product of the liberal trading order established under the leadership of the USA.

The eventual collapse of outright US hegemony spelt the end of the old system that had so favoured Asia. Although there is debate over the extent of this collapse (Nau, 1990) and optimistic predictions that accommodations can be made to sustain the major working features of the old system (Keohane, 1984; Maswood, 1989), the reality is that the world faces a phase of reconstituting its organizational forms; and no-one can know for certain just what structures will emerge. Investment and production will therefore, at least for the short term, face an unscripted, unknown and dramatically changing environment. It is uncertain whether Pacific Asia will be better placed than its competitors to succeed in the fluid atmosphere of the 1990s.

The reasons for this conclusion lie in two somewhat paradoxical features of the new environment: the emergence of what Ohmae (1989) and others have called Triad economics; and the compelling pace of technological change, which challenges the corporate organizational forms that were adequate for the previous period. The regionalization of markets, and perhaps production systems, favours Europe and North America, while Asian structures have adapted more quickly to the demands of the new technologies.

Regionalism and globalization pose immense challenges to corporations seeking to succeed in the 1990s. The secret of Japan's successful assault on world markets lay in an emphasis on standard products and components rather than on a range of items tailored to different markets. Cost savings, through standardization and concentration and quality control and reliability became the weapons. Integrated industrial policies designed to move companies in a timed sequence to improve consistently the value-added component became the setting. The giant trading companies (*sogo shosha*) provided the marketing edge.

The structures of the old global system allowed Japan to concentrate its production processes at home. When cost pressures or environmental concerns pushed manufacturing offshore it was generally to neighbouring Asian nations which acted as extensions of the export platform Japan had created at home. Of the top dozen Japanese firms producing motor vehicles, earth-moving equipment and electronic goods in 1986, half relied on exports for over 50 per cent of their sales and the remainder for over 25 per cent (James, 1989, p. 36).

Japanese, and other Asian, corporations now have to develop organizational forms that will work in the era of Triad economics. European and North American consortia have developed more flexible forms and more spatially distributed operating systems. US and European corporations have been investing in each other's regions over a long period.

INTERNATIONALIZATION AND CORPORATE STRUCTURES

Japanese Corporate Structures

As its firms captured ever larger shares of world markets it became accepted that Asia had produced the most effective corporate structures for successfully operating in the global arena; the *kigyo keiretsu* of Japan and the *chaebols* of Korea are outstanding examples. The *kigyo keiretsu* are large industrial groupings normally formed around a bank or a general trading company (*sogo shosha*). Surrounding the core are complex webs of subsidiaries and affiliates with large degrees of cross-shareholding; they operate in a wide variety of enterprises but do not possess an integrated corporate structure in the American or European sense. Around the members of the *kigyo keiretsu* cluster a very large number of small firms engaged through sub-contracting arrangements in providing components of the production process. The total picture is of a dense interrelated manufacturing complex in Japan, replicated nowhere else in the world.

For more than 30 years after the Second World War relatively open markets were dominated by standardized products that were enormously responsive to cost reductions through scale economies. This system allowed fairly simple global strategies to be effective. It favoured the preferred option of Pacific Asian manufacturers: to rely on national (or regional) production bases which could exploit the special benefits of the *kigyo keiretsu* and *chaebol* systems. They then exploited weaknesses in the marketing and distribution systems of other countries.

The Japanese corporate form invoked the traditional concept of *ie*, or perpetual descent group. From this flowed the cornerstones of Japanese governance: implicit contracting founded on trust, extensive reciprocal shareholdings and implicit reciprocal trade agreements, managerial incentives aligned to overall corporate growth and early selective intervention in the case of problems by key stakeholders (Kester, 1991, p. 12).

The system has received critical support from government. At least from 1952, when the Enterprises Rationalization Promotion Law was introduced, the government has provided subsidies, dialogue and direction to promote technological change. The choice always revolved around those technologies likely to transform the established existing pattern.

The other significant factor in giving Japanese companies a competitive edge was the close association with banks, and the banks' association with government. Without a real securities market from the late nineteenth century through the early twentieth century the company–bank relationship became particularly intimate. The *zaibatsu* banks that emerged provided long-term credit and emergency funds as well as day-to-day banking needs.

The central role of the banks as credit providers continued in the postwar period. The banks are also at the centre of the information-sharing system that so marks the *kigyo keiretsu.* The banks help to stabilize corporate performance over time, and when problems appear they can make selective interventions. The special links between companies, banks and government are also cited as a reason why Japanese companies have been able to operate with a less stringent focus on profit maximizing.

The Japanese exploited four approaches to competitive advantage in their global advances. They concentrated on building layers of advantage, they searched out weak spots ('loose bricks') in their competitors' armoury, they kept changing the terms of the engagement and they competed through collaboration (de-skilling competitors through alliances and out-sourcing deals).

Japan as Exemplar

Japan charted its own industrial course and so became a model for others to follow. Korea, a land with even fewer resources than Japan, adopted much of the Japanese methodology and by the mid-1980s was widely acclaimed as having reproduced the Japanese miracle. By the mid-1980s within Asia the expectation, indeed the certainty, was that it was only a matter of time before nations such as Malaysia, Thailand and Indonesia repeated the lesson and achieved the desired status of a NIC (newly industrializing country).

Korea maintained a strong import-substitution regime for target industries and consumer goods organized through the *chaebols* which dominate the economy (in 1989–90 the four largest had sales equivalent to half Korea's GNP and 40 per cent of its exports). As early as the 1970s, Korea was alarmed by the decline of US hegemony and made a deliberate thrust towards long-run future international competitiveness and industrial transformation (Woo, 1991, p. 11). The state shouldered the risk of investment in lumpy projects with a long gestation period. Crucial to the process was a credit-based financial structure controlled by the state.

Taiwan, the other major success story of East Asia, contrasts with Korea in that the export base is dominated by small to medium firms (in 1985 accounting for 65 per cent of manufacturing exports and 40 per cent of total production). In this more differentiated environment the state has still played a key role. In many sectors public enterprise was used for the initial push (fuels, chemicals, mining, fertilizers, food processing) and in the 1950s and 1960s public enterprise played a large part in setting the synthetic fibres, metal and shipbuilding industries on their way. Established large-scale private firms were often exposed to administrative guidance, and the structure at large was supported by import controls, tariffs, entry requirements, domestic content requirements, investment incentives and concessional credit.

Japan has had another significant, and recent, influence on the region, both as a producer of information technology and as a pioneer of new technology. The beginnings of the industry in the region came with the dispersal of various stages of semiconductor assembling from the USA (and to a lesser extent

Europe). [B]y 1986, Japan had come to dominate the semiconductor industry and Korea and Taiwan were making great strides in developing their own industries.

In the 1980s Korea and Taiwan both moved to become major players in information technology. In 1990 Korea was the world's sixth largest producer of electronics; its US$24.1 billion was well behind the USA (US$199.1 billion) and Japan (US$188.9 billion) but was expected soon to overtake both France (US$26.1 billion) and Britain (US$26.1 billion). Only Germany (US$40.6 billion) seemed to be securely ahead among the European producers.

The history of the electronics industry in Taiwan has some distinct similarities to Korea's. From very early days the government fostered industrial technology through national science and technology development plans, and research and development institutes in strategic areas. In 1990 Taiwan ranked eighth in the world among electronic producers (US$14.1 billion). Singapore (US$13.2 billion) was ninth, Hong Kong eleventh (US$8.1 billion) and Malaysia fifteenth (US$6.1 billion).

THE SWEEP OF PACIFIC ASIAN GROWTH

With good cause most attention has been paid to Japan and the larger Asian NICs in the considerable literature seeking to explain the growth of manufacturing and exports in the region. Less attention has been given to how the region as a whole is structured and to the growth of levels of intraregional trade in the 1980s.

At the centre of this regional change is the Japanese economy, and the dominating agent of change has frequently been Japanese investment. Japan in the 1980s replaced the USA, the pioneer investor in the region, as the major force (Tables 1, 2 and 3).

There are two important aspects of Japanese investment in Asian manufacturing (see Table 2). Within ASEAN, Japanese companies are producing a spatially specialized system of production; for example, there is an intraregional distribution of auto parts established by Japanese car-makers whereby production of different parts of certain models is shared among different countries.

Table 1 Japanese FDI, 1969–1989 (US$ billion)

Region	1969–73	1978–84	1986–89
World	8.3	49.2	170.3
North America	1.9	16.1	82.1
Asia	2.0	12.0	21.0
Europe	1.8	6.0	34.0

Source: JETRO (1991, p. 10).

Table 2 Japanese FDI in Asia, 1988–1990

Region	US$ billion		
	1988	1989	1990
Taiwan	0.37	0.49	0.45
Hong Kong	1.66	1.90	1.79
Singapore	0.75	1.90	0.84
Thailand	0.86	1.28	1.15
Malaysia	0.39	0.67	0.73
Indonesia	0.59	0.63	1.11

Source: Japanese Ministry of Finance.

Table 3 Japanese FDI in manufacturing: small and medium firms, 1980–1989 (percentages)

Region	1980	1985	1986	1987	1988	1989
World	100	100	100	100	100	100
Asia	57.6	63.5	64.9	72.7	65.6	64.7
NICs	36.4	30.7	47.0	46.3	26.8	20.4
China		21.1	10.8	6.4	8.1	7.9
Other Asia[a]	21.2	11.7	7.1	20.0	30.7	36.4

[a] Includes China 1980.

Source: JETRO (1991, p. 32).

The second aspect has been the spread of small to medium Japanese firms throughout the region (see Table 3). The number of such investments made throughout the world averaged 300 in the first half of the 1980s but then doubled each year to peak at 1625 in 1988. Manufacturing is the major part of this investment and most (67.4 per cent) of it occurs in Asia. There has been a pronounced shift to ASEAN countries: in 1986, 47 per cent of total investment was in Asian NICs, 10.8 per cent in China and only 7.1 per cent in the rest of Asia. By 1989, investment in the NICs had fallen to 20.4 per cent, in China investment fell to 7.9 per

Table 4 South Korean FDI, 1980–1989 (US$ billion)

Region	1980	1985	1986	1987	1988	1989
South East Asia	1.6	17.7	7.2	131.1	41.5	124.1
North America	11.1	10.8	76.1	71.4	41.6	31.7
Latin America	2.9	40.8	2.4	4.5	9.9	58.9
Europe	0.7	0.9	5.6	6.8	18.8	18.3
World	21.1	117.8	172.0	397.3	212.9	492.5

Source: Bank of Korea.

cent and investment in the remainder of Asia (principally ASEAN) climbed to 36.4 per cent of the total.

The NICs entered periods of solid surplus in the second half of the 1980s. Taiwan's current account surplus was US$7.0 billion in 1984, peaked at US$18.0 billion in 1988 (when foreign reserves reached US$76.7 billion) and was US$11.3 billion in 1990. Korea's current account as US$0.9 billion in deficit in 1985, climbed into a surplus of US$4.7 billion in 1986, peaked at US$14.2 billion in 1988 then fell, recording a deficit of US$2.1 billion in 1990. Korea's external debt fell from US$46.7 billion in 1985 to US$29.4 billion in 1989, while reserves grew over the same period from US$7.7 billion to US$22.5 billion. The Asian NICs became exporters of capital.

In 1989 Korean FDI rose by 93 per cent; there were 369 projects valued at US$0.927 billion, 49 per cent of which were in manufacturing (see Table 4). Investments in South East Asia grew at three times the rate of general investments.

The fear of trade blocs engendered by the EC single market and the North American Free Trade Agreement has raised the question of defensive retaliation within Asia. Loosely, there has been talk of a yen bloc, given Japan's role as the centre of gravity of the region's economies. There is reluctance about this because of old fears concerning Japanese dominance, and the practical consideration that less than 12 per cent of the region's trade (excluding Japan's own trade) is denominated in yen and less than 15 per cent of the region's reserves are held in yen (*Far Eastern Economic Review*, 11 October 1990, p. 73).

Australia led the way in establishing the APEC (Asia–Pacific Economic Cooperation) framework. This was viewed suspiciously by ASEAN nations, largely because of the presence of the USA and Canada in the group. Its aims of providing an economic forum where issues such as trade can be discussed represents a vague response to the realities of the northern blocs.

In December 1990 Malaysia proposed the establishment of an East Asian Economic Grouping (EAEG), a trade bloc containing ASEAN with the East Asian economies and China that would exclude the USA, Europe, Australia and New Zealand. In October 1991 Thailand produced a modified proposal that would create within ASEAN an ASEAN Free Trade Area (AFTA); this would be phased in over 10 to 15 years, and would incorporate the Indonesian proposal of a Common Effective Preferential Tariff (CEPT) in the transition period. The problem with the idea is one that had always dogged ASEAN: the countries tend to produce similar items and have little trade with each other. The proportion of exports from countries in ASEAN to other ASEAN members has never exceeded 20 per cent, and in 1991 was only 10 per cent (*Far Eastern Economic Review*, 24 October 1991, p. 64).

Pacific Asia, which appears so threatening to Europe and North America, is not well equipped to retaliate against increased protectionism. However much individual countries are reluctant to embrace the idea, and however reluctant the region's giant is to accept the leading geopolitical role, the essence of a regional response to world trends lies with Japan.

INTO THE FUTURE: NATIONS, REGIONS AND THE GLOBAL SPHERE

The Financial Mould

The struggles across the global geopolitical scene reflect the resurgence of nationalism (sprung from the collapse of the dual superpower administration of the world system) and the emergence of protectionist trade blocs. The contrasting fact is that production and marketing systems integrated over vast geographic areas are technologically more possible than ever before. Three factors in particular condition the ability of nations or regions to succeed in this situation:

- their place in the international financial markets and their ability to prosper from or control the effects of those markets;
- their ability to produce within the information-intensive production systems;
- their ability to produce and integrate a range of new technologies.

A Deregulated World

The global financial system was profoundly transformed after the final collapse of the Bretton Woods agreement in 1973. The primary change was a philosophical shift towards deregulation; while being accepted in principle, the fact of deregulation was not nearly as widespread as its advocates insisted. Asian nations in general resisted opening up their systems.

Instability in exchange and interest rates, mobility of funds, great liquidity, successive streams of new products to take advantage of the 'deregulated' markets and an emphasis on short-term profits were the trademarks of the new financial system.

From 1973 to 1988 the assets of the top 300 international banks increased seven-fold to US$15.43 trillion, a rate of increase that was double the growth of world GDP (De Carmoy, 1990, p. 111). By the end of the 1980s international trade was running at a level of around US$3000 billion per year while spot and forward exchange operations were between 50 and 75 per cent of that level *each day*. The financial world began to dominate the industrial world.

There were substantial shifts in the geographic division of financial power over two decades of change. Japan emerged as the clear winner. Japan became the world's major creditor nation and in the final years of the 1980s seven (and at one time eight) of the world's ten largest banks were Japanese. Japan became the major force in the principal money market centres, especially London.

Japan as a Financial Force

Japan had conquered the world's financial markets, just as before it had triumphed in so many of the world's manufacturing export markets. A long Japanese reign in the financial sphere seemed inevitable, and contributed substantially to the conclusion, drawn inevitably by so many, that the twenty-first century would belong to Pacific Asia. The proposition needs careful examination; the future shape of the Japanese financial system is critical not only for the future prospects of Asia but to the behaviour of the global financial system.

Japanese finance, immensely strong to the world at large, has a degree of brittleness within. Property investment, both at home and abroad, became a Japanese obsession. The stock market soared to extraordinary levels. Manufacturing companies turned to *zaiteku*, financial engineering, as a major source of profit.

Urban property prices increased two and a half times over the 1980s. In 1990 Japan placed a theoretical value of US$14,000 billion on its property, land in Tokyo was selling at US$400,000 a square metre and Tokyo suburban houses were selling at 37 times average earnings. Land in Japan was valued as being four times as great as in the USA, a country 25 times larger.

Rising to its peak on 29 December 1990, the Nikkei stock market index stood at 38,915; the index had risen by 285 per cent between 1982 and mid-August 1987, and it then climbed a further 46 per cent to 1990 in defiance of the falls in other parts of the world. It fluctuated around a declining trendline throughout 1990 and 1991, reaching a low of 21,457 in 1991. In April 1992 it had fallen to 16,598, 57 per cent below the 1989 peak.

The problem in Japan is that all aspects of the financial system are intimately related. When Yasushi Mieno, head of the Bank of Japan, pushed up interest rates in 1990 and 1991 (eventually to a peak of 6 per cent) and restricted the money supply, alongside Ministry of Finance directives limiting lending for real estate, all segments of the Japanese system came under pressure.

The Tokyo Stock Exchange (TSE) has a very different market from those in other countries. Because of the immense web of cross-shareholdings associated with the *keiretsu*, 70 per cent of shares are normally never traded. The overheated markets pushed firms into issuing equity in the form of convertible bonds and warrant bonds; this, as well as the great earnings of the corporations, weakened their traditional dependence on the banks. Warrants are Eurobonds that give the buyer the option to purchase the bonds at fixed prices in four to five years time. By issuing dollar-denominated warrants in London and then using swaps to convert the money into yen the corporations were able to raise capital at what appeared to be very low (0.5 per cent) or negative rates of interest. The warrants were like time-bombs (and US$115 billion worth were issued between 1987 and 1989): harmless in a rising market but deadly in a collapsing market unless the options are mispriced in the company's favour.

Much of the capital raised on the stock market did not go into new production facilities but into the greater rewards offered in the stock and property markets. The fall in the stock market has occurred; and the property market has slipped significantly over recent years. In 1991 bankruptcies in Japan reached record levels, and most of these were property-related. By mid-year 1992, 70 per cent of property developers were reported as no longer paying the interest on their loans (*The Economist*, 11 April 1992, p. 72).

At the heart of both the shaky stock and property markets were the banks. When the stock market fell it wiped out US$20 billion of bank capital, leading to a scramble to make it up.

The property downturn affected the entire Japanese banking industry. In June 1990 the trust banks had 52 per cent of their total loans held against property collateral. The 30,000 lending institutions in the non-bank sector had a US$158 billion exposure to property (*The Economist*, 16 February 1991, p. 67). The biggest fears surround the 131 regional banks and the 452 *shinkin* (credit union) banks who married an over-exposure to property with a vigorous setting-up of *tokkin* accounts.

In 1991 banking frauds of the order of US$5 billion were exposed and security compensation deals worth US$1.3 billion came to light. Japan's major banks and security firms were involved. Ritual resignations were made by the heads of some organizations and eventually by the Minister of Finance. A deep malaise remained, however, at the heart of the Japanese financial system.

Japan cannot be isolated from the world financial system. Curiously, despite the number of Japanese institutions among the leading banks, they are relatively no more powerful than the American banks that preceded them. What is more, the Japanese banks have not been very profitable. They adopted similar tactics to their industrial cousins: seeking 'loose bricks' in the fabric of world banking and then exercising their financial clout by pushing margins down to very slim levels.

When in late 1987 the Bank for International Settlements (BIS) introduced new regulations that required banks to hold capital-to-assets ratios of 8 per cent by March 1993, the entire complexion of Japanese banking was changed.

Japanese banks were caught in a web of debt at home and abroad. They faced great challenges in meeting international capital requirements. Their traditional links to the industrial sector were punctured by the very success of *zaiteku*.

Japan will remain a significant force in the world financial scene, but the challenges both at home and abroad are such that it will be as much controlled by as controlling the system.

Asia in the 1990s

Asia's prospects for the next decade rest with Japan itself, and the way in which the regional economies can be integrated to withstand the threats of Europe and North America.

The key thing in Japan's favour is its commanding drive into high-tech industries. It has achieved this by various means.

(1) It has developed clear strategic intents, and built core competences that enable them to succeed.
(2) Japanese companies have been brilliant exponents of competing through collaboration: they have used licensing, outsourcing and joint ventures to gain technology, to calibrate competitors' strengths and weaknesses, and to penetrate markets.
(3) When necessary, Japanese companies have made strategic acquisitions; Sony's celebrated takeover of CBS is an example. Such acquisitions are generally not financially motivated (as were the majority of US and European takeovers during the mergers and acquisitions wave of the 1980s). They usually involved firms with whom the Japanese had prior production and marketing arrangements, and they then often proceeded through joint venture arrangements and finally led to equity positions.

Asia is at the competitive forefront of the next generation of information products. Japan's competitors have followed with some alarm Japan's progress in moving from being a follower in the technological race to becoming a leader.

Japan and other Pacific Asian nations are likely to continue their high-performance record in R&D because the importance of research is accepted throughout the whole community: governments assist and reward research developments, and firms direct enormous resources to the process.

Japanese corporations have hundreds of different projects under way at the same time, and this leads to introductions of new products at regular intervals. Resource allocation is not made on the basis of single products, however, but a whole series of products and extensions and variations. The Japanese method is to run multiple projects with short lives (at NEC the average life of a project is under six months). Thus, cultivating core competences does not involve vastly outspending rivals on individual elements. Rather, it is the structure and environment of the R&D that counts. The corporation, in Kodama's terms (1989, p. 201) shifts from being a place for production to being a place for thinking.

While Japan and other Asian nations will maintain a high profile at the cutting edges of technology their success will be dependent on transforming the 'export-island' corporate structures into production systems that can overcome the barriers of a more protectionist world. The ability to do this is greatly complicated by the problems facing the finance industry.

Japan's attitudes to finance were in tune with its overall view of the world. As it became the world's biggest creditor nation and the world's largest earner of trading surpluses Japan sought to protect its own finance system from the rest of the world. This was impossible and the strains inevitably became too great. The stock market and banking crises of the early 1990s were the result.

The collapse of the stock market has challenged the delicate web of cross-shareholdings that hold together the *kigyo keiretsu*. Banks are less capable of performing the salvage operations that they performed in the past while the collapsing share market and falling profits place each element of the groupings under strain. The special features of the system that served the Japanese export island so well in the open trading decades from the 1950s begin to look precarious in the restricted era of the 1990s.

Japan might be the heart of the Pacific Asian economy but it is not the whole. Might the more highly integrated regional economy be capable of relating to external challenges and the problems of Japan itself? The answer is probably no.

There is no clear shape to the emerging regional structure. Japanese, Korean and Taiwanese investment and manufacturing growth have created an East Asian industrial core with strong linkages to the ASEAN periphery. This represents the most obvious form of regional cohesion but there are many other complicating regional forces. China is a conundrum.

At another level other forces are at work. North and South Korea, China and Russia have explored a giant free trade area near their common borders. Singapore has taken a leading role in linking the economies of Malaysia and Indonesia with itself. Thailand has become a major investor from Burma in an arc across Indo-China. And to the north Central Asia, with its array of newly independent countries (Azerbaijan, Kazakhstan, Uzbekistan, Kirghizia and Tajikistan) introduces another complication to the maze that is regional Asia.

Without doubt Pacific Asia, with the world's largest population, the world's youngest population, the world's fastest growing population and the fastest growing incomes, will remain a vital and dominant segment of the world's economy. Its economic passage in the 1990s, however, will not be as comfortable or as predictable as the seers of the early 1980s believed.

Alison Stenning and Michael J. Bradshaw

'Globalization and Transformation: The Changing Geography of the Post-Socialist World'[1]

INTRODUCTION

The focus of much economic geography in recent years has been an attempt to reconcile the study of increasingly global processes of change with a continuing, even strengthening, attention to the difference that space makes. Until very recently, however, this work has concentrated on the western world (primarily North America and Western Europe), complemented by a body of work dealing with issues of underdevelopment, aid and most recently, post-colonialism in the so-called Third World. The states of the former Soviet bloc[2] – the second world – have been for many years largely absent from mainstream economic geography, and this exclusion has rested on two processes.

On the one hand, a preoccupation with the uneven development of *capitalism* within western economic geography permitted writers and researchers to produce otherwise excellent work which almost entirely omitted any discussion of the Soviet states. Implicitly it seems many western geographers have practised a regional geography very much centred on Western Europe and North America, and all but ignored structures and processes in a part of the world which is home to nearly 400 million people (see Table 1). On the other hand, study of the Soviet states developed in relative isolation from the theoretical and philosophical developments in the social sciences more generally. As the sociologist Michael Burawoy writes, 'Soviet studies constituted the Soviet Union [and, to a lesser extent, east central Europe] as "special" and so cut themselves off from developments in other areas of social science' (Burawoy, 1992, p. 778). Sovietologists 'set the Soviet Union [and eastern Europe] apart from other countries' (*ibid.*) and considered Soviet-style communism 'a unique phenomenon in the history of mankind [*sic*]' (Fleron and Hoffman, 1993, p. 3). This theoretical isolation and separatism was reinforced by the practical difficulties of doing research in the Soviet states – the availability of material, the ability to travel, etc. – so that much of the very rich work produced on the Soviet states had a statistical bias which worked against conceptually integrating the Soviet states into broader economic geographies.

The events of recent years however have forced the countries of the former Soviet Union and east central Europe onto the economic geography agenda, not least because the map of this part of the world has now been redrawn and there are many new states to be integrated into the world economy (see Figures 1 and 2). The uneven development of capitalist social and economic relations – the central focus of economic geography – is being played out day by day in Russia, Tajikistan, Slovakia and Bulgaria, for example. The post-socialist context sets the processes of change long studied by western economic geographers in even greater relief. In few places are contemporary

Figure 1 East Central Europe and the Baltic States

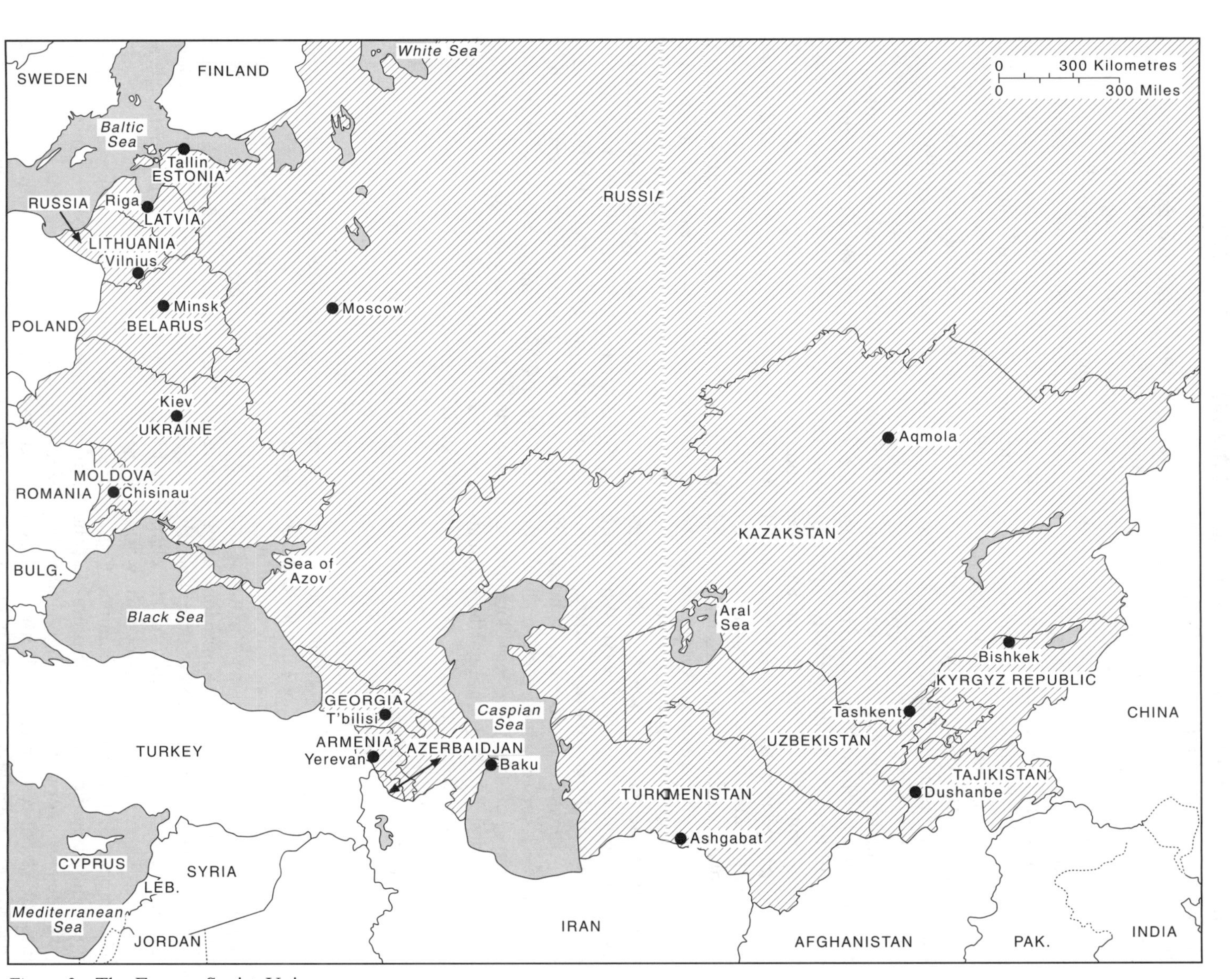

Figure 2 The Former Soviet Union

debates about the importance of institutions, embeddedness, the transfer of knowledge and the cultural construction of the economy being better played out 'on the ground' than in the process of the contemporary construction of 'capitalism' in the former Soviet states. Out of this emerges a considerable body of geographical and sociological work on networks, on regulation and on embeddedness rooted in the experiences of the former Soviet Union and east central Europe (see, for example, Pickles and Smith, 1998; Grabher and Stark, 1997).

Bringing together work on globalisation and work on the contemporary transformations of the post-socialist world is important for three principal reasons. From an 'area studies' perspective, the ongoing processes of globalization and the 'reality' of an increasingly global world (however contested and uneven) must be increasingly recognized as the wider context within which post-socialist transformations are taking place (see the special edition of *Regional Studies*, 1998). While both qualitatively and quantitatively different, post-socialist transformations have much in common with the economic, social and cultural restructurings of many western localities in recent decades. Working through these similarities and differences has the potential to contribute to the elaboration and sophistication of theories and concepts created to understand changing economic practice, institutions and structures across the world. Such work also contributes to less exceptionalist, more integrated understandings of change in different parts of the world and aids in the identification of links and connections – material and discursive – between and within regions of the world. A geography of globalisation is clearly incomplete without consideration of the changing geographies of the post-socialist world. The impact of the 'Russian Economic Crisis' upon global financial markets is proof that the post-socialist world is increasingly integrated into the global economy; even if it is only as a location for speculative investment in government debt.

COMMUNISM AND CAPITALISM

From the 1960s, and the early debates over 'convergence', onwards, social scientists have been engaged in theorising the relationship between communist and capitalist social systems (see Lane, 1978, for more on convergence theories of the 1960s and 1970s). Some, such as Tony Cliff, argued that communism in fact represented capitalism in a statist form (Cliff, 1974) whilst others insisted that state socialism and capitalism were (are) entirely different systems (see, for example, Davis and Scase, 1985). More recently, debates have centred on the relationship between Fordism and communism, and between post-Fordism and post-communism (Murray, 1992; Joffe, 1990; Smith, 1998; Altvater, 1993; see also Peck and Tickell, 1994). These debates are far from resolved, but they bring to the fore the idea that post-socialist transformations must at least be considered in the context of western restructurings. The author who makes this point most explicitly is the Pole Grzegorz Gorzelak, who writes 'the restructuring processes that dominate in the post-socialist transformation very strongly resemble the phenomena which shaped economic life in more advanced Western countries since the 1960s and especially during the 1970s' (Gorzelak, 1996, p. 3). He adds that post-socialist transformations should be 'de-mythologized' and regarded as 'normal' processes of technological and organizational change, linked to the wider processes of globalization, but unfortunately fails to develop these ideas further.

Dominant analyses of the transformation of the economies of the former Soviet Union and east central Europe have utilized the concept of four 'pillars' – liberalization, stabilization, privatization and internationalization (see, for example, Gros and Steinherr, 1995). These four pillars can be broken down into a number of structural and institutional features, which include the freeing of prices from state control; currency convertibility; the extraction of the state from economic ownership (in all but 'strategic' fields); self-management of enterprises; the development of a private entrepreneurial sector; the liberalization of foreign trade; accession to a wide variety of international organizations (such as the EBRD, the World Bank, the IMF, the World Trade Organization, etc.); the reduction of soft credits and subsidies, etc. These processes are linked through to organizational, technological and institutional changes, which in their turn are related to cultural, political and behavioural shifts which encourage an understanding of transformation not simply as a reprofiling of enterprises and a remaking of macroeconomic structures, but as a 'profound restructuring of regional economic life' (Smith, 1998, p. 3). In an abstract sense, these processes of change are not unlike the changes detailed, for example, by Cooke in *Localities*, whose focus was the local experiences of economic restructuring, social adjustment and political change during Britain's deindustrialization (Cooke, 1989). Thatcher's Britain was characterized by a hegemonic position of neo-liberal, monetarist discourses which demanded stable, balanced budgets and low inflation, internal competition in the private sector and the opening-up of the public sector (industry and public services) to performance and efficiency indicators and fully-fledged privatization. All this was set within the context of an increasingly international and open world economy. From this perspective, economic restructuring in the West has been characterized, like that in the former Soviet Union and east central Europe, by pressures to liberalize, stabilize, privatize and internationalize.

GEOGRAPHY MATTERS

One of the fundamental tenets of economic geographies of restructuring, however, is the conceptualization of the interplay of 'big' processes of global, historical change with local conditions, and a recognition that geography matters – a process such as privatization is structured by and structures the localities and regions in and through which it takes place. As Duncan (1989) notes, contextual conditions – that is particular moments of institutional relations, patterns of historical development – exist on every spatial scale to mediate the development of capitalist social and economic relations. So, whilst on a general, almost abstract level, it may be true to suggest that processes of transformation in the former Soviet Union and east central Europe are leading to a 'shifting kaleidoscope' of change (Massey, 1994, p. 110) in much the same way as restructuring in the West, work within the region brings out the difference that place makes.

The first argument to recognize is the difference of sheer scale. No society has ever made the transition to the market on such a vast level of magnitude. Other countries have privatized entire sectors of the economy, and many 'Third World' countries have implemented huge programmes of stabilization and liberalization in response to global restructuring. However, transformation in the former Soviet Union apparently involves the dramatic change of all sectors and regions of the economy from heavily subsidized, distorted state ownership to liberal, rational private ownership, submitting them not just to lower tariffs, but to a surge of imports and an increasing flow of direct investment. The deindustrialization of Britain, as studied by Western geographers, pales in significance to the potential for restructuring created by transition in the former Soviet Union and east central Europe.

The difference of scale relates not only to the breadth of transformation, but also to the depth. Many post-Soviet firms are so uncompetitive that they are actually 'value-subtracting' such that the materials entering the production process are worth more than the finished product at the factory gate. The necessary scale of closure and rationalization if firms were to be subjugated fully to market rationale, with no protection, would be immense (see Gaddy and Ickes, 1998, on the concept of the 'virtual economy'). Even those that are competitive will have to restructure significantly, particularly to divest of vertically and horizontally integrated parts of enterprises, most noticeably the divestment of social infrastructures on which towns and cities depend.[3] Many firms were structured in ways that could not be justified under a market economy, with machine building enterprises making their own nuts and bolts, for example. Less voluntary enterprise divestiture will take place through asset stripping, as only the profitable parts of enterprises are privatized and sustained with new investment whilst loss-making sectors remain in state hands and are not restructured.

Beyond the generalizations listed above, it is also necessary to note some of the specific characteristics of the Soviet-style economy, which pose a particular problem for the restructuring of regional economies. The most fundamental is perhaps the planned nature of Soviet industrial location, which, though proclaiming an interest in balanced development (see Hamilton, 1967, and Pallot and Shaw, 1981), resulted in the concentration of certain sectors of the economy in certain regions. Treyvish *et al* (1993a, b) describe the pattern of primary, secondary and tertiary employment across the former Soviet Union, for example, showing clear regions of manufacturing and agricultural predominance. Disaggregating further to particular productive branches, it is clear to see that certain regions 'contain' particular concentrations of single industries. The extreme of this tendency is the existence of 'company towns' whose survival depends entirely on the survival of a single enterprise. Enterprises of the defence industry are notable examples of this phenomenon since many large and medium-sized towns rely almost exclusively on one or more related defence industries. The service sector was notoriously underdeveloped even in major urban centres and local economic development and planning were more often than not rendered secondary to national economic gain. These problems of hyper-industrialization have been further aggravated by the legacies of central control that have served to reduce inter-enterprise linkages within individual cities and regions. With the collapse of centrally-orchestrated trade, enterprises have had to find new sources of supply in the local economy.

A PLURALITY OF TRANSFORMATIONS

These economic changes are confused and complicated by the simultaneous (and inherently interlinked) processes of social, political and cultural change. This point is made most clearly by Stark who argues that the countries of east central Europe and the former Soviet Union are experiencing:

> a plurality of transitions in a dual sense: across the region, we are seeing a multiplicity of distinctive strategies; within any given country, we find not one transition but many occurring in different domains – political, economic and social – and the temporality of these processes are often asynchronous and their articulation seldom harmonious. (Stark, 1992, p. 301)

The post-socialist era has also played witness to the reconfiguration of centre-province relations, mass migrations and other demographic transformations and the (re)construction of new state formations, identities and citizenships (see, for example, Regulska, 1997; G. Smith, 1996; Pilkington, 1998). Local resource endowments have been revalued and reassessed in the context of world prices. Local élites have worked to regionalize patterns of economic and political change through the consolidation of their authority in the market context, establishing regional associations and creating the space to manage the implementation of national and international reform policies (see Bradshaw and Hanson, 1998, for a review of research on Russian regions). Finally, and critically, post-socialist transformations have been characterized by a shifting political economy of scales as different functions of governance and control are exercised at different scales. The emphasis on national strategic goals under the Soviet system has given way to an increasing role for both the local and the international. As the states of the former Soviet Union and east central Europe are opened up to the material and discursive influences of globalization, so too can we witness processes 'reducing the level of central control and empowering regional authorities and individual enterprises, as well as individual citizens' (Bradshaw, 1997b, p. 4).

This plurality of scales and variability in transformation can be clearly identified with empirical examples from the region. As Smith notes, echoing Stark, the '"transition to capitalism" is working its way out in a diverse group of some 28 countries. This complexity is further compounded by the great variety of transitions found at the subnational level . . .' (Smith, 1998, p. 1). The increasing wealth of information about the post-socialist states and their attempts to construct capitalism has shattered the image of the Soviet bloc as monolithic and homogeneous. Popular notions of winners and losers amongst the 'transition economies' highlight the uneven and contested nature of post-socialist transformations. States adopt and adapt the four pillars of liberalization, stabilization, privatization and internationalization in different ways, following national paths to integration in the world economy, and, for this reason, it is increasingly apparent that there is no single path from plan to market (see Table 1). In large part, this is because each of the 'transition economies' started from different points. Some had a history of a market-oriented democratic system in their none-too-distant past; but others had never existed as independent states. A number of cleavages are now becoming apparent. For example, a fast reforming core, including the central European states of Hungary, Poland, Czech Republic and Slovenia and the Baltic State of Estonia, is now seeking membership of the European Union. In contrast, the peripheral states of the post-Soviet world are slipping into the ranks of the 'Third World'. Finally, there are those states that have been damaged by warfare and have been unable to implement reform policies.

Yet, analysis of 'transition progress' at the national level hides the fact that marketization has tended to aggravate pre-existing spatial inequalities and has also generated new types of deprivation. The increasing accessibility of information on these states, new opportunities to travel to and work in the former Soviet Union and east central Europe, often collaborating with local colleagues, and the explosion of interest in their economic development has encouraged intensive research exploring local patterns of transformation, getting beyond the national generalizations and explaining subnational responses to global changes. Again, winners and losers are identified and this is often married with incredibly rich, well-researched stories of the lived experiences of 'the transition to capitalism'.

As the largest state in the world, and the economy most heavily influenced by the Soviet system, the Russian Federation has posed the greatest challenge to 'transitologists'. Because of the scale of the economy, and the fragility of the federal system, the regional dimension of transition has become a major focus of research on transition in Russia (Bradshaw and Hanson, 1998). Research by political geographers and political scientists has focused upon challenges to Russia's federal system, the emergence of regional elites and the changing nature of centre–region relations whilst economic geographers and economists have focused upon trying to explain patterns of regional economic change. As Figures 3 and 4 illustrate, economic transition in Russia has generated new economic geographies which defy simple explanation (Bradshaw *et al*, 1998). Extensive statistical analysis has highlighted the importance of inherited economic structure (Sutherland and Hanson, 1996); while case study analysis has often revealed a development strategy focused on subsidizing bankrupt formerly-state owned enterprises, rather than promoting new economic activity. It is now recognized that many enterprises, under new political and economic conditions, are simply doing the wrong thing in the wrong place.

CONCLUSION

These geographies of transformation and globalization testify to a 'shifting kaleidoscope' of change as the interaction of 'geographical surfaces' and 'the demands of industry' (Massey, 1994, p. 110) are played out in new ways. As Giddens argues, 'the structuration of

Table 1 Socio-economic indicators for transition economies in 1996

	Population, 1996 (millions)	Land area (thousand sq. km.)	GNP ($ billions)	GNP/per capita ($)	GDP growth (% 90–96)	Energy use (kg oil equiv. per capita)	Urban pop. (%)	Ag. (%)	Trade (as % GDP)	Human Develop. Index	Rank out of 174
Albania	3	27	2.7	820	1.5	314	38	55	52	0.633	93
Bulgaria	8	111	2.4	1190	–3.5	2724	69	13	127	0.773	62
Croatia	5	56	18.1	3800	–1	1435	56	16	95	nd	nd
Czech Republic	10	77	48.9	4740	–1	3918	66	11	117	0.872	37
Hungary	10	92	44.3	4340	–0.4	2454	65	15	79	0.855	46
Macedonia, FYR	2	25	2	990	–9.1	1308	60	22	86	nd	nd
Poland	39	304	100.9	3230	3.2	2448	64	27	49	0.819	56
Romania	23	230	36.2	1600	0	1941	56	24	60	0.738	74
Slovak Republic	5	48	18.2	3410	–1	3272	59	12	126	0.864	41
Slovenia	2	20	18.4	9240	4.3	2806	52	6	111	nd	nd
East/Central Europe	107	990	292.1	3336	–0.7	2262	59	20	90	na	na
Armenia	4	28	2.4	630	–21.2	444	69	18	86	0.680	93
Azerbaidjan	8	87	3.6	480	–17.7	1735	56	31	62	0.665	96
Belarus	10	207	22.5	2070	–8.3	2305	72	20	96	0.787	61
Estonia	1	42	4.5	3080	–6.5	3454	73	14	159	0.749	68
Georgia	5	70	4.6	850	–26.1	342	59	26	44	0.645	101
Kazakhstan	16	2671	22.2	1350	–10.5	3337	60	22	65	0.740	72
Kyrgyz Republic	5	192	2.5	550	–12.3	513	39	32	86	0.663	99
Latvia	2	62	5.7	2300	–10.7	1471	73	16	102	0.820	55
Lithuania	4	65	8.5	2280	–6	2291	73	18	115	0.719	81
Moldova	4	33	2.5	590	–16.7	963	52	33	118	0.633	104
Russia	148	16889	356	2410	–9	4079	76	14	42	0.084	57
Tajikistan	6	141	2	340	–16.4	563	32	41	228	0.616	105
Turkmenistan	5	470	4.3	940	–9.6	3047	45	37	nd	0.695	90
Ukraine	51	579	60.9	1200	–13.6	3136	71	20	93	0.719	80
Uzbekistan	23	414	23.5	1010	–3.5	2043	41	35	69	0.679	94
Former Soviet Union	292	21950	525.7	1339	–13	1982	59	27	98	na	na

Sources: The World Bank (1998), World Development Indicators. Washington, DC: The World Bank and UNDP (1997), Human Development under Transition: Europe and the CIS. New York: UNDP.

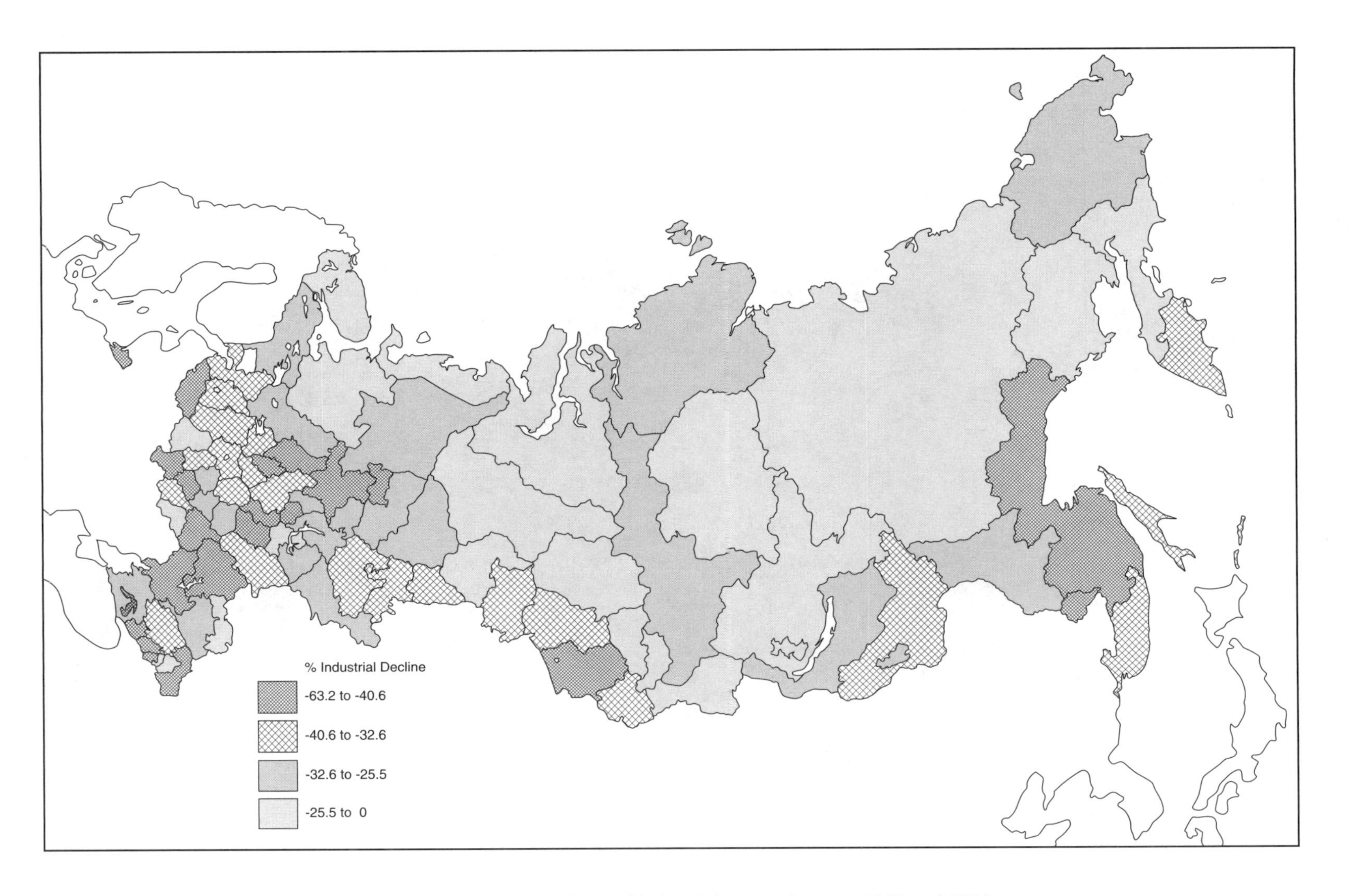

Figure 3 The Russian Federation: decline in the physical volume of industrial output between 1992 and 1994

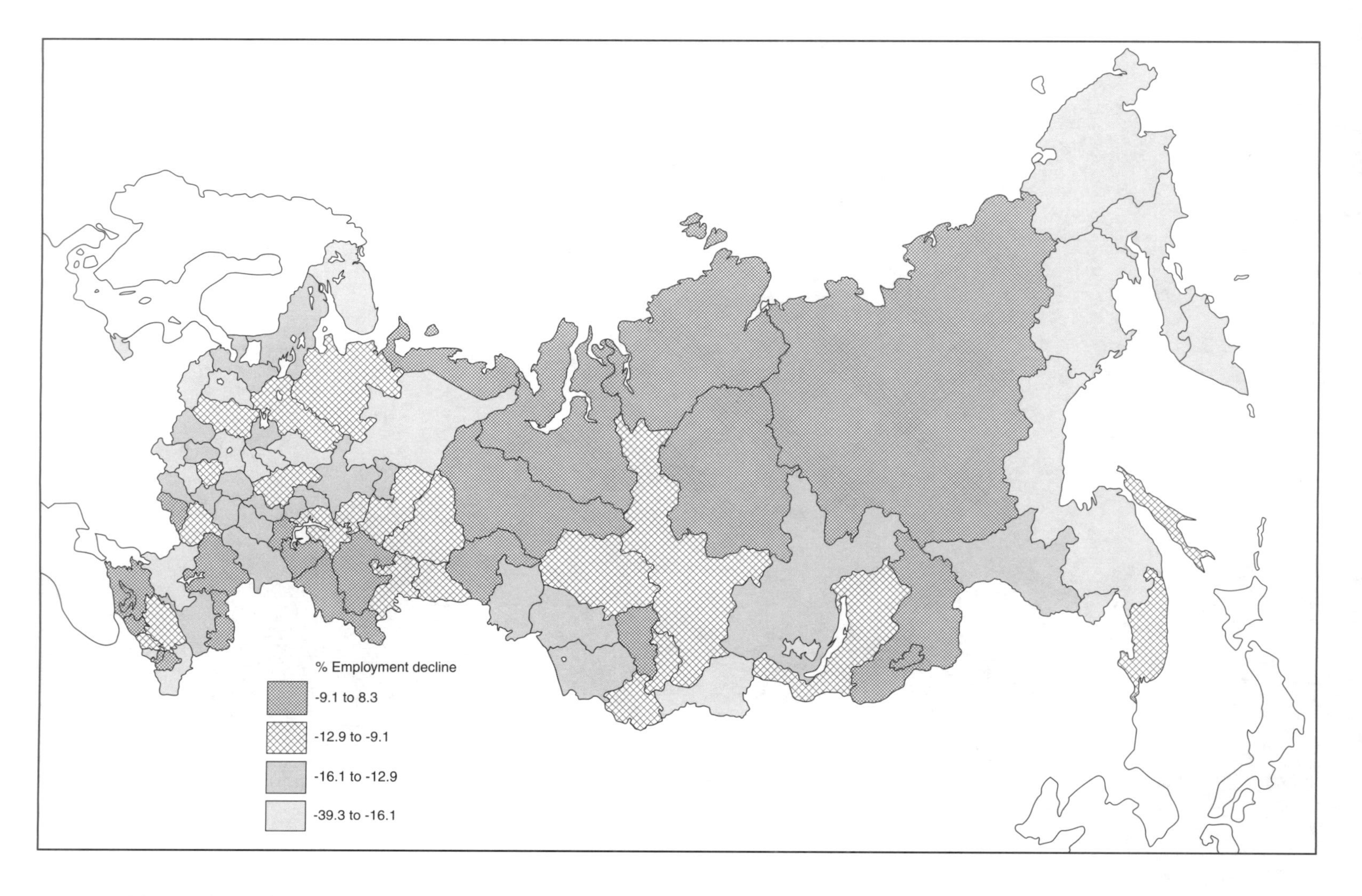

Figure 4 The Russian Federation: decline in industrial employment between 1991 and 1994

every social system, however small or large, occurs in time and space' (Giddens, 1981, p. 91) – the dismantling of the communist system and the construction of capitalism do not occur in 'uncharted territories' but in and through existing geographies which remake and are remade by the processes of restructuring. A failure to take account of the very real geographies of the region – the cultural, social, political and economic contexts of change – leads to a failure of reform policies – as events in the Russian Federation in late summer 1998 have aptly demonstrated. Despite, and because of, the centralized nature of the Soviet system, different places in the former Soviet Union and east central Europe find themselves with different social and spatial legacies. When articulated with the uneven processes of globalization and transformation, the evidence is that different post-socialist countries, regions and localities occupy very different places in the world.

The events in Russia have also clearly shown this part of the world's integration into the wider global economy – though the Russian economy is very small in global terms, financial crises there have hit Western banks and investment funds hard, have raised the threat of regional political instability (including the fear of nuclear weaponry) and have begun to challenge the hegemony of neo-liberal reform policies worldwide. Notwithstanding the success or failure of reforms, the last ten years have seen the countries and regions of east central Europe and the former Soviet Union become increasingly subject to and participant in the processes of globalization and economic restructuring. The general concerns of reconciling globalization with the difference that space makes would seem to be particularly pertinent in the study of post-socialist transformations.

NOTES

1. This chapter draws heavily on previous work by the authors – see, for example, Bradshaw (1990, 1991, 1997a, b, c); Bradshaw and Lynn (1994); Bradshaw *et al* (1998); Stenning (1996, 1997, 1998).
2. What to call the part of the world that is now facing a transition away from socialism is a difficult question. No one name is unproblematic – 'post-socialist' and 'post-communist' make assumptions about the earlier presence of socialism or communism; 'post-Soviet' or 'former Soviet' define a set of very diverse states primarily in terms of their relationship with the former Soviet Union and 'the second world' invokes ideas about a hierarchical world system which may not today be appropriate, whilst an apparently simple geographical descriptor such as 'the former Soviet Union and east central Europe (FSU/ECE)' shifts attention away from other transforming states in Africa, Latin America and East Asia. This paper uses each of these terms, but has as its central focus the central and east European states which adopted some form of the Soviet model in the post-war era and the successor states to the Soviet Union (see Figures 1 and 2). It is recognized that there are other post-socialist states, but these are only dealt with implicitly in this paper.
3. The vital role played by enterprises, not only in job provision, but also in health, educational, cultural and social activities serves to increase the devastation that enterprise restructuring may lead to within communities (see Shomina, 1992).

Katharyne Mitchell

'Flexible Circulation in the Pacific Rim: Capitalisms in Cultural Context'

Economic Geography (1995)

The notion of a general restructuring of international capitalism is often taken as the starting point for theoretical debates concerning changes in the global built environment. Areas of agreement include the increasing internationalization of the economy, widespread financial deregulation, the accelerating importance of new information technologies, and a growing integration of various networks of production and circulation. There is, however, serious disagreement regarding just about everything else. Given a general shift in the nature of capitalism, what exactly is the shift from? In the realm of production, how extensive was Fordism to begin with? Is it being transformed everywhere, or just in some sectors and some geographic regions? Are the transformations manifested in a new, flexible specialization or in other kinds of productive systems? How universal is the logic and the impact of the shift? Can we discuss it in the abstract? What are the spatial manifestations of the changes?

Much of the disagreement arises from different conceptions of the degree to which economic practice is socially embedded (see Granovetter, 1985). In a spectrum from economism to embeddedism theorists of global restructuring and urban development debate the autonomy of the economic sphere. At one end, there is a tendency to view economic activity with a universal, rational, or abstract 'logic'.

Although few geographers would locate themselves at the far end of this spectrum, many draw heavily from the abstract logic of economic transactions. In this framework, economic conditions act to propel change: the resulting changes are then *conditioned* by institutions, states, agencies, and sociospatial environments.

Toward the other end of the spectrum there is a stronger emphasis on the embedded quality of economic activity. Here, historical and cultural context is given greater theoretical weight and economic and social activity are viewed as intertwined rather than causal. In examining productive organization and new spatial patterns in the context of rapid capitalist restructuring, geographers emphasize the importance of mutual interdependence and trust (**Gertler**, 1988) (see Part Three, Section 3.3); preexisting spatial structures (Sayer, 1989); institutional atmosphere, local value systems, community, and family (Becattini, 1990; Trigilia, 1990; Amin, 1991). The focus here is not merely on the particularities of local places and cultures and how they are transformed upon contact with the capitalist mode of production. It is rather how specific local traditions actively coproduce and rework global systems. Capitalism is thus theorized as capable of infinite variation – arising, combining and transforming in odd and unique ways around the globe.

Here, Harvey's [1988] discussion of the 'fundamental propositions' of capitalism serves as a useful reminder of the ways in which structural principles and local, socio-spatial particularities must be considered concurrently. In [this paper] I remain attentive to Harvey's general 'organizing principles', but also show how specific historical and cultural factors produce extremely diverse forms within this broad systemic framework.

[The] paper examines some of the social and spatial structures informing contemporary Hong Kong Chinese business society. Although this obviously encompasses a widely diverse group, the focus here is on the extremely wealthy business families, such as the Li family, which control vast amounts of capital in Hong Kong and abroad. The particular form of capitalism employed by members of this financial elite is embedded in sociocultural relations, yet is extremely successful in the contemporary global economy – a period wherein financial deregulation, deterritorialized systems of credit, quick and flexible decision making, subcontracting, and the effective exchange of information are of crucial importance. In this paper the connections between these features of contemporary global restructuring, and the current success of the particularist form of capitalism employed by families such as those of Li Ka-shing (1991), are enumerated through a case study of investment in Vancouver real estate during the late 1980s.

HONG KONG BUSINESS PRACTICE AND GLOBAL RESTRUCTURING

Familial and Regional Ties, Networks, and Business Trust

Ownership and management of Chinese businesses usually remains in the family. When large-scale operations are extended overseas, a prominent pattern is for the spin-off project to be controlled by a son or son-in-law – even though professional managers may be hired to manage various aspects of the business operations (Redding, 1990; Montagu-Pollock, 1991). An important element that helps to cement these extended family ties across spatial divides is business trust. The ethic of trust is central to the business success of Chinese entrepreneurs in Hong Kong, Taiwan, and in many overseas communities (see Silin, 1972; Barton, 1983; S. Wong, 1991; Kao, 1991). In order for the established norms of reciprocity to operate effectively, business people must trust that the other players will uphold certain social norms.

Credit and Capital Accumulation

Many contemporary Chinese businesspeople rely on mutual aid associations (*hui*) for venture capital. In these groups, formal laws or administrative agencies are rarely used to enforce behaviour; trust remains an important element in bonding the business transaction. By pooling funds and obtaining capital through informal channels, the bulk of the principal necessary for a business venture can often be accumulated long before the business venture is initiated. Chinese businesses that borrow from savings pools, or from within the extended family network, or even from personal connections within the banking system, are thus in an advantageous competitive position in comparison with other businesses whose loans are constrained by the necessity to return the capital plus interest within a specified time period.

[I]n the Ming and Qing dynasties trust was not placed in the constancy of money or the financial system, but in various binding yet elastic relationships between people. Rather than focusing on the creation of legal institutions to protect currency and trade, the emphasis of the state was to maintain and supervise self-regulating associations that strengthened the hierarchical relational ties characteristic of Chinese social culture at that time (Liu, 1990; Mann, 1987; Hamilton, 1991). This type of trust based on relational ties remains critical in contemporary Chinese credit practices, functioning as an assumed constant with respect to time and space.

Indeed, even in the case of international transactions, where geographic distance might operate to sever particularist ties of this nature, the 'personal' character of credit transactions has been carefully maintained. As will be shown in the case of Hong Kong and Vancouver in the latter part of this paper, the painstaking establishment and maintenance of long-term, extended family ties remains a critical ingredient in the contemporary circulation of finance capital overseas.

Flexible Credit in the Contemporary Global Economy

Why are the socially embedded economic practices described above particularly successful in the current period of global restructuring? What are the features of late capitalism that have articulated well with the contemporary business practices of an elite Hong Kong society? In the past 15 years, qualitative shifts have occurred in the international financial system, with major implications for the use of money and credit worldwide.

Through the innovation of new financial instruments, the ability of nation-states to control the production and circulation of money and credit through traditional forms of regulation, including compartmentalization and the restriction of financial institutions to prescribed areas, has been greatly reduced (Thrift, 1990, p. 1136; Leyshon and Thrift, 1992, pp. 51–2). As financial flows have become unmoored from national space economies and increasingly global in focus, there has been a move toward a general 'deterritorialisation of credit' (Leyshon and Thrift, 1992, p. 54). With the continued internationalization of accumulation processes via advanced technologies of computerization and communications, the semblance of regulation and control in the banking industry has all but disappeared.

This leads to an interesting contrast with the Chinese business practices discussed earlier. In Hong Kong, most businesses are owned and managed by the extended family. In this scenario, conflicts between owners and managers are unlikely, and conflict between money capital and productive capital will rarely occur.

In terms of the worldwide deregulation of finance in the 1980s, Chinese businesses were also at an advantage. The 'deterritorialization of credit' and decreasing control by financial agencies served to increase the general reliance on informal credit channels and direct, unmediated contact between creditor and debtor. This type of credit system, reliant to a greater degree on personal contacts both within and outside the banking industry, was akin to the type already familiar to many Chinese businesspeople. In the practice of seeking informal channels for credit accumulation, Chinese businesses were far in advance of their Western counterparts, for they were able to rely on previously established ascribed and achieved social relations that already formed the core of their informal credit networks. This gave them a distinct advantage in a number of business situations in Asia and, increasingly, worldwide.

Information and the Conversion of Values in a Mixed Economy

In contemporary Chinese business networks, the exchange of information remains an integral part of the financial network and is *institutionalized* in that network in a manner fundamentally different from the exchange of information in most Western business practices. As with capitalism itself, there are fundamental propositions of information concerning new technologies and the ability to access and interconnect with those technologies. Unquestionably, the new advances in telecommunications – the greater speeds and distances covered within the latest systems – are of central importance to the business practices of *all* societies. At the same time, however, information itself is culturally specific: who says what to whom and how it is received and employed depend on the location and positioning of the individuals who are conducting business in any given society.

In most contemporary Chinese business societies, the creation and manipulation of information performs an important function within the mixed system of commodity and 'gift' exchange. Gift exchange operates within systems characterized to some measure by interlocking and functions as a transaction between persons that ordinarily leads to economic profit or gain, but which does not function as a purely economic exchange.

The binding of gifts to the persons involved in the exchange is the primary quality of the *guanxi* economy. This binding quality means that the receiver of the gift is indebted to the donor, has lost stature through the acceptance of the gift, and must reciprocate the gift in order to regain face and remain within the social network. Conversion of values in the gift economy (repayment of the debt) is predicated on this understanding of moral obligation in the context of ongoing social relationships; it can take place in a number of ways – all of which are distinctive from both barter and simple commodity exchange.

The quality of the personal relations integral to the gift economy is also integral to contemporary Chinese business networks and works in perfect symbiosis; obligations cannot be exactly or completely repaid, and thus the binding quality of the relationship continues. At the same time, although different in form from the symbolic capital of a state distributive economy, the potential profitability of symbolic capital conversion for the creditor may be equally lucrative within the context of the contemporary global economy.

Rather than monetary repayment of debt, which may threaten the social fabric of personal relations, gifts of 'information capital' can be employed to save face and to extend the business relationship. In the context of the international financial system, good information can ordinarily be converted into major profits quite easily. Good information, such as insider or early news, can be shared because of the position of those who acquire and disseminate it. In Chinese society, one's 'position' refers to a literal, geographic position, as well as to a position in a social hierarchy.

Information shared between capitalists is 'good' because it is contextually specific to both initiator and receiver; in other words, it is useful because of the *positioning* of both individuals within a broader context that is both literal geographic and social-hierarchical. The passing of data from initiator to receiver in a manner that permits understanding and effective manipulation depends upon a shared system of meanings between the two parties. Information, as all interworked system of signs, is culturally inscribed, and thus must be understood in terms of culture as well as of technology.

In many Western business practices, the *technology* that enables the collection and transmission of information incorporates information's exchange value within itself – a process that often leads to the fetishization of information technologies. In many Chinese business societies, in contrast, information is exchanged in the 'gift' economy – one that is particularly conducive to the elastic and indeterminate quality of information.

THE HONG KONG–VANCOUVER CONNECTION

Hong Kong from Colonialism to 1997: British Decline and the Growth of the Chinese Multinational

In Hong Kong, the British colonial government and the powerful British trading houses (*hongs*) have exerted control over local business practices since the middle of the nineteenth century. Until recently, the majority of Hong Kong Chinese residents had little or no access to employment in the public sector, to political representation, or to business activity within the giant British *hongs*. In the post-World War II period, rapid and highly profitable capitalist expansion was the acknowledged mandate of government agents, through the tacit alliance between business interest and the bureaucracy. This alliance was composed of an elite group of British businessmen and politicians and a small number of powerful Chinese businessmen.

As recently as 1976, the dominance of British *hongs* in the economic and political life of Hong Kong was extensive. Most of the major business groups were controlled by non-Chinese families or organizations, with long histories in Hong Kong. They dominated the hotel, property, and public utilities industries. In the last decade and a half, however, power has shifted from British businesses to Chinese business groups. By 1986, the rapid rise of large, multicompany Chinese businesses and the exodus of many British businesses transformed the economic profile of Hong Kong, and Hong Kong capitalists pondered their business strategies in light of a new geopolitical order.

Geopolitics and the Global Real Estate Market

Numerous Chinese capitalists in Hong Kong insured against negative political repercussions from the pending 1997 changeover by increasing their investments in new locations around the globe.[1] Although most of the large business groups have massive amounts of fixed capital in the infrastructure of Hong Kong and southern China and plan to remain firmly established in Hong Kong, they nevertheless also invested heavily in foreign companies.

In addition, Hong Kong's highly successful export industry of the 1960s produced a tremendous accumulation of capital, much of which went into real estate speculation during the 1970s. Although property development continued to be highly lucrative throughout the 1980s, the incredible overagglomeration of wealth in this sector by players like Li Ka-shing forced a movement into new geographic arenas.[2] Much capital investment in property was directed toward countries such as Canada.

Deregulation and Privatization: The Role of the State in Canada

From the early 1980s through the early 1990s, a number of measures were initiated at all three levels of conservative government that transformed Vancouver's business and urban environment. These measures, including the deregulation of the financial sector and the privatization of land, aided a general shift of western and eastern Canadian capital into real estate development and property acquisition. Due to timing and deliberate government strategies, the measures also proved particularly favourable for Hong Kong business groups interested in Vancouver's burgeoning property markets.[3]

Vancouver's rapid entrance into the elite circles of Hong Kong big business occurred largely through joint ventures within the private banking industry. Banks with a previously established relationship, such as the long-standing tie between the Canadian Imperial Bank of Commerce (CIBC) and Li Ka-shing's family, had a tremendous advantage. CIBC set up an office in Hong Kong in 1968 and was introduced to Li Ka-shing through a personal connection. In 1974, Li's property company, Cheung Kong, formed a joint venture with CIBC. Li is currently the largest single shareholder in CIBC, holding the legal limit of just under 10 percent.[4]

The government's deregulation of the Canadian banking industry also facilitated a move toward greater market share by foreign banks. The Hong Kong and Shanghai Banking Group (Hongkong Bank), in particular, became a major player on the Canadian financial scene. This bank, under the direction of Li Ka-shing after 1978, took over an ailing Bank of British Columbia in 1986. The takeover, which greatly expanded the role of Hong Kong's financial presence in British Columbia, was facilitated by the Canadian government.

The Expo Land Development and Extended Family Networking

In the late 1980s, Vancouver's built environment was an attractive investment site for foreign investors. In downtown Vancouver, the former Expo lands were especially enticing owing to their ongoing devalorization by the government of British Columbia. The process of decline initiated by the government's lack of confidence in the area furthered the flight of actual or potential capital investment, hastening the area's overall devalorization (cf. Zukin, 1982, p. 15).

Li Ka-shing's native place is Chao'an, Guangdong, but he made his first fortune in the plastics industry in Hong Kong. He purchased property in Hong Kong in the late 1960s, when prices were relatively low, and shifted his main business to property development in the 1970s (G. Wong, 1991, p. 142). Li currently holds 34.9 percent of Cheung Kong Property Development Investment Holdings.

Li's interest in the Expo property may have originated from any of his numerous sources of information around the world. His long-term relationship with Colliers Macaulay Nicolls Inc., a major commercial real estate firm in Vancouver, undoubtedly enabled him to consider the project in a useful investment context.

Following the purchase of the former Expo lands from the provincial government in 1988, Li made his son, Victor, senior vice-president in charge of the development of the land by Concord Pacific.

After the land purchase in 1988, Li Ka-shing invited two Hong Kong Chinese friends to join him in the deal. Billionaires Cheng Yu-tung (and his son, Henry) and Lee Shau-kee became partners soon after the site was obtained, as did the Canadian Imperial Bank of Commerce. [I]nformal relationships and trust [were of key importance] in the formation of these business ties. Interlinking of the top business echelon as a form of extended family is shown in Victor Li's references to Hong Kong's business elites as 'uncles'. With these 'uncles' Li shares information about investment projects around the world.

Reputation, Information, and Business Trust

Victor Li's status as 'nephew' among the business elite of Hong Kong stems from his family relation to his father, whose good reputation is widely known. The importance of reputation is evident in both the willingness to become partners in various business ventures and also in the readiness to follow the financial experiments of a well-respected investor such as Li Ka-shing. By the late 1980s, Li's stature was so great that his investment decisions were followed with interest by Chinese communities worldwide; he is widely recognized as the best and most trustworthy developer in Hong Kong. When he invested in the Expo lands, Vancouver real estate could not have become hotter.

Management and Family Control

As with many large Hong Kong conglomerates, Cheung Kong claims to use 'the best of both Western and Chinese management'. [M]ost of the Chinese Pacific Rim companies [Li] has assisted have been 'family controlled, owned and managed'. [A]s a result of this family control, 'the nature of business practice differs':

> In the West, [in larger companies] you tend to have hired management, and because of the stock system, the ownership and the management gradually gets slightly divorced. The management controls the company, while the investors control the capital. In the East, the owners are the managers. [T]hat's really the difference. All the big companies in Hong Kong or Taiwan, they're still controlled by individuals. And therefore the management style is different. (Interview, Stanley Kwok, Vancouver, January 1991)

After Li Ka-shing brought new partners into Concord Pacific following the purchase of the Expo lands, Victor Li reassured jittery Vancouver politicians that his father would maintain major control of the property. He emphasized that Concord Pacific is not a subsidiary or related to the massive Hong Kong conglomerates owned by the other partners, but a private company 'financed mainly by family money'. Victor Li's remarks were intended to placate local Vancouver citizens, politicians, and journalists, who were becoming increasingly vocal against the perceived sell-out of the Expo lands by the Canadian government. The strategy of winning acceptance on the local level was bolstered by the hiring of Stanley Kwok, a highly respected Chinese-Canadian architect in Vancouver. At the time of his hiring, Kwok was friends with numerous influential Vancouver officials, including the former director of urban planning, Raymond Spaxman. He was also personally acquainted with all three of the major Hong Kong shareholders in Concord Pacific.

Kwok's personal contacts on both sides of the Pacific enabled him to serve as an effective mediator between the business and urban planning circles of both Hong Kong and Vancouver. In his position as intermediary, Stanley Kwok facilitated the articulation of Hong Kong and Vancouver capitalists as they interconnected globally.

CONCLUSION

[T]his paper [has] examined the social embeddedness of economic practice within elite Hong Kong Chinese business society. The focus was on extended family networking, business trust, and the importance of sociocultural connections in the use of credit and information. Although numerous Western theorists have argued that these particularist features of economic behaviour cannot be effective in an international or 'modern' capitalist setting, large, family-run businesses such as Li Ka-shing's Concord Pacific have proven extraordinarily successful in recent global business ventures. Clearly many elements of recent global restructuring – especially the increased flexibility in capital circulation through widespread financial deregulation and privatization – have articulated well with the business practices of contemporary Chinese companies.

Geographers have worried that the increasing reference to the specific attributes of local sites and traditions will result in a plethora of empirical case studies void of theoretical claims (Scott, 1988a; Scott and Storper, 1987). Theory, however, must not be limited to an economistic understanding of causally determined interactions. Capitalism's own dynamism should be mirrored in its theorization – with local–global contact seen as organic and dialectical, operating not merely on the transformation of local places, but also on the continual reworking of global systems.

Differentiation in the capitalist mode of production occurs on many levels. By focusing on the sociocultural specificities of flexible circulation in Hong Kong Chinese business society, it is possible to theorize not just the manner in which capitalism is differentiated across time and space, but also, with successive comparative analyses, how these various forms compete and articulate in the new global economy. Examining these international convergences enriches our understanding of capitalism's 'abstract' organizing principles, while at the same time yielding greater insight into the manifold ways in which subjects and cities participate in and are disciplined by its multiple births and transmogrifications.

NOTES

1. In 1997, Hong Kong [changed] hands from British political control to control by China.
2. In the late 1980s, Li's companies controlled more than 10 percent of the entire Hong Kong stock exchange. As a result of this dominance, he was discouraged from further local investment by securities regulators.
3. Deliberate government strategies of attracting investment from Asia, particularly Hong Kong, included business immigration visas for the wealthy, various tax incentives, and a concerted advertising campaign.
4. This financial connection has proven invaluable for the Li family. Over the years, the bank has lent 'hundreds of millions of dollars' to Li Ka-shing.

Section 2.5 Regulating the New Capitalism

Peter Dicken

'International Production in a Volatile Regulatory Environment: The Influence of National Regulatory Policies on the Spatial Strategies of Transnational Corporations'

Geoforum (1992)

INTRODUCTION

In face of the globalising strategies of transnational corporations (TNCs) it has become conventional wisdom in some quarters to regard the nation-state as increasingly irrelevant to economic activity at the global scale. TNCs, it is often implied, slice their way at will through national boundaries, rendering virtually all state economic policies ineffectual. Although there is a kernel of truth in this viewpoint it is a gross misrepresentation of current reality. In fact, there is a renewed recognition, among writers across a broad spectrum of ideological positions, that the state's role, though much changed by the increasing complexity of both its internal and external relations, remains highly significant as one of the primary shapers of the global economy.

Differential state regulatory (and deregulatory) policies in the areas of trade, industry and foreign investment, therefore, remain highly significant. They are exceptionally important to the strategic behaviour of TNCs which, of course, attempt to take advantage of national differences in regulatory regimes in their pursuit of global competitive advantage. Current developments in the international political economy, notably the strengthening of regional economic integration in Europe and North America and the continuing GATT negotiations in the Uruguay Round, are transforming the geographical scale and nature of the 'regulatory surface' on which TNCs operate.

The result is a dynamic bargaining process in which each party attempts to capture the greatest gains. The notion that the changing structure of the global economy is the outcome of a complex combination of processes involving both TNCs and states is increasingly widely accepted across the political spectrum. Thus, Gordon (1988, p. 64) argues that 'TNCs are neither all-powerful nor fully equipped to shape a new world economy by themselves' whilst Ostry (1990, p. 1) asserts that 'the international environment of the coming decades will be shaped not by governments or international institutions but by the *interaction* of the two main actors, governments and global corporations – especially in the Triad: the United States, Europe and Japan'.

REGULATORY STRUCTURES IN THE GLOBAL ECONOMY

It is widely accepted that the mode of international economic regulation, and especially the general stability of that mode, is driven primarily by the existence of a hegemonic power which has the political and economic clout both to initiate, and to ensure the maintenance of, the regulatory mechanisms. The role of the United Kingdom and, subsequently, of the US as the global hegemon has been extensively documented. Today, however, the US hegemonic role has been eroded (Gilpin, 1987; Kennedy, 1987; MacEwan and Tabb, 1989b). We appear to be in a phase of transition to multipolar economy captured, at least crudely, by Ohmae's (1985) concept of the global triad.

For the TNC, the two most critical aspects of state regulatory policy are, first, *access* to markets and/or resources (including human resources) and, second, *rules of operation* for firms operating within particular national (or supranational) jurisdictions. The primary regulatory mechanisms governing access to markets is, of course, trade policy. This is both the longest-standing mechanism of state regulation of international economic activity and also the only one which is set within an international institutional framework, the GATT.

National regulation of access to markets via imports – the traditional mode of entry before the development of direct presence by TNCs – is achieved through the use of either tariffs or nontariff barriers (NTBs). In recent years, however, although tariff levels have continued to fall overall, NTBs have proliferated, both in the frequency with which they are being introduced and in their diversity of form. A plethora of import barriers has been invented, the most important of which are the various forms of import quota. Governments have displayed considerable ingenuity in devising other more subtle, and often less transparent, trade restrictions. It has been estimated that NTBs now apply to more than a quarter of all industrialised country imports. In addition, there has been a surge in anti-dumping regulations by several states, notably Australia, the US, Canada and the European Community (EC).

A second group of regulatory policies, those relating to inward foreign direct investment (FDI), are concerned with both access and rules of operation. In controlling access, that is regulating the entry of foreign firms, governments may operate screening mechanisms to filter out those investments which do not meet national objectives, whether economic, social or political. Foreign firms may in fact be excluded entirely from particularly sensitive sectors of the economy or the degree of foreign involvement may be restricted. The operations of those firms which have been allowed a physical presence may be regulated in a whole variety of ways including, in particular, the operation of local content requirements, export levels, rules governing the export of profits or capital and the levels and methods of taxing locally-derived profits.

Although few states operate a totally closed policy towards inward FDI, the actual degree of openness varies considerably. In general, developed market economies tend to be less restrictive in their policies towards foreign investment than developing market economies. One obvious reason is that the developed economies are the major sources of, as well as the dominant destinations for, the world's FDI.

A third group of policies which directly or indirectly impinge upon the spatial strategies of TNCs are industry policies. Most governments have some kind of policy whose aim is to regulate and/or stimulate industrial activity within its jurisdiction and, therefore, to enhance its internationally competitive position. The tool box of industry policies is extremely diverse, ranging from the obvious one of financial and fiscal incentives, through state procurement policies, technology policies, policies to encourage industrial restructuring, to regulate mergers and influence industrial structure, to regulate the labour market, and so on. The list of possibilities is very long indeed. Such policies (some of which are specifically designed to stimulate industrial activity, others explicitly to regulate such activity) may be applied either generally across the whole of a nation's industry or, more commonly, applied selectively by specific sector, type of firm or designated geographical area.

Johnson (1982, pp. 19–20) draws a distinction between two different 'ideal type' models of state policy stance:

> A regulatory, or market-rational, state concerns itself with the forms and procedures of economic competition, but it does not concern itself with substantive matters. For example, the United States government has many regulations concerning the antitrust implications of the size of firms, but it does not concern itself with what industries ought to exist and what industries are no longer needed. The developmental, or plan-rational, state, by contrast has as its dominant features precisely the setting of such substantive social and economic goals. The government will give greatest precedence to industrial policy, that is, to a concern with the structure of domestic industry and with promoting the structure that enhances the nation's international competitiveness.

It is far easier to find examples of developmental states in today's world than of pure regulatory states, although countries like the US and the United Kingdom are more regulatory than developmental in their current policy stances. But, certainly, virtually all the newly industrialising countries (NICs), including those in East and South East Asia, and, *a fortiori*, Japan itself, are undoubtedly developmental states. Despite some prevailing views to the contrary, there is no doubt that the role of the state in the NICs has been a central driving force in their economic development, although not in a uniform manner. Even though it may be more valid to see the developmental state and the market-rational/regulatory state as two ideal types at the far ends of a continuum it is, nevertheless, not unreasonable to interpret many of the current politico-economic tensions in the global economy as being as least partly a reflection of a clash between these two different models of state behaviour.

Of course, these three areas of regulatory activity – trade, FDI and industry policies – are not, in reality, set in totally separate boxes. Indeed, the fundamental feature of the developmental state is its use of all three policy strands as part of a developmental strategy. However, one of the most significant developments of the past few years is the growing demand within such market-rational states as the US for a *strategic* policy stance. Rather misleadingly, the term used to express this orientation is 'strategic trade policy' (STP). But, by definition, STP involves far more than just trade policy. In the US case, for example, it encompasses issues of strategic industry policy and, by extension, FDI policy as well as trade policy.

A second major development in the last 12 years or so has been the trend towards the deregulation of certain economic activities. As Cerny (1991, p. 173) observes, 'deregulation' emerged as a major phenomenon in many industrialised countries during the 1980s:

> The concept of deregulation – as it has been used by most commentators, whether politicians, journalists or academics – is deceptively simple. It means the lifting or abolishing of government regulations on a range of economic activities in order to allow markets to work more freely, as in classical capitalist economic theory.

However, as Cerny points out, the issues are, in fact, extremely complex. In particular, deregulation at the national scale has to be seen in its wider, international, context. In fact 'deregulation cannot take place without the creation of a new regulation to replace the old' (Cerny, 1991, p. 174). In effect, what is often termed deregulation is really 'reregulation'.

This interpretation is especially apparent in the context of the third major development in the international regulatory environment: the strengthening of regional economic integration. The final years of the 1980s witnessed two extremely significant developments in this regard. The first was the process to complete the Single European Market by the end of 1992. The second was the signing of the Canada–United States Free Trade Agreement and the initiation of negotiations to create North America Free Trade Area involving, in the first instance, Mexico. Some observers see these developments as a kind of concerted move to restructure the global economy into three major regional blocs. [T]he emergence of changed and strengthened regional blocs greatly alters the nature of the international regulatory surface on which TNCs operate. But it does so in complex, and often uncertain, ways.

In summary, the changing regulatory structures in the global economy in the areas of trade, foreign investment and industrial policy are a reflection of the increasingly intensive competitive position in which nation-states now find themselves. In an increasingly integrated and interdependent global economy nation-states are engaged in fierce competitive rivalry. Of course, a major component of the competitive struggle between states is their relationships with TNCs whose spatial strategies, in turn, are strongly influenced by state regulatory behaviour.

REGULATORY STRUCTURES AS CONSTRAINTS AND OPPORTUNITIES: INTERACTIONS BETWEEN TNCs AND NATION-STATES

[I]n considering TNC attitudes and responses towards different regulatory structures in the global economy we need to keep in mind the fundamental nature of the relationship between TNCs and states.

In the specific case of regulation, the most obvious assumption would be that TNCs will invariably seek the removal of all regulatory barriers which act as constraints and impede their ability to locate wherever, and to behave however, they wish. Removal of all barriers to entry, whether to imports or to direct presence; freedom to export capital and profits from local operations; freedom to import materials, components and corporate services; freedom to operate unhindered in the local labour markets – these would all seem to be the ultimate preference for TNCs. Certainly, given the existence of differential regulatory structures in the global economy, TNCs will seek to overcome, circumvent or subvert them. Regulatory mechanisms are, indeed, constraints to a TNC's strategic and operational behaviour.

Yet it is not quite as simple as this. The very existence of regulatory structures can be perceived as an *opportunity* available to TNCs to take advantage of regulatory differences between states by shifting activities between locations according to differentials in the regulatory surface. Where foreign regulatory differences do exist, TNCs may well use them in their negotiations with governments, particularly where the latter are anxious to attract specific TNC investments. Such competitive bids between states (and between regions and communities within states) to attract TNC investment has undoubtedly intensified in recent years (Dicken, 1990).

Such general comments should not create an impression of a uniform response among TNCs to regulatory differentials although there are some kinds of regulation which will tend to produce a fairly general reaction. But, for the most part, responses to regulatory differences are contingent upon specific circumstances in which particular TNC strategies are especially significant.

Firms operating in so-called multi-domestic industries must take account of global forces while, conversely, firms operating in global industries must be responsive to national and local differences. The intensification of global competition in a world which still retains a high degree of local differentiation creates, for all TNCs, an internal tension between globalisation forces on the one hand and localisation forces on the other (Bartlett and Ghoshal, 1987).

These pressures greatly complicate the nature of the responses of TNCs to regulatory differences between states. Such complications are intensified by the fact that the global–local tension applies at each stage of the firm's production (or value-added) chain. Spatial variations in regulatory structures will surely influence (though are unlikely to determine) the extent to which the production chain is integrated or disintegrated geographically. Whether or not a specific regulatory mechanism is a constraint or an opportunity will depend, at least in part, on whether the firm is pursuing a nationally-responsive or globally-integrated strategy. For example, the primary requirement for a firm pursuing a nationally responsive marketing strategy is unhindered access to its target markets. Regulatory barriers, such as tariffs or NTBs, will usually stimulate a response by which the firm sets up operations within the regulated market. The picture is complicated in regional economic groupings, such as the EC prior to the completion of the Single Market, where there are differential regulatory barriers affecting a firm's spatial strategy but in which access to the larger regional market may be achieved. In these circumstances, firms may well be able to play off one member country against another on the basis of regulatory differences.

A more complex situation arises in the case of TNCs attempting to pursue a globally-integrated strategy. The essence of such a strategy is the ability to locate each part of the production chain in optimal conditions (for example, to use cheap labour for basic assembly operations). The integrated structure, therefore, consists of a complex network of managerial, research, production, marketing, distribution and service nodes linked together by intricate cross-border flows. On the one hand, such global networks provide TNCs with the potential ability to minimise regulatory constraints and maximise regulatory opportunities because they may be able to switch and reswitch their operations between alternative locations. In this sense, global firms may be able to engage in 'regulatory arbitrage', that is, to benefit from the differences in regulatory structures between states at a global scale.

On the other hand, states may well have the ability to inhibit the achievements of globally-integrated strategies. [W]hat matters is the relative bargaining power of the state on the one hand and the TNC on the other. The relative bargaining power of TNCs and host countries is a function of three related elements (Figure 1): (1) the relative demand by each of the parties for the resources which the other controls, (2) the constraints on each which reflect the translation of bargaining power into control over the outcomes, (3) the negotiating status of the participants.

Conventionally, the activities of TNCs have been equated with the actual ownership of assets in more than one country. Indeed, this is the only criterion on which comparative statistical data are available at an international scale. However FDI, whether by greenfield investment or by acquisition and merger, is only one possible mode of foreign investment which firms might adopt. Recognition of a diversity of modes implies a broader definition of the TNCs; one which is not confined to direct equity ownership. This, indeed, has been the position of the UN Center on Transnational Corporations whose definition has broadened substantially during the past 15 years. Cowling and Sugden (1987, p. 60) suggest a far more flexible definition of the TNC: 'A transnational is the means of co-ordinating production from one centre of strategic decision making when this co-ordination takes a firm across national boundaries'.

Thus, a firm's international involvement may well take the form not only of FDI of the conventional variety but also (and increasingly) various forms of strategic alliance and various forms of networking. At least in part, although to varying degrees, these modes are a response to regulatory differences in the global economy. Strategic alliances, for example, are very often a means for firms to get a foothold in a tightly regulated market by joining up with domestic firms.

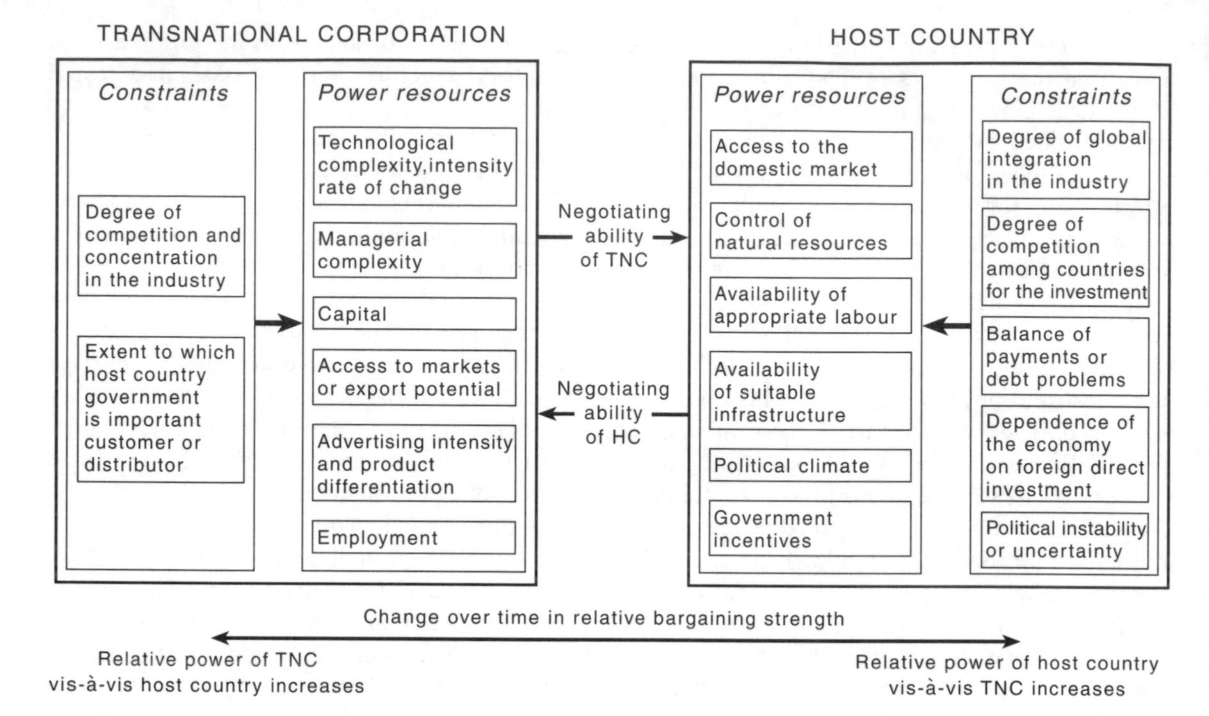

Figure 1 Elements in the bargaining relationship between TNCs and host countries. *Source*: Dicken (1992, Figure 12.5). Based on material in Kobrin (1987)

CONCLUSION

[T]he influence of national regulatory policies on the spatial strategies of transnational corporations is a two-way process. On one hand, TNCs respond to differentials in the regulatory surface: on the other hand, the changing nature of the regulatory surface itself is, at least partly, a state response to the strategies of TNCs. The scope of this paper has been limited to just three aspects of state regulation: trade, foreign investment and industry policy. But these, of course, need to be set within the broader regulatory framework at the macro-scale. Unfortunately, the geographical literature has tended to pay scant attention to these complex, but highly significant, politico-economic interactions. Indeed, much of the literature on the new international division of labour gives very little emphasis to the role of the state as a significant actor in shaping the contemporary global economy. There is a real need for geographers to engage in research into the interactions between states and TNCs, into the dynamic impact of changing regulatory environments, and into the bargaining processes involved.

Adam Tickell and Jamie A. Peck

'Social Regulation after Fordism: Regulation Theory, Neo-Liberalism and the Global–Local Nexus'

Economy and Society (1995)

INTRODUCTION

Critical emphasis in regulation theory is placed on the 'structural coupling' between the system of accumulation (a macro-economically coherent production–distribution–consumption relationship) and the ensemble of state forms, social norms, political practices and institutional networks which regulationists term the mode of social regulation (MSR). Our proposition here is that much post-Fordist speculation is based on a series of abstractions from changing conditions in *production*, themselves based on empirical generalizations. For all its use of regulationist jargon, a great deal of the post-Fordist literature falls considerably short of the requirements of a regulation approach in failing to specify either how the putative post-Fordist economy might be socially regulated or how it might be pieced together in macro-economic terms. We cannot, therefore, speak of a post-Fordist *regime of accumulation* because such a system has yet to be comprehensively identified.

For many, regulation theory has in effect become a theory of post-Fordism. This is a mistaken perception, though in many ways an understandable one. Regulation theory has been caught up in post-Fordist debates despite the fact that the theory is not predictive. Its terminology has been used to link tendentious developments in flexible production to posited systemic shifts in the regime of accumulation, despite the fact that the theoretical criteria for the establishment of a regime have not been satisfied. While an important strand of thinking about post-Fordism placed flexible production at the economic heart of the new regime (compare Scott, 1988b, with **Pollert**, 1988 (see Part Five, Section 5.2), and Gertler, 1992), political variants of neo-liberalism were heralded by some as being functional for post-Fordism (compare Jessop, 1989b, 1991b, with Hay, 1994, and Peck and Jones, 1995). As the dysfunctionality of both flexibility and neo-liberalism has become apparent even to their strongest advocates, regulation theory itself has become tarred with the post-Fordist brush. Yet what is at stake here is not just a defence of the integrity of a particular theoretical position, but the much broader question of how we interpret, theorize and respond politically to the present political–economic disjuncture.

In this paper we deploy a regulationist framework to raise some critical questions about certain post-Fordist propositions, focusing on the question of space and social regulation. Contrary to the central role of the nation-state under Fordism, it has been suggested that the post-Fordist era will be characterized by a 'hollowed out' nation-state, eroded from both above and below as supranational and subnational state structures come to play an increasingly important part in shaping systems of social regulation and, more generally, the course of accumulation (see Jessop, 1994a; Mayer, 1994). Such claims are in some senses just as controversial as those which have been made about post-Fordist production and politics, because they suggest that not only are the defining features of the ascendant regime already identifiable, but also that the incipient 'spatial logic' of this system can be discerned. Continuing to be faced, as we do, by conditions of *systemic instability* at the level of the global economy, such claims are premature.

REGULATION THEORY AS METHOD

Regulation theory provides a conceptual framework for understanding processes of capitalist growth, crisis and reproduction. Systems of accumulation and MSRs which are coupled together in a stable fashion are the basis for what regulationists term *regimes of accumulation*. These refer to particular capitalist development paths – defined in terms of historical phases and patterns of development – which are characterized by economic growth and under which (immanent) crisis tendencies are contained, mediated or at least postponed. [T]he regulation approach is not simply about medium-term growth and how to get it. Rather, its focus is on the institutional forms and practices within the MSR which guide and stabilize the accumulation process and which provide a *temporary* resolution to the crisis tendencies which are seen to be endemic in the accumulation process.

Regulation theory, then, confronts the paradox that capitalism has proved rather more durable than envisaged in marxian theory, that crises may not *only* be way-stations on the path of terminal decline, but that – in terms of the actualities of capitalist development – they may also play a rejuvenating role, 'brutally restoring the contradictory unity' of the accumulation process.[1] Central to this entire process, for regulation theorists, is the role played by the MSR.

Regulationists insist that, in order to understand capitalist reproduction in its integral sense, it is necessary to understand the wider social and institutional context in which accumulation occurs (Jessop, 1992c). This is accomplished through an analytical focus on the nexus of the accumulation process and the MSR. The MSR is defined as a set of codified social relations which have the effect of guiding and sustaining the accumulation process (Aglietta, 1979, p. 382). Comprised of a complex ensemble of social norms and habits; state forms, structures and practices; customs and networks; and institutionalized compromises, rules of conduct and enforceable laws, the MSR defines 'the social context in which expanded economic reproduction occurs' (Jessop, 1992a, p. 50). The establishment of a broadly functional coupling between the accumulation system and the MSR is, for regulationists, a prerequisite for the formation of a regime of accumulation.

While critical for the theory, however, the MSR is often rather poorly defined within regulationist writing. By definition, MSRs are not determined functionally by the requirements of the accumulation process. The complex processes through which MSRs are formed around particular accumulation trajectories, are in essence a political process (see also Lipietz, 1992a), one which is reducible neither to some functionalist response to the 'needs of capital' nor to conscious action on the part of capitalist states.

According to Jessop, for example, the MSR provides the most sound basis on which a regime of accumulation should be determined (rather than the labour process or the pattern of accumulation) because it specifies:

> the institutional and organizational conditions which secure[d] Fordism as a national accumulation regime and is especially helpful in defining the peculiarities of different Fordist regimes. But [MSRs] cannot be properly understood without considering how [they] modify and yet remain subject to the general laws of capital accumulation. (Jessop, 1992a, p. 50)

Thus, while it may be possible to make both abstract claims and empirical generalizations

Table 1 Variants of Fordism

Type of Fordist regime	Characteristics	Examples
'Classic Fordism'	Mass-production and consumption underwritten by social democratic welfare state.	USA
'Flex-Fordism'	Decentralized, federalized state. Close co-operation between financial and industrial capital, including facilitation of interfirm co-operation.	West Germany
'Flawed Fordism'	Inadequate integration of financial and productive capital at the level of the nation-state. Archaic and obstructive class politics identified by some authors.	United Kingdom
'State Fordism'	State plays a leading role in creation of conditions of mass production, including state control of industry. 'L'etat entrepreneur'.	France
'Delayed Fordism'	Cheap labour immediately adjacent to Fordist core. State intervention has key role in rapid industrialization in 1960s.	Italy, Spain
'Peripheral Fordism'	Local assembly followed by export of Fordist goods. Heavy indebtedness. Authoritarian state structures coupled with movement for democracy, attempts to emulate Fordism accumulation system in absence of corresponding mode of social regulation.	Mexico, Brazil
'Racial Fordism'	Dualistic workforce. Privileged minority has North American style working conditions and remuneration levels which rely on authoritarian state structures and 'super-exploitation' of majority population.	South Africa
'Primitive Taylorization'	Taylorist labour process with almost endless supply of labour. Bloody exploitation and huge extraction of surplus value. Dictatorial states and high social tension.	Malaysia, Bangladesh, The Philippines
'Hybrid Fordism'	Profit-driven expansion based on modified Taylorism. Truncated internal market, societal segmentation and under-developed welfare state. Indirect wage indexation.	Japan

Source: updated from Tickell and Peck (1992).

about globally hegemonic regimes of accumulation, regulationist accounts have pointed to a considerable degree of institutional (and geographical) variability *within* individual regimes. So, while global Fordism could be characterized in general terms as a global system, the building blocks from which this system was constructed were in effect a whole series of *national Fordisms* (Lipietz, 1987; but see Aglietta, 1985; Leyshon and Tickell, 1994), [t]he specificities of which are sketched out on a country by country basis in Table 1.[2]

THE POLITICS OF POST-FORDISM

For some critics, however, the entire regulationist project has been undermined by speculative work on post-Fordism. Thrift (1989b), for instance, has warned of the perils inherent in the use of 'stage' or 'transition' models. But the regulationist insistence that periodization (and for that matter regionalization) provides a useful approach for understanding the historical geography of capitalist development should not be read to imply that the rules do not change, that capitalism will progress infinitely through some regular succession of alternating phases of growth and crisis. Regulation theory differs from conceptualizations such as long-wave theory on this crucial point – it offers no guarantees that a successor regime will come along, that durable growth will be restored (Kotz, 1990). It also does not say that the new regime will be the mirror-image of the last (for example, as a rigid Fordism is replaced by a flexible post-Fordism). The entire burden of the theory is that history matters, institutions matter, geography matters, that crises and regimes are unique events (albeit simultaneously expressing structural processes).

This said, regulation theory *does* have things to say in the interpretation of developments and experiments *after* Fordism. The uncertainties created by the breakdown of Fordism–Keynesianism are bound to stimulate critiques of existing production systems and institutional structures, as well as experimental searches for new solutions. Needless to say, such conditions pose immense interpretative problems for social scientists, particularly those concerned with separating the durable wheat from the ephemeral chaff. More broadly, there are real analytical dangers in attempting to read regime-wide and predictive conclusions from the flux of crisis.

Post-Fordist Social Regulation?

A critical silence in much of the post-Fordist literature – perhaps *the* critical silence – is the issue of social regulation. For all the audacious claims about flexible production, flexible labour systems, new industrial spaces and so on, relatively little has been said about the political and social institutions needed to sustain these nascent economic trends. While some point to an emerging consensus around certain of these (sometimes empirically definable) developments, the questions of their durability and of how they should be interpreted remain wide open (Gertler, 1992). From a regulationist perspective, these features cannot be seen to be emblematic of a new regime until their sustainability can be demonstrated. This in turn implies that (some kind of) post-Keynesian MSR must be identifiable.

Because the rise of neo-liberalism in the 1970s and 1980s coincided with the breakdown of Fordism and the apparently terminal collapse of Keynesian social regulation, it has in some accounts been afforded the status of an MSR-in-waiting. This issue has been confronted most explicitly by Jessop (1992b, pp. 31–2), who defines the 'emerging post-Fordist mode of regulation', based on the twin principles of flexibility and supply-side innovation. For Jessop, these developments 'could create a distinctive ensemble of regulatory practices which is reproducible in the medium term' (1992b, p. 32) and 'could prove structurally congruent and functionally adequate to post-Fordist accumulation regimes' (1994b, p. 27).

Contrary to Jessop's position, we would argue, first, that neo-liberalism and the post-Fordist MSR (as he defines it) are related; second, that failures of the former raise real questions about the durability of the latter; and, third, that political dangers lurk in separating the two. The contradictions and problems of neo-liberalism cannot be put on one side while the post-Fordist MSR is theorized. On the contrary, they imply that it is inappropriate to label this ensemble of regulatory practices as an MSR because it is internally crisis-prone and therefore unstable.

Lipietz (1992a) has suggested that four flaws exist in the neo-liberal model. First, the model is associated with a tendency for social polarization. Second, it does not resolve the contradictions of the Taylorist labour process, central to the breakdown of Fordism. Third, it tends to fuel, rather than contain, swings in the business cycle, with the result that macroeconomic crashes are a constant threat. Fourth, the deregulation of international trade does not lead unproblematically to structural adjustment but to the exacerbation of structural imbalances and to forced deflations.

The patterns of growth exhibited in regions such as Britain's M4 corridor and the Californian technopoles, and for which so much is claimed by proponents of (and apologists for) neo-liberalism, both faltered badly in the recession of the early 1990s. In the south east of England, chronic and simultaneous overheating in labour, housing, finance and commercial property markets contributed to the suffocation, in the late 1980s, of the regional growth pattern. This was subsequently to pull Britain as a whole into a deflationary spiral which led to the longest recession of the post-war period. Meanwhile, the downturn in California triggered a state-level fiscal crisis.

Interestingly – and in stark contrast to the ideology of neo-liberalism – the mode of growth exhibited during the 1980s in both California and the M4 corridor was significantly underwritten by state defence expenditure (Lovering, 1991; Markusen, 1991). These areas were consequently the beneficiaries of what was in effect a repackaged version of regional Keynesianism (coupled with regressive income redistribution), presented through the rhetoric of enterprise and self-help.

Locating Neo-Liberalism

The crisis of Fordism–Keynesianism contributed to the growing ascendancy of neo-liberal ideologies, implicit in which was a rejection of state regulation – at least in terms of established forms in intervention. [N]eo-liberals hold that markets enter crisis as a result of the distorting actions of state regulation (see, for example, Hayek, 1976; Dowd, 1988). While the ideology of neo-liberalism suggests a 'hands-off' approach to economic management, the task which confronted neo-liberal nation-states during the 1980s was that of dismantling (parts of) the Keynesian welfare state. This fostered a series of neo-liberal *political projects*, such as Thatcherism, Reaganism and Nakasoneism, variants on the underlying (neo-liberal) ideology which differ from one another as much as did the national variants of Fordism–Keynesianism shown in Table 1.

It is no coincidence, though, that such a rash of neo-liberal projects should have emerged after Fordism. This is because neo-liberalism reflects a deeper set of currents in capitalist development. There are essentially two ways of interpreting the current grip of neo-liberalism. One interpretation is that, because there are so many neo-liberal projects (and few presently working alternatives), they must be part of a post-Keynesian institutional fix. Here, neo-liberalism is portrayed as a regulatory solution. Another interpretation – and the one which we favour – is that the ascendancy of neo-liberalism represents a regulatory hole, one which has elements of market regulation but which represents the *absence* of a new institutional fix. Market regulation is, as **Lash and Urry** (1994) (see Section 2.2) emphasize, simply inadequate to the task of regulating capitalism. Contrary to Jessop, then, we suggest that, far from providing the basis for a new regulatory fix, neo-liberalism represents the *source of the problem*. It is, as a result, tautological to argue that neo-liberalism is promoting a 'market-guided transition towards the new economic regime' (Jessop, 1994a, p. 16).

SOCIAL REGULATION UNDER FORDISM AND BEYOND

The process of social regulation under Fordism was anchored in the Keynesian welfare state, under which collective bargaining and monopoly pricing were institutionalized; policy instruments were deployed to maintain and manage aggregate demand; and norms of mass consumption and 'American ways of life' were generalized (see Jessop, 1992a, b). By implication, these processes of social regulation were rooted, first and foremost, in and around the *nation*-state.

Regulationist research now identifies as one of the fundamental tensions of the Fordist regime the uneasy interface between *national* forms of regulation and the *globalizing* dynamic of accumulation. The regulatory 'logic' of the Fordist system may have been rooted in the nationally constituted Keynesian welfare state, but this regulatory system was *itself* predicated on a specifically configured international order.

Essential to the viability of Keynesian welfare state systems in core Fordist countries was the institutionalization of international economic and political relations under the *Pax Americana*. The establishment of the US as global financial hegemon – backed up by the country's military, economic and political might and institutionalized under Bretton Woods – provided a means of regulating the international system in a way compatible with the requirements of Keynesian regulation at the level of the nation-state (Corbridge, 1988; Altvater, 1992).

However, the degree of control exercised by Bretton Woods over *monetary* circulation could not be exerted over the emerging global *credit* system. Lubricated by the growth of TNCs and with the development of unregulated 'Euromarkets' in money, the global credit system became, in effect, a privatized system, beyond the reach of political–institutional control (Gilpin, 1987; Walter, 1991; Swyngedouw, 1992). As private capital began increasingly to circuit globally on a deregulated basis, Keynesian nation-states progressively lost control of one of the most important macro-economic levers – the setting of interest rates. 'Unregulated global credit was [consequently] a factor of erosion of the (political institutional) regulation of the whole Fordist system' (Altvater, 1992, p. 37).

There is an important sense in which Fordism was undermined by its emergent geographical contradictions, or, alternatively, that the contradictions of Fordism took on a geographical form. The erosion of interest-rate sovereignty undermined the basis for the regulation of aggregate demand at the nation-state level. In the absence of effective demand management, the system of Fordist regulation was itself ruptured. The continuing process of globalization, moreover, implies that the current crisis is *qualitatively* different from earlier crises (see Martin, 1993).

REGULATION THEORY, NEO-LIBERALISM AND THE 'NEW SPATIAL DISORDER'

Regulation theory has essentially three things to say about space: first, the nature of the coupling between accumulation and regulation varies from nation-state to nation-state; second, for a regime of accumulation to stabilize, this coupling must be functional at the level of the nation-state; and, third, capitalism is associated with endemic uneven development, as expressed through the shifting nature of core–periphery relations. A limitation of the theory, however, is that it contains no explicit conception of uneven spatial development, at either the subnational or the supranational scales. So, while regulation theory raises some interesting questions about uneven development, it is unable at the moment to provide the answers.

If regulation theory is to be spatialized, it must loosen its exclusive grip on the nation-state: while the nation-state will no doubt continue to be one of the key arenas of struggle, this may not be the scale at which future institution-building is rooted. The question of functionality in accumulation–regulation relationships consequently needs to be opened up at other spatial scales (see Peck and Tickell, 1992, Table 1; Marden, 1992).

Table 2 summarizes the spatial constitution of regulatory relations during and after Fordism. The table illustrates the spatial contradictions of Fordism and the ways in which they contributed to the unpicking of the Fordist regime of accumulation. It also illustrates the spatial relations which have emerged in the unstable period since Fordism's demise. This new geography is not, of course, the polar opposite of its predecessor, although there are fundamental differences between them. Important here is the process of 'glocalization' (the simultaneous development of globalization and localization: Swyngedouw, 1992). Although the national arena remains important as the site for *discursive* struggle, Jessop argues that nation-states have become 'hollowed out': power has been ceded both to supranational institutions and to local states.

Restructuring in the international financial system has been a key stimulus for change at the supranational scale. These changes have not only led to a structural shift in accumulation, where the returns on the fictitious churning of financial capital can now far outstrip the potential returns from material production (Walter, 1991; Altvater, 1992), they have destabilized the spatial strategies employed by transnational corporations. Such instability has strengthened pressures to expand the role of supranational state institutions – such as the European Union.

It has recently been argued that local states and regions have assumed a heightened importance as a result of the hollowing out of

Table 2 Spatial constitution of regulatory relations under and after Fordism

	Fordism		After-Fordism	
Spatial scale	Characteristics	Contradictions	Characteristics	Contradictions
Global	Bretton Woods financial system and GATT underwrite financial stability and global trade, acting as mechanisms which 'transmit' Fordist features internationally. An international 'regulated space'.	USA acts as governor and guarantor of regulatory order at same time as exploiting system for its own economic interests. US ideology of market undermines efficacy of international regulatory discourse.	International financial system operates without control of regulators, while market 'logic' dominates the new GATT. 24 hour global financial markets enable 'regulatory arbitrage' further undermining regulation.	Increasingly volatile and unstable financial system. Rapid systemic transmission of economic cycles, accentuating growth and decline and undermining basis for stable development.
Global-national relations	Nation-states have capacity to set independent monetary policy within the context of US hegemony.	Later stages of Fordism witness progressive internationalization of capital which undermines national autarchy. Transmission of US 'domestic' problems through global economy.	Emergent supranational bodies and 'pooled sovereignty' in attempt to regain control over internationalized capital (e.g. BIS, EU). TNCs engage in regulatory arbitrage.	National economies become further absorbed into global circuits of capital, necessitating further supranationalization of power (not always pursued). Nation-states further undermined.
National scale	Central regulatory functions dispensed by Keynesian welfare state, securing conditions for mass production and consumption.	Fiscal crisis of nation-state triggered by deindustrialization, rise of mass unemployment and loss of interest rate sovereignty.	'Hollowing out' of nation-state, as national governments cede power to supranational and local bodies.	State loses control over accumulation process and becomes more responsive to demands of capital. Less able to meet social welfare objectives, further undermining cohesion of national social formation.
National–local relations	Centralization and consolidation of nation state powers as governments attempt to control economies and introduce social welfare systems. Nation states seek to ameliorate worst effects of uneven development via regional policy.	Political and economic contradictions of uneven development within nation state. Failure of regional policy following deepening peripheralization.	Unstable and uneasy relationship. Geographically specific political responses. Targeted local interventionism replaced by 'competitive localism'.	Market rules foster local regulatory undercutting and zero-sum local-local competition. Exacerbation of spatial inequality.
Local scale	Key regulatory functions around social reproduction dispensed through local welfare states.	Fiscal crisis of national state transmitted to local state, undermining local welfarism.	Claims that local states have enhanced economic role. Supply-side local state manages, e.g., training policy.	Local states powerless in global economy, reacting to external economic forces. Few degrees of local freedom.

the nation-state (see, for example, Stoker, 1990; Mayer, 1992; Swyngedouw, 1992; Jessop, 1994a). In some variants of the localization thesis, local states are seen to have an enhanced role in the world economy because they have been able to bypass national states. Mayer, for example, argues that neo-liberal states have devolved significant, material powers to cities and regions which:

> have become direct players in the world economy. The particular location of a place within the international division of labour under conditions of heightened inter-urban competition thus not only sets certain constraints, but itself becomes an asset to be exploited *on the basis of locally-determined priorities.* By identifying the particular strengths and assets a city or region has to offer (to investors etc.), *local political actors can exact payments and concessions, and can exert leverage over supra-local actors.* (Mayer, 1992, pp. 263, 269, emphasis added)

Such claims are, to say the very least, debatable. While cities and regions *may* be competing with each other, it is difficult to see that they have significantly greater *power* now than during the Fordist period. If *national* states are insufficiently powerful to set their economic policy or to prevent transnational companies from engaging in regulatory arbitrage, local states will surely have even less success.

As Jessop (1994a) argues, in many ways the most significant change over the past twenty years has been the collapse of the Keynesian welfare state and its regulatory infrastructure. This poses questions about the *significance* of the process of glocalization described above. There clearly have been significant formal increases in local economic intervention across both Europe and North America (see Moulaert *et al*, 1988; Davis, 1993). These have, however, surely been the result of the vacuum which has emerged at the level of the nation-state following the breakdown of Fordism–Keynesianism, rather than the granting of *new* powers to the local state. As *nation-states* are losing the capacity for demand-side intervention, it is difficult to see what significant levers are available at the local level *vis-à-vis* the more powerful process of globalization.

As Table 3 suggests, the emergent regulatory problems of post-Fordism will demand putative solutions based at different spatial scales. In the words of Dunford and Kafkalas (1992, p. 29), 'without supra-regional regulation and organisation, decentralisation does not offer real prospects of macro-economic stability or a full realisation of the productive potential of new technologies'. It is difficult to see how a new local order can precede a new global order. The geographies of *after*-Fordism sketched in Table 2, then, are intrinsically unstable. They are geographies of crisis.

CONCLUSION

Given the existence of severe reservations about the sustainability of flexible accumulation (see Pollert, 1988; Gertler, 1988; Tickell and Peck, 1992), it is impossible to make any conclusive statements about the spatial logics of post-Fordism. There are, to be sure, substantial shifts in the spatial ordering of the world economy under way at present. However, in theorizing these changes, there has, as Martin (1993) has argued, been an over-eagerness to read off spatial stereotypes from simplified macro-economic characterizations and to afford such stereotypes a conceptual status that they do not deserve. The geography of glocalization is the geography of the *after*-Fordist crisis, not the geography of stable post-Fordism. Such stability cannot be attained until neo-liberalism is defeated. In order to meet the essential prerequisite of reproducibility, the new pattern of social regulation must be an *anti*-liberalist one.

NOTES

1. There is a danger of teleology in regulation theory. Crises do not occur in order to rejuvenate capitalism, they arise out of the contradictions that remain inherent within capitalism. Most regulationists maintain that there are a number of different forms of crisis, ranging from

Table 3 Taming the *after*-Fordist tiger

Spatial scale	Regulatory 'problem'	Putative solution
International	Unstable and volatile neo-liberal financial system undermines national economic intervention and global stability	Supranational institutions which reassert control over financial capital? Potential in existing institutions but needs to be realized by democratization and eschewing of neo-liberalism. Financial co-operation and common currencies diminish advantages of speculative global finance system but render weak countries more vulnerable to external economic conditions.
	Neo-mercantilism and worsening terms of trade for Third World countries induces significant risk in trading system. International Trading Organization heralds neo-liberal trading regime. Regional trading and political blocs organized along neo-liberal lines. Creates supranational instability and exacerbates uneven development.	Formation and enhancement of regional trading blocs? Provide some protection for those in strong blocs but these are likely to include only the already economically powerful, leaving the poor to get poorer. Spatially redistributive policies to ameliorate worst effects of uneven development (at national and supranational level).
	Regulatory 'arbitrage' and undercutting (e.g. British opt-out of the European Social Chapter).	Supranational institutions to assert common minimum standards across range of areas (e.g. minimum capital adequacy standards). Development of high-skill rather than low labour cost national base.
National	Mass unemployment	Renewed national Fordism-Keynesianisms are unlikely to emerge. Supranational Keynesianism allied to new 'flexible production systems'?
	National neo-liberalisms unable to contain geographical contradictions.	Policies to stimulate growth in lagging regions and contain growth in cores. Regional policy organized at national and/or supranational levels?
	'Hollowing out' undermines nation-states' welfarist objectives.	Supranational regulation to prevent pressures to minimize standards. International neo-liberalism stimulates regulatory undercutting and needs to be countered. Need for progressive fiscal structures to forge new social compromise.
Local	Zero-sum competition between localities and regions encourages geographically uneven undermining of social standards and fragmentation.	Embedding of capital within localities – using technology or training policy? – to stimulate a spatial fix. Stimulation of regional development. National and supranational regulation of wasteful competition.
	Local growth coalitions are inherently unstable and short-termist.	Democratization of growth for all. Reduced emphasis on growth coalitions based on business. Enhanced power for regional and local governments.
	Local state seen as increasingly central to economic regeneration but with limited powers to intervene	Increased political autonomy and power within wider structural frameworks. True 'subsidiarity'.

a crisis for a unit of capital through to an (as yet unrealized) crisis of capitalism itself (Dunford, 1990; Tickell and Peck, 1992). Furthermore, as Hay (1994) emphasizes, the notion of crisis in regulation theory underestimates the extent to which crisis is socially and politically mediated.

2. Significantly, this geographical variability within regimes of accumulation requires a conception of spatially uneven development to be located with the regulationist framework, one absent from the original theorization (Peck and Tickell, 1992).

PART THREE

Spaces of Production

Carhenge, Alliance, Nebraska. A collection of 38 cars from the 1950s and 1960s painted grey and welded in place to replicate Stonehenge.

PART THREE SPACES OF PRODUCTION

Section 3.1 Introduction: Spaces of Production – Towards New Economic Geographies

> Different areas come to specialize either on manufacturing or on certain services and there is no particular reason why these will coincide within national boundaries.
>
> (Lash and Urry, 1994, p. 195)

INTRODUCTION

Part Two of this Reader explored the changing forms of capitalism as economic activities become increasingly influenced by the forces of globalization. These global changes have important implications for the production of goods and services both in advanced and developing economies. The resulting new economic landscape in countries like the US and the UK is one in which paid work is likely to be a form of service employment, and in which production becomes a combination of manufacturing and service activities. The papers in this section explore the alteration in the division of labour and the resulting shift towards service employment, as well as the rise of new spaces and forms of production.

Since the 1960s employment has shifted away from manufacturing towards a variety of service activities and occupations (**Daniels; Bryson *et al***). This shift has important implications for the nature of work, production and the structure and operation of economies from the local through to the global scale. Employment statistics show that the advanced economies have experienced a rapid process of deindustrialization (Rowthorn and Wells, 1987). This process of deindustrialization has been particularly focused on older – often coalfield-based – industrial regions, many of which have suffered acutely since 1970 from wholesale factory closures and huge manufacturing job losses (**Florida**). Regional deindustrialization has reflected previous over-specialization by these regions on primary and secondary sectors – coalmining, iron and steel, shipbuilding, textiles – which have been in crisis, and drastically reducing employment, because of a collapse in demand (the shift from coal to oil, for example), intensified competition from low-cost or more-efficient foreign competitors, or the adoption of new computer-controlled production techniques requiring only a fraction of the former labour force.

This said, however, a select few of these older regions have since the 1980s been experiencing a rejuvenation of manufacturing industry either because of successful restructuring and renewal of innovative capacity, after great trauma, of some of their traditional industries (as with the US Mid-West: see **Florida**) and/or because of the attraction to the region by government policy incentives of external, often foreign, branch plant investment (as with Wales, whose manufacturing employment record 1981–97 was better than any other UK region). Equally, national deindustrialization in terms of manufacturing employment losses has often gone hand-in-hand with continuing manufacturing output growth. Thus, while UK manufacturing employment fell between 1981 and 1993 by 31.3%, UK manufacturing output rose by 24.4%. This suggests that manufacturing is not becoming less important, but that since the 1960s there have been quantitative and qualitative changes in the workings of the capitalist economy. These changes reflect alterations in the relationship between manufacturing and service employment and in the ways in which manufactured products are produced, sold and consumed.

When investigating these alterations, it is usual in economic geography to separate service activities from manufacturing. Most economic geographers concentrate research on one of these sectors. This part of the Reader does not make this distinction. Service functions are an integral part of the production system. Manufacturing could not occur without service activities and services would be unable to function without manufactured products. The shift towards service employment is not a shift away from manufacturing activities, but a realignment of the types of skills and expertise required to produce a finished commodity. In certain instances, places (regions, cities, etc.) come to specialize in particular skills or expertise.

REWORKING THE DIVISION OF LABOUR

The transition from an industrial to an advanced industrial economy is associated with changes in the production process and in the nature of the added value contained in physical products (**Bryson *et al***; Bryson, 1996, 1997). In an industrial society the transformation of raw materials during the production process was predominantly the result of the activities of direct labour on raw materials (Walker, 1985). The division of labour in the workplace was concerned with specialization of activity to increase productivity, or to transfer the control of the production process away from the shop floor to management. Constant technological and organizational innovations led to de-skilling of parts of the workforce and the creation of a middle management tier. This transition and transformation of skills created a division of labour based on the relationship with the production process. This development of managerial activities created an indirect labour force whose task was to service, support and control the production process.

During the twentieth century, production has become dominated by the increasing concentration of control into the management layers of organizations and the creation of a management division of labour as well as by the development and reworking of a new international division of labour (**Lakha**; **Daniels**; see Pearson's paper in Part Five, Section 5.3). The former development represents the creation of a whole series of complex internalized and externalized service occupations and industries (**Bryson *et al***). The latter development involves the interplay between local and global forces (**Lakha; Roberts**).

At the same time an escalation in the knowledge content of both goods and services has occurred. The concept of a knowledge or information economy (Castells, 1996), however, distracts attention from the relationship between knowledge and the physical production of goods. Knowledge is either contained in physical products, or is indirectly involved in their production, circulation or consumption. The increase in the knowledge content of goods is both a product of the increasing division of labour, and of increases in the design and technological content of both goods and services. It is also part of a shift in consumer behaviour away from the consumption of mass-produced goods towards the consumption of customized products and lifestyle shopping (see **Harvey**, and the papers in Part Four of this Reader).

At the heart of the information revolution, according to the paper by **Daniels**, is the computer, or more precisely the combination of computers with telecommunications which has become known as information technology (see **Warf**'s paper in Part Two, Section 2.3). Dramatic increases in computer power and data-handling techniques have enabled the development of giant computer databases and associated surveillance, by government and private firms, of individuals' and household behaviour. Such surveillance is not new, but the impact of the new information technology makes the collection of personal data more efficient as well as expanding both the geographic extent of the collection area and the nature and scale of linkages between areas.

Externalization or the outsourcing of service functions and the creation of new types of service occupations represents an increase or extension of the division of labour. An increasing division of labour reflects both increasing specialization of activity with a resultant increase in the complexity of production, and alterations in the way in which production is organized. Here the important point is made in the paper by **Sayer and Walker** in which they explore the notion of an *extended labour process*. This is the work that occurs before and after goods and services are physically produced. Thus, research and development, design, market research, trial production, product testing, marketing, customer care and sales are all essential parts of the production process. The fact that they can be separated in both time and space from the actual production process does not necessarily imply that they are not an integral part of the manufacturing sector of the economy. Ultimately, this means that the dramatic growth in service employment reflects alterations in the way in which manufacturing and primary production is organized, rather than the development of a service or knowledge economy.

The reworking of the division of labour is also a spatial process. As outlined in Section 3.4 on new spaces of production, new technologies (air, telecommunications) and falling tariff barriers from the 1960s enabled the development in certain manufacturing industries of a new international division of labour, involving a split between production in low-cost developing countries and control, research and marketing in advanced economies. The paper by **Lakha** makes the important point that this new international division of labour now extends beyond manufacturing and into services. Thus, India has developed a software industry on the back of low-cost professional labour linked to the global demand for software expertise.

Lakha's paper deals with a relatively highly paid workforce of professional employees. Service employment, however, is also characterized by low-paid, part-time work. This aspect of service work is explored by **Christopherson** (see also **Waldinger and Lapp; Pinch** in Part Five, Section 5.4). This paper highlights three important characteristics of service industry growth in advanced economies.

First, it is associated with substantial growth in numbers of small and medium-sized enterprises (SMEs), for the simple reason that notwithstanding some giants, most service sectors are dominated (in numbers at least) by small businesses (**Bryson *et al***). In 1997, for example, firms with fewer than 50 employees accounted for 99.6% of businesses and 56.6% of jobs in the UK computer services industry, and 98.3% and 60.0% of UK firms and jobs, respectively, in the management consultancy, accountancy, legal and market research services sector (DTI, 1998a). Very high proportions of small firms reflect low barriers to entry in terms of the need for capital or premises, and numerous niche market opportunities (Keeble *et al*, 1992). Small service firm growth has in fact been most remarkable in professional and business services, the UK stock of businesses in this sector growing by an astonishing 22.4% or 69 000 between 1994 and 1997, compared with a decline of 5.8% or 77 000 in the stock of businesses in all other sectors of the UK economy put together (DTI, 1998b).

Second, the shift towards services is reflected in the growth of part-time, temporary and self-employment in the advanced economies. This trend can be explained by reference to the work of Hart (1987) who suggests that employers seek flexibility by varying the number of people employed (flexibility on the extensive margin), or by varying the number of hours people work as well as the times at which the work occurs (flexibility on the intensive margin). **Christopherson** notes that between 1973 and 1979, 12.5 million jobs were created in the US, and another 14.5 million were added in the 1980s. Nearly one in four of these jobs was part-time.

The third trend is the increasing feminization of the workforce (see **McDowell**'s paper in Part Five, Section 5.3). This is associated both with the shift towards service employment and the rise of part-time jobs, many of which are undertaken by women. These employment trends are especially pronounced in those service industries, such as retailing, secretarial work, cleaning and care-related occupations, that provide standard rather than customized services.

RETHINKING THE SPATIAL MOSAIC

The papers in this section of the Reader explore different ways of conceptualizing the geography, or spatial mosaic, of economic activity. They highlight two significant shifts in the focus of economic geography: first, the return of the region, or local agglomeration production systems, as a meaningful unit of analysis, and secondly, an awareness of the relationship between culture and economic activities.

Amin and Thrift explore the localization thesis drawing upon two examples: the City of London and the Italian leather industry. This analysis echoes the early work of the Cambridge economist Alfred Marshall (1909) on small-firm districts in Lancashire and Yorkshire during the nineteenth century. To Marshall an agglomeration economy functioned on the basis of both social and economic processes. The former involved a common culture and dense social networks, the latter intense specialization and close inter-firm linkages. Such processes are central to the success of the City of London and Italy's leather industry. To Amin and Thrift both these places are 'neo-Marshallian nodes'; they are 'neo' because they are nodes which are part of a complex global network of economic interaction. Such networks provide these places with a competitive position in the global economy. However, highly localized production systems drawing upon certain basic structures whilst highly successful are very difficult to replicate.

Economic activity has always had an uneven geography. The paper by **Allen** explores the uneven character of growth across the UK suggesting that it can be seen as the product of distinctive geographies that are laid down, one over another. This notion of 'regionalized modes of growth' provides an alternative way of thinking about economic activities. Not all service or manufacturing activities operate internationally and not all operate locally. The important point is that different parts of a country's economy will be connected to the world economy in different, and maybe distinctive, ways. Thus, the City of London's financial institutions are increasingly more connected to the economies of New York and Tokyo than Newcastle and Manchester. Allen's paper allows us to conceptualize an economy as a spatial mosaic of economic activity with each activity having its own unique set of local, national and global linkages.

Markusen argues that a focus solely on new production spaces of the Italian 'industrial district' (**Amin and Thrift**'s neo-Marshallian nodes) or US high-technology types ignores the success, resilience and growth of other very different types of production location. Indeed, she points out that the great majority of recently successful manufacturing regions in the US do not conform to what she calls the 'new industrial district' model. Instead, and by inductive research, she identifies three additional and distinctive spatial types, namely 'hub and spoke' industrial districts (Seattle, Toyota City), 'satellite platforms' (Wales, North Carolina's Research Triangle), and 'state-anchored districts' (Santa Fe, Denver). These different district types or 'sticky places', she argues, all manage to capture and anchor intrinsically mobile production capital within increasingly 'slippery' space, but in different ways and with different regional consequences.

For geographers such as **Gertler**, national cultural differences are powerful influences on firm behaviour and competitiveness, and hence very important in understanding variations in manufacturing growth in different locations, especially globally. The spread of new production methods will be problematic. Their adoption will depend on the existence of similarities in the social relations of production in the region of origin and in the region to which they are transferred. To **Gertler** 'closeness' between users and producers may be becoming more important as technology (machines and production systems) are embedded in local cultures, for example work practices and training cultures.

Closeness and regional business cultures are also central to **Storper**'s seminal paper on the resurgence of regional economies. In a highly uncertain world, the development of a regional 'world of production', characterized by common cultural conventions, rules, practices and institutions which enable firms to learn and benefit from other local firms and organizations, may be a major source of sustained competitive advantage. This paper has had an important impact on economic geography because it introduced geographers to another conceptual tool for understanding regional economies, namely untraded interdependencies. Companies are situated in a web of traded and untraded interdependencies. Untraded interdependencies take the form of 'conventions, informal rules and habits that co-ordinate economic actors under conditions of uncertainty' (Storper, 1997, p. 5). This concept can also be traced back to the ideas of Marshall (**Amin and Thrift**). Writing in 1909 (pp. 152–3) about agglomeration economies he noted that:

> The mysteries of [a] trade become no mysteries; but are as it were in the air, and children learn many of them unconsciously. Good work is rightly appreciated; inventions and improvements in machinery, in processes and the general organization of the business have their merits promptly discussed; if one man [*sic*] starts a new idea it is taken up by others and combined with suggestions of their own, and thus becomes the source of further ideas.

NEW SPACES OF PRODUCTION?

All the changes described above have spatial expression, at both global and intra-national/regional scales. At the global level, economic geographers initially focused on the way in which some US giant manufacturing corporations in labour-intensive industries – clothing, textiles, electronic consumer goods – in the 1970s and 1980s restructured their activities so as to decentralize low-skilled assembly production to new previously unindustrialized and cheaper developing country locations. The development of this 'new international division of labour' was one factor in the rapid growth of certain manufacturing-intensive Asian economies, such as Malaysia, Singapore and Taiwan, although the growth of indigenous companies with their culture of strong government support and close workforce–management relations has been equally if not more important. More recently, as noted earlier, the global division of labour has also developed in service industries such as software (**Lakha**) and financial services, the latter involving both the global nodes of New York, London and Tokyo (**Sassen**) and new offshore financial centres such as the Cayman Islands (**Roberts**).

Geographers have also shown that global restructuring by large manufacturing companies is actually taking a variety of forms, rather than conforming to one over-arching 'new international division of labour', with **Fujita and Hill** contrasting American-influenced global Fordism with Japanese multi-national global Toyoatism. They argue that toyotaism has different and more positive implications for local development in host nations. Their work again illustrates the growing recent trend for economic geographers to stress the importance of cultural differences between firms and countries in influencing manufacturing competitiveness, location and growth, culture being interpreted widely as embracing the mix of social value systems, institutional structures, and both unwritten and legal codes of behaviour and organizational relations characteristic of particular countries, and even regions.

Most geographical interest has however focused on understanding the fascinating changes in the location of production within advanced industrial – and 'post-industrial' – countries which have occurred since the 1960s, at the regional and local levels. Much of this work has investigated the apparent rise of 'new industrial spaces', often conceptualized as the geographical manifestation of a new production system of 'flexible specialisation' (Scott, 1988; **Storper**). The nature of these new production spaces varies considerably, two major and in many ways different types being the 'industrial districts' of north-east and central Italy and the high-technology regional clusters that have developed in the US (Silicon Valley, Boston's Route 128) and some European countries (Cambridge, Grenoble). As Lash and Urry (1994, p. 195) note, many do specialize on manufacturing, as with the Prato (Tuscany) textile district (Dei Ottati, 1998), or the Orange County aerospace and electronic equipment cluster in California (Scott, 1986). But the increasing blurring of manufacturing and service functions discussed earlier is also evident in the evolution of these clusters, as with the Cambridge high-technology complex where rapid 1990s employment growth has been dominated by high-technology services, not manufacturing (Keeble *et al*, 1999; **Massey**).

In studying the sometimes remarkable rise of these new industrial spaces, economic geographers have increasingly stressed the importance of local 'networking' (Yeung, 1994) and 'untraded interdependencies' (**Storper**) between the flexibly specialized firms, often small firms, and organizations which constitute them. **Henry *et al*** provide a fascinating account of just how important such local interdependencies have been in the rise and current competitive success of Britain's 'Motor Sport Valley', a crescent-shaped area focused on Oxfordshire which contains the world's largest single concentration of technologically-dynamic motor racing companies.

While global linkages and markets are crucial to the success of these agglomerations (**Amin and Thrift**), so too appear to be localized formal and informal networking relationships, the movement of 'embodied expertise' in the form of highly-skilled workers and entrepreneurs, and the associated inter-organizational exchange and development of new knowledge. A regional collective learning capacity (Keeble *et al*, 1999) is arguably especially important both for technology-based new industrial spaces and specialized service agglomerations such as the City of London (**Amin and Thrift**). As **Bryson *et al*** show, many small specialized business service firms, such as management consultancies, in cities like London benefit greatly from membership of networks of expertise through the flexibility conferred by close links with other specialized service firms and associate consultants. Such firms perhaps epitomize 'flexible specialization' in action.

Most research on service employment and activities has been focused on an analysis of Europe or the US (Bryson and Daniels, 1998). The paper by **Selya** explores Taiwan's growing service economy. Service activities are highly concentrated in Taiwan's major cities because these contain large pools of highly skilled labour and developed telecommunications (**Daniels**). Selya's paper can be read using Allen's concept of regionalization with different regions in Taiwan developing distinctive relationships with the national as well as the global economy. Thus the concentration of service activities in Taipei City is a function of its connections to the global economy. Taipei is a global city, but a much less important global city than **Sassen**'s London and New York. **Selya**, however, does not fall into the trap of dismissing manufacturing enterprises as relict features of Taiwan's economic landscape. Manufacturing is still important for Taiwan's economy and service employment and exports will be unable to replace manufacturing's contribution to the national economy. However, Taiwan's new service jobs may be able to assist textile and clothing manufacturers to discover and capture new niche markets. This brings the analysis back to **Sayer and Walker**'s extended division of labour, with Taiwan's economy developing new ways of production based on a combination of manufacturing and service activities.

As the London case indicates, however, recent processes of uneven regional development and production restructuring have also powerfully affected many localities which are neither 'new' nor 'industrial' spaces. The case of specialized 'service spaces' is highlighted by **Sassen**'s work which eloquently addresses the apparent paradox of the continuing if not enhanced importance of 'place-boundedness' to advanced financial and business services in an era of increasingly 'place-less' global telecommunications. As she points out, competitive success in such services often demands not only global 'real-time' contact with international markets and capital, but also local availability of sophisticated telecommunications infrastructure, a large and diverse labour market, and specialized local services and business cultures, which global cities such as London, New York and Tokyo are uniquely well-placed to provide (**Amin and Thrift**). This is not to deny, however, that new global forces have also created entirely new service spaces, as with the rise of small offshore banking centres such as the Cayman Islands, a process analysed by **Roberts**. This new regulatory and financial space exemplifies the contradictory yet simultaneous movement towards global integration and local fragmentation, in which the local helps shape the global.

Global forces are also active in the very different case of rejuvenation of an 'old industrial space' described in **Florida**'s case study of the US Mid-West, where one key renewal process has been the attraction of external manufacturing investment and foreign-owned branch plants. Equally important here, however, has been the adoption of radically new forms of production organization by local, especially larger, manufacturing firms, often involving cooperative relationships with their smaller local suppliers.

CONCLUSION

Continual change and evolution is one of the primary features of both a capitalist economy and of organizations. Change is driven by technological innovation and by the introduction of new forms of management and work. All of these are driven by the forces of competition. One of the most recent alterations in capitalist economies has been the transformation of employment away from manufacturing to service activities. However, this does not imply that capitalism is shifting away from a manufacturing to a service driven system. What is occurring is a continual evolution of the division of labour so elegantly described by Adam Smith in 1776 (Smith, 1977). To Smith '[t]he greatest improvement in the productive powers of labour, and the greater part of the skill, dexterity, and judgement with which it is anywhere directed, or applied, seem to have been the effects of the division of labour' (Smith, 1977, p. 109). The development of capitalism has stimulated the creation of new support functions which feed into the manufacturing production process. Initially many of these support functions were undertaken within manufacturing companies, but there has been a gradual shift towards employing the services of independent service providers.

The increasing division of labour in capitalist economies, however, is complicated by a recent blurring of the boundaries between manufacturing and service activities. Many manufacturing companies are concerned that the added value associated with the purchase, customization, management and servicing of their products is being acquired by other companies. A good example of this process is the relationship between the computer company IBM and companies that specialize in the design, installation and servicing of computer installations. One such service company, Admiral plc, was established in 1979, and now employs over 2000 people and had a turnover of £90.8 million in 1996. The relationship between the customer and a manufacturing company is very different to that between a client and a service company. The difference is one of depth and frequency of interaction. The client/service relationship is more likely to provide further sales opportunities. The supply of an initial product to a customer thus provides opportunities for the sale of services. The development of a strong relationship between a client and a supplier also provides opportunities for the exchange of information that might stimulate new product innovation. Many manufacturing companies have failed to capture the service market for their products primarily because of differences between the way in which manufacturing and service companies operate. In contrast to manufacturing companies, service companies are engaged in relationship marketing. Service companies are actively involved with the management of the client–supplier relationship as well as the management of customer expectations of the supplied service.

The distinction between the manufactured and service component of a product is breaking down. The implications of this process are that the boundaries between service and manufacturing activities are disappearing. This will involve a reconceptualization of the ways in which social scientists identify boundaries between economic activities. In other words, the terms 'service' and 'manufacturing' may inhibit understanding of economic processes as it is becoming increasingly difficult to distinguish manufacturing activities from service functions.

Finally, these changes are also being worked out in an increasingly varied and complex 'regional mosaic' of production locations, shaped by global forces interacting with culturally, politically and historically-specific local economies to generate a diversity of regional economic trajectories. These involve both new and mature industrial spaces, specialized manufacturing and service regions, global cities and off shore locations, and advanced and developing countries. The contemporary geography of production is thus a dynamic and evolving palimpsest, which reflects ongoing changes in the division of labour, the organisation of production, and the balance between and blurring of manufacturing and service activities.

REFERENCES

Bryson J R (1996) Small business service firms and the 1990s recession in the United Kingdom: Implications for local economic development, *Local Economy Journal* **11**: 221–36.

Bryson J R (1997) Business service firms, service space and the management of change, *Entrepreneurship and Regional Development* **9**: 93–111.

Bryson J R and Daniels P W (eds) (1998) *Service Industries in the Global Economy*, Cheltenham: Edward Elgar.
Castells M (1996) *The Information Age: Economy, Society and Culture, Volume 1: The Rise of the Network Society*, Oxford: Blackwell.
Dei Ottati G (1998) The remarkable resilience of the industrial districts of Tuscany, in H J Braczyk, P Cooke and M Heidenreich (eds) *Regional Innovation Systems*, London: UCL Press.
Department of Trade and Industry (1998a) *Small and Medium Sized Enterprises (SME) Statistics for the United Kingdom, 1997*, Sheffield: SME Statistics Unit.
Department of Trade and Industry (1998b) *Business Start-Ups and Closures: VAT Registrations and De-registrations in 1997*, Sheffield: SME Statistics Unit.
Hart R (1987) *Working Time and Employment*, London: Allen and Unwin.
Keeble D, Bryson J R, and Wood P (1992) Entrepreneurship and flexibility in business services: The rise of small management consultancy and market research firms in the UK, in K Caley, F Chell, C Chittenden and C Mason (eds) *Small Enterprise Development: Policy and Practice in Action*, London: Paul Chapman, pp. 43–58.
Keeble D, Lawson C, Moore B and Wilkinson F (1999) Collective learning processes, networking and 'institutional thickness' in the Cambridge region, *Regional Studies*, **33**, 4: 319–32.
Lash S and Urry J (1994) *Economies of Signs and Space*, London: Sage.
Marshall A (1909) *Elements of Economics of Industry: Being the First Volume of Elements of Economics*, London: Macmillan.
Rowthorn R E and Wells J R (1987) *De-Industrialisation and Foreign Trade*, Cambridge: Cambridge University Press.
Scott A J (1986) High technology industry and territorial development: the rise of the Orange County complex, 1955–1984, *Urban Geography* **7**, 1: 3–43.
Scott A J (1988) Flexible production systems and regional development: The rise of new industrial spaces in North America and western Europe, *International Journal of Urban and Regional Research* **12**, 2: 171–86.
Smith A (1977) *The Wealth of Nations*, Harmondsworth: Penguin.
Storper M (1997) *The Regional World: Territorial Development in a Global Economy*, New York: The Guilford Press.
Walker R (1985) Is there a service economy? The changing capitalist division of labour, *Science and Society* **49**, 1: 42–83.
Yeung H W (1994) Critical reviews of geographical perspectives on business organizations and the organization of production: Towards a network approach, *Progress in Human Geography* **18**, 4: 460–90.

SELECTED FURTHER READING

Allen J, Massey D and Cochrane A (1998) *Rethinking the Region*, London: Routledge.
Hudson R (1997) Regional futures: Industrial restructuring, new high volume production concepts and spatial development strategies in the new Europe, *Regional Studies* **31**, 5: 467–78.
Malecki E J (1997) *Technology and Economic Development: The Dynamics of Local, Regional and National Competitiveness*, Harlow: Longman.
Marshall N and Wood P (1995) *Services & Space: Key Aspects of Urban and Regional Development*, Harlow: Longman.
Sassen S (1994) *Cities in a World Economy*, London: Pine Forge.
Scott A J (1988) *New Industrial Spaces*, London: Pion.
Storper M (1997) *The Regional World: Territorial Development in a Global Economy*, New York: The Guilford Press.

Section 3.2 Reworking the Division of Labour

Andrew Sayer and Richard Walker

'Information and Substance in Products . . . The Extended Division of Labor

from ***The New Social Economy: Reworking the Division of Labour*** **(1992)**

INFORMATION AND SUBSTANCE IN PRODUCTS

Confusion over the changing physical nature of goods has led some observers to see services where none exist. *The Economist*, in calling services 'everything you cannot drop on your foot', manifests the antiquated notion of goods derived from the mechanical age. This view fails to see that computer software, which consists of electronic signals on a tape or disk, can be every bit as much a material good as a chair. A customized program written for one customer would be a labor-service, but a packaged program such as Lotus 1-2-3 served up on the shelf at ComputerLand is unquestionably a good. Even a custom program written on my machine and transferred to yours by diskette is a good: it has a discrete, tangible, and fungible form (which the carrier disk helps to emphasize), unlike a true labor-service. Yet software is generally misclassified as a business service. The real distinction here is between things that are easily seen and grasped, and those that are not.

There is a coordinate but broader problem of seeing written, informational products as goods rather than services. A legal brief or an environmental impact statement is simply intellectual craft applied to paper, as a chair is woodcraft applied to lumber. As long as the brief remains in the lawyer's head it may be used as a labor-service, such as advice-giving, but once it is on paper and takes a material form, it is potentially useful to anyone. Hence many technical consultants produce goods rather than direct, personal advice to their (usually business) consumers.

Some service theorists think we have entered an age of information and communication, leaving the world of industrial goods behind (e.g. Hepworth, 1990). The information explosion in the contemporary economy is readily apparent, but information is not a free-floating ether; it must be pinned down. Information can be either part of industrial products or, as we shall see, part of their production, circulation, or consumption. We consider its role in outputs here.

All goods communicate symbolic information, if only in their appearance: chairs carry the utterly conventional meaning that they are things to be sat on – though in the case of a throne, a great deal more is implied. In other words, human beings speak through their objects and actions as well as through their throats. Only certain goods, however, have as their principal use-value an ability to store, transfer, and interpret information. Similarly, not every labor-service is informational: some have relatively mute purposes, such as trimming a hedge, while others are meant to impart wisdom, as in engineering consultancies. The electronic age has unquestionably increased the ability to package greater amounts of information in more sophisticated products, and industrial output has become more information-rich, but there is no convincing evidence for a large, distinct information-producing sector in the economy as a fourth sector beyond agriculture, manufacturing, and services.

A handful of sectors loosely designated as 'the media' sell goods and labor-services particularly laden with information, and are for this reason commonly classified as services. These include such activities as printing and publishing (of books and newspapers), film-making, and recording which were in the past considered branches of manufacturing. The shift in label can be justified, because the media remain industries producing chiefly goods that store, transmit, and impart information, whether in the form of music on records, stories on film, or news programs on tape. These forms of storage and transmission raise additional problems of circulation, to which we shall return.

A group of business labor-services is also information-heavy. They fall under three main headings: software, technical consultancies, and data banks. Software is a necessary adjunct to information processing by machine, including communications systems; technical consultancies sell specialized dollops of advice in areas such as product engineering, marketing and finance; data banks offer massive amounts of unprocessed information for use by those with particular expertise; armed with computers, usually in media companies, management, or financial investing.

Communication networks, such as telephone and telegraph systems, have almost a pure information-handling function, with no additional output, as a rule. They act as a special type of infrastructure, which is used collectively and often state-run, though private systems are becoming more common. For business consumers, communication has always been an integral part of production (and circulation). Within the fabric of production communication acts in a slightly different way: a division of labor requires that information pass between workers and workplaces to coordinate social labor processes. Intermediate goods cannot be silently passed along the production chain in hope that workers at the next step, on the shop floor or at the next factory, will understand what is to be done, so there is communication, in the form of speech, *kanban* boards, packing labels, etc. Similarly, distributors and final consumers need information about the goods they are handling or buying. Of course, communication networks are also utilized for purposes of final consumption, as in personal phone calls, but these are secondary to business uses.

In sum, while the informational aspects of goods is increasing, the use-value (information content) of a commodity should not obscure the materiality of physical goods, thereby leaping over the tangible labor involved in their production. We are still talking about production and industries. We do not wish to ignore the transition to a more electronic or informational age; we do want to emphasize that such a transition still lies within the bosom of industrial capitalism.

INTERMEDIATE OUTPUT: PRODUCER GOODS AND SERVICES

Not all products are meant for final consumption. The general division of labor in modern capitalism creates differentiation and specialization across immensely complex production systems. Economic growth rests squarely on the multiplication of the division of labor represented by a deepening web of intermediate inputs and outputs. Although it is possible to carve the economy into sectors with discrete commodity outputs (whether goods or labor-services), they must ultimately be joined together into systems involving intermediate as well as final outputs. Intermediate outputs serve as materials and means of production for other sectors, and goods and labor-services sold from one business to another can ultimately contribute to other goods or other services of any kind. A steel bar can ultimately appear in diverse forms: as part of a car, a tin can filled with processed food, or a bank building, and tracing intermediate commodity and labor flows is the staple of modern input–output analysis.

Traditionally, the most important intermediate outputs were capital goods (i.e. machinery), but producer services have been the most important intermediate outputs and the fastest growing segment of the service economy in recent years. The growth of producer services does not alter the fact that final output remains largely in goods, as with an engineering consultant who helps design a factory to produce diskdrives. The expansion of business services does indicate that the social division of labor in the production of all outputs is steadily enlarging.

All goods require labor-service inputs and, conversely, all labor-services are produced with the use of goods. Ultimately, all production requires both labor and goods (materials and machines): it is production of commodities by means of commodities. The regress is infinite, and as we shift from the realm of outputs to that of inputs into production, we arrive eventually at the labor process. It is possible with the right assumptions in the end to reduce all productive inputs to quantities of labor (Walker, 1988). Individual jobs almost never have a discrete product, so all work becomes service work.

THE PERSISTENCE OF INDUSTRIAL PRODUCTION

The dominance of goods production in the modern industrial economy is simply not in question. There has been no large or sweeping shift toward a different sort of output, direct labor-services; instead, many classic industries, such as electric power, construction, food processing, and transportation have been swept into the capacious bin of the service sector, falsely enlarging its size in national accounts.

At the same time, the social division of labor has been changing. First, new sectors have arisen whose output is atypical of earlier industries: electronic devices and plastics embody new technological principles and are materially different in ways that can be profoundly dislocating. With the shift toward electronic miniaturization and information storage, especially, the tangible material substratum diminishes radically while the meaningful content soars. Second, the division of labor has not only widened, but deepened as the layers of intermediate inputs have multiplied. Mostly this involves unfinished goods or components, but there has been a notable growth in business services, which are both intangible and closely related to more sophisticated production technologies, systems of management, and circulation functions. These are considered in depth below, where we shall see that distinguishing between goods and labor-services as the final outputs of production, which is central to many discussions of services, does not take one very far in analyzing the economy, its division of labor, and its developmental tendencies.

An important tendency in industrialization is that intermediate outputs are increasingly directed toward expanding areas of labor and away from traditional manufacturing. Compare the shift from agriculture to industry: the agrarian economy was once the principal source of industrial materials and the chief user of manufactured goods, but the manufacturing economy soon became its own largest market and source of inputs. Today, steel and concrete are going the way of wheat and corn, which – while still much in demand and produced in large volume occupy only a small fraction of the modern labor force. Agriculture is no longer the heartland of the economy, and neither are traditional heavy industries; electronics, software, and optics have moved into this position. The leading edge of product development is increasingly directed away from manufacturing altogether: the largest use of computers is not in running steel furnaces or metal-cutting machines but in accounting, design, and retailing.

Because the industrial base is continually replenished with new products, there is no discernible tendency for goods production to fade away. True, manufacturing has been shrinking slightly as a share of national output and employment in the advanced countries; even if we include business services, output and employment have not grown in recent years. But this trend is complicated by the overall slackening of economic growth in the last two decades, relative to the quarter century after the Second World War, and by the uneven performance of national economies: after all, the two top-performing industrial countries in the world today, Japan and West Germany, are also those with the largest manufacturing components (Cohen and Zysman, 1987). Nonetheless, overall, there has been a shift from manufacturing toward other parts of the economy, in particular toward circulation and social consumption activities.

THE EXTENDED DIVISION OF LABOR: THE TEMPORAL AND SPATIAL DIMENSIONS OF PRODUCTION

To deal with the modern division of labor, it is necessary to break with the fiction that everything happens in an instant and begin to grapple with *extended labor processes*, that is, work that takes place before and after products are actually formed by direct, hands-on labor, as well as work complementary to the immediate labor process. Extended labor, from R&D to janitorial work to automobile repair, is regularly mislabeled as services, especially where it is sold in a commodity form, when it is, in fact, nothing more than work that can be separated in time and space from the core of direct labor.

Pre-production labor usually includes research on materials, processing techniques, and product uses as well as development work on process engineering, prototype design, trial production runs, and product testing. This should not be confused with the early steps of a sequential process; for example, the initial clearing of a field for planting or the breeding of experimental livestock are pre-production labor but each year's plowing and calving are part of the ordinary cycle of agricultural labor. The dividing line between short developmental runs and regular production can be imprecise, and one should not assume that design ends at the moment production is regularized. Product design and process engineering were long the province of skilled workers or capitalists themselves, while research was an esoteric speciality of outside scholars. In the 20th century, research and development and engineering departments grew to be major divisions at the large corporations.

Auxiliary labor comprises those tasks which take place on a regular basis during actual production, but which back up or complement direct labor without ever coming in contact with materials or products. Auxiliary work has expanded in step with the growing technical division of labor within the workplace. Three general types may be discerned: tasks that feed into and out of the production line, such as the work of the stockboy or quality checker; routine repair and maintenance of machines and buildings; and gathering and exchanging information: instructions, feedback on performance, and interaction among workers. There is a strong spatial element here because workers have to communicate across factories, between departments, and even between far-flung workplaces.

Post-production labor is work applied to, or an adjunct to, a good after the immediate production process. Such labor begins within the factory with inspection, labeling, and packaging; continues with transport (delivery) of a good away from its point of origin; and includes the additional work required in wholesaling and retailing, such as minor assembly and clean-up of furniture or automobiles. Even after sale to the consumer there can be further work involved in delivery, installation, adjustment, and instruction for use – especially with equipment sold to businesses. The product is useless without such labor, so it cannot be considered only part of the sales effort.

Finally, repair and maintenance can continue for years after a product has been put into use. This work is a regular part of the litany of the service economy, but such labor merely restores goods to proper order, as with car repairs, road work, or house painting. In other words, the labor process has been extended because the original good has a long life and is cheaper to maintain than to throw away or remake. The need for maintenance and repair has grown along with the growing stock of fixed capital, infrastructure, and consumer durables, and therefore represents an enlargement of the world of industrial goods and their further penetration into personal life. No personal service is given except to soothe the customer's nerves.

Salim Lakha

'The New International Division of Labour and the Indian Computer Software Industry'

***Modern Asian Studies* (1994)**

The literature on the new international division of labour (NIDL) highlights the rapid growth of the electronics industry in East and Southeast Asia (Henderson, 1986; Rasiah, 1988; Salih *et al*, 1988). By contrast, the Indian electronics industry has received less attention (Joseph, 1989). Nevertheless, in computer software India is emerging as a competitive location for software development and exports.

Indian software producers are confident that in the 1990s they will command a status in software comparable to that of South Korea and Taiwan in hardware. Over the last few years many foreign corporations have contracted out their software development to Indian companies or alternatively set up export units in India.

In response to rising costs, combined with a shortage of software professionals in the advanced capitalist countries, the transnational corporations (TNCs) are increasingly resorting to offshore development of computer software, especially in Asia (Nalven and Tate, 1989). Amongst the Asian locations – Hong Kong, Singapore, South Korea, Taiwan, Thailand and Philippines – India is at the forefront because of its low production costs and large reservoir of competent scientific and technical personnel. Apart from low cost labour, some offshore locations offer additional benefits to the TNCs such as higher productivity, efficient management, and a higher rate of project completion.

This paper examines the growth of the computer software sector in India in the context of the discussions on the NIDL. It confirms some of the propositions of NIDL theorists by arguing that the strategy of expanding the Indian software industry through exports is promoting its integration into the global division of labour. However, the industry is not merely an export enclave of the TNCs but is articulated with the local economy. To that extent, the paper supports the critics of the NIDL who argue that the newly industrialized countries (NICs) cannot be regarded as simply extensions of the NIDL (Castells, 1989a). Importantly, the offshore development of software by the TNCs indicates a broadening of the NIDL which now extends beyond low value-added manufacturing.

THE NEW INTERNATIONAL DIVISION OF LABOUR

According to NIDL theorists, the contemporary world economy is shifting away from the traditional division of the globe consisting of a minority of industrialized nations and many primary producers, that is, the developing countries (Froebel *et al*, 1980). Instead the NIDL is characterized by a fragmentation of manufacturing processes which are globally dispersed. It involves the relocation of certain industrial activities from the advanced countries to the less advanced ones, and is motivated by the need for cheap labour, both unskilled and semi-skilled, which is employed to perform non-complex, routine operations in factories catering for the world market. The underlying logic of this process is explained by the 'valorization and accumulation of capital' which requires three conditions: the availability of a global reserve of industrial labour, the possibility for the fragmentation of production processes, and an efficient transport and communications network. Whilst NIDL theorists were mainly concerned with changes in manufacturing, comparable developments in computer software are currently underway.

Subsequent review of NIDL propositions has considerably refined the original claims and shed light on the complexities connected with the whole process of the globalization of manufacturing. In particular, there is increasing recognition of the broadening of export-oriented industrialization in Asia rather than its mere confinement to enclaves as claimed by NIDL theorists. Nevertheless, this manufacturing growth is based on an international division of labour where the labour intensive and technologically less sophisticated production processes are located in the NICs and other developing nations (Higgot *et al*, 1985). Whilst low labour costs are an important factor in the investment decisions of TNCs, they are not the sole attraction. The availability of an efficient infrastructure, relatively educated and compliant labour force, and the host nations' attitude towards foreign investment are significant considerations too, [as are] the wide-ranging tax and tariff concessions offered by Free Trade Zones in many Asian countries (Salih *et al*, 1988).

However, in contrast to the semiconductor industry of Southeast Asia where the labour force is mostly engaged in assembly type operations, the labour requirements of the software sector demand professional skills. This skills differential is a major factor influencing the international mobility of computer personnel. Computer professionals, whose skills are widely sought after in the advanced countries, enjoy substantial financial rewards. Therefore, in the software industry the NIDL is characterized by transnational upward wage mobility for at least some sections of the labour force.

Technological upgrading within the hierarchical international division of labour is a major concern of NIDL analyses. It is widely recognized that the NIDL is not static as NICs attempt to upgrade their technologies. Such mobility, however, is not automatic and neither is success guaranteed. Singapore's attempts from 1979 onwards to promote the 'Second Industrial Revolution', which aimed to upgrade skills and enhance existing technology, proved less successful than anticipated by the government. Despite increased foreign investment in computer hardware, the emphasis in electronics still remained on assembly work (Rodan, 1987). The pertinent issue is not whether there is any technological upgrading at all, but how far, and under what conditions, the technological gap between the advanced capitalist countries and the less advanced ones is bridged (Sivanandan, 1989).

Linked to the question of technology is a concern with the mobility of capital and the permanence of employment offered by the TNCs. It is feared that in response to automation in the advanced countries, the production processes located in NICs and elsewhere could become redundant. [T]he likelihood of a massive relocation of facilities is considered unlikely, partly owing to the high costs associated with relocating the established patterns of sourcing, production and marketing (Ernst, 1985). Instead, in response to rising wage levels, there has emerged a sub-regional division of labour. Low wage countries in Southeast Asia have become locations for 'assembly plants for large-batch standardized low grade outputs' and relatively high wage centres like Hong Kong and Singapore are increasingly undertaking critical testing tasks for which they have sufficient numbers of well qualified scientific and technical personnel (Henderson, 1986).

Also, the ability of Asian countries to improve their infrastructure and education facilities in the face of automation, makes it possible to accommodate some of the technological changes at existing sites without the need to relocate. Thus within the NIDL, the NICs, and others less developed, possess some flexibility which allows them to modify the consequences of technological changes emanating from the advanced industrial centres.

Despite some pessimistic prognosis the NICs have extended their technological base to a stage where their current capabilities provide substantial opportunities for further technical gains. Countries such as South Korea and Brazil have made determined attempts to move beyond clones in computer production and develop indigenous capability for the manufacture of sophisticated computer systems (Evans and Tigre, 1989). Similarly, India is extending its technological capabilities in electronics by acquiring technical know-how through foreign collaboration; it is also expanding its computer manufacturing industry through encouraging exports.

The active involvement of the government of India through the establishment of public sector computer companies, controls over TNCs domestic market share, and the imposition of foreign exchange regulations is reminiscent of the early actions of MITI in Japan. The Indian government's involvement goes well beyond tariff controls. It has engaged in charting the economic direction as well as participating in the production of consumer and essential non-consumer goods.

The extent to which countries progress within the NIDL will depend on a variety of factors besides the dominant interests of TNCs. Whilst the NIDL is underpinned by the global strategies of TNCs, development policies pursued by individual states are important, too. The role of the state is not confined to the provision of infrastructure facilities but extends to other spheres such as the mediation of different class interests, including those of foreign capital. In the case of India where the state has traditionally assumed an interventionist role, it is in a relatively strong position to direct the development process.

In addition to an interventionist state, Castells (1989a) attributes the competitiveness of NICs to the external orientation of their economies, an educated work force, and a progressive push for technological enhancement through government policies and technological transfer from the TNCs. Variations in the industrial structures of individual countries can also have a substantial impact on the pattern of technological development. The industrial structure of the NICs in Asia exhibits varying connections between the state, the TNCs and local capital, both small and large.

The structure of the computer software industry in India is distinctive because of the significant involvement of local capital and the existence of a large pool of human resources. Though the presence of foreign companies in software is rapidly growing, the state retains an important role in shaping the development of the software sector.

ROLE OF THE STATE

The government has actively intervened in the development of the computer software industry through policy initiatives and the provision of the necessary infrastructure facilities. [Its] commitment to software development was stated in its policy document of November 1986, which enunciated the main objectives, including the promotion of software exports to obtain a sizeable proportion of the global market in software, and an integrated development of software for national and export markets (GOI, 1986). [This document] provided guidelines on imports of computer hardware and software [and] foreign collaboration and foreign investments. [In line with] the economic liberalization in India since 1984–85 (Girdner, 1987; Kohli, 1989) for 100% export-oriented projects, 100% foreign equity was permitted.

The government has [also] established export-oriented Software Technology Parks in the cities of Bangalore, Bhubaneswar and Pune.

Bangalore is a major electronics centre in the state of Karnataka, whereas Pune is situated near Bombay, a leading location for banks and electronics. The software park in Bangalore is aimed at attracting small to medium sized businesses through the provision of various services which include centralized airconditioning and power, international telecommunications links, financial and marketing support, leasing arrangements for personal computers and minicomputers, and fast approvals from both the state and central governments. A similar park in Pune provides direct satellite connection with Boston from a base station owned and run by the government's *Videsh Sanchar Nigam* which controls India's overseas communications.

A software technology park can also be established by private investors who have obtained government approval; the US corporation, Texas Instruments, has operated (since 1986) a technology park in Bangalore with its own satellite link to the corporation's main facility in Bedford, UK. These developments in communications are integral to the incorporation of the Indian software industry into the NIDL, and complement other infrastructural facilities such as the Export Processing Zones which offer a wide range of benefits to software exporters.

Other major initiatives [are] aiding computerization across the nation (Raj, 1990). Through Nicnet, over 400 district capitals in India are now integrated into a government computer network linked by satellite. By 1994 all the 5,500 blocks (responsible for community development) will be connected. Another major network, Indonet, includes 70 corporations and links nine cities through telephone lines. Indonet's 'international gateway' links users to many overseas networks.

[These] government initiatives have created an environment and an infrastructure for software development that has attracted both local and foreign investors. The software policy, predicated on the principle of 'Flood-in Flood-out', was intended to allow into India a range of US developed software that could be utilized by Indian programmers to improve their own software which could then be exported to the US and Western Europe (Mukhi and Chellam, 1988). This strategy, whilst aiming to expand software development and exports, is also reliant on the TNCs for the inflow of technology, and especially access to overseas markets.

GROWTH OF COMPUTER SOFTWARE

[In] the growth of computer software in India the state and local capital have also played an important role. The software sector in India [began] during the early 1960s with the introduction of commercial computers by big companies able to afford the high costs of operating such technology. [This created] a pool of skilled programmers. However, the pressure to earn foreign exchange through exports [led to a] government-sponsored 'software export scheme' in June 1976 whereby software producers willing to undertake exports would receive priority for import[ing] computers (Grieco, 1984). Software exports are essential to counteract the high costs of imports triggered by economic liberalization and the anticipated growth of the computer industry. By the end of the century, the computer industry is targeted to reach the production level of US$8,000 million. Since exports of computer hardware products are unlikely to compensate for the substantial inflow of imports, foreign exchange earnings [will depend] upon software exports.

The software industry also [benefited] in 1978 when IBM withdrew from India in response to pressure from the Indian government to indigenize part of its equity holding. Since IBM systems were widely used in the country [this] forced the local industry to fill the void left by IBM. The public sector company, Computer Maintenance Corporation (CMC), [which] took over the task of servicing IBM equipment and software, subsequently has emerged as one of the most successful Indian computer firms in software development (Harding, 1989). Further, many of the Indian professionals previously employed and trained by IBM set up small software businesses.

Besides the foreign exchange imperative and the conflict with foreign capital, namely IBM, the development of the software industry has also been aided by the government's belief that India has a comparative advantage in software because of its large pool of scientific personnel that is proficient in English, relatively low-cost and capable of high productivity. Software is labour intensive and therefore compatible with the country's human resource endowment. Global trends in electronics have, moreover, favoured the promotion of software as it is expanding more rapidly worldwide than the hardware sector.

In recent years, the growth of software in India has been considerably aided by the expansion of the electronics industry which received priority as a result of the government's attempts at economic and technological modernization. Within electronics, the computer industry has also grown rapidly at an annual compounded growth rate of 60% from 1984 to 1987. The installed computer base expand[ed] from 3,500 systems in 1983 to 26,560 in 1987. Similarly, software exports expanded from only [US$22million in 1984] to US$117.5 million [in 1990–91].

Computer software is one of the single biggest export items and in 1989 its export value in the electronics sector was only exceeded by computers and electronic components. [Exports] are based upon capability across a wide range of activities including computer aided design, computer aided manufacturing, and expert systems, all of which demand a high skills input since they represent the more specialized and sophisticated branches of software.

HUMAN RESOURCES AND THE INTERNATIONAL DIVISION OF LABOUR

India possesses a vast source of scientific personnel which can be employed at a relatively low rate of remuneration. Estimates reveal that India has technical personnel numbering around 2.5 million and 225 million high school graduates, [while] the country's work force of engineers and skilled technicians ranks third highest in the world; in 1989, computer personnel numbered 80,000 professionals [with] 40,000 software professionals (Sridharan, 1989). To augment the pool of computer personnel, the government is encouraging the spread of computer education. Almost 350 institutions offered computer courses at degree and diploma levels in 1988–89, whereas in 1983 there were only 30 such institutions. Despite these efforts, personnel requirements are exceeding the current supply.

Other limitations have surfaced too. The level of computer education is criticized for failing to meet international standards [while] the best graduates migrate overseas (Mukhi and Chellam, 1988). This outflow of professionals is also facilitated by their proficiency in the English language which in India is still an important medium of communication in business, education and the government.

Nevertheless, the Indian computer professionals' command of English is a major asset for the software industry since it is the main language spoken in parts of the world that collectively provide the largest market for computer software. Combined with this advantage, the Indian software developers are conversant with all the major computer systems. Further, through overseas exposure some professionals are familiar with foreign scientific developments, and especially British and US business practices.

The rapid growth of the Indian software sector, however, is not based solely upon the availability of scientific-entrepreneurial expertise since low labour costs are a major consideration too (Nash, 1989). Computer professionals in India earn only about one-sixth to one-eighth of an average salary of a British or US professional (Mukhi and Chellam, 1988). [This] assume[s] considerable importance since producing software is a labour intensive task involving labour input of between 70 to 80% of the cost. Prevailing cost ratios, therefore, favour Indian software producers.

International cost differences explain, to some extent, why some advanced countries are shifting their software production offshore to Asian locations, leading to a worldwide division of labour in the software industry. The lower costs offshore, coupled with a shortage of skilled labour in the UK, has led certain British businesses to locate the most labour intensive stages of their software development in India. [Indian] companies like TCS have provided tailored financial software to their British clients with savings of between 40–50%. Since coding, the actual writing of the program, is the most labour intensive activity, it is cost efficient to locate it in India where labour costs are lower. Many foreign firms including American Express, Swissair, Yamaha, National Westminster Bank (UK) and others are now relying on Indian software companies for coding (Nalven and Tate, 1989).

Foreign corporations also derive other benefits in India. The standard of work is comparable to that offered by some companies in Europe and the USA, and organizations like TCS obtain productivity levels as high as 150% compared to their competitors in UK (Nash, 1989). Importantly, the advantage of cost savings is reinforced by prompt delivery, quality and the prevailing work practices. Those familiar with the work environment in India claim that Indian programmers are subjected to longer work hours; commit themselves more readily to urgent job schedules, and have fewer leisure opportunities. An average working week of 60 hours, including overtime, is regarded by some as common practice.

Whilst currently India enjoys several advantages in software production, certain limitations could over time counteract the benefits derived by the industry. [These include a low] level of research and development in Indian companies compared to US corporations which allocate up to 15% of their sales revenue on R and D activity. By contrast, Indian firms spend only between one to two per cent. [In addition,] quality computer personnel [are] being lost to its technologically superior competitors who have better research facilities, working conditions, and offer higher remuneration. In an industry where human skills are integral to market strength, the drain of human resources presents a serious loss, especially when the corresponding gains accrue to competitors dominating the markets that the Indian companies seek to penetrate further.

MARKETS AND FOREIGN COLLABORATION

The software market in India reveals a high degree of concentration with only a few companies exercising dominant control both in the domestic and export spheres. The overseas market [exhibits] a heavy reliance on the US. India's market orientation highlights some of the structural weaknesses of the software industry which is currently confined to a subordinate status within the NIDL.

In the international market for software products, India is still a very minor contender. The world software market has experienced massive growth reaching US$130 billion in 1991. Within this huge international market, India's share of software exports is just 0.1%.

Within the international software market, Indian companies are constrained by their lack of specialization in packaged software; half the world market in software is based on packaged software sales. This, however, accounts for only one per cent of the country's software exports. India's lack of concentration on packaged software is attributed to its unfamiliarity with the most recent market trends and the environment in US and Europe, differences in computer systems, [and] severe competition in the software industry with high costs in marketing and promotion. Experts suggest that 70% to 80% of the final price of a software package is accounted for by marketing costs. [These] impose severe constraints on particularly the smaller firms as they are unable to afford an extensive sales network. Whilst there are around 700 software companies operating in the country, most (550) have only a small annual turnover under Rs10 lakh. Only a few companies dominate the domestic and export markets with TCS and Tata Unisys Ltd (TUL) in the forefront; their share of India's total software export income in 1990–91 was 43%.

Owing to their small size, many software companies are unable to compete internationally because they cannot afford the cost of high-speed international communication necessary to transmit packages and communicate with overseas clients [or of] more advanced computer hardware and software. The lack of venture capital is yet another major impediment to the successful marketing of packaged software in overseas markets.

Other [obstacles to such marketing include] the limited utilization of computers in India [which] restricts the volume of software production, thus preventing the necessary price reductions. Another disincentive is the readily available supply of pirated software. It is estimated that about 90% of the software for many of the internationally popular packages used in India is pirated (Mukhi and Chellam, 1988). The high price of software in India is an important reason for the widespread piracy.

As a result of the weakness of the packaged software sector, Indian software exports are largely based on customized production, that is, software produced to cater for the requirements of particular clients. Between 50% to 80% of the exports are destined for the US, mainly in the form of customized software which has shown a strong growth in the US market. The leading status of the US market explains why many Indian companies have sought collaborations with major American companies. [Thus] in its initial phase, TCS collaborated with Burroughs Corporation, [though] it subsequently moved away, appointing its own marketing agents in the major US cities. This trend might reflect its stronger base as part of the powerful Tata Group, [and a] staff of 3,000 employees.

[Other] Indian software exporters are closely tied to the software requirements of their foreign partners or overseas based parent companies. For example, in 1988 amongst the top five software exporters in India, the two fully foreign owned companies, namely, Texas Instruments and Citicorp Overseas Software Limited (COSL), were both mainly engaged in developing software for their parent organizations. The other export leader, TUL, a joint venture between the Tata group and Unisys, produced most of its software for the US partner, Unisys. TUL's domestic sales were only five per cent of its entire worldwide software income in 1989. Under the joint venture, Unisys provided the specifications, undertook project management, and reaped the benefit of cheap professional labour in India. In 1988, the top five software companies contributed over 74% of the total exports, and the above-mentioned three exporters accounted for 37.5%.

In view of the further liberalization of the Indian economy, continuing interest by the TNCs in the Indian software industry, and growing ties with the US, foreign capital is likely to exercise an even greater influence in the future over India's software sector. Links between India and the US will serve to integrate further the Indian software industry into the NIDL.

CONCLUSION

Even though, currently, India is a very minor player in the international market for software, [its] emergence as a software exporter reveals a broadening of the NIDL which now extends beyond manufacturing. The growth of the software industry contributes to a higher level of skills formation since it requires the professional skills of computer programmers, software engineers and other scientific personnel. By contrast, many manufacturing processes relocated in the developing countries have largely sought semi-skilled labour.

Whilst India has actively pursued the promotion of software exports, the development of software for local application as in the case of the communication network, Nicnet, is [also] a significant achievement

These achievements underline the major role of the state in both the promotion of exports and the broadening application of software for local economic development. The state's involvement in infrastructure development, support for computer education, and the setting up of public sector computer enterprises are indicative of the priority accorded to the development of knowledge-intensive industries. It also suggests that the state is not entirely constrained by the NIDL, but is capable of extending the technological base beyond the limited requirements of the NIDL.

The development of the software industry in India is the result of an interplay between both local and international forces. Globally, software has emerged as a growth industry which, as a consequence of labour shortages and rising costs, has increasingly sought offshore locations to overcome the constraints on growth. Countries like India with their relatively low-cost professional labour and burgeoning computer industries have proved competitive locations for investments by TNCs engaged in software development. Locally, as stated above, the state in India has intervened to enhance the competitive capability of the country's software industry. Additionally, local capital has provided the bases for collaboration. This conjuncture of local and international forces confirms that the NIDL can be fully comprehended only by taking into account the articulation between the state, local capital and the TNCs.

Whilst India's considerable supply of low-cost scientific and technical personnel is perceived by many as a major advantage for the promotion of the software industry, it is not on its own sufficient to sustain further development. Over time other factors such as R and D, the level of computerization in the country, availability of capital, and an affordable as well as efficient communications infrastructure will prove crucial in upgrading India's status within the global software industry.

Peter Daniels

'Services in a Shrinking World'

Geography (1995)

The last quarter of the twentieth century has been remarkable for the inexorable expansion of service industries (Bryson and Daniels, 1998). Consequently, there are few aspects of activity in contemporary economies that are not dependent upon, or in some way influenced by, the involvement of service industries. For many of us services are something that we consume without realising that if they were not available our economic and social milieu would be rather different; although this is not to say that it would be any better! The administrative functions that provide the framework for the organisation and day-to-day functioning of the economy and of society, the infrastructure of roads and railways that enable the distribution of goods or the completion of the journey to work, the education, health or recreation facilities that sustain the quality of the labour force or the capital markets that provide the finance for enterprises of all sizes and complexity are all vital to the smooth operation of contemporary economies. The list is endless.

In OECD countries such as the USA and the UK the strong performance of the service sector is observed in terms of: (a) the higher average annual compound growth of employment in the sector since 1960 (Table 1); (b) the even faster expansion of producer or intermediate services such as investment banking, management consultancy or advertising, and (c) the positive contribution of services to the national trade balance (Figure 1). Since the mid-1970s imports of goods (merchandise) to the OECD countries have exceeded exports while the value of services exports has consistently exceeded imports since 1970. Although the size of the positive balance of trade in services has fluctuated, the overall trend throughout the period has been upwards.

This kind of evidence suggests that services in general, and especially those services incorporated in the production of goods and other services (the so-called producer services) are now a key component in the economic and spatial development of advanced, developing and post-socialist economies (Allen, 1988b). This paper focuses on the producer services which in the UK, for example, have experienced significant absolute employment growth during the 1980s (Figure 2). Producer services have become so intertwined with other aspects of the production process that they are not just *dependent or parasitic* activities in the way suggested by Marx or by Adam Smith. Producer services are *self-propelling*, they do not just respond to the requirements of manufacturing industry, they themselves develop new 'products' which are indispensable to the competitiveness or survival of other economic activities. They contribute to gross domestic product (GDP) through invisible earnings. They comprise a significant proportion of the value added

Table 1 Services as a percentage of total employment, selected countries, 1960–1987

Country		Type of service			
		Producer	Distributive	Personal	Social
France	1960	3.5	16.8	7.9	16.0
	1987	9.0	20.1	7.9	26.4
Germany	1960	3.4	17.5	7.4	10.3
	1987	7.7	18.1	8.1	21.6
Japan	1960	3.3	18.5	7.5	8.2
	1987	10.2	25.1	10.2	13.0
Netherlands	1960	4.2	20.4	8.5	14.7
	1987	10.8	21.3	6.5	28.4
Sweden	1960	3.5	19.4	8.4	16.3
	1987	7.2	19.2	5.9	35.1
UK	1960	4.4	20.6	8.0	15.8
	1987	10.4	21.3	10.1	25.3
USA	1960	6.4	22.2	11.3	21.2
	1987	13.6	21.5	12.5	26.0
Average	1960	4.1	19.3	8.4	14.6
	1987	9.8	20.9	8.7	25.1

Source: After Elfring (1988)

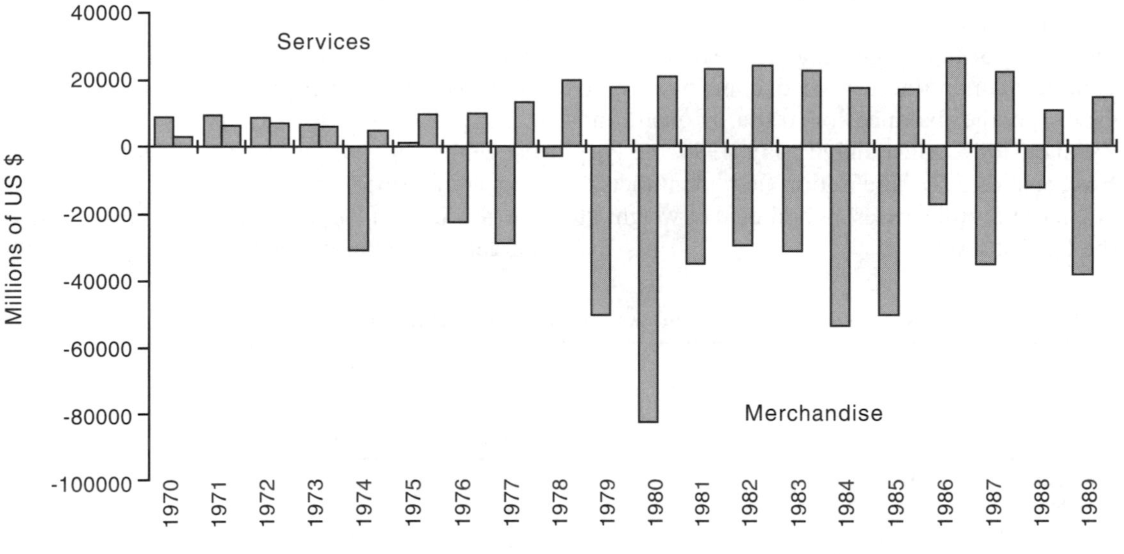

Figure 1 Balance of trade in merchandise and services, 23 OECD countries, 1970–89. *Source*: OECD (1992)

to a good or service and there is evidence indicating that, in the same way as manufacturing industry, they have become part of the export base of local economies. Thus, between 5 and 15 per cent of the turnover of a sample of more than 1,300 business service firms in the UK in 1991 was generated by clients in other EU countries and some 5 to 10 per cent from outside the EU (Table 2).

Despite the evidence above, there continues to be a good deal of antipathy among economic geographers both to the concept and the significance attached to producer services for industrial restructuring, regional develop-

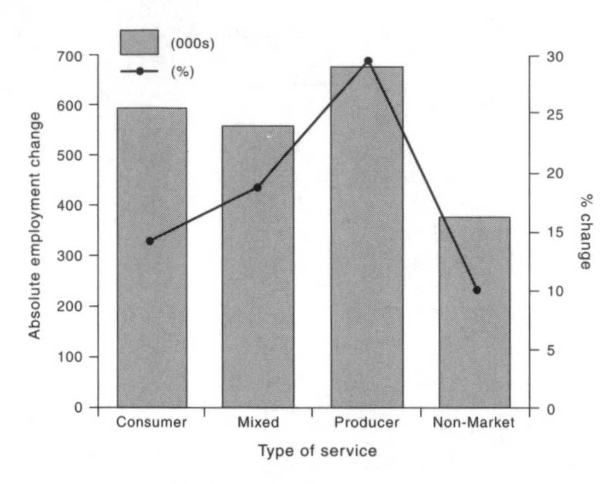

Figure 2 Growth of service employment, UK, 1981–89

ment and the evolution of metropolitan systems at national and international level. This uncertainty is not, of course, confined to geographers. Many economists would argue that it is inherently dangerous for national economic development policies to rely on the dynamics of the service sector to compensate for the diminishing volume of manufacturing-led contributions to GDP or the balance of trade (Cohen and Zysman, 1987; **Allen and du Gay**, 1994, see Part Five, Section 5.2). The notion that 'manufacturing matters' still carries a good deal of weight at the highest level.

SOME EXPLANATIONS FOR THE EMERGENCE OF PRODUCER SERVICES

Whatever the justification for the manufacturing fixation, we cannot ignore the part played by producer services in contemporary economic and social restructuring. It is important to recognise the role they play in affecting both *what* is being produced and *how* production occurs. Clearly the factors involved are likely to be wide ranging (given the diversity of activities within the producer services). We can usefully distinguish between *supply side and demand side* factors along with the mediating role performed by technology (information, telecommunications, transport, computing) and by the *business organisational environment*.

At the simplest level technology has enabled producer services to deliver (supply) their products more rapidly, reliably, over greater distances, at lower cost. It has also assisted demand by improving accessibility (which may be measured in a number of different ways) and by improving customer awareness of alternative sources of services. The *business and organisational environment* has been changing with respect to, for example: the size and distribution of organisations; almost frenetic merger and take-over activity during the 1980s

Table 2 Business service turnover by market and type of client, selected UK cities, 1991

City[a]	Turnover (%)					
	UK	EU	Non-EU	Households	Public sector	Private sector
Birmingham (10)	77	17	7	2	33	65
Bristol (14)	93	6	2	0	18	82
Cardiff (6)	87	10	3	1	7	92
Edinburgh (8)	97	2	1	7	15	78
Glasgow (19)	92	4	4	7	30	63
Leeds (12)	97	1	3	8	16	76
Liverpool (9)	88	5	7	5	20	76
London (excluding City) (365)	85	7	7	6	14	76
City (50)	79	8	13	5	11	84
Manchester (19)	88	3	9	1	8	85
Nottingham (9)	79	11	10	1	12	87
Reading (19)	77	11	13	0	34	66
UK Total (1303)	87	7	6	6	18	76

[a] Number of business service firms in brackets.

creating multifunctional, multi-product service firms; intensification of competition within increasingly extensive geographical limits (indeed global competition is commonplace for certain producer services such as securities dealing, trading in futures, engineering consultancy); an acute awareness of the need to increase productivity and to lower costs in order to remain competitive; and the ever-present pressure to innovate, to devise new products, or to 'fine tune' existing products.

In order to survive effectively under these circumstances, businesses require the appropriate *knowledge and information*. Producer services are primarily (not exclusively, if we include blue collar producer services) purveyors of this commodity or enable users to obtain access to it through, for example, providing the appropriate financial instruments for raising capital to purchase competitors or to obtain a controlling interest. The key point is that *knowledge is capital* and producer services are a key source of this knowledge.

Thus, on the *supply* side, competition between producer service firms has promoted diversification or specialisation (depending on the size of the firm and its business strategy). Firms have become market makers by extensively advertising the quality/diversity/complementarity of the services they offer or by devising new services which they must then persuade clients that they need in order to advance their objectives. Producer service firms must ensure that their delivery systems take maximum advantage of the opportunities offered by IT (using LANS, global communications facilities, state-of-the-art operating systems for desk-top microcomputers, desk-top publishing).

In relation to the *demand* side the increased specialisations of knowledge inputs (whether for manufacturing or service firms) often makes it necessary for firms to look outside the resources of their own organisation. The resulting demand for specialised inputs has also been stimulated by the internationalisation of production of many goods and services. Products are becoming more differentiated in an effort to appeal to more specific market niches; product lives are shortening and product lines broadening. Therefore greater emphasis is given to planning and product development, to flexibility in production, to advertising, to getting the product onto retailers' shelves or into office administrative practices.

GLOBAL INDICATORS OF SERVICES EXPANSION

Service industries have therefore infiltrated production and consumption at all geographical scales. The focus here is on their internationalisation. Travel and tourism and advertising services provide two examples. The increasingly global scope of their activities in part reflects the changing scope of consumer tastes which, as a result of media activities such as satellite television, are shaped by an awareness of international rather than national or local products or environments. Travel and tourism supported 183 million jobs (10.2 per cent of those in employment worldwide) in 1991 and generated more than one-tenth of the world's GDP (Table 3). They provided employment for one in five workers in Spain, one in six in Portugal, Belgium, Luxembourg and the Caribbean, and one in ten in the United States. In a recent report the Brussels based World Travel and Tourism Council therefore argues that travel and tourism are a powerful driving stimulus for economic development, especially for post-socialist countries such as those in Eastern Europe that are making the transition to a market economy.

Finding the right balance between the positive and the negative impacts of tourism is a delicate matter, especially for recently discovered tourist venues that are often more 'exotic' and fragile environments which are suddenly made accessible by relatively low cost air travel. Nevertheless, on the positive side, travel and tourism act as an economic catalyst by injecting hard currency directly into the economy; by reducing the time lag before new jobs and purchasing power filter into the rest of the economy; and by generating two and a half times more economic activity than their own direct output by

Table 3 Travel and tourism, major world regions, 1991

Region	Gross output		Employment	
	US$ (million)	Proportion of world total	Millions	Proportion of region total
Africa	41	1.4	4	11.6
North America	765	26.5	13	10.1
Latin America	68	2.4	13	10.1
Caribbean	25	0.9	1	16.0
Asia and Pacific	558	19.2	106	10.0
Europe	1,316	45.4	39	10.6
Total	2,773	100.0	176	

stimulating infrastructure development and other public works projects, which in turn boost other investment in building, transport, telecommunications and engineering. In addition, travel and tourism stimulate investment by small and medium-sized enterprises, thus generating the service and entrepreneurial drive necessary for a market economy. They also enhance protection of the environment, since sound environmental planning and the latest environmental technology are an integral part of the tourism development process.

Internationalisation, which may take the form of cross-border trade or organisational strategies directed at achieving direct representation in overseas markets, has played an important part in raising the global profile of services such as advertising, engineering or management consultancies, banking, insurance and computing. Many products and services can, in principle, be used to equal effect in different parts of the world but user preferences or culture, for example, will vary considerably. Thus, globalisation of markets for services such as advertising (Levitt, 1983) is a useful concept but difficult to operationalise because of the need to take account of the nuances of local (national) conditions, including the regulatory environment (Daniels, 1995). Advertising in newspapers and in the form of handbills and billboards was taking place in Britain as early as the seventeenth century (Fraser, 1981). Its subsequent transformation into a US$ multi-billion, multi-media activity advertising goods and services in markets extending from the local to the global scale has taken place in close concert with the marketing strategies of manufacturing and services firms, advances in the technology of communications and information management and related developments in the media used by advertisers. The worldwide growth of advertising expenditure has been substantial throughout the 1980s (Figure 3).

The market for advertising services can be divided into local (domestic) and international (products, services produced and marketed outside the country of origin). Changes in the organisational structure and location of the advertising industry since 1945 largely reflect the behaviour of its clients, the advertisers and the media industries, that have looked beyond national markets. In an effort to extend or to protect market share many users of advertising services have engaged in mergers and takeovers that have created large multinational enterprises (MNEs) spanning the globe. This has made it necessary for advertising agencies to follow their clients, led by the large US agencies who expanded into Europe, for example, both before and after the last War as US manufacturers such as Ford, General Motors, Kodak or IBM established plants there. The peak of this activity was reached during the 1960s – a total of 59 overseas offices of US advertising agencies in 1960 had increased to 246 by 1970 (Weinstein, 1977). The initial impetus was sustained by the prospect of more open and less demanding markets outside the US (Sinclair, 1987), tax concessions for firms engaging in overseas activities and the rapid diffusion of television.

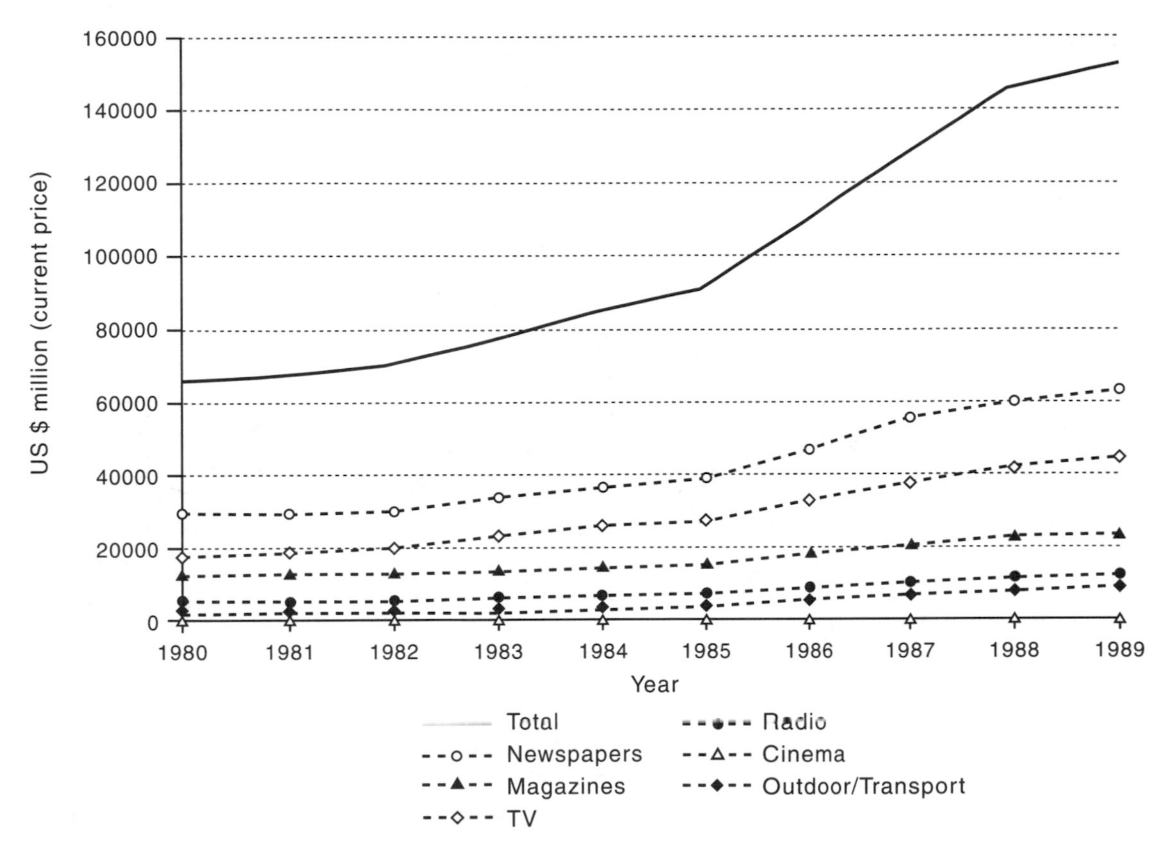

Figure 3 Total advertising expenditure, Europe, USA and Japan, 1980–89. *Source*: Data reproduced in Waterson (1992)

A dichotomised structure has therefore emerged with a small number of large advertising agencies and a large numbers of small firms. The top five firms in Europe, for example, attracted a disproportionate share of the total billings received by the top twenty in 1991 and 1992. In France the top five agencies accounted for 57 per cent of the total billings. Even in Spain, where the proportion was lowest, it amounted to well over one-third of all billings. If the top three agencies alone are considered they attracted one-third of total billings across Europe in 1992 with France (40 per cent) and Spain (24 per cent) again at opposite ends of the range.

The geographical distribution of the leading 100 agencies in 1982 and 1990 has also changed dramatically (Figure 4). In 1982 France, Great Britain, the Netherlands and Germany dominated the distribution, with 64 per cent of the total; by 1990 all except Great Britain, with 54 per cent, recorded a decreasing number of top 100 agencies. This is a powerful illustration of the impact of internationalisation during the 1980s on agency location decisions. As one apex of the so-called 'Golden Triangle' with New York and Tokyo, London was one focus of the intensive investment, transactional and employment growth that accompanied the internationalisation of advertising. Employment in the top 100 advertising agencies increased by a factor of 15 and turnover by a factor of 12. Whereas in 1982 Paris accounted for the majority (33 per cent) of the top 100 agency employees compared with less than 16 per cent in London, by 1992 it had just 27 per cent while London was the location for some 63 per cent of the total. The status of Paris as the leading location for advertising agencies in 1982

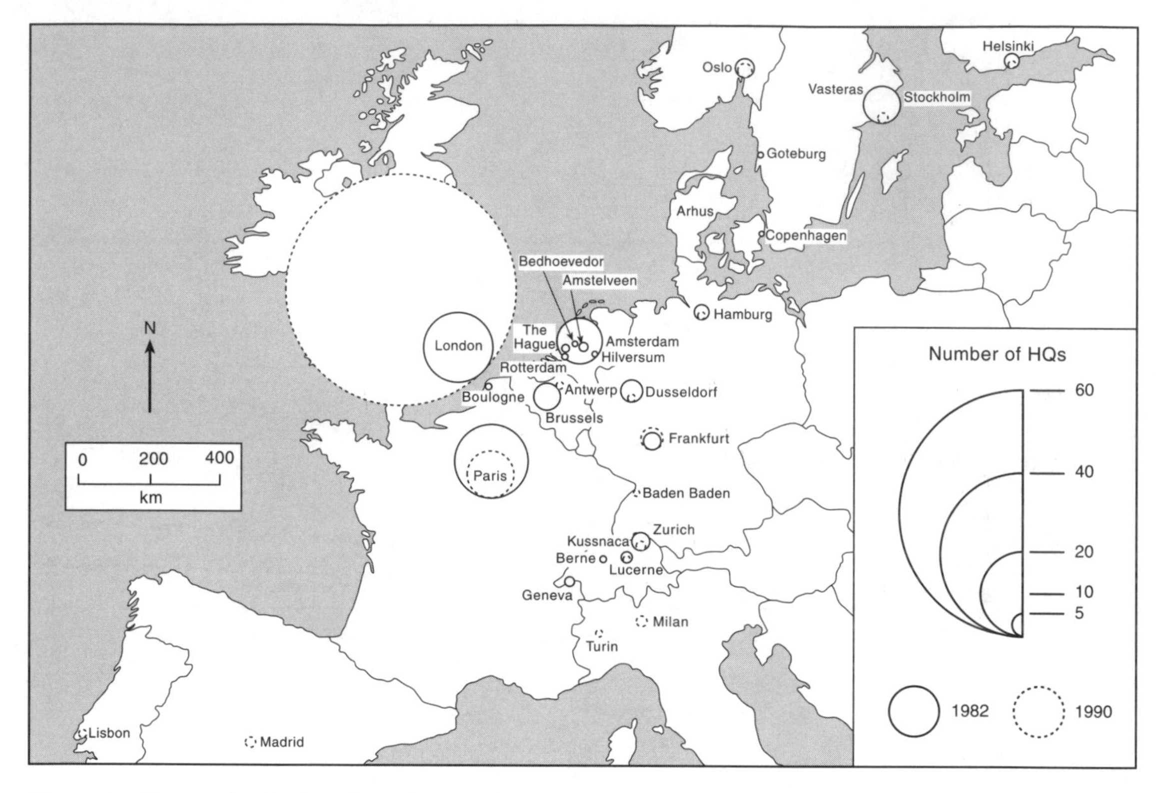

Figure 4 Changes in the location of HQs of the top 100 advertising agencies, Europe, 1982 and 1990

therefore declined sharply, along with Amsterdam, Dusseldorf and Stockholm. This geographical concentration of agencies has been accompanied, however, by some dispersal of top 100 agency locations. Spain, Italy and Portugal are included in the 1990 list for the first time with agencies in Madrid, Milan and Lisbon. Relative to London the scale of activity is however very low.

More recently, growth has focused less on the establishment of a presence in a wider range of locations and more on a restructuring of the industry into major transnational groups (networks) that are co-ordinated by holding companies. The networks mainly involve the leading firms (Figure 5). Many are relatively new with British and French firms especially prominent (Euromonitor, 1993). The networks comprise wholly owned agencies and a mix of joint ventures, minority shares, majority shares and associate status and most span all the EU and some, such as BBDO, also extend into Eastern Europe. The largest European network (EuroRSCG, France) employed 6,250 in 1992 and claimed billings in excess of US$5 billion. The smallest, Bozell, employed 650 and claimed billings of US$0.5 billion. Networks thrive on a mix of local and multinational accounts and are the most effective way to service clients who have globalised product development but still sell locally, and therefore need local advertising. Toyota, for example, manufactures centrally but sells locally. Such arrangements favour the formation of agency networks. Other advantages of a network include market coverage, risk diminution, promotion of agency profile and an increased volume of business. In order to spread the effects of fluctuations in demand, most networks would like to achieve a 50/50 split between local and multinational accounts, but for the present, the former remain in the majority (Nicholls, 1993).

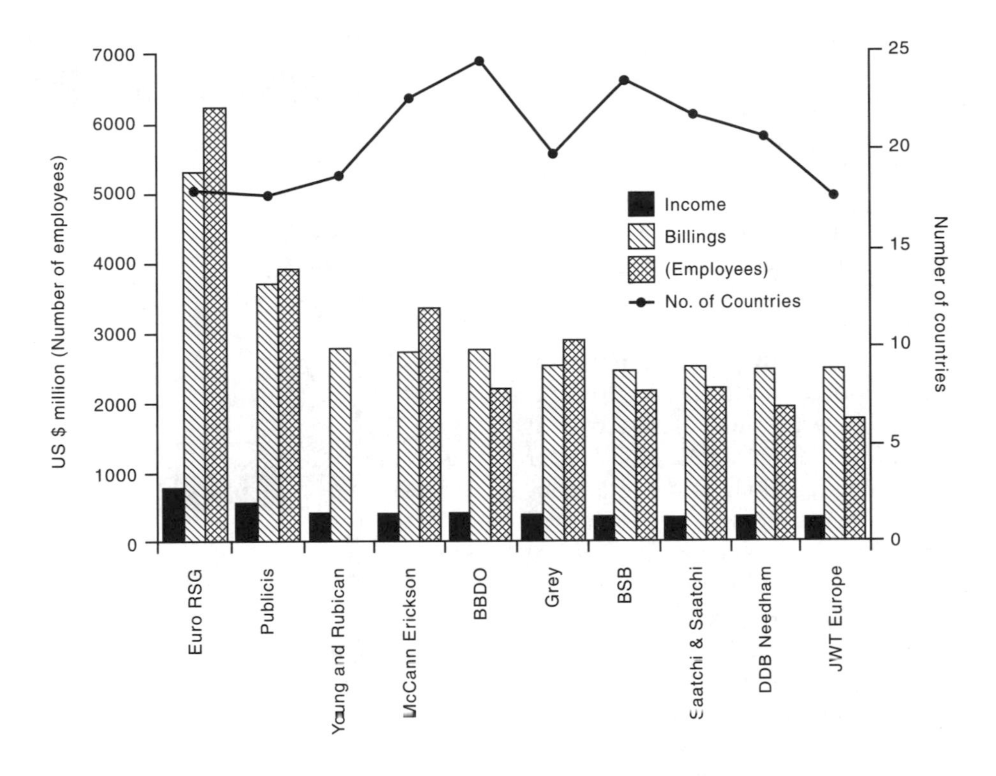

Figure 5 Top ten advertising agency networks, Europe, 1992. *Source*: Nicholls (1993)

IMPACT ON LOCATION OF PRODUCER SERVICES

[A]dvances in IT have been widely assumed to increase the footlooseness of producer services. In practice their effect has been to facilitate increased concentration of ownership and of location, although there are also some examples of the converse effect. In order to be able to compete in both national and global markets, knowledge and information-intensive services must deliver state-of-the-art advice to increasingly sophisticated knowledgeable and discriminating clients. They must deploy the best IT and the most creative human resources. These are the most readily accessible in what are often referred to as the transactional (Gottman, 1983), world (Friedmann, 1986) or global cities (Sassen, 1991). The association of telecommunications-dependent information binds together a global system of interdependent cities. The organisational development and locational strategies of producer-service multinationals are both a cause and effect of the interdependence which, together with the multiplier effects that are generated, underwrites the process of cumulative causation (Pred, 1977; Dunning and Norman, 1987). Thus, the growth and diversification of economic activities in metropolitan areas as a result of the international transactional activities of MNEs has a tendency to generate further innovation that promotes further growth involving the same select group of major cities. As a result, decision-making functions accumulate to a disproportionate degree in a relatively few locations along with their high level, specialist staff and officials whose presence makes these cities attractive for further rounds of locational decision making and investment (Budd and Whimster, 1992). With growth impulses more likely to be transmitted between cities at the top end of the global hierarchy, the prospects for downward filtering of economic and social restructuring to second or third tier cities are rather limited. The fact that producer services are becoming more active participants in this process is likely to exaggerate further the differences between the 'core' and the 'peripheral' cities.

While on the one hand there is clear evidence that IT and telecommunications have promoted centralisation of both the organisation and location of advanced producer services, we should not overlook examples of their distance-substitution effects. Agglomeration of services in large cities carries with it penalties in the form of high rents, high labour costs, difficulties of recruiting suitable labour, congestion and other symptoms of overcrowding such as crime and poor environmental quality. In recent years a number of service firms have sought to use telecommunications technology to escape the diseconomies of metropolitan location. The establishment of offshore or satellite offices (sometimes described as 'clerical telecommuting') has been made possible, especially for the conduct of more routine functions leaving high-level decision-making activities in front offices at the key locations. Thus, Ireland is increasingly seen as an electronic back office for firms headquartered in Europe's core cities or in the leading US cities (Figure 6). A leading American data processing firm employs more than 600 in the Shannon area who are on line each day to its mainframe computer in Colorado via a satellite link leased from AT&T. Some 20,000 items of electronic mail are sent daily and in a 20-minute bulk transmission at the end of the working day a quarter of a million items of information go through subscribers to 200 American magazines. More than 50 workers are employed by a New York life insurance company to process claims that are flown in overnight from the US and the decisions relayed back by satellite on the same working day (New York is five hours behind Shannon). American Airlines has a back office providing secretarial services in Barbados, while within the US, companies like American Express have moved major back offices from New York to small cities in the Mid-West.

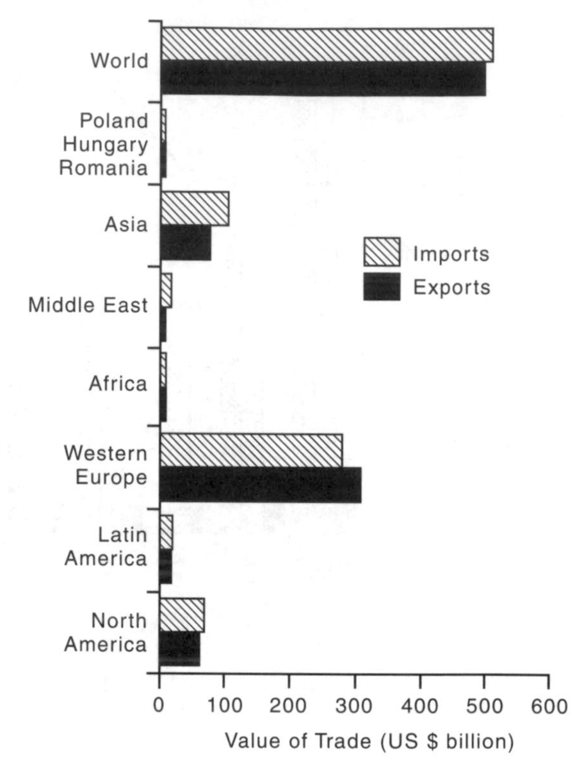

Figure 6 Commercial services trade, by world region, 1987

Telecommuting or teleworking is not a new idea and has received much attention from futurologists and telecommunications advocates for at least 15–20 years. There are some 1.2 million teleworkers (4 per cent of the working population) in Britain (see for example Huws, 1993). Claims assessors for Prudential Assurance work from home and the company is now one of the leading exponents of teleworking. The potential for employees to operate from home is considerable in a wide range of service jobs from professional white-collar work to routine data processing activities. But the process of changing attitudes to the workplace, conditioning individuals to work independently without supervision or even to go without the daily grind of commuting takes much longer than the available technology allows for. It is interesting that the concept of telecommuting and the isolated work and life style that it represents has recently been modified by the concept of the telehamlet or the televillage; places where people come together to live even though they can retreat to the privacy of their own workspace (home) when there is serious work to be done. A related idea is the telecottage, a room in a village hall or community centre, equipped with IT and telecommunications facilities. There are perhaps 15–20 examples in the UK at present (five opened in Wales during the summer and autumn of 1993 alone); often started as community projects financed to help people, especially women returning to work, to upgrade their skills.

CONCLUSION

It was suggested a few years ago that service industries were the Cinderella of economic and urban geography (Daniels, 1991). Perhaps things have moved on a little since then, but we should not underestimate the obstacles that still confront effective analyses of the pivotal role of services in advanced, less developed or former centrally planned economies. National and international statistical agencies continue to use classifications of economic activity that largely overlook the restructuring of production and the emergence of completely new occupations during the last 25 years. The IT revolution has played a large part in effecting these changes, as well as making it difficult to keep track of service transactions, and unless there is a complete overhaul of data-gathering practices, the mismatch between what is available and what is needed will widen further.

It is suggested that the rise of the service industries is both the cause and the effect of a shrinking world. Telecommunications services have greatly reduced the tyranny of distance for all kinds of service-related interactions. In real terms the costs of telecommunication have been falling, together with the costs of the hardware. In theory, this should mean more equitable access, but it is clear that the emerging geography of services is far from uniform. Some countries (Figure 6) and some cities attract a disproportionate share of the growth and associated benefits with all that this means for economic development, standards of living and quality of life. This is certainly true for the knowledge- and information-intensive services that now have a key role in shaping corporate and national competitiveness. It is less significant in the case of services such as tourism that enable places to capitalise upon physical rather than human resource endowments.

Susan Christopherson

'Flexibility in the US Service Economy and the Emerging Spatial Division of Labour'

***Transactions of the Institute of British Geographers* (1989)**

RATIONALIZATION AND SPECIALIZATION: COMPLEMENTARY PATTERNS IN SERVICES

A wide range of evidence demonstrates that manufacturing is being carried out in smaller enterprises producing a wider variety of outputs. US service sector firms, however, are characterized by quite different trends in production organization. One tendency is that of systemic rationalization and a drive to achieve scale economies. The other tendency is toward specialization of service products. Even more confusing, these two patterns may exist within the same firm.

In contrast with trends in manufacturing, employment in the fastest growing US service industries is concentrated in very large firms which have expanded by increasing the number of establishments they control. The general direction toward production in smaller units (that has been noted as occurring across industrialized countries) is more substantially attributable to the increasing number of small units in service firms than the vertical disintegration of manufacturing. If we look across the full range of service industries, they have more establishments at all firm employment size categories than goods producing industries.

The fastest growing industries in the US, business services, retail and health are clearly dominated by large firms. Over 50 per cent of retail employment in 1984 was in firms with over 100 employees. In consumer, health and business services 66 per cent of employment was in firms of this size and 46 per cent in firms with more than 500 employees.

This pattern of concentration in large multi-establishment firms is most evident in the retail sector where the number of establishments in firms with the highest volume of sales has significantly increased. Although retail is still dominated by small establishments, the proportion of establishments in the lowest total firm sales category declined from 92 per cent of total establishments in 1972 to 85 per cent in 1982. This reorganization in large, higher sales volume multi-establishment firms is even more apparent in employment figures. Between 1984 and 1986 employment in retail firms with over 500 employees increased 16 per cent compared to an increase of only 2 per cent in firms with less than 100 employees.

In the health care industry, there is only indirect evidence but the same apparent pattern. For example, 28 'for-profit' hospital chains managed, owned, or leased 809 hospitals in the US in 1983 and 958 hospitals in 1984. Their profits increased 30 per cent during that year. In both these industries, the trend is toward achieving scale economies by producing and distributing services through multiple market proximate units.

In addition to multiplying units through which to produce and distribute services, large US service firms are taking a series of steps to standardize and rationalize production. As in manufacturing, there has been significant capital investment (primarily in computer technology) aimed at reorganizing the work process (Stanbach, 1987). Secondly, the workforce has been redeployed between routine (low value-added) and non-routine (high value-added) work. For example, routine nursing care has been shifted out of hospitals to different types of routine nursing care facilities. As a consequence, the non-hospital health care sector grew 70 per cent in the 1970s. This new pattern of health care provision is exemplified by Beverly Enterprises, a nursing home chain which employs 116 000 people (more than Chrysler Corporation at 115 000) and deploys their labour in 1200 different locations.

The emerging spatial form of the service firm is quite different from that which we associate with manufacturing enterprises. First, these firms are characterized by the physical separation of labour among a set of worksites with similar functions (the retail outlet, the nursing home, the rental office). Second, even within the corporate headquarters, distinct worksites are emerging. In the health care industry, activities which once took place in the same building, such as out-patient care or laboratory testing, are being physically separated to allow separate billing. In addition, as financial management and marketing have become more important to the modern hospital, administrative staff has increased relative to health care workers. This work, too, is more frequently carried out under separate administration and frequently in a location apart from the hospital itself.

Although employment in consumer and distributive services is concentrated in large firms, there is a parallel trend toward small firm growth in producer services or what might be called services to services. Case studies of vertical disintegration in service industries suggest that much of this growth is an aspect of the service rationalization process. For example, large firms more frequently subcontract activities which they formerly carried out (i.e. food services; laboratory tests; building maintenance). Whereas, in the large-firm, small establishment model, there is little specialization across units, subcontractors provide specialized services. These building maintenance and catering firms, too, are currently growing in size as a consequence of mergers, occasionally financed by the firms to which they subcontract their services.

Although they originate in two different processes – the vertical disintegration of production and the proliferation of small establishments within large firms – the downsizing of the workplace and physical separation of production activities are among the most prominent aspects of contemporary service employment in the United States. Fifty-five per cent of all workers work in an establishment with 100 employees or less. What is important to recognize is that the large firm does not disappear in either of these processes. It merely assumes a different role, emphasizing finance and distribution, and distancing itself directly or indirectly from the actual production and provision of the service. The drive to achieve scale economies in service provision suggests that an expanded use of flexible labour in these industries is associated primarily with rationalization and cost reduction. It is only marginally a consequence of the need to produce multiple differentiated outputs. This supposition is further supported by an examination of the sectoral distribution and social characteristics of the flexible workforce in the United States and of some of the more significant developments in US patterns of work.

THE FLEXIBLE WORKFORCE IN THE UNITED STATES

From 1973 to 1979, 12.5 million jobs were added to the US economy. Another 14.5 million jobs were added in the 1980s. Nearly a quarter of these jobs were part-time and approximately 66 per cent were filled by women. One out of every six US jobs or about 19 million total is a part-time job. This yearly average figure understates the dimensions of the part-time work experience, however, for a much larger proportion of the workforce is employed part-time at some point during the year. In 1985, for example, the number of people who worked part-time for a portion of the year was double that of the annual average number of part-time workers. Eighty-nine per cent of the part-time workforce is employed in the service industries. Within service industries, the most important employer of part-time workers is the wholesale and retail trade where 30 per cent of the workforce is composed of part-time workers.

Part-time jobs are not only a sizable portion of total employment, they are growing faster than full-time jobs. Of the 10 million jobs created since 1980, one quarter have been part-time. Part-time work is divided among two groups of workers, those who work part-time by choice and those who work part-time because full-time work is not available to them. Thirteen and a half million workers work part-time voluntarily while a growing portion of the part-time workforce, 5.6 million, are involuntary part-time workers.

Since the 1950s, part-time work in the US shows two trends: a secular increase over time and a tendency to fluctuate more with respect to the business cycle. Part-time work has increased very slowly as a proportion of the total work force (from 16 per cent in the 1950s). The increased cyclical sensitivity of the part-time workforce is attributable to the larger portion of this workforce which is involuntary. In the mid-1970s the rate of growth in voluntary part-time employment began to slow and that of involuntary part-time employment to increase. Of the 2.9 per cent increase in part-time employment between 1969 and 1987, 2.4 per cent is attributable to growth in involuntary part-time work.

In the 1940s, men part-time workers outnumbered women because of the concentration of part-time work in primary sector industries. (Twenty-four per cent of the workforce in agriculture currently works part-time but agriculture is much less significant with respect to the sectoral distribution of employment.) The contemporary part-time workforce in the US is disproportionately composed of younger and older workers and of women in comparison with the workforce as a whole. Fifty-four per cent of part-time workers are wives or children in married couple families.

Although part-time jobs are the most numerous of flexible jobs other forms of flexible employment are expanding more rapidly, particularly temporary work. The use of temporary workers is especially prevalent in those situations, such as general clerical work and data entry, where the pattern of demand is not predictable and in work where generalized rather than firm-specific skills are required. Temporary work contracts can take a number of forms including:

1. a short term job;
2. a long term job with no employment security, lower pay or no benefits;
3. a structured internal temporary worker pool (most common in large public institutions, such as universities and hospitals).

In most cases, recruitment and hiring of temporary workers is carried out independently of the hiring of permanent personnel. Temporary workers differ from part-time workers in two important respects. Part-time workers more frequently work variable hours (rather than generally full-time as do temporary workers) increasingly geared to peak transaction periods. Secondly, in contrast with temporary workers, part-time workers are frequently required to have firm specific knowledge of rules and procedures while temporary workers have non-industry-specific skills.

In the US temporary 'industry' close to a million workers are employed as temporaries at any one given time but, more significantly, at least 3 million people work as temporaries at some time during any given year. Average annual employment in the industry increased from 340 000 in 1978 to 944 000 in 1987. The temporary supply industry is growing at 3 times the growth rate of service industries and 8 times the rate of all non-agricultural industries. Between 1988 and 1995, the temporary help industry is projected to grow 5 per cent annually in comparison with a 1.3 per cent growth rate for all industries.

Occupationally, 52 per cent of Temporary Help Service Firm workers are employed in technical, sales and administrative support occupations, most of these in clerical occupations. The concentration of clerical workers is 21 times their concentration in all industries. Sixty-six per cent of these workers work full-time. The second largest group is composed of operators, fabricators and labourers. These workers are more likely to be men, more likely to be Black and more likely to work part-time. In addition, non-office temporary help appears to be growing faster than the clerical component of the industry. Agencies specializing in non-office temporary help accounted for only a third of the total temporary help service employment in 1972 but for 45 per cent of the total by 1982.

The temporary industry employs only a small portion of temporary workers. The largest portion are 'direct hires', employed as on-call workers in large firms and more and more frequently, in the public sector. The US federal government is one of the largest employers of temporary workers and under revised regulations can hire 'temps' up to four years without providing benefits or job security. Approximately 300 000 workers in the executive branch, including the postal service, are currently employed as temporary workers. Among the private firms with their own 'in-house' temporary labour services are Standard Oil, Hunt-Wesson, Beatrice Foods, Hewlett Packard and Atlantic Richfield (ARCO).

The role of the temporary industry and the temporary employee has changed over time. As the temporary agency has become more established as a labour market institution, more firms are restructuring work to use a 'permanent' temporary labour force to do certain jobs. Rather than a part-time phenomenon, temporary workers are more frequently employed on long assignments, for weeks and even months.

Who makes up the temporary workforce? Temporary workers are similar to the part-time workforce – they tend to be young and female. The best available information on the characteristics of this workforce, from the May 1985 Current Population Survey, indicates that 64 per cent of temporary workers are women and that one of three of them is between 16 and 24 years of age. Blacks are also over-represented in the temporary workforce. They constitute 20 per cent of the temporary workforce in comparison with 10 per cent of the workforce across all industries.

Temporary and part-time workers are overwhelmingly employed in large service and retail firms as a buffer workforce and to reduce labour cost. Their use has little to do with flexibility as it is conceived in the manufacturing model of flexible specialization but instead reflects the ability of service firms to rationalize and standardize their operations and to deploy work hours as needed. The other major form of flexible work, independent subcontracting, plays a more ambiguous role. In some cases independent subcontractors reduce costs for the firms buying their services. In other cases they provide a specialized service input.

Approximately 7.5 per cent of the employed in the US are self-employed in unincorporated businesses and an additional 2.6 per cent operate incorporated businesses. Businesses conducted in addition to full-time work are operated by between 2–3 per cent of the workforce and over 4 per cent of self-employed business owners own more than one business. Changes in the role of self-employed workers in the American economy are closely related to the development of the service economy over the past 15 to 18 years.

Self-employment declined steadily between 1950 and 1970, led by losses in retail trade. Self-employment in agriculture has always been high but continues to decline (from 67 per cent of all full-time equivalent employment in 1950 to 50.1 per cent in 1986). Despite continuing declines in sectors such as agriculture where self-employment has been the predominant employment form, self-employment began to grow after 1970, particularly in the service sector.

Self-employed independent contractors are an important source of high-skilled professional workers for industries needing short term specialized services. Independent contractors are prevalent in electronics, chemicals, and business services and among a set of professional occupations, including graphic design, engineering, technical writing, systems analysis and programming. These occupations have some common characteristics that make them amenable to independent contracting. They are highly skilled but their skills are not industry specific. Self-employed workers frequently work on projects that are non-routine and carried out within a definite time frame. The expanding use of self-employed workers is presumably related to the increasing project orientation of business, for example the temporary employment of specialized teams to market a new product or to develop a specialized software application.

Self-employed independent contractors are also used to reduce both direct and indirect labour costs. This type of contracting, also subject to abuse of working conditions and 'off the books' payment, is typified by homework in electronics and apparel, but is also represented in services by clerical homeworkers. Clerical 'independent contractors' currently number between 5000 and 10 000 workers in the United States. There is reason to believe, however, that this form of work will expand. The companies which have homework programmes, including New York Telephone, American Express, Walgreens, Investors Diversified Services and Blue Cross–Blue Shield, are very large firms which hire large numbers of clerical workers. Of the approximately 250 companies with home-based work programmes, between 20 and 30 are known to be in the process of expanding their programmes. The programmes now in operation are essentially pilot programmes which will be evaluated and redesigned to facilitate homework productivity and supervision. The attractions of home-based work for the firm are substantial. They include increased productivity, elimination of non-wage benefit costs, reduction of turnover, and reduced costs related to off-hours computer utilization as well as decreased office space needs. Only about 10 per cent of the home-based subcontractors are full-time workers.

As with other emerging forms of employment, there are significant differences in the self-employed workforce by sex and race. For example, among the self-employed, more women than men are sole proprietors (70 per cent to 60 per cent). More men than women own businesses in addition to full-time employment (17 per cent to 15 per cent) and more women than men own casual businesses. Casual businesses are defined in terms of their total earnings ($1224 average in 1983) but 37 per cent of 'casual' business owners reported working full-time at them. As a consequence of this distribution, self-employed women had annualized earnings of $3767 in 1983 while men's annualized earnings were $13 520 (higher than the $12 079 earned by female paid employees). Median earnings for male self-employed incorporated workers (according to 1987 CPS data) were $35 114 and for female, $16 669. For unincorporated self-employed workers, the medians are $17 942 for men and $7930 for women. Despite these differentials, women are the fastest growing portion of the self-employed workforce. Between 1979 and 1983, the number of unincorporated self-employed women increased 5 times faster than men and more than 3 times faster than wage and salary women. More than 9 per cent of working women own a business in the US.

These flexible work patterns are only the most visible expression of a structural transformation of work and a redistribution of work-time among the US population. The extent of this transformation is demonstrated by broader evidence concerning the redistribution of work in the US. For example, if one looks at the relationship between the number of hours worked per capita and the number of hours worked per working age adult between 1950 and 1985, per capita work hours were 14 per cent higher in 1986 than in 1965 while hours worked per working age adult (between 16 and 65) declined 4 per cent during the same period.

Almost all this increase in work hours is attributable to increased work hours by women. Women across all age groups increased their hours of paid work by about 6 hours a week between 1975 and 1985. Overall, the evidence indicates that a higher gross national product per capita has been achieved by adding more people to the workforce.

Another aspect of the redistribution of work in the American workforce is evidence of a decreasing unemployment rate among the employed population but a continuing hard core population of 'discouraged workers', people who say they want work but are not seeking jobs or who work less than half the year and earn under $10 000. This group is estimated at between 10 million and 20 million people and excludes the majority of non-workers who remain out of the workforce for reasons of health, education, or retirement. Although the US unemployment rate has fallen from 9.5 per cent in 1983 to under 6 per cent in 1988, this statistic measures only the status of those who are actively in the workforce. Despite apparent labour shortages and a declining unemployment rate, the number of discouraged workers has remained sizable and stable since the recession of the early 1980s.

Within the employed workforce, patterns of work and the distribution of worktime are also changing. The 40 hour work week is becoming less common in the United States with considerable growth in both longer and shorter work weeks. For example, the number of women working more than 49 hours per week increased 50 per cent between May 1979 and May 1985.

The year to year variation in work hours is also increasing. The average year to year change in work hours during the 1970s was nearly 320 hours (the average for men workers). Variation in work hours ranged from 280 hours per year for white women workers to 350 hours for Black workers. Thus, fewer adult Americans hold stable full-time jobs and receive pay for 40 hour work weeks.

On virtually every measure, the US workforce stands out as being more 'flexible', that is more responsive to changes in supply and demand in the market, than its counterparts in other industrialized countries. Labour turnover, as measured in the percentage of job holders who hold jobs for two years or less, is higher in the US than in thirteen other industrialized countries with which it was compared in an OECD study (OECD, 1986, p. 51). Wage flexibility as measured in the dispersal of wages within sectors, is also greater in the United States than in industrialized countries generally (OECD, 1986). And, although the US has a lower percentage of workers in part-time jobs than the UK or Sweden, the share of total labour input (in terms of hours) by part-time workers is higher in the US. Part-time work is increasing faster in other industrialized countries but the US is unique with respect to the multiple forms and extent of flexibility that characterize its workforce.

SOME IMPLICATIONS FOR THE SPATIAL DIVISION OF LABOUR

A politically-informed analysis of the type I have sketched in this paper can potentially affect the way we approach international comparisons of emerging patterns of work. It could, for example, encourage a second look at what appear to be patterns of convergence, say, in the proportion of part-time workers or in service sector employment. Although there has been a general trend across industrialized countries for part-time work to increase substantially and for this form of work to employ predominantly married women, there continue to be significant differences among countries in the age and gender composition of the part-time workforce: in how part-time workers are used; in how they are compensated; and in how the State regulates part-time work (Standing, 1986). These differences are constituted through a political process, within the workforce and potential workforce, as well as between capital and labour.

If the allocation of work is politically determined, we cannot assume that services, including childcare, food preparation, and health care as well as business services, will be provided in the same way in all societies. In this regard, Urry makes the valuable point that empirical evidence does not support the notion that there is a natural trajectory toward ever greater levels of service employment in modern societies. The US, for example, has had a larger service sector throughout the twentieth century than many European countries, and in the second half of the century it has grown at the expense of the extractive sector rather than manufacturing (Urry, 1987). So, the allocation of many types of service work in an individual society is an open question and a political question (Warde, 1986). Despite a general trend toward private provision throughout industrialized societies, we can expect that the allocation of work among home, State and private enterprise will take different forms.

A second set of implications for the spatial division of labour is drawn from the narrower sphere of industrial politics. Clark (1986) has demonstrated how the limits of union influence in the United States were constructed as a consequence of the geographic concentration of 'lead' industries, such as automobiles and steel and the consequent local community orientation of industrial politics. Decentralized bargaining was eventually codified in US labour law. The labour institutions constructed within this context may be incapable of responding to the needs of a workforce in flexible service jobs. The new forms of work and the new workforce are foreign to the major US labour unions which continue to operate in the 'mass-production' mode, for example opposing part-time work rather than organizing part-time workers. The combination of a tradition of exclusionary practices with respect to women and minorities and a history of localized industrial politics makes it very difficult to regulate flexible work carried out in national firms composed of many small dispersed establishments. It is this 'legacy' which is, at least in part, responsible for the burgeoning of flexible forms of employment in the US.

The trends in US service industries also raise questions about the way geographers view processes which effect change in spatial patterns. The standardization and rationalization of services is eroding local differences in production and consumption. More people do much the same thing. Should this be of interest to geographers, who have traditionally been concerned with what makes places different from one another? The answer is yes, if only because underlying an interest in spatial difference is a more basic concern for the alteration of spatial patterns in and across places. In some service-dominant economies, the systemic rationalization of production and distribution in retail, health, insurance, and banking firms is altering local economies and employment possibilities to make them more similar from place to place. Thus, geographers need to examine not only those processes in production that create new forms of local difference but also processes the intent of which is to reduce local difference. Systemic rationalization intersects with both old and new sources of differentiation, including the place-based transformation of manufacturing production. It is in this emerging interaction that place is being redefined.

CONCLUSION

The introduction of two new elements could significantly alter the way we approach the emerging spatial division of labour. One is a concern for the role of labour in the transformation process and more particularly the legacy of mass-production trade unionism in shaping responses to the issues raised by new patterns of work, including 'flexible' work.

The second is attention to changing forms of production in sectors other than manufacturing. If one looks at the organization of production in services, in which the majority of workers in industrialized countries are, in fact, employed, a significantly different set of patterns emerge than are captured in the manufacturing-oriented flexibility model. To understand new patterns of labour use, we need to examine how the organization of production in services differs from the manufacturing model. And, since service workers have worked, for the most part, outside the traditional wage–work-time bargain, we need to assess what effect that has had on tendencies toward flexibility in these sectors. Only an enlarged vision of economic change, one that incorporates an understanding of production in services and of industrial politics, will allow us to adequately interpret the emerging spatial division of labour.

Section 3.3 Rethinking the Spatial Mosaic

Ash Amin and Nigel Thrift

'Neo-Marshallian Nodes in Global Networks'

***International Journal of Urban and Regional Research* (1992)**

This paper take[s] the emergence of new localized industrial complexes seriously, but set[s] them firmly within a context of expanding global corporate networks via consideration of the history of two industrial districts, Santa Croce in Tuscany [and] the City of London.

THE LOCALIZATION THESIS

The most powerful case for a major return to the regional economy comes from writers speculating on the rise of locally agglomerated production systems out of the crisis of mass production (Piore and Sabel, 1984; Sabel, 1989; Scott, 1988b; Storper, 1989; Hirst and Zeitlin, 1991). The concepts of 'flexible specialization' or 'flexible accumulation'[1] describe the transition to a new era of vertically disintegrated and locationally fixed production.

The key argument is that the irreversible growth in recent decades of consumer sovereignty, market volatility and shortened product life-cycles requires production to be organized on an extremely flexible basis, [including] the 'deverticalization' of the division of labour between independent but interlinked units; numerical and task flexibility among the workforce; [and] greater reliance on innovation.

Such a change is said to be particularly evident in industries which face pronounced volatility and product innovation in their niche markets. Examples include electronics [and] craft products.

This change, it is argued, implies a return to place – a dependence on locational proximity between different agents involved in any production filiere. Echoing Marshall['s] work on small-firm districts in Lancashire and Yorkshire during the nineteenth century, [agglomeration] advantages are said to include the build-up of a local pool of expertise and know-how and a culture of labour flexibility and cooperation resulting from dense social interaction and trust; lowered transport and transaction costs; and the growth of a local infrastructure of specialized services, distribution networks and supply structures.

[Protagonists argue] that the most dynamic and competitive examples of [recent] industrial restructuring have been 'Marshallian' in their spatial dynamics. Examples include high-tech areas such as Silicon Valley, [and] industrial districts in semi-rural (the Third Italy) and inner-city environments (e.g. motion pictures in Los Angeles), in which networks of specialist small firms produce craft or quality consumer goods.

The novel conceptual aspect of the thesis is the (re)discovery of the locational importance of inter-firm relationships, notably in relation to the exchange of information and goods between buyer and seller and its influence on linkage costs.

[However,] to anticipate a pervasive return to local production complexes in the postfordist economy is nonsensical. First, the conditions and areas cited by the localization thesis [are not] the only examples of success. Others include the reconsolidation of major metropolitan areas such as London and Paris as centres of finance, management, [and] business services, the resurgence of major provincial cities such as Birmingham [and] Turin, [and] growing wealth in certain rural areas characterized by in-migration by commuters looking for a pleasant lifestyle. These have little in common with the logic of flexible specialization.

Second, the very areas cited as examples of postfordist growth are evolv[ing], and perhaps fragment[ing] internally. Silicon Valley [is] now being drawn into a wider spatial division of labour [because] of intense inward investment by overseas multinational corporations, and the export of assembly and intermediate production functions respectively to areas of cheap labour and growing market demand. Some Italian industrial districts [are experiencing] increase[d] international linkages (Amin, 1989; Harrison, 1990).

Third, the ingredients for local success identified by the localization thesis are not readily transferable to other areas. The dismal failure of strategies to promote technopoles in European less favoured regions, [or] local networking between small and large firms within depressed industrial regions, bears witness to this difficulty.

A final problem is that there is no conclusive evidence of the demise of fordist principles of mass production and consumption. The idea of a clean break is too simple a caricature of historical change (Gertler, 1988; Sayer, 1989; Thrift, 1989c). Sensitivity to diversity is particularly essential when analysi[ng] the geography of production.

A REFORMULATION: GLOBAL NETWORKS

In this paper we retain the notion of 'localization', but consider local complexes as the outgrowths of a world economy which is still rapidly internationalizing and [characterized by] global corporate power. We [also] want to provide a reformulation of the significance of local networking [via] Marshall's work on 'industrial atmosphere'.

An important shift [since] the 1970s is a move from an international to a global economy. This global economy has four particularly important [characteristics]. First, industries increasingly function on an integrated world scale, through the medium of global corporate networks. Second, corporate power has continued to advance, so that the new global industries are increasingly oligopolistic, progressively cartelized. Third, today's global corporations have become more decentralized through increased 'hollowing out', new forms of subcontracting, new types of joint ventures, strategic alliances and other new 'networked' forms of corporate organization. However, power and control [remain] in the hands of global corporations, rather than [there being] a genuine spread of authority to smaller or local players in corporate networks (Amin and Dietrich, 1991).

Further, increasing corporate integration may well be accompanied by increasing geographical integration, as more and more places are drawn into, or excluded from, the web of global corporate networks. Sites within these networks might be more autonomous than [previously], [but] they are still locked into a global corporate web.

Fourth, there is a new, more volatile balance of power between nation-states and corporations. The result is the increasing prominence of cross-national issue coalitions.

The net result has been the growth of increasingly integrated global production filieres coordinated by large corporations. But, because these filieres are more decentralized and less hierarchically governed, there are considerable problems of integration and coordination.

Three of these stand out. The first is *representation*. Information has to be gathered and analysed about what is happening in these filieres. But [notwithstanding] advances in global communications and a massive increase in the quantity and quality of information, the problem of how that information is interpreted remains. There is a *growing interpretive task*.

The second problem is social *interaction*. Global production filieres are sociable structures. Social interaction is still needed to gather information and to tap into particular knowledge structures, to make agreements and coalitions, and continually to cement relations of *trust*, of implicit contract (Marceau, 1989). New forms of corporate interaction, like joint ventures and strategic alliances, have made social interaction more rather than less central to corporate life. The third problem is tracking *innovation*. [It] is how to [maintain] product and process innovation in a decentralized system and to successfully market products in the early customized stages when they can succeed or fail.

Thus the world economy may have become more decentralized, but centres are still needed. Centres, first, are needed to represent, that is to generate and disseminate discourses, collective beliefs, stories about what world production filieres are like. These discourses [influence] the direction in which industries and corporations can go – whether we are talking about new fashions in design or products or new management trends (like strategic alliances). Centres are also needed as points at which knowledge structures, many of which carry considerable social barriers to entry, can be tapped into.

Secondly, centres are needed to interact, that is to act as centres of sociability, so gathering information, establishing or maintaining coalitions and monitoring trust and implicit contracts. Thirdly, they are needed in order to develop, test and track innovations.

These have to be *geographical* centres, that is, place-bound communities in which the agglomeration and interaction between firms, institutions and social groups acts to generate and reinforce that 'industrial atmosphere' which nurtures the knowledge, communication and innovation structures required for retaining competitive advantage in a given global production filiere.

To be sure, this form of localization is quite different from the older habit of the vertically integrated, hierarchical firm of concentrating its strategic functions, its 'head', in headquarters in major cities. In contrast, neo-Marshallian nodes in global networks act, as it were, as a collective 'brain', as centres of excellence in a given industry, offering for collective consumption local contact networks, knowledge structures and a plethora of institutions underwriting individual entrepreneurship.

Such [localization] appears to be of particular relevance in industries characterized by knowledge-based competition, rapidly changing technological standards and volatile markets. Marshallian nodes offer an industrial atmosphere and infrastructure which firms, small and large, isolated or interconnected, can dip into as and when required.

TWO EXAMPLES

Two examples of centred places in a global system [are] Santa Croce, a vertically disintegrated small-firm industrial district of the classical Marshallian type [and] the City of London, which still has some Marshallian features but now relies on global networking to cement these features in place.

These districts may appear dissimilar at first sight. For example, the City has a much larger employment base than Santa Croce. But such dissimilarities are outweighed by notable similarities. First, they are in global industries with similar market conditions (product volatility, reduced product life-cycles, design intensity, flexibility of volume, etc.). Second, in both districts, despite intensifying external linkage formation, many needs can be met locally. Third, both districts rely on strong knowledge structures. Fourth, both districts have strong traditions of 'thick' social interaction and 'collective consciousness', the result of distinctive institutional mixes. Fifth, both districts are under threat. Santa Croce [faces] the threat of incorporation back into global networks through vertical reintegration. The City [faces] an American-forced change from implicit to explicit contracts [which] might increase transaction costs and reduce the need for specialized knowledge to the point where many crucial intermediaries are no longer necessary or viable.

Marshall in Tuscany: Leather Tanning in Santa Croce sull'Arno

Santa Croce is a small town in the lower Arno Valley, 40 kilometres east of Pisa, which specializes in the production of bovine leather for predominantly the 'fashion' end of the shoe and bag industries. The lower Arno Valley accounts for about 25% of the national employment in the leather and hide tanning industry.

In Santa Croce, an area no larger than 10 square kilometres, are clustered 300 artisan firms employing 4500 workers and 200 subcontractors employing 1700 workers. 15% of its sales revenue [comes] from exports, [but] the industry is still heavily dependent on the Italian market, particularly upon buyers in Tuscany.

Twenty years ago, Santa Croce was not a Marshallian industrial district. There were many fewer firms, production was more vertically integrated, the product was more standardized (albeit artisanal) and power [rested with] the older and larger tanneries. Today, Santa Croce is a highly successful 'flexibly specialized' small-firm industrial district, which derives its competitive strength from specializing in the seasonally based fashionwear niche of the industry. Typically, market conditions in this sector – e.g. product volatility, a very short product life-cycle, design intensity, flexibility of volume – demand an innovative excellence and organizational flexibility which Santa Croce has been able to develop.

The boom in demand for Italian leather fashionwear in the 1970s and 1980s [enabled] area-wide specialization and growth in the output of cured leather. Multiplication of independent small firms [and] task-specialist subcontractors was more a result of specific local peculiarities than of the new market conditions. Opposed to the highly polluting effects of the tanning process – [in] Santa Croce you can recognize the Marshallian 'industrial atmosphere' by its smell – the local Communist administration was unsympathetic to factory expansion applications. This, together with the strong tradition of small-scale entrepreneurship in rural Tuscany, led to a proliferation of independently owned firms. Further encouragements to fragmented entrepreneurship were the preference of local rural savings banks to spread their portfolio of loans widely but thinly as a risk-minimization strategy, and [government] fiscal incentives to firms with less than 15 employees.

This initial, and somewhat 'accidental', response to rapidly expanding demand was gradually turned into an organizational strength capable of responding with the minimum of effort and cost to new and rapidly changing market signals. The tanners – many of whom call themselves 'artists' – became more and more specialized. Firms were able to keep costs down via different mechanisms of cooperation [such as] the joint purchase of raw materials [or] pooling of resources to employ export consultants. The main device for cost flexibility, however, has been an elaborate system of putting-out between tanners and independent subcontractors (often ex-workers). The production cycle in leather tanning is composed of 15–20 phases, of which at least half are subcontracted to task-specialist firms. Constantly at work, and specializing in operations which are most easily mechanized, the subcontractors have been able to reduce drastically the cost of individual tasks while also providing the tanners with numerical flexibility. This division of labour among and between locally based tanners and subcontractors, combining simultaneously the advantages of complementarity between specialists and competition between the numerous firms operating in identical market niches, is perhaps the key factor of success.

But other factors have also played their part. One is area specialization. Santa Croce is a one-product town which offers the full range of agglomeration and external economies to leather tanning firms. The area contains the warehouses of major international leather traders as well as the offices of independent import agents, brokers and customs specialists. There are the depots of the major multinational chemical giants as well as locally owned companies selling paints, dyes, chemicals to the tanners. There are at least three savings banks which have consistently provided easy and informal access to finance. There are several manufacturers of plant and machinery, tailor-made for the leather tanning industry, and there is a ready supply base for second-hand equipment and maintenance services. There are numerous independent sales representatives, export agents and buyers of finished leather. The local Association of Leather Tanners, the mayor's office, the bigger local entrepreneurs and the Pisa offices of the Ministry of Industry and Trade also act as collective agents to further local interests at national and international trade fairs. There are several international haulage companies and shipping agents capable of rapidly transporting goods to any part of the world. There is, at the end of the value-added chain, a company which makes glue from the fat extracted from the hides and skins. Finally, there is a water purification depot collectively funded by the leather tanners, the effluence of which is sold to a company which converts the nontoxic solids into fertilizer. All in an area of 10 square kilometres!

The entire community in Santa Croce, in one way or another, is associated with leather tanning. Through spin-off, along the value-added chain, this guarantees the local supply of virtually all the ingredients necessary for entrepreneurial success in quality based and volatile markets. This 'valorization' of the milieu is a product of the progressive deepening of the social division of labour (vertical disintegration) at the local level. The area not only produces specialized skills and artisan capability, but also powerful external economies of agglomeration and a constant supply of industry-specific information, ideas, inputs, machinery and services – Marshall's 'industrial atmosphere'.

A third factor in safeguarding the success of the area is the institutionalization, at the local level, of individual sectional interests (e.g. the Association of Leather Tanners, the Association of Subcontractors, savings banks, trade union branches, etc.), as well as a sense of common purpose [reflecting] the intricate interdependences of a vertically disintegrated production system. Not only has this prevented the growth of rogue forms of individual profiteering which may destabilize the system of mutual interdependence, but it has also created a mechanism for collectivizing opportunities and costs as well as ensuring the rapid transmission of information and knowledge across the industrial district.

The 'collectivization of governance' has been of particular importance for the industrial district in recent years, as it has tried to cope with new pressures. By the mid-1980s, a honeymoon period of spectacular success was coming to an end [because] of growing competition in international markets from Southeast Asia, declin[ing] demand from the Italian footwear industry, big price increases [and] shortages of uncured skins and hides, and new costs [from] environmental controls on effluence discharge. These are problems which different interest groups have not been able to resolve individually. Resulting collective responses [include] joint funding by the tanners of an effluence treatment plant and multi-source funding (involving tanners, subcontractors, a local bank and the regional authorities) of an [advice] centre on market trends, management skills and information technology. How successful these efforts will be is not known. What matters is that Santa Croce continues to possess a local institutional capability to respond collectively and swiftly to new market pressures.

This said, evidence suggest[s] that, into the 1990s, the organization of industry in Santa Croce will be 'post-Marshallian', that is, less locally confined and less vertically disintegrated. Tanners [are increasingly] import[ing] semi-finished leather, to the detriment of locally based hide importers, subcontractors and chemical treatment firms. There is also a threat of 'forward' internationalization. The oldest and most powerful [companies] have begun to open distribution outlets overseas as well as tanneries, in countries either producing hides and skins or promising growth in the leather goods industries.

The risk, then, is that Santa Croce will come to perform only specific tasks in an internationally integrated value-added chain, [and hence] become less vertically disintegrated. Such a narrowing runs the risk of threatening the institutional synergy and richness of activity which hitherto has secured the area's success as an industrial district. With functional simplification and larger premises, the larger tanners [may also] seek to internalize individual production tasks more than before.

If the twin processes of internationalization and vertical integration at the local level become the dominant trend, Santa Croce will lose its current integrity as a self-contained 'regional' economy. But, it will remain a central node within the leather-tanning industry. Twenty years of Marshallian growth have made Santa Croce into a nerve centre of artisan ability, product and design innovation and commercial acumen within the international fashion-oriented leather goods filiere. This unrivalled expertise will guarantee its survival as a centre of design and commercial excellence.

Marshall in London: The City and Global Financial Services

[Until] well into the twentieth century, the City of London could have been characterized as a classical Marshallian industrial district. During this period it consisted of a network of small financial service firms, and a set of markets and market-clearing mechanisms, in close contact with and close proximity to one another, employing as many as 200,000 people. These firms and markets consistently minimized transaction costs through social and spatial propinquity, creating an industrial 'soup' rather than an industrial atmosphere, an upper-middle-class craft community of quite extraordinary contact intensity (Thrift and Leyshon, 1992).

This 'old City' was run on 'mesocorporatist' (Cawson, 1985; Moran, 1991) lines: that is, it was a largely self-regulating system of collective governance with the Bank of England acting to protect it from pluralist regulatory systems and politics. Second, the City was tightly socially integrated. All its key workers were drawn from highly specific social backgrounds, reinforced by various socializing processes (the firms themselves, often partnerships, the markets, the livery companies, the Masonic lodges, the clubs and so on). This generated a 'collective consciousness' and afforded quick assessments of character. Thus the old City was a trust-maximizing system and the knowledge structure was, in effect, the social structure.

Third, the City's markets were large and liquid, [such that] to participate in these markets it was often important to know the right people, the right gentlemanly discourses. And fourth, the City was tightly concentrated in space, [reflecting] the ability of institutions like the Bank of England and the various markets to demand propinquity, and the peculiar importance of face-to-face transactions, an importance which [was] both a cause and a result of the dense spatial arrangement (Pryke, 1991).

Thus, the old City of London was a protected, self-regulating, socially and culturally specific enclave. It was, of course, tied in, in the strongest possible way, to the world economy of the time, but here was a case of 'the local going global' (Thrift, 1990; Pryke, 1991).

Since the early 1960s there has been a sea change in the way that the City of London [operates], [leading] to the emergence of a 'new City'. At least five major changes have threatened the City's integrity as an industrial district. The first has been the emergence of larger and larger oligopolistic financial service firms, such as pension funds, insurance companies, securities houses or banks. Many of these firms are substantial multinational corporations. Second[ly], financial markets have become increasingly international in both space and time, often operating around the clock. They have also become, through a massive investment in telecommunications, increasingly electronic; more and more communication is taking place at a distance, with some markets becoming almost entirely decentralized. [And] many new fictitious capital markets are increasingly based on securitized products which demand new 'disintermediated' relationships between financial service firms.

The third change has been in the mode of regulation of the City. Neither 'deregulation' nor 'reregulation' adequately describes what has occurred (Moran, 1991), [namely] the progressive accretion of a more carefully codified, institutionalized and legalistic mesocorporatism, forced on the City by a state succumbing to American pressure for change and by the diffusion of an American ideology of correct and incorrect practice. Formal contracts have replaced implicit contracts.

Fourth[ly], the City's social structure has become more open in terms of class, gender and ethnicity and more keyed into the knowledge structure in financial services. The fifth change, facilitated by the investment in telecommunications technologies, [is that] innovation in products is now much more rapid than before, but they are also more likely to fail (de Cecco, 1987).

Given the scope of the changes, why [has] the City of London persisted at all? The answer to this question can be related to the imperatives of the new global financial services production filiere. In its old incarnation, the City was the result of the local going global. In its new incarnation, the City is a result of the global going local. In particular, the City survives for three related reasons. First, it is a centre of representation, of surveillance and scripting of the global financial services filiere. Thus, much of the *world's* financial press operates from or near to the City; so does much research analysis [and] information processing. The concentration of this activity in one place [enables] the City [to] watch and script the global financial services filiere industry by watching itself. Equally, the City represents an important part of the knowledge structure of world financial services. It is still convenient to have this massive body of knowledge, much of which is highly specialized, easily accessible in one place.

The second reason for the city's persistence as an industrial district [is] its role as a centre of interaction. The City is a social centre of the global corporate networks of the financial service industry, with a large throughflow of workers from other countries. [It] is still a vital meeting place in which important deals can be made, issues marketed, coalitions formed, syndicates established and trust/implicit contracts cemented.

The third reason for the City's persistence is as a proving ground for product innovation. [In] financial service[s] product innovation and marketing are very closely tied together. A product cannot take off unless it is aggressively marketed, usually to a quite specific set of people in large investment institutions and companies. The City allows new products to be evolved and tested quickly and efficiently, [while] the availability in one place of many investment institutions and companies means that products can be easily socialized and customized.

To conclude, the City is still an industrial district, but it is an industrial district which depends on fewer, larger firms, which still need a place from which and with which they can represent and analyse their world, a place where they can meet in order to add flesh and trust to electronically mediated personae, and a place which can be counted on to continue to invent new products and markets. Clearly, the City faces challenges from other financial centres, especially European financial centres such as Frankfurt and Paris (Leyshon and Thrift, 1992). In part, this will clearly continue to depend on its constitution as a local and global social space.

CONCLUSION

This paper [argues] that there does not appear to be any inexorable trend towards the localization of production. Vertical disintegration

[and] 'networking' may [be] encouraging the resurgence along Marshallian lines of some regions as self-contained units of economic development. [But] in this age of intensifying global hierarchies and global corporate networks, local economic prospects are becoming more dependent upon global corporate organizational forces.

In such a context, efforts to encourage Marshallian growth in other areas through the formation of highly localized production systems are likely to fail. Santa Croce and the City of London illustrate only too clearly that the conditions for such growth are difficult to capture through even the most innovative policy measures, *unless certain basic structures are already in place*. These include a critical mass of know-how, skills and finance in rapidly evolving growth markets, a sociocultural and institutional infrastructure capable of scripting and funding a common industrial agenda, and entrepreneurial traditions encouraging growth through vertical disintegration of the division of labour.

The majority of localities may [therefore] need to abandon the illusion of self-sustaining growth and accept the constraints laid down by the process of increasingly globally integrated industrial development and growth. This may [involve] pursuing those interregional and international linkages (trade, technology transfer, production) which will be of most benefit to the locality in question, [and] upgrading the position of the locality within international corporate hierarchies by improvements to [its] skill, research, supply and infrastructure base in order to attract 'better quality' branch investments.

NOTE

1. Flexible specialization describe[s] transformations in the production process stimulated by new technological, skill and market developments. In contrast, flexible accumulation, drawing upon the Regulation Approach [and] Harvey (1989), refers to a broader macroeconomic design for the twenty-first century, transcending the 'fordist' regime of accumulation [based on] mass consumption, mass production and Keynesian regulation of the economy.

John Allen

'Services and the UK Space Economy: Regionalization and Economic Dislocation'

***Transactions of the Institute of British Geographers* (1992)**

THE REGIONALIZATION OF SERVICES

In adopting the term *regionalization*, I wish to convey the idea that the uneven character of growth across a country or parts of a country can be seen as the product of the distinctive geographies that are laid down, *one over another*, by the changing spatial structures of production and circulation (Allen, 1988c; Massey, 1988). In this account of uneven growth, the intention is to break with growth models that work within the framework of the national economy, by drawing attention to the social relations of production and circulation that operate *across* national economies as well as those that shape development *between* them. A minimal aim will be to show how a highly 'open' economy such as the UK is characterized by regionalized modes of economic growth, of which the most significant in recent times is service sector growth across the London City region and beyond.

I wish to contrast those kinds of input/output relations with a wider range of social relationships. I hasten to add that this is not to deny the significance of functional relationships within a national economy; on the contrary, it is to recognize them as *one* kind of intra-national relation. In addition to these market relations within a country, I wish to draw attention to the production and circulation relations that structure an industry across national and international space. This involves a consideration of a variety of spatial structures. Within *production* it involves tracing the labour geographies, the spatial hierarchies of labour associated with particular industries across the regions, as well as the pattern of labour migration for different occupational classes. Within *circulation* it involves a grasp of the spatial structuring of markets for different industries, including that of international trade flows, as well as an understanding of the geography of supply relationships. It also involves charting the financial flows that cut across national boundaries in order to gauge the volume and nature of these transactions, and how and where this translates into economic growth.

In broad terms, it is the different *networks* of production and circulation relations that circumscribe firms, industries and their workforces which provide the key to their distinctive geographies and which, in turn, carry a range of economic consequences for the generation and transmission of growth across the regions. Attempting to trace this network of overlapping relations is obviously a complex affair, yet this framework of enquiry does suggest a way of thinking about service growth which does not habitually link it to manufacturing, nor routinely assume a series of sectoral linkages internal to the UK economy. Service industries can and do act as a catalyst of growth for other service industries, within and beyond the confines of the national economy. This is not to suggest, however, that services or certain services are now the 'motor' of the modern economy; that is, representing some kind of 'leading edge' development which will, over time, spread across the country. On the contrary, in the UK recent service growth has taken distinctive regionalized forms.

We shall attempt to illustrate this line of argument by focusing upon the network of social relations associated with three types of service industry: finance and commerce, high technology services, and tourism – each of which represented an important source of economic growth in the UK over the last decade, especially in the late 1980s. Although much of the empirical detail may be familiar, it is the framework of enquiry that I hope will be of interest.

If we start with the financial services industry, it has frequently been observed that the UK economy historically has displayed a functional divide between manufacturing and financial sectors, with the latter more orientated towards international rather than domestic markets. Yet the general nature of this observation has often obscured the discontinuous nature of this divide (Barrett Brown, 1988). Whilst it is reasonable to suggest that financial institutions within the UK, in particular banks, have maintained an arm's length relationship with manufacturing industry, the form of that relationship has altered over time. For much of the nineteenth century, long term lending and finance for capital investment was the norm, only switching to short term financing of industry with the amalgamation and centralization of bank capital at the turn of the century. This arrangement remained intact until the late 1970s, when the arrival of foreign banks in the City led the shift in financing from overdrafts to wholesale bank loans (Fine and Harris, 1985). Until recently, the international role of the UK's financial sector had been locked into the framework of the Sterling Area and Empire, but with the rise of the euromarkets this orientation was to change in a rather dramatic way. At the risk of constructing a further generalization, if we separate for the moment the domestic from the international components of the UK's financial sector (the former refers to the traditional overdraft markets, whereas the latter refers to the wholesale money markets and the corporate banking markets), it is possible to talk about the *supranational character* of the latter in the contemporary period (Harris, 1988). Based in the City of London and the south east of the UK the international sector of finance has been reshaped by a series of recent developments.

The first development is the internationalization of bank lending which, via the euromarkets, has grown spectacularly since the 1970s. Following Coakley and Harris (1983) London essentially acts as a clearing house for the massive flow of funds that move through London – coming from and going to other countries. Apart from UK multinationals, much of this market remains separate from the rest of UK industry and little of the money lent derives from 'local' funds (unlike Tokyo and New York). The importance of this activity can be gauged by the number of foreign banks attracted to the City, which virtually doubled in the late 1970s and early 1980s, and also by the amount of international lending advanced, which rose sixfold over the same period. Overseas banks in London dominate international lending, as they do the foreign exchange business. Indeed, developments in the eurocurrency markets have enabled London to remain the foremost foreign exchange market, enjoying a turnover in the late 1980s one third greater than its nearest rival, New York. London is also the lead eurobond centre, gaining a considerable boost in recent years from the development of new financial instruments such as swaps and equity warrants. Finally, the internationalization of London's equity markets brought about through the 'Big Bang' reforms of 1986, rapidly increased the turnover of foreign equities, although overall turnover remains small in comparison with New York or Tokyo. The volume of London's business, however, especially in international banking and foreign exchange, does represent a massive flow of capital through the City and in consequence a considerable source of income for the financial services provided. Indeed it could be argued that the rise of euromarkets, together with the growth of international banks in London have effectively tied London and parts of the south east into a network of international relations which are very different from growth patterns across other UK regions.

International finance, however, is only one aspect of the City's activities. Consideration should also be given to the commercial activities which dominate the City, in terms of both employment and income generated (Thrift, 1987). Dealing in commodities, as well as mediating the kinds of monetary transactions that characterize the euromarkets represents the prime activity of the City. Other international services which contribute towards the earning power of the City include insurance, fund management, shipping, corporate consultancy, accountancy, advertising and legal services. Whilst it is difficult to measure the value of international trade in such services, as it is to disentangle flows of capital from the fees and commissions received, the awesome global transaction figures do provide some indication of the significance and strength of London's financial and commercial network. To register, however, that London's daily turnover in foreign exchange at the end of the 1980s amounted to $187 billion or that the yearly turnover in foreign equities stood at $71 billion is perhaps less important to note than the network of global relations which sustains this activity.

Moreover, this network is not confined to the flows of finance and commerce that lock the south east into a set of international structures. Such circulation flows act as a dynamic of growth for the region as Thrift and Leyshon (1992) have recently shown, yet their spatial reach goes beyond the boundaries of the south east planning region. As a spin-off of growth in the financial services sector, various provincial cities in the south of the UK have benefited directly in employment terms, as the major banking and insurance institutions decentralize their back office routine activities, and indirectly through the income multiplier. The spatial structures which lie behind the growth of provincial financial centres are not necessarily the same as those which sustain the London City region, however (Leyshon *et al*, 1989). The growth of local financial service firms may arise from provincial demands, whereas the back office locations of financial institutions will very likely be in response to the availability of a relatively stable, educated, cheap labour supply. The latter is an example of transmitted growth from the City, the former is not. Likewise, the regulatory change and corporate reorganization which has led to a shift in the spatial structures of domestic financial institutions, reinforcing an existing southern bias as well as benefiting the larger provincial centres, would form part of a different growth dynamic from that of London's financial and commercial complex.

In focusing upon the financial services industry, therefore, the geography of both production and circulation relations is central to an understanding of how different patterns of growth are laid down across the country. In this case London and the south east represent the hub of this growth, not simply because they dominate the pattern of employment in the financial and commercial services sector in the UK (with over half the total jobs in the region), but rather because of the type of economic activities performed in the London City region which tie it into the flows of the global financial markets. The kind of interdependencies that are of interest here are indeed functional, but they refer to a network of relations that takes place *across* countries and which go beyond the narrow confines of input–output relations. Equally, this is not to suggest that no important financial or commercial activities take place outside the south of the UK, merely that the majority are not part of the same structures of interdependence that have helped to shape the changing fortunes of the south east in recent years.

A similar kind of exercise – equally truncated – can be adopted to trace the network of relations which have shaped the growth of a high technology sector across the UK. As with the financial services sector it is important to discriminate between those parts of a sector which generate growth and those which represent part of the growth transmission process. Not all industries within the high technology sector for instance may be considered propulsive. Indeed, much high technology activity involves relatively routine production tasks such as assembly, as opposed to the tasks of innovation. Broadly speaking, within the UK it is the high technology *services*, namely telecommunications, computing services, and research and development which act as the bearers of innovation rather than high-tech manufacturing. Moreover, within the UK almost half of all employment in the high technology sector, around one half of a million workers, is to be found in the above three service industries (Butchart, 1987).

The uneven geography of high-tech activity within the UK is well known, particularly its strong southern bias (Keeble, 1989). What is particularly striking however is the concentration of the high technology service industries in southern provincial cities and the south east in general (Begg and Cameron, 1988). For example, sixty eight per cent of total employment in research and development is in the three southern regions, along with sixty per cent of jobs in the telecommunications industry. In computer services, the south east dominates the industry with just under two thirds of total employment (Cooke *et al*, 1992). Overall, over half of high-tech service employment is in the south east, of which a quarter is in London. This contrasts sharply, for example, with the manufacturing side of the computing and communication industries which saw a dramatic reduction in its south east employment base in the 1980s, especially in London (Cooke *et al*, 1992).

In terms of market relations, much of this concentration can be traced to state involvement in the location of government research establishments and Ministry of Defence expenditure, with recent changes in the defence industry reinforcing a south east bias (Lovering, 1991; Morgan and Sayer, 1988). Also the demands of an information intensive financial services sector in London have provided a growing market for 'state of the art' information and communication technologies, indicating once again the relative importance of service industries as markets for the products of other service industries. These, however, are not the only kind of relations that account for the dynamism of high-tech service activity in the region. The spatial hierarchy of skills and qualifications, with the greatest number of highly qualified, highly paid staff to be found in the south east, represents a spatial structuring of production that also contributes towards the region's dynamism (Keeble, 1989). Moreover, such groups are increasingly part of global labour markets which take in London and the south east, though little further in terms of their spatial reach (Fielding, 1992). Other parts of the UK do possess a high-tech presence, but the nature of their activities and their position in an international spatial division of labour does not mirror that of the south generally. As Begg and Cameron (1988) among others have shown, the routine production and less skilled high-tech activities are more widely dispersed, with different regions tending to specialize in one or more high-tech manufacturing activity.

Again, the spatial networks in the case of the high technology service industries which are of interest include input–output relations, but they also include production as well as market relationships, international as well as national structures of independence. Alongside the financial and commercial sector, it is also interesting to note how this network of relations has contributed towards the regionalized character of much of the private service sector growth across the southern half of England.

Finally, with the disparate tourist industries in the UK it is more difficult to discern a spatial network of relations. If we follow Urry's argument (1990a) then tourism clearly operates as a dynamic of growth across the regions, providing an independent source of income. Indeed, in his more extensive treatment of the tourist industry (1990b), he draws a connection between the 'openness' of the UK economy and the tourist industry in particular. Moreover, he refers to the emergence of an international division of tourist sites, with different countries providing particular kinds of attraction. In the UK, it is history and heritage which are seen to constitute the major attraction, with London dominating the market for overseas visitors. According to international passenger survey data (1988), London receives almost sixty per cent of all expenditure by overseas visitors, close on £3½ billion in 1988 (BTA/ETB, 1989). In fact, the flow of overseas visitors has been highly localized with London and the south east capturing over two thirds of all trips and around one half of all overnight stays (Jeffrey and Hubbard, 1988; BTA/ETB, 1989).

What is less clear from this account are the differences between kinds of 'heritage' and the relations of ownership that circumscribe them. The diverse ways in which different cities and regions of the UK seek to promote their areas through the reconstruction and 'sale' of the past and the different scale of the enterprises involved make generalizations risky. Nonetheless, the work of Townsend (1992) points to the limited employment potential of some of the more recent 'industrial' attractions that have appeared in northern cities in comparison with the relatively longstanding attractions of country house and 'historic' buildings in the south. Beyond a certain level, it would appear that the presence of a wide range of tourist attractions, including that of museums, galleries, theatres, and opera, act as a spiral of growth reinforcing one another, as is the case with London.

As with tourist markets, tourist-related employment in the hotel, travel, and entertainment industries have their own pattern of local and international interdependencies. In the UK, estimates of the total number of people dependent upon tourism for their livelihood range from one to one and a half million, many of whom are working on a part-time and/or self-employed basis. Of this figure, the largest provider of jobs is the hotel trade, and in regional terms just over one third of all jobs in this industry are located in the south east, with nearly one half of that figure in London. As a dynamic of growth, however, few of the jobs are of the high wage, high skill variety found within parts of the financial, commercial and high technology service industries. Yet as a source of foreign currency and local multipliers, the hotel trade is regarded by Medlick (1985) and others to be a major generator of growth. And it is the London City region which is seen to have benefited most from the shift to history and heritage tourism in the UK.

We have briefly considered the network of relations associated with three of the most significant forms of contemporary service growth in the UK – finance and commerce, high technology, and tourism – each of which possess their own distinctive geography. The lines of interdependence that shape each sector – the relations of production and circulation that are organized over national and international space – differ considerably between them. Yet their geographies, that is, the pattern of uneven development which is laid down by each, *overlap* in more than a diagrammatic sense – taking in London, parts of the south east, as well as locations in adjoining regions. This set of overlapping geographies, as it were, although the outcome of different sets of social relations, have contributed towards what may be referred to as a regionalized mode of service growth *across* the south of the UK (Allen, 1988c).

This is not a description of a simple north–south divide however. Nor does it refer to an 'international' south and a 'local' north. The pattern is obviously more complex than that, with for example, parts of manufacturing in the north tied to the global economy in terms of both markets and labour hierarchies through foreign direct investment, parts of the south bypassed by the service growth described here, and both manufacturing and service firms across the north connected to the capital city as a locus of strategic economic control. Nonetheless, it is possible over the last decade and a half to discern a varied and somewhat cumulative mode of service growth which has moved across London and much of the south, but little further. Across the rest of the UK space economy, other forms of growth undoubtedly have occurred based upon combinations of private and public service growth as well as local and international manufacturing. These forms of growth co-exist and cut across one another, although their relative weight in the reorganization of the national space economy is likely to be of unequal value. Of course, it is quite possible that the fragility of service growth across the London City region witnessed in the early 1990s signals the contraction of this growth pattern. Even so, the structural causes of this mode of growth remain largely intact, especially those that lie behind the financial and commercial services sector. What is more, such causal relations are suggestive of more than the coexistence of regionalized modes of growth, they also point to a number of dislocations in the UK economy which have taken a spatial form.

THE PECULIARITIES OF THE UK

The idea of a national economy exhibiting signs of structural dislocation is not unique to the UK, nor is it a particularly novel description of the UK economy. Lash and Urry (1987), for example, have referred to the UK historically as a 'Makier' (middleman) economy, with an imbalance of industries and markets which were unfavourable to heavy manufacturing industry and orientated towards the 'interests' of the City and commerce. Moreover, there is a longstanding debate on the 'exceptional' character of Britain's political class which resurfaced in the late 1980s with an even stronger emphasis upon the pre-industrial stamp of the nation's ruling bloc and the hegemonic role of the City. At issue is the distorting effect of the City's international orientation upon the rest of the UK economy, especially its disappearing manufacturing base. We have touched upon this sense of dislocation in our discussion of developments in the UK's financial services sector. Here, however, following Laclau (1990), I wish to give a more explicit sense to the term dislocation by using it as a synonym for uneven developments.

On this view, development is uneven precisely because an existing set of social arrangements is disrupted by the introduction of elements or forces external to it. Any set of social arrangements, be they wrecked manufacturing regions attempting to turn themselves around, public sector service cities attempting to broaden their economic base, or successful regions bent on achieving further success, may experience dislocation as events which have their origin outside of them disrupt their development. So, for example, it could be argued that the arrival of foreign banks in the City of London from the 1970s on, especially US and Japanese banks and their dominance of the new euromarkets, effectively dislocated the 'old' commercial regime of the City. A potential consequence of this disruption, if I have understood Laclau correctly, would be a re-combination of elements – not simply of the 'old' and the 'new' – as if the overseas banks represented a more advanced stage of development which has been grafted onto the more traditional practices of the City's institutions, but rather a combination of national and international elements which together have produced the supra-national character of the City.

Dislocations, however, need not be solely economic or international in their origin; arguably the neo-liberal political strategies adopted by the UK government in the 1980s also represent a source of dislocation. Moreover, their uneven geographical impact, perhaps more as an unintended consequence, appears to have exacerbated spatial and economic divides around service sector growth. It is possible to cover only a limited range of political developments here, but the broad suggestion is that the regionalized mode of service growth outlined in the previous section should not be seen as the inexorable outcome of local and global economic forces. Rather it may be understood as the result of a succession of economic *and* political dislocations which have tended to *reinforce* this regionalized growth pattern.

Among the more important economic strategies associated with neo-liberalism are the liberation of markets, the deregulation of markets, the privatization of services, and the positive stance adopted towards the internationalization of the UK economy (Jessop *et al*, 1988). Each of these strategies has an impact upon the structure of particular service industries and, in a number of cases, they also influenced their spatial organization. We have already noted the increased openness of the UK's financial sector, but as a consequence of the liberalization and deregulation of the sector's activities, the effect has been to further enhance its predominance in the UK economy as a whole and to intensify the concentration of its activities in the London City region. The abolition of exchange control, the radical reorganization of the stock market as part of the 'Big Bang' reforms, and the relaxation of barriers to trade within the financial markets have further emphasized the supra-national quality of London's markets, as well as lending support to the argument that London and parts of the south east economy have detached themselves from a national economy (King, 1990).

The latter is perhaps too strong a claim, especially as it understates the significance of the economic relationships that bind cities and regions outside of the south east to London's growth pattern. Nonetheless, even in the sphere of domestic banking and finance, the recent liberalization measures which have promoted the integration of personal financial markets and stimulated competition between the clearing banks and the building societies have also tended to emphasize the distinctiveness of the London City region. Losses in banking employment in the north during the 1980s were mirrored by growth in London and across the south as the major institutions rationalized and centralized their activities. Although, as noted earlier, the organizational changes introduced by the domestic financial institutions are the outcome of a different set of growth dynamics from that of the City and its euromarkets, the increased dominance of the London City region in domestic financial control is an indirect consequence of the *political* shake up of the international financial markets.

Similarly, attempts to liberalize a range of public services through the introduction of market competition in certain areas of service provision have also, again perhaps unintentionally, worked to the advantage of the south east. In the cleaning and catering sectors, for example, which were among the main activities contracted out by central government departments, local authorities and the National Health Service, contracts awarded to the private sector have tended to favour national and in some cases multinational, companies based in the south east (Ascher, 1987; Mohan, 1988a). The significance of this is felt not so much in terms of job numbers, as in their ability to shape the relations between service jobs across their branch networks. Similarly, employment growth in the south east in the private health care and welfare industries (as well as private education) owes much to the opening up of traditional public sector service areas to private capital (Mohan, 1988b).

The geographical impact of the privatization of public service corporations such as British Telecom and Cable and Wireless PLC is more difficult to unravel, especially in comparison with the heavy job losses in the north that followed the privatization of nationalized manufacturing industries (Champion and Townsend, 1990). The main areas of service activity affected by privatization are the telecommunications and transport industries. Along with attempts to deregulate the networks of both industries, privatization has led companies like British Telecom to become more internationally orientated, seeking a share of overseas markets (Cooke *et al*, 1992). In general, the increased international stance adopted by the telecommunications sector has merely served to consolidate the concentration of its activities in the south east noted earlier. Indeed, Howells and Green (1988) point to the limited development of telecommunication services in less profitable, peripheral areas of the UK and anticipate a more restricted network.

Pulling these few political strands together does not provide evidence to support the view that a political bias in favour of the south east was apparent in the 1980s, however. On the contrary, it could be argued that the qualified success of the neo-liberal strategy rests, in part, with the adaptation of civil society in the south east to the shift from public to private regimes of growth. In one sense, neo-liberalism has dislocated the relation between the public and the private in south east civil society. The line between the public and the private has been largely redrawn, or rather, a new combination of public and private elements has produced a civil society which could be said to have its core within the region.

What is peculiar about this set of circumstances is not that the City and the political class whose interests are represented through it are divorced from the rest of the UK, socially and economically; rather, it is the unintended outcome of national political strategies working alongside intra- and inter-national economic interdependencies to reinforce a regionalized mode of service-growth across London and parts of the south. However, the novelty of the present situation – which is now passing – does not necessarily entail an eventual return to national modes of growth. The idea of *one* overall direction, *one* 'engine' of growth, to an industrial economy is a distinctively *modern* notion and one which is called into question by today's 'open' economies.

CONCLUDING REMARKS

[G]eography, or rather the conceptual framework of uneven development, opens up a way of thinking about the discourse of services rather than thinking with it. The notion of regionalized modes of growth which take their character and shape from a combination of production and circulation relations, including functional market relations, operating across and between 'open' national economies, provides an alternative way of conceptualizing service sector growth, or indeed other modes of growth. This is not to imply that all service activities have taken a global turn. Clearly, some service industries have become international in many economic aspects, whilst others remain predominantly local concerns. Rather, it is to suggest that we should now be thinking beyond a conception of the modern economy which routinely emphasizes the links between services and manufacturing within the national space economy.

Ann Markusen

'Sticky Places in Slippery Space: A Typology of Industrial Districts'

***Economic Geography* (1996)**

THE PUZZLE OF STICKINESS IN AN INCREASINGLY SLIPPERY WORLD

In a world of dramatically improved communications systems and corporations that are increasingly mobile internationally, it is puzzling why certain places are able to sustain their attractiveness to both capital and labor. Movement is, of course, costly and disruptive to both. David Harvey's (1982) work on capital's need for 'spatial fix' and Storper and Walker's (1989) work on labor and reproduction suggest generic reasons why hypermobility cannot completely obliterate production ensembles in space. But neither account explains why certain places manage to anchor productive activity while others do not.

The problem is most acute in advanced capitalist countries, where wage levels and standards of living are substantially higher than in newly incorporated labor-rich and increasingly technically competent countries (Howes and Markusen, 1993). Production space in these countries has become increasingly 'slippery', as the ease to capital of moving plants grows and as new competing lines are set up in lower-cost regions elsewhere. Often the only alternative for the region of exit or any other aspirant appears to be matching local production conditions to those in the competitor place, lowering wages and reproduction costs to the lower common denominator.

Alarmed by the welfare implications of such a strategy, economists, geographers, and economic development planners have sought alternative models of development in which existing activities are sustained or transformed in ways that maintain relatively high wage levels, social wages, and quality of life. One extensively researched formulation is that of the 'flexibly specialized' or 'new industrial district' (NID), based on Italy [and] the role of small, innovative firms, embedded within a regionally cooperative system of industrial governance which enables them to adapt and flourish despite globalizing tendencies.

In this paper, I argue that there are at least three other types of industrial districts, or 'sticky places' that have demonstrated resiliency in the postwar period in advanced industrialized countries. Stickiness connotes both ability to attract as well as to keep, like fly tape, and thus it applies to both new and established regions. Based on an inductive analysis of the more successful metropolitan regions in the United States, I show that structures and dynamic paths quite different from those captured in the NID formulation have enabled both relatively mature and up-and-coming regions to weather heightened capital mobility. Contrary to the emphasis on small firms in the NID formulation, these alternative models demonstrate the continued power of the state and/or multinational corporations under certain circumstances to shape and anchor industrial districts, providing the glue that makes it difficult for smaller firms to leave, encouraging them to stay and expand, and attracting newcomers into the region. These models exhibit greater propensities for networking across district lines, rather than within, and a much greater tendency to be exogenously driven and thus focused on external policy issues than do NIDs. From a welfare point of view, the four types perform quite differently with regard to income distribution, permissiveness toward labor organization, short-to-medium-term cyclicality, and longer-term vulnerability to secular change.

IDENTIFYING AND ANALYZING STICKY PLACES

The three alternative models of sticky places developed in this paper were constructed through a process of inductive inquiry similar to that used in researching NIDs.

First, we surveyed metropolitan growth since 1970 for [the United States, Japan, Korea, and Brazil]. We then selected at least one case with apparent conformity to the NID formulation and three to five others whose industrial structure and organization appeared to be quite different.

Conceptually, we inquired into the presence or absence of features specified in the NID formulation [such as] firm size, networks among district firms, [and] districtwide governance structures. In addition, we examined the role of the state as rule maker, as producer and consumer of goods and services, and as underwriter of innovation (Christopherson, 1993; Saxenian, 1994), the role of large firms (Dicken, 1992; Harrison, 1994), [and] the embeddedness of firms both within their districts and in nonlocal networks (Granovetter, 1985; Markusen, 1994). Fourth, we investigated the longer-term developmental dynamic of major industries, to determine their resiliency and/or vulnerability to longer-term atrophy. Fifth, we assessed the long-term dynamic potential of each region, including the ability of [its existing industrial ensemble] to release locally anchored resources, human and physical, into new, unrelated specialized sectors. Finally, we searched for connections between district structure and social welfare metrics, including employment growth rates, cyclical stability, associated income and wealth distribution, trade union presence, and political diversity.

Like Storper and Harrison (1991), we opt for an extensive connotation of industrial districts, which does not confine it to the most common usage, called here the Marshallian (or Italianate variant) district. Elsewhere, we offer the following definition: an industrial district is a sizable and spatially delimited area of trade-oriented economic activity which has a distinctive economic specialization, be it resource-related, manufacturing, or services (Park and Markusen, 1994).

In what follows, I present four distinctive industrial spatial types: (1) the Marshallian NID, with its recent Italianate variety; (2) the hub-and-spoke district, where regional structure revolves around one or several major corporations in one or a few industries; (3) the satellite industrial platform, comprised chiefly of branch plants of absent multinational corporations; and (4) the state-centered district, a more electic category, where a major government tenant anchors the regional economy (a capital city, key military or research facility, public corporation). The hypothesized features of each are summarized in Table 1. Schematic visual models of each of the first three are offered in Figure 1. Here, firm relationships within the region are depicted inside the circle versus those outside of it – suppliers to the left, customers to the right. A real-world district may be an amalgam of one or more types, and over time districts may mutate from one type to another.

Central to the differences among sticky places and their ability to persist are presence (or absence) of distinctive and lopsided power relationships, sometimes within the district and sometimes between district entities and those residing elsewhere.

Table 1 Hypothesized features of new industrial district types

Marshallian industrial districts

- Business structure dominated by small, locally owned firms
- Scale economies relatively low
- Substantial intradistrict trade among buyers and suppliers
- Key investment decisions made locally
- Long-term contracts and commitments between local buyers and suppliers
- Low degrees of cooperation or linkage with firms external to the district
- Labor market internal to the district, highly flexible
- Workers committed to district, rather than to firms
- High rates of labor in-migration, lower levels of out-migration
- Evolution of unique local cultural identity, bonds
- Specialized sources of finance, technical expertise, business services available in district outside of firms
- Existence of 'patient capital' within district
- Turmoil, but good long-term prospects for growth and employment

Italianate variant (in addition to the above)

- High incidence of exchanges of personnel between customers and suppliers
- Disproportionate shares of workers engaged in design, innovation
- Strong trade associations that provide shared infrastructure-management, training, marketing, technical or financial help, i.e., mechanisms for risk sharing and stabilization
- Strong local government role in regulating and promoting core industries

Hub-and-spoke districts

- Core firms embedded nonlocally, with substantial links to suppliers and competitors outside of the district
- Scale economies relatively high
- Low rates of turnover of local business except in third tier
- Substantial intradistrict trade among dominant firms and suppliers
- Key investment decisions made locally, but spread out globally
- Long-term contracts and commitments between dominant firms and suppliers
- High degrees of cooperation, linkages with external firms both locally and externally
- Moderate incidence of exchanges of personnel between customers and suppliers
- Low degree of cooperation among large competitor firms to share risk, stabilize market, share innovation
- Labor market internal to the district, less flexible
- Disproportionate share of blue-collar workers
- Workers committed to large firms first, then to district, then to small firms
- High rates of labor in-migration, but less out-migration
- Evolution of unique local cultural identity, bonds
- Specialized sources of finance, technical expertise, business services dominated by large firms
- Little 'patient capital' within district outside of large firms
- Absence of trade associations that provide shared infrastructure-management, training, marketing, technical or financial help, i.e. mechanisms for risk sharing and stabilization
- Strong local government role in regulating and promoting core industries in local and provincial and national government
- High degree of public involvement in providing infrastructure
- Long-term prospects for growth dependent upon prospects for the industry and strategies of dominant firms

Satellite industrial platforms

- Business structure dominated by large, externally owned and headquartered firms
- Scale economies moderate to high
- Low to moderate rates of turnover of platform tenants

continues overleaf

Table 1 (continued)

- Minimal intradistrict trade among buyers and suppliers
- Key investment decisions made externally
- Absence of long-term commitments to suppliers locally
- High degrees of cooperation, linkages with external firms, especially with parent company
- High incidence of exchanges of personnel between customers and suppliers externally but not locally
- Low degree of cooperation among competitor firms to share risk, stabilize market, share innovation
- Labor market external to the district, internal to vertically integrated firm
- Workers committed to firm rather than district
- High rates of labor in-migration and out-migration at managerial, professional, technical levels; little at blue- and pink-collar levels
- Little evolution of unique local cultural identity, bonds
- Main sources of finance, technical expertise, business services provided externally, through firm or external purchase
- No 'patient capital' within district
- No trade associations that provide shared infrastructure-management, training, marketing, technical, or financial help, i.e. mechanisms for risk sharing and stabilization
- Strong local government role in providing infrastructure, tax breaks, and other generic business inducements
- Growth jeopardized by intermediate-term portability of plants and activities elsewhere to similarly constructed platforms

State-anchored industrial districts

- Business structure dominated by one or several large government institutions such as military bases, state or national capitals, large public universities, surrounded by suppliers and customers (including those regulated)
- Scale economies relatively high in public-sector activities
- Low rates of turnover of local business
- Substantial intradistrict trade among dominant institutions and suppliers, but not among others
- Key investment decisions made at various levels of government, some internal, some external
- Short-term contracts and commitments between dominant institutions and suppliers, customers
- High degrees of cooperation, linkages with external firms for externally headquartered supplier organizations
- Moderate incidence of exchanges of personnel between customers and suppliers
- Low degree of cooperation among local private-sector firms to share risk, stabilize market, share innovation
- Labor market internal if state capital, national if university or military facility or other federal offices for professional/technical and managerial workers
- Disproportionate shares of clerical and professional workers
- Workers committed to large institutions first, then to district, then to small firms
- High rates of labor in-migration, but less out-migration unless government is withdrawing or closing down
- Evolution of unique local cultural identity, bonds
- No specialized sources of finance, technical expertise, business services
- No 'patient capital' within district
- Weak trade associations to share information about public-sector client
- Weak local government role in regulating and promoting core activities
- High degree of public involvement in providing infrastructure
- Long-term prospects for growth dependent on prospects for government facilities at core

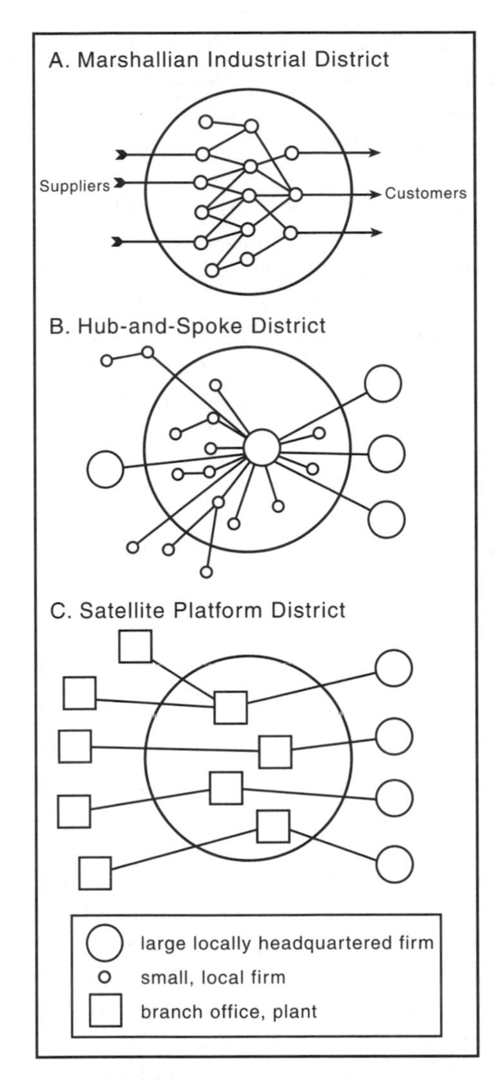

Figure 1 Firm size, connections, and local versus nonlocal embeddedness

Very few of [the fastest-growing industrial cities in the United States] can be characterized as NIDs, but many of them reproduce the conditions present in the other three models of 'sticky places'.

MARSHALLIAN AND ITALIANATE INDUSTRIAL DISTRICTS

In his original formulation of the industrial district, Marshall envisioned a region of small, locally owned firms. Scale economies are relatively low, forestalling the rise of large firms. Within the district, substantial trade is transacted between buyers and sellers, often entailing long-term contracts or commitments (Figure 1), for eventual export from the region.

What makes the industrial district so special and vibrant, in Marshall's account, is the nature and quality of the local labor market, which is highly flexible. Individuals move from firm to firm, and owners as well as workers live in the same community, where they benefit from the fact that 'the secrets of industry are in the air'. [This] enables the evolution of strong local cultural identity and shared industrial expertise.

The Marshallian district also encompasses a relatively specialized set of services tailored to the unique products/industries of the district. These services include technical expertise in certain product lines, machinery and marketing, and maintenance and repair services. They include local financial institutions offering so-called 'patient capital', willing to take longer-term risks because they have both inside information and trust in [local] entrepreneurs.

In Marshall's formulation, it was not necessary that [local] actors should be consciously cooperating with each other. But more recent researchers have argued that concerted efforts to cooperate among district members and to build governance structures to improve districtwide competitiveness can improve prospects – that is, increase the stickiness of the district.

Features characterizing Italianate districts are articulated in intensive case studies on the Italian (Piore and Sabel, 1984; Bull *et al*, 1991) and American cases – Orange County (Scott, 1986; Scott and Paul, 1990) and Silicon Valley (Saxenian, 1994) – though not without debate (Malecki, 1987; Florida and Kenney, 1990). Few cases have been identified outside of Europe or the United States.

Unlike the passivity of Marshall's firms, Italianate districts exhibit frequent and intensive exchanges of personnel between customers and suppliers and cooperation among competitor firms to share risk, stabilize markets, and share innovation. Disproportionate shares of workers are engaged in design and innovative activities. Activist trade associations provide shared infrastructure – management, training, marketing, technical, or financial help – as well as providing forums to hammer out collective strategy. Local and regional governments may be central in regulating and promoting core industries. Trust among district members is central to their ability to cooperate and act collectively (Harrison, 1992; Saxenian, 1994), although critics argue that the power of large corporations to shape Italian industrial districts has been understated (Harrison, 1994, Chap. 4).

In assessing the growth, stability, equity, and politics of Italianate industrial districts, the Italian variety must be distinguished from the Silicon Valley and Orange County cases. In terms of growth and stability, both Marshallian and Italianate industrial districts retain good long-term prospects. Although more standardized functions may be driven elsewhere by inflated regional costs, innovation (so the theory goes) will ensure the revitalization of these 'seedbeds of innovation'. But other hypotheses have been advanced. Agglomerative specialization can actually impede the development of other sectors, whose presence might diversify the economy and counteract maturation in the original sector. Pittsburgh in the late nineteenth and Detroit in the early twentieth century resembled Italianate districts and Silicon Valley, but the evolution of oligopoly and the crowding out of other sectors left both quite vulnerable to the inevitable maturation and decentralization of those industries (Chinitz, 1960; Markusen, 1985).

On the equity front, the high-tech Silicon Valleys and Orange Counties depart strikingly from the Italian industrial districts. Italian districts often [exhibit] resilient cultures, [based] politically on long-standing communities, unions, and the Italian communist party. [There are] strong leadership roles for unions and relatively good income distributions. In the California cases, in contrast, district cooperation is purely between entrepreneurs and firms, [within] a non-union environment. Income distribution tends to be highly dualized [and] politics tends toward the conservative.

The 'new industrial district' approach has much to offer and has deservedly captured the imagination of scholars and local economic development activists alike. But many of the faster-growing regions of the world turn out not to be primarily characterized by these same features. Furthermore, other structural forms may be associated with superior welfare and political cultures.

HUB-AND-SPOKE INDUSTRIAL DISTRICTS

Another quite different type of industrial district is where a number of key firms and/or facilities act as anchors or hubs to the regional economy, with suppliers and related activities spread out around them like spokes of a wheel. A simple version of this form is depicted in Figure 1, where a single large firm (e.g. Boeing in Seattle or Toyota in Toyota City) buys from both local and external suppliers and sells chiefly to external customers. Intensive case studies of hub-and-spoke districts include Seattle (Gray *et al*, 1996) [and] San Jose dos Campos and Campinas, Brazil (Diniz and Razavi, 1994).

The dynamism in hub-and-spoke economies is associated with the position of these anchor organizations in their national and international markets. Other local firms tend to have subordinate relationships to them. If over time the anchors evoke a critical mass of agglomerated skilled labor and business services around them, they may set off a more diversified developmental process.

Hub-and-spoke districts are thus dominated by one or several large, vertically integrated firms, in one or more sectors, surrounded by smaller and less powerful suppliers. The large player(s) may be oligopolists in a single industry, as in Detroit, or loosely linked hubs in several industries may coexist in a region. In

Seattle, for instance, Weyerhauser [is] the dominant resource-sector company, Boeing the dominant industrial employer (commercial aircraft and military/spacecraft), Microsoft the leading services firm, the Hutchinson Cancer Center the progenitor of a series of biotechnology firms, and the Port of Seattle the transportation hub. Core firms are embedded nonlocally, with substantial links to suppliers, competitors, and customers outside the district. Internal scale and scope economies are relatively high. Key investment decisions are made locally, but their consequences are spread out globally.

Intradistrict cooperation will generally be on the terms of the hub firm. Substantial intradistrict trade among suppliers and hub firms [is] often embodied in long-term contracts and commitments. Markedly lacking is cooperation among competitor firms to share risk, stabilize the market, and share innovation.

The labor market in hub-and-spoke districts is internal to both large hub firms and to the district. Workers' loyalties are to core firms first, then to the district, and only after that to small firms.

Hub-and-spoke districts do evolve unique local cultures related to hub activities. Detroit is known as Motor City, and sports teams of many cities have been named after dominant sectors – the Steelers, the Brewers, the Millers (the old Minneapolis team). They develop considerable expertise in the labor pool in specialized industrial capabilities, and engender specialized business service sectors tailored to their needs.

Districts of this sort lack some of the more celebrated governance structures of the Italianate industrial districts [such as] 'patient capital'. The few trade associations that exist are relatively weak. Hub firms will try politically to ensure that area politicians represent the interests of their firm at national and international levels. They may also be actively involved in improving area educational institutions and the provision of infrastructure.

In the long run, hub-and-spoke districts are quite dependent on their major industries and firms within them for their stickiness. Growth and stability can be jeopardized by portability of plants away from the region, by the long-term decline of the industry, or by poor management of the principal firms. But stickiness also depends on the degree to which mature sectors can release local resources into new, unrelated sectors. A sobering historical example of the vulnerability of hub-and-spoke districts is Detroit, [whose] vitality was severely taxed by the oligopolistic rigidity of the locally headquartered auto industry. [The latter's] control over the Detroit area's resources prevented the diversification of its economy (Chinitz, 1960). A counter example is Seattle, where several unique features of Boeing as the undisputed anchor to the regional economy have contributed to (or at least not prevented) the region's diversification into other sectors – port-related activities, software, biotechnology (Gray *et al*, 1996).

Hub-and-spoke industrial districts may be characterized by relatively good income distributions. Market power, natural economies of scale, and high levels of labor productivity [may result in high] wages. [This] is [however] dependent on the presence of unions or the threat of their emergence.

SATELLITE PLATFORMS

A third variant of rapidly growing industrial districts [is] the satellite platform – a congregation of branch facilities of externally based multiplant firms. Often these are assembled by national or provincial governments as a way of stimulating regional development in outlying areas. Tenants of satellite platforms may range from routine assembly functions to relatively sophisticated research, but they must be able to more or less 'stand alone', detachable spatially from [the parent] firm or agglomerations (Glasmeier, 1988).

Satellite platforms may be found in almost all countries, example[s including] the United States Research Triangle Park, a collection of unrelated research centers of major multinational corporations (Luger and Goldstein, 1990), [and] Elkhart, Indiana, where auto-related branch plants have been attracted by relatively low-wage labor. In Brazil, a remarkable case is the state sponsored expansion of Manaus as an import/export zone (Diniz and Borges Santos, 1995). Satellite platforms [are] dominated by large, externally situated firms that make key investment decisions. Scale economies are moderate to high, and minimal intra-district trade takes place among platform tenants. Orders and commitments to local suppliers are conspicuously absent. Heterogeneity and remote control mitigate against cooperative ventures among resident plants, [unlike the] hub-and-spoke district, where the large multi-locational firm is locally based. The most conspicuous feature [of Figure 1] is the absence of any connections or networks within the region and the predominance of links to the parent corporation and other branch plants elsewhere.

Relationships external to the facility are common but are with the headquarters firm, not locally with other branch facilities. To buttress this nonplace embeddedness, the labor market cuts across district boundaries; it is internal to the vertically integrated firm, rather than to the district [especially] at managerial, professional, and technical levels. Only blue- and pink-collar labor will be hired locally.

A number of features of the satellite platform constrain its development into a better-articulated regional economy. The main sources of finance, technical expertise, and business services are external to the region, namely corporate headquarters. Satellite districts have little 'patient capital', and because activities are diverse, they lack industry-specific trade associations that would provide shared infrastructure and help with management, training, and marketing problems.

Satellite platforms' future growth is jeopardized by the intermediate-term portability of plants elsewhere to similarly constructed platforms. Higher-end activities, depending on skilled personnel, will be less vulnerable, purely low-cost districts more so. Satellite platforms do not engender local cultural bonds. Thus they may be less sticky, especially if less skilled, than other types of district. They do remain sticky, however, to the extent that [they involve] large capital investments.

The record on income distribution is mixed. Entry of such platforms into previously depressed regions does contribute to higher incomes. However, satellite platforms artificially cordon off employment in some operations of a corporation from those in other regions, spreading income inequality out spatially. Somewhat better jobs for rural Japan obscure the concentration of top-paid corporate jobs elsewhere and deterioration in inner-city Tokyo, especially for blue-collar workers. Implications for politics in satellite platform regions are also mixed.

STATE-ANCHORED DISTRICTS

A fourth form of sticky place is the state-anchored industrial district, where a public or nonprofit entity, be it a military base, a defense plant, a weapons lab, a university, a prison complex, or a concentration of government offices, is a key anchor tenant in the district. Here, the local business structure is dominated by facilities whose locational calculus and economic relationships are determined in the political realm. This type of district is much more difficult to theorize. It is apt to look much like the hub-and-spoke district in Figure 1, although a facility can operate with few connections to the regional economy, resembling the satellite platform case.

Many of the fastest growing industrial districts in the United States and elsewhere owe their performance to state facilities. Military bases, military academies, and weapons labs explain the phenomenal postwar growth of Santa Fe and Colorado Springs, while defense plants contributed dramatically to the growth of Los Angeles, Silicon Valley, and Seattle (Markusen *et al*, 1991). State universities and/or state capitals explain the growth of cities like Madison, Austin, and Boulder. In Brazil,

Campinas owes much to its top-ranked university, while San Jose dos Campos's growth is based on the government-owned, military-oriented aerospace complex (Diniz and Razavi, 1994).

In general, scale economies are relatively high in such complexes [and] because [they] are so large, supplier sectors do grow up around them. [For] state capitals and universities, high degrees of cooperation may exist [with] suppliers, and activity will be relatively immune from the threat of exodus. This is less true for national facilities, especially in times of fiscal stringency or redundancy of function (e.g. the closing of military bases). [Here] decisions are made external to the district and may be indifferent to regional development impacts.

Government contracting may encourage the development of long-term supply relationships, based on trust and cooperation. However, these ties need not be localized; they may span thousands of miles between Los Angeles and Washington, D.C., for instance. For state capitals, the labor market will tend to be relatively local or regional. For universities and national facilities, labor markets will operate externally for the higher-skilled occupations as also for blue-collar and unskilled positions in military bases. Indigenous firms will play less of a role in these districts than in Marshallian or hub-and-spoke districts although some may emerge out of specialized technology transfer (universities).

In state-anchored industrial districts, long-term growth prospects depend on two factors: the prospects for the facility, and the extent to which the facility encourages growth within the region by spawning local suppliers, spinning off new businesses, or supplying labor or other factors of production to the local economy. Often, the mammoth size of the facility – New Mexico's Los Alamos Laboratories, for instance, with an annual budget of $1.4 billion, mostly for personnel, or New London, Connecticut's Electric Boat submarine manufacturing facility, with its 20,000 workers – overwhelms any contribution, real or potential, that may be made through second effects.

STICKY MIXES

Although the presence of Marshallian industrial districts can be confirmed in a number of American instances, the claims made for the paradigmatic ascendancy of this form of new industrial space (Scott's rubric) do not square with the experience of most rapidly growing agglomerations in industrialized and industrializing countries. Most rapidly growing metropolitan areas owe their performance to hub firms or industries, satellite platforms, and/or state anchors, or some combination thereof.

Many localities, especially larger metropolitan areas, exhibit elements of all four models. Silicon Valley, for instance, hosts an industrial district in electronics (Saxenian, 1994) but also revolves around several important hubs (Lockheed Space and Missiles, Hewlett Packard, Stanford University), as well as hosting large 'platform' type branch plants of US, Japanese, Korean, and European companies (e.g. IBM, Old, NTK Ceramics, Hyundai, Samsung). Furthermore, Silicon Valley is now and has been the fourth largest recipient of military spending contracts in the nation, a fact that shapes its defense electronics and communications sector (Saxenian, 1985; Markusen *et al*, 1991; Golob *et al*, 1995).

An intriguing question is whether regions can maintain their stickiness by transforming themselves from one type of district to another. Historically, Detroit made the transition from a Marshallian district to a hub-and-spoke district. Satellite platforms may be able to encourage backward and forward linkages that transform them into more Marshallian or hub-and-spoke type districts; scholars are debating whether this is occurring around large Japanese auto transplants in the United States. A state-centered district might do the same.

Many localities with stable or slowly declining growth patterns are also struggling to be sticky places, and many are succeeding in stanching their losses by remaking their industrial structures. New England, for instance, began as early as the 1950s to transform itself into a diversified military–industrial complex, escaping the deeper displacement that occurred post-1970 in the Industrial Midwest (Markusen *et al*, 1991). Although New England has not posted above-average long-term growth rates, it deserves study as a sticky place. Midwestern cities like Chicago, Milwaukee, and Cleveland are trying to make themselves more sticky by anchoring and upgrading existing expertise in industries like metals, machining, and automobiles.

RESEARCH AND POLICY IMPLICATIONS

This exercise in distinguishing among types of sticky places illustrates the diversity in spatial form, industrial complexion and maturity, institutional configurations, and welfare outcomes found in contemporary regional economies. It cautions that the singular enthusiasm for flexibly specialized industrial districts, especially the high-tech American variant, is ill-founded on both growth/stability and equity grounds. In large part, the problem here lies in the limits of the research strategy used in the NID literature, which intensively studies particular localities extracted from their embeddedness in a larger global economy.

Furthermore, the study of industrial districts and networks within them has generally been confined to smaller firms in particular industries; their links to larger firms and to other firms and institutions outside the region have been ignored. Nor is the zero-sum nature of much of this growth acknowledged – that certain places grow at the expense of other places, that high-wage employment in some regions is linked to low-wage employment in others, and that only a few places can possibly aspire to become Silicon Valleys of the future.

In reality, sticky places are complex products of multiple forces: corporate strategies, industrial structures, profit cycles, state priorities, local and national politics. Their success cannot be studied by focusing only on local institutions and behaviors, because their companies, workers, and other institutions (universities, government installations) are embedded in external relationships – both cooperative and competitive – that condition their commitment to the locality and their success there.

Improving relationships and building networks that reach outside of the region may prove more productive for some localities than concentrating on indigenous firms. Furthermore, our work on hub-and-spoke and satellite platform structures suggests that large firms can be significant contributors to regional development, albeit posing problems of dominance and vulnerability, and that recruitment of an external firm or plant may be a good strategy for a region at a particular developmental moment.

The prominence of hub-and-spoke and satellite platforms among US sticky places suggests that economic development strategies built on cross-regional alliances might be as important to localities as purely local networking approaches. More sophisticated and pluralistic profiles of industrial districts and how they operate, both internally and externally, must be joined with more intensive study of multinational corporations and state institutions if a more powerful geographic contribution to progressive strategy is to emerge.

Meric S. Gertler

'"Being There": Proximity, Organization, and Culture in the Development and Adoption of Advanced Manufacturing Technologies'

Economic Geography (1995)

[A] growing number of geographers have chronicled the apparent rise of post-Fordist economic systems (Schoenberger, 1988; Harvey, 1989; Storper and Walker, 1989). These systems are said to employ a flexible approach to production, reflected in employment relations, the organization of work within firms, and the broader social division of labor (Cooke and Morgan, 1991). In particular, [workers] point to a new set of flexible process technologies whose programmable properties offer producers prospects of great versatility, limited downtime, unparalleled precision, and superior quality. The same technologies are said to hold the potential to unleash the creative potential of workers and to compel manufacturers to establish a new regime of cooperation on the shopfloor (Florida, 1991).

Despite the popularity of such arguments, the pervasiveness of such practices, especially in locations outside the 'paradigmatic' flexible production regions, [has been questioned] (Gertler, 1988, 1992; Sayer, 1989; Pudup, 1992). [F]or example, while rates of adoption of flexible technologies such as computerized numerical control (CNC) are reasonably high among manufacturers in the United States, Great Britain, and Canada, many firms in these countries have experienced great difficulty in trying to implement such technologies effectively.

Furthermore, many of these implementation difficulties seem to arise in older, mature industrial regions, far removed from the major production sites of the new flexible production technologies (Gertler, 1993). Increasingly, the leading producers of these process technologies are to be found in countries like Germany, Japan, and Italy, while once-dominant American machinery producers have seen market shares drop significantly (Graham, 1993).

At the same time, case studies suggest that the production and use of such advanced machinery frequently occurs within the very same region, whether it be textiles in Germany's Baden–Wurttemberg (Piore and Sabel, 1984; Sabel *et al*, 1987), leather goods, ceramic tiles, and knitwear in Emilia–Romagna (Brusco, 1986; Russo, 1985), or semiconductors in Japan (Stowsky, 1987). The same literature suggests that the spatial coincidence of machine production and use in the recently successful industrial districts is not merely coincidental. The importance of spatial context in determining success in technology implementation [is] the central subject [of] this paper.

MACHINE PRODUCTION AND USE FROM AN INTERACTIVE PERSPECTIVE

[Recent] work [on] industrial organization and technological change (Lundvall, 1988; Porter, 1990) has suggested that 'closeness' between the users and producers of advanced machinery is important for a variety of reasons. Capital goods differ in important ways from other kinds of inputs purchased by manufacturers. [They are] of central importance to a manufacturer's operations and long-lived, [while] uncertainty surrounds [their] future use qualities. Because of these properties, the wise firm tends to purchase advanced capital goods through extensive interaction and communication [with] the producer of the machinery.

This mode of purchase offers benefits to prospective machinery users and producers alike, so that complex production equipment is not only more likely to be *adopted* successfully when there is close and frequent interaction between producer and user, but to be *produced* more successfully as well. Benefits and closeness between users and producers are most likely to be important when the technology involved is expensive, complex, and rapidly developing.

Significant implications flow directly from this literature. First, much of the success enjoyed by manufacturers in the 'canonical' industrial districts of Europe and Asia may be attributed to their close and constructive relationship with nearby producers of innovative machinery. Second, it should come as no surprise that machinery producers in these same regions have become highly successful competitors in international markets in recent years. Third, this literature suggests that many of the problems arising when manufacturers in the mature industrial regions of the United States, Britain, or Canada attempt to implement new, technologically advanced process technologies stem from the greater difficulty these users have in developing and maintaining a 'close' relationship with advanced machine producers, as the latter are currently (and increasingly) more likely to be located at a considerable distance.

Despite these arguments, it is possible that certain countervailing forces might reduce or qualify the attenuating influence of simple physical distance between user and producer. First, users and producers might be able to communicate by using modern telecommunications (especially the fax and telephone) and rapid air transportation of technical personnel (or key parts). In addition, many European and Asian producers contract with distributors, sales representatives, and maintenance firms in North America to perform on-site service functions on their behalf.

Second, large, multilocational (including multinational) firms may serve as highly effective, distance-transcending vectors of (intrafirm) technology transfer. [Thus] production regions that might be viewed as peripheral may nevertheless [contain] advanced machinery that is being used effectively by the local branch plant operations of such large, multisite firms. According to this argument, then, what matters more than simple physical distance is what one might call *organizational distance*.

Third, another interpretation of the difficulties encountered by the users of advanced machinery in 'peripheral' locations has more to do with differences in cultures, institutions, and the legacy of past industrial practices than with the problems caused by distance between users and producers. For example, Lundvall (1988) argues the importance of a common culture and language shared by users and producers, to facilitate the transmission of highly encoded information concerning users' needs and the capabilities and proper operation of complex and rapidly changing process technologies (see also Storper, 1992). Others point to differences in training cultures and attitudes toward technology as the crucial issues (Stowsky, 1987; Gordon, 1989).

According to this latter view, the typical American, British, or Canadian firm regards technology as something embodied entirely within the physical properties and design of machinery and production systems themselves. This stands in sharp contrast to European and Japanese manufacturers, who appreciate the necessity of social interaction for effective machine production and use, [and] regard the technological capabilities of a production process as [reflecting] the interaction between machines and skilled workers who have built up a wealth of knowledge and problem-solving abilities through many years of training and learning by doing.

In consequence, Anglo-American users of advanced machinery tend systematically to undervalue the importance of training while an advanced machine designed and built in, for example, Germany will be considerably more difficult to implement successfully in a North American user plant than in a German user plant because the 'culture' of industrial practices peculiar to Germany (high skill levels of factory workers, stability of employment relations, cooperative decision making on the shopfloor, strong emphasis on training) have been incorporated into the design of the German-made machine. According to this view, then, physical distance is really just a proxy for *cultural distance*, where 'culture' refers to a set of dominant workplace practices shaped in large part by legislative definitions of employment relations and the nature of the (public and private) industrial training system. Furthermore, this approach would seem to ascribe continuing importance to nation-states and the economic and social institutions created by them (see Gertler, 1992).

RESEARCH QUESTIONS AND METHODOLOGY

This empirical study investigate[s] the relationship between users and producers of advanced process technologies in order to shed more light on the source of users' difficulties in implementing complex new machinery.

Specifically, the study seeks to answer the following questions: (1) Is a high-quality (i.e. interaction-intensive) relationship between user and producer necessary to support the successful implementation of advanced manufacturing technology? (2) Must users and producers be 'close' to one another for such a high-quality relationship to develop? (3) Under what circumstances is 'closeness' likely to be most important? (4) How close is 'close', and is this to be understood only in terms of physical distance, or are organizational or 'cultural' conceptions of distance more important?

The study investigates advanced machinery implementation among manufacturers in Ontario using (1) a postal questionnaire survey of users of advanced manufacturing technologies, and (2) follow-up plant visits and personal interviews with a subset of firms. The survey targeted establishments in four sectors: transportation products (automotive parts and aerospace), electrical and electronic products, fabricated metal products and plastic and rubber products. Plants in these industries [are] among the most frequent users of complex new process technologies. Prior to mailing, plants were contacted by telephone to determine if they did in fact use any advanced manufacturing technologies [e.g. computer-aided design, robots, etc.]. The sample of establishments was stratified by both plant size (employment) and ownership (Canadian or foreign-owned). The resulting framework sought a balanced sample in six different cells – three size categories (small, medium, large) by two ownership categories (domestic, foreign).

The survey instrument solicited information for up to three advanced process technologies in use. Consequently, the final data base contained information describing 407 distinct technology implementation cases. The questionnaire sought information on the type of advanced process technology used, where it was produced, the extent and nature of difficulties encountered during technology implementation, the perceived importance of spatial proximity between user and producer, and the frequency and type of interaction between the user and producer. Plant visits and personal interviews were conducted with 30 survey respondents to allow more detailed probing of the kinds of relationships addressed in the questionnaire.

USER–PRODUCER INTERACTION AND THE IMPLEMENTATION OF COMPLEX PROCESS TECHNOLOGIES IN ONTARIO MANUFACTURING INDUSTRIES

Does Distance Matter?

Users [were first asked to] assess the importance of having their major machinery producers located at varying distances from them. There is an obvious and consistent pattern to the responses. Just under 40 percent of plants judged a very close physical link (20 kilometers or less) to be somewhat or very important. This figure rises steadily, however, as the spatial scale increases. More than half responded positively on colocation within the same region (within a 75 kilometer radius) with over two-thirds responding positively at the national scale.

At the continental scale, more than half of all firms judged colocation to be very important, and three-quarters very or somewhat important. The general conclusion is that having the producer of one's major machinery on the same continent is quite important to most users and, for many users, even shorter distances between them and producers are preferred.

Respondents were also asked to explain the answers they gave to the prior question. For plants indicating that closeness to producers was important, the overwhelming consideration appeared to be the ability to obtain good service or spare parts quickly in the event of breakdowns, especially from the original producer of the machinery. Distributors and sales representatives were often seen as distinctly inferior substitute sources of service, repairs, modifications, and technical information.

Other considerations underlying the importance of close contact between user and producer included problems caused by different time zones or the simple cost of transportation (for parts, service personnel, or trainers). Intervening international borders were also identified as problematic [because of] duties, tariff payments or difficulty of entry by foreign service personnel. Finally, communications problems when producers were not 'close by' included both the problems arising when the producers' first language was not English, and problems associated with communicating complex technical information over long distances.

Continental colocation of user and producer may thus be especially important, since overseas interaction is likely to be more difficult for all of the physical, logistical, and cultural reasons noted above. For many users even location of their producer on the same continent may not be good enough to suit their needs. However, in certain circumstances, closeness between user and producer is deemed unimportant, for example when user firms compensate for their 'peripheral' location relative to machine producers by developing and exploiting strong technical expertise within the firm itself.

When Does Distance Matter Most?

What are the conditions under which closeness between user and producer is likely to be most important? First, closeness may be more important to smaller user plants since they are more likely to lack financial or organizational resources to overcome the attenuating influence of distance. This expectation is confirmed [by the survey results]. Here one sees two notable patterns. First, the importance of closeness generally increases with spatial scale. Second, for the most part, small and medium-sized plants attach greater importance to closeness to machine producer at every spatial scale examined. The largest plants seem to be the least concerned with the need for spatial proximity, particularly at subnational scales.

Larger establishments can thus exploit their greater spatial reach to overcome the problems associated with long distances from machine user to producer. However, this reach is more accurately described as continental than global. Furthermore, larger plants constitute more important customers for producers and hence are considerably more likely to receive good service, no matter where they are located.

Of course, plant size may be less important than firm size. The survey results confirm the suspicion that organizational status is indeed important. At all spatial scales, single-plant establishments (which are by definition Canadian-owned) are by far the most likely of the three plant types to judge closeness to machinery producer as very important. However, within the multiplant category, ownership type (foreign versus domestic) does not appear to make much of a difference to the importance of closeness.

The Influence of 'Cultural' Distance

Cultural commonality between user and producer as a result of both a shared code of communication as well as a common legacy of industrial practices and institutions was investigated by asking user plants to gauge the degree of difficulty they have had in operating their advanced machinery. This was then related to user plant ownership and the geographic origin of the process technology.

The results are very revealing. In both Canadian and foreign-owned plants, only about 40 percent of users acknowledge some difficulty in operating complex machinery obtained from Canadian producers.

However, when the origin shifts to overseas (i.e. outside North America) machinery producers, the 60/40 easy/difficult split for Canadian users is almost reversed: now 57 percent of Canadian users report difficulties, and the share increase is especially notable in the 'very difficult' category (from 13 to almost 24 percent). However, foreign-owned users indicate only a modest increase in difficulty, with 57 percent of their experiences being rated as unproblematic. In other words, when the source of advanced machinery is overseas, foreign-owned plants are substantially less likely to encounter difficulty in operation than are Canadian-owned establishments.

Cultural commonality thus appears to exert an important influence on technology implementation for both large and small plants and single- and multiplant firms alike. When this area was probed in conversation with production managers and engineers, lack of common language was usually the first consideration mentioned. However, many also indicated that the problem was deeper than one of language. After all, most European and Japanese machine-producing firms know that their sales, engineering, and service personnel have to be reasonably proficient at speaking English if they wish to make a foreign sale.

More subtle forces at work became evident when Canadian user-plant personnel described the often difficult implementation processes they had experienced. Different business cultures, compared with European machine producers, necessitated establishing a relationship over time, with repeated personal interaction, in order to obtain the kind of product and service they were seeking. Even after persevering, many commented on an apparent chauvinism on the part of (especially northern) European machine builders, who were resistant to altering their designs to suit North American users' requests because they felt simply that 'they [the producers] knew best' and that North American users, if only they were more technologically sophisticated, would see that they were right. Implementation problems were thus common amongst Ontario plants where users have attempted to implement northern European technology while lacking the necessary 'cultural' affinity.

'Being There': Site Visits

Closeness between users and producers of advanced technologies allegedly permits greater ease of interaction and communication before, during, and after installation of machinery, especially through direct, face-to-face contact. Our interviews revealed that site visits were absolutely crucial for ultimate success when the technology being implemented was new, complex, and expensive. They were also seen as the most useful medium by which to deliver training to the user's technical and operative personnel.

Many users complained of the difficulties that arose when their contact with producers was strictly remote. In many cases, 'training' might be provided in the form of printed manuals, while in other cases, the producer might send the user a videotape. These were regarded as poor substitutes for 'being there'.

These observations raise an important question – namely, what kinds of plants do (and do not) receive site visits? And what kinds of producers are more likely to provide this service to users? From the survey data four major patterns are evident. First, while the probability of a user plant receiving a site visit is notably greater in the earlier stages of technology acquisition than in the latter stages, the declining frequency of site visit reports as one moves from installation to operation seems to be greatest when the machinery in question originates overseas (from nearly 80 percent during installation to roughly 64 percent during operation). In this sense, then, distance does appear to 'matter'.

Second, the likelihood of receiving a site visit from the machinery producer declines as the distance from the machinery source increases. Third, the attenuating influence of distance seems to be especially strong for smaller plants. Fourth, small plants are less likely to receive site visits than are large plants, no matter where the machinery producer is located.

Distance decay in the likelihood of receiving a site visit is especially marked in the case of Canadian plants (both single- and multiplant firms). Indeed, foreign-owned plants experience almost *no decline in the likelihood of receiving a site visit* as the distance from the source increases. This pattern may be one more manifestation of the 'cultural affinity' arguments made earlier, or may also provide evidence of intrafirm, though international, transfer of production technologies and assistance to foreign-owned plants located in Ontario.

The pattern suggest that plants that are small and Canadian-owned single-establishment operations are less likely to receive site visits from machinery producers. Furthermore, this discrepancy appears to widen as the source of machinery becomes more distant, and this 'distance decay' effect is especially marked during the operation phase of technology acquisition.

Vendor-Provided Training: Who Provides It, and Who Gets It?

While these patterns concerning site visits are quite revealing, one of the most important potential benefits of a site visit is the opportunity for the producer to provide explicit, hands-on training for the employees of the user plant.

An analysis of training reveals that Canadian single-plant establishments and small plants are the least likely to receive vendor-provided training. As for the vendors themselves, US producers are the least likely to provide training, although they do seem to serve the needs of the foreign-owned and larger plants better than the rest of their market. On the whole, it is the overseas vendors who appear to be serving their users' needs best, in terms of the proportion of users receiving training from them. This may reflect the recognized unwillingness of their North American customers to purchase complex machinery made overseas without training support.

However, many users were unhappy with the quality of the training support they have received from distant vendors – again, particularly from those located in Europe. And this dissatisfaction was particularly strongly felt among the smaller, independent Canadian-owned plants.

Further evidence of a possible 'cultural affinity' effect is that Canadian machinery producers seem to receive the highest overall satisfaction ratings for their training, particularly from Canadian-owned users. At the same time, foreign-owned users are almost twice as likely as Canadian users to claim satisfaction with training provided by overseas machinery producers. Interestingly, Canadian users also give a high satisfaction rating to vendor-provided training from US producers, suggesting that the commonality of language and work practices may facilitate the transmission of technical knowledge to engineers and shopfloor workers in the Canadian user plants. And, following the same pattern, foreign-owned users are notably less likely to express satisfaction with training provided by a Canadian machinery vendor than are Canadian user firms.

IMPLICATIONS FOR REGIONAL–INDUSTRIAL POLICY

We preceded our discussion of empirical findings by posing four research questions, to which we now return. First, our research appears to lend strong support to the idea that a high-quality (interaction-intensive) relationship between the user and producer of advanced capital goods is of major importance in facilitating implementation by the user. Second, 'closeness' between producer and user does indeed seem to facilitate the formation and maintenance of a high-quality, interaction-intensive relationship in which there is a more open flow of information. Where large distances intervened, problems frequently arose. Third, lack of 'closeness' appears to be especially onerous for smaller enterprises and for domestically owned single-plant establishments.

Fourth, even large and multiplant establishments indicate overwhelmingly that, given the choice, they would rather do business with a producer of advanced machinery located on the same continent. In fact, many users would prefer to have such producers located very close by. The research shows that 'closeness' is to be understood in an organizational and cultural sense, as well as in the more traditional physical sense of the term. Considerations here run from the relatively banal (problems with time zones, border crossings) to the interesting (communications problems related to language differences, technical complexity of subject matter) to more fundamental concerns (problems arising from contextual differences in workplace practices, training cultures, and conceptions of 'technology').

Of particular interest here are the findings concerning the difficulties arising from a mismatch in work cultures between machinery users and producers The research reported here suggests that what shows up in Canada or the United States as a 'training problem' may in part be symptomatic of a 'cultural' gulf between machinery users and the (overseas) firms producing their process technologies. Where advanced machinery is largely imported from countries whose industrial culture is based to a much greater extent upon a *socially* determined notion of technology (compared to the 'embodied' conception dominant in the Anglo-American economies), implementation problems may be expected. Of course, physical distance only serves to exacerbate implementation difficulties.

This suggests that a greater emphasis on training must be accompanied by a sea change in Anglo-American manufacturers' understanding of technology: they must acknowledge that productivity can best be generated from production processes in which skilled employees work closely with advanced machinery over extended periods of time. This amounts to espousing an employment relation based on stability, trust, and the free exchange of information between workers and managers. Only within the context of such a relationship can workers build up the kinds of skills and experience necessary to engage in effective learning by doing.

CONCLUSION: IMPLICATIONS FOR THEORETICAL DEBATES

A central issue in the ongoing debate within economic geography regarding the end of the Fordist era and the nature of its replacement concerns the extent to which 'flexible' production processes have already diffused throughout the industrial world, as well as their potential to become more widely adopted, particularly in more mature industrial regions. The machinery and systems in question here constitute the very heart of these flexible technologies.

First, the analysis presented here indicates that the spread of such production methods to the industrial 'diaspora' will be anything but unproblematic. Predictions of the 'diffusion' of flexible production methods to mature industrial regions may be premature and overly optimistic, underestimating the difficulties involved in making this transition. This 'diffusionist' perspective is guilty of conceiving of the problem of manufacturing renewal in mature industrial regions as one of insufficient rates of uptake (or 'adoption') of process technologies whose productive attributes are assumed to be wholly contained within a set of inanimate objects (Gertler, 1993). A more Marxian conception of capital as a social relationship [suggests that] difficulties (or what I have called 'implementation pathologies') will arise if the social relations surrounding the use of such technologies in their regions of origin are not substantially present in the region of implementation.

Second, the findings hold important implications for our understanding of the 'success stories' – the canonical industrial districts of Europe, Asia, California, and elsewhere. A common interpretation of the success of such regions emphasizes the beneficial effects of transaction cost reductions stemming from the physical proximity of interacting firms in these industrial agglomerations (Scott, 1988b). However, this paper's findings strongly suggest that physical proximity alone does not constitute a full explanation of the success of users (and producers) in such districts. Nor are common language, shared codes of communication, trust, or 'embeddedness' (Harrison, 1992), on their own, sufficient to explain the success with which users are able to implement advanced process technologies. Underemphasized in this literature is the importance of work practices and training cultures (and the broader regulatory and institutional framework) that are shared by both users and producers. In the absence of this commonality, it is difficult to imagine the advanced machinery users in these districts attaining the levels of success they have been alleged to enjoy.

Third, the similarity between the arguments made in this paper and those offered by Rosenberg (1976, 1982a,b) in his analysis of user–producer relations in British and American industrial clusters in earlier historical periods suggests that this thesis may have broader applicability to capitalist industrialization in general. Despite tremendous improvements in transportation and communications technologies, the quantum leap in complexity of production technologies since the early 1800s (in particular, the recent transition to *electronic control* of mechanical systems – what Florida (1991) has dubbed 'mechatronics') suggests that 'closeness' between user and producer is more important today. Given the increasingly global nature of production systems and corporate organization, there is more than a little irony in this insight.

Michael Storper

'The Resurgence of Regional Economies, Ten Years Later: The Region as a Nexus of Untraded Interdependencies'

***European Urban and Regional Studies* (1995)**

THE REDISCOVERY OF THE REGION AND ITS CRITICS: POST-FORDISM, DISTRICTS AND ALL THAT

Something funny happened in the early 1980s. The region was rediscovered by a group of heterodox political economists, sociologists, political scientists and geographers. These workers asserted that the region might be a fundamental basis of economic and social life 'after mass production' as the centre of 'post-Fordist', 'flexible', 'learning-based', production systems. This debate over regionalization in contemporary capitalism continues to generate fascinating propositions.

Three main 'schools' have participated in the debate: those interested in institutions; those focusing on industrial organization and transactions; and those who concentrate their attention on technological change and learning. I claim that there is good reason for including the region as an essential level of economic co-ordination in capitalism. But none of the main schools in the resurgence of regional economies in post-Fordism debate has come up with the correct formulation of why this is the case. The general, and necessary, role of the region is as the locus of what economists are beginning to call 'untraded interdependencies' between actors; these untraded interdependencies generate region-specific material and non-material assets in production. These assets are the central form of scarcity in contemporary capitalism, with its fantastic capacity for production of standardized outputs, essentially because they are not standardized. The region, in this analysis, is important as an underpinning for these interdependencies, which allow actors to generate technological and organizational change; hence, the region is a key source of *becoming* – of development – in capitalism.

INSTITUTIONS AND INDUSTRIAL DIVIDES AND THE ILL-FATED DEBATE OVER SMALL FIRMS

From the mid-1970s, Italian scholars called attention to the different development model which characterized the Northeast-Centre (NEC) of their country, dubbed the 'Third Italy' by Bagnasco (1977). Piore and Sabel (1984) were the first to capture this as a model of flexibility plus specialization. Generalizing from Italy to certain other cases (notably German), they then placed the success of such forms of production in macro-economic and historical context and postulated the possibility of an 'industrial divide' separating a putative era of flexible specialization from that of post-war mass production.

Their account was both empirically rich and theoretically powerful. Nonetheless a number of serious criticisms [have been made] of the flexible specialization–industrial districts thesis.

1. Production systems dominated by small firms, especially the exceptionally small firms of the Third Italy, are few and far between in this world; it is wrong to base the possibility of an industrial divide on small firm examples.
2. A model of an industrial divide needs to cover a wide sectoral composition, and the Italian and German examples concentrated on specialized industries.
3. There are deep historical roots to the Italian and German examples. If collaboration–competition was a particularity of the conventions of certain regions, what about other regions with (Anglo-American) competitive norms?
4. Although Piore and Sabel did assert clearly that there was a convergence between big and small firms, they overdefined their model in terms of small firm industrial systems.
5. [The Third Italy is] the extreme case of localization. Most competitively successful production systems do not approach that level of regional closure (Gordon, 1990).
6. The most important criticism is that the flexible specialization model did not define, in analytical terms, precisely what it was that distinguished a technologically dynamic regionally rooted system of firms from those systems of firms that did not share these characteristics, but still appeared to be flexible and specialized.

[Nonetheless], the fundamental contributions of the flexible specialization school consist of four points which remain unchallenged.

First, technologies and divisions of labour in production are not dictated by a movement towards a globally optimal, foreseeable 'best practice' for each sector. They are, rather, the outcomes of institutional pressures and choices made at critical points in the histories of products and their markets, and the direction of development is thus not necessarily towards greater scale and integration, but can be the reverse. Second, the flexible specialization school got something basically right in identifying flexibility and specialization as fundamental alternatives to mass production. Third, though the original examples of regionalism were much too pure, it seems clear that the dynamic forces in contemporary capitalist development are both localized and territorially specific. Fourth, appropriately institutionalized networks are essential to successful ongoing adaptation of a regional economy in the face of uncertainty (technological, market, and so on).

INDUSTRIAL ORGANIZATION, TRANSACTIONS, AGGLOMERATION: THE CALIFORNIAN SCHOOL OF EXTERNAL ECONOMIES

The 'Californian School' (Scott, 1988b; Scott and Storper, 1987) rooted flexibility in the division of labour in production, and linked that to agglomeration via an analysis of the transactions costs associated with inter-firm linkages. It assumed that certain market conditions gave rise to uncertainty which is met by deepening the division of labour. In turn, vertical disintegration of production raises the transactions costs of traded interdependencies (input–output relations). Agglomeration is an outcome of the minimization of these transactions costs. The external economies which attach to interdependent production systems are maximized by agglomeration, for without agglomeration, the advantages of interdependence – flexibility, risk minimization, specialization – are reduced.

This analysis seemed to have several advantages over the flexible specialization school. First, it did not seem to depend on thick and historical institutional contexts. Indeed, we argued that new industries emerging after technological branching points enjoy what we labelled 'windows of locational opportunity', in that they are not attached to old stocks of external economies. But once a group of firms begins to get ahead, the proliferation of external linkages gives them advantages which rapidly attract new entrants and hence, only a few major new agglomerations form in a given new industry such as Silicon Valley.

Second, older industries, analogous to the European cases, could be accounted for via the process of externalization and interlinkage of firms – the story of Hollywood, experiencing vertical disintegration and re-agglomeration – was a case in point. Indeed, we defined the model around three groups of sectors – high technology, revitalized craft production, and producer and financial services.

As the debate proceeded, however, a serious critique did emerge. The Californian school's analysis does apply to certain modernized craft- or traditional industries as well as certain labour-intensive manufacturing and service sectors: but in other industries, dense local input–output relations are not present in sufficient quantity to account for the existence of agglomeration. Examples [include] parts of high technology and mechanical engineering.

INNOVATION, HIGH TECHNOLOGY, AND REGIONAL DEVELOPMENT

From the late 1970s on, students of regional development investigated the regionally uneven distribution of high technology industries (Malecki, 1984; Breheny and McQuaid, 1988). The American School of high technology regional development sought the conditions for growth in Silicon Valley and Route 128 (Markusen *et al*, 1986). The work identified many different factors, the single most discussed [being] the research university–spin off process. Secondary factors [included] a 'high quality of life', good infrastructure, even climate. The problems come when [university links] is taken to be a universal logic of new technology based infant industry development. There are many research universities, but a much smaller number of Silicon Valleys.

A second branch of the American school is the 'regional politics' approach (Markusen *et al*, 1986). It holds that regional coalitions push for the transfer of high technology resources: thus, Silicon Valley got ahead partially because its early industrialists were clever enough to commandeer resources from the military–industrial complex. Here again are interesting observations, but they fall far short of a coherent theory.

An alternative, European, approach has been developed by the GREMI group (Groupement Européen des Milieux Innovateurs). Their central theoretical notion is that of the *milieu* (Aydalot, 1986; Aydalot and Keeble, 1988; Camagni, 1991). The milieu is essentially a context for development, which empowers and guides innovative agents to be able to innovate and to coordinate with other innovating agents. The milieu is something like a territorial version of what the American economic sociologist Mark Granovetter (1985) labelled the 'embeddedness' of social and economic processes. The milieu is described, variously, as a system of regional institutions, rules, and practices which lead to innovation and as a network of actors: producers, researchers, politicians, and so on, in a region.

[However] the GREMI group has never been able to identify the economic logic by which milieu fosters innovation. There is a circularity: innovation occurs because of a milieu, and a milieu is what exists in regions where there is innovation. Nonetheless, the GREMI group [rightly] argue [that] the economic process is fundamentally about creation of knowledge and resources, and the context for this is likely to have territorial boundaries and specificities.

TECHNOLOGY, PATH DEPENDENCY, AND UNTRADED INTERDEPENDENCIES

How then can we identify the intangible aspect of a regional economy that underlies innovative, flexible, agglomerations? The first insight came from evolutionary economics (Nelson and Winter, 1982), refined for technology by Dosi, Pavitt [and] Soete (Dosi *et al*, 1990; Arthur, 1989). They claimed that technologies develop along pathways or trajectories, which are characterized by strong irreversibilities. Technologies are the products of *interdependent* choices, and interdependency means uncertainty. There are also significant *technological spillovers* in the economy: knowing how to do one thing is frequently consequent upon knowing how to do another (Romer, 1990) such that technological excellence comes in packages or ensembles. Since such excellence relies frequently on knowledge or practices which are not fully codifiable, the particular firms who master it are tied into various kinds of networks with other firms, both through formal exchanges and through untraded interdependencies. The latter include labour markets, public institutions, and locally- or nationally-derived rules of action, customs, understandings, and values (Dosi and Orsenigo, 1985). The western world is now a vast 'learning economy'.

The argument can be summarized now, as the following:

1. Technological change is *path dependent*.
2. It is path dependent because it involves interdependencies between choices made over time – choices are sequenced in time, not simultaneous, and often irreversible.
3. These choices have a spatial dimension, which is closely tied to their temporal uncertainty and interdependence. Some inter-organizational dependencies within the division of labour, that is input–output or network relations, involve some degree of territorialization. But in all cases where organizations cluster together in territorial space in order to travel along a technological trajectory, they have interdependencies which are *untraded*, including labour markets, and 'conventions', or common languages and rules for developing, communicating and interpreting knowledge.

I began to use the evolutionary paradigm to explain the appearance of new high technology centres (Storper, 1985, 1986) [and] subsequently to try and explain the 'industrial divide' crossed in the film industry, as a path-dependent phenomenon (Storper, 1989). I more recently tried to show (Storper, 1992) that globalization of economic activity was linked to regionalization, through the mechanism of localized technological learning.

FROM TECHNOLOGY TO ACTION: THE REGION AS A NEXUS OF UNTRADED INTERDEPENDENCIES

The evolutionary school of technological change opened up the question of economic development as one of learning, or *becoming*, and of untraded interdependencies as a major feature of this process. But certain limitations remained. What defines the trajectory? And why limit trajectory to 'technology'? Surely the evolutionary properties of production systems involve more than hardware. Learning certainly concerns all dimensions of production: the design of products, processes, know-how, and the evolution of organizational skills

But perhaps even this does not go far enough. There are other dimensions of the evolution of production systems which cannot easily be captured via the word 'technology'. All production systems involve uncertainty. The main way that such uncertainty is resolved is through *conventions*, which are taken-for-granted rules and routines between the partners in different kinds of relations defined by uncertainty. The conventions that producers attempt so as to cope successfully with uncertainty constitute 'frameworks of economic action', and different frameworks are possible for different kinds of products, markets, and kinds of labour: these are 'possible worlds of production', defined by the ensemble of their conventions. Such conventions are always different for fundamentally different kinds of products, but they are also often different from region to region and especially from nation to nation. Some such 'worlds of production' are more successful than others in competition, of course. But the evolution of these production systems is strongly dependent on the worlds of production, rooted in their underlying conventions, constructed in given times and places, and to which producers, workers, and consumers are subject.

This definition of the problem permits us to analyse more of the regional economies identified by recent scholarship than any of the preceding perspectives on regionalism and post-Fordism, and to do so in a dynamic manner. We may say that, in technological or organizational spaces, there are moments when economic assets are more general and others when they are more specific. For example, an old, well-developed, highly codifiable, and standarized technology rests on both knowledge and physical inputs that are widely diffused, but highly specific. The elaboration of the technology over time, its differentiation into many different products and using more and more differentiated inputs, makes the assets of the industry often highly specific to its firms and products. In the late 1940s and early 1950s, product and process technologies were quite specific for vacuum-tube based electronic products in what was then an old technology. But then miniaturization came along to replace the vacuum tube, first via the transistor, and then via the silicon-based semiconductor. When new, these technologies had few established inputs that were specific to them; they had to 'invent' their own input chains, and the knowledge going into them, which had not yet become highly applications-specific. In other words, they began as generic assets which, over time, evolved into more specific assets. This is one reason why the semiconductor industry was not attached, geographically, to its parent industry – radio and television equipment – on the East Coast, and instead found its centre in Silicon Valley. It had to reinvent its own input chain thus creating assets specific to the emerging technological space. It also had to convert generic electronic engineers into labour which had skills specific to semiconductors. [And] technological spillovers created cognate fields of knowledge.

In three ways – the labour market, the input-output system, and the knowledge system – there is a process of becoming specific.

But these three levels are, in my view, not the deepest levels at which this analysis can be formulated. All three, I would propose, evolve from generality to specificity on a foundation of conventions which make possible communication, interpretation, and co-ordination among the actors who are making them become specific. In other words, frameworks of action which govern the production system evolve from a position of generality to specificity, as they are sedimented into conventions, practised by actors, and sometimes embodied in formal institutions and rules. They are also highly regionalized. Certain dimensions of this culture have diffused, globally; but there is only one Silicon Valley if one wants to be 'in the know' for the most advanced innovations in semiconductor technology.

We know that input–output relations alone have costs associated with uncertainty in infant industries that are high enough to bring agglomerations into existence in the early days of existence of many industries. Such geographical input–output constraints [often however] disappear rapidly in many industries, including high technology industries, as inputs become more standardized and are produced at higher output levels. [Yet] in many cases, less decentralization (often associated today with 'globalization') takes place than would seem possible. Silicon Valley, as an agglomeration, show[s] essentially no sign of weakening even to this day. Might this be because the *geographically-constrained untraded interdependencies outlive geographically-constrained input–output linkages*? And could this especially be the case when there is [*sic*] high levels of technological or organizational learning? Here, the rules of action that permit participants in the production system to develop, communicate and interpret information [and] knowledge may have greater geographical constraints than the process of trading inputs and outputs. The conventions which underlie the collective activity known as learning are a principal form of untraded interdependency, along with labour markets. Thus, regional economies constitute nexuses of untraded interdependencies which become specific but public assets of production communities, and which underpin the production and reproduction of other specific assets such as labour and hardware.

I propose, then, that an answer to the principal dilemma of contemporary economic geography – the resurgence of regional economies, and of territorial specialization in an age of increasing ease in transportation and communication of inputs and outputs – must be sought in two sorts of analysis. One is the tension between respecialization and destandardization of inputs and outputs which, all things being equal, raises transactions costs. The other is the association of organizational and technological learning with agglomeration, which in turn has two roots. The first, and more limited case, is that of localized input–output relations, which constitute webs of user–producer relations essential to information development and exchange, and hence to learning (Russo, 1985; von Hippel, 1987). The other, and more general case, is the untraded interdependencies which attach to the process of economic and organizational learning and coordination. Where these I–O [input–output] relations or untraded interdependencies are localized, [which] is quite frequent in cases of technological or organizational dynamism, then we can say that the region is a key, necessary element in the 'supply architecture' for learning and innovation. It can now be seen that theoretical predictions that globalization means the end to economies of proximity have been exaggerated by many analysts because they have deduced them only from I–O analysis.

A few additional observations are:

1. These nexuses of interdependencies are not territorially concentrated in all industries or at all times. Agglomeration is a part of the supply architecture for innovation and efficiency in only some cases. But because scholarship has concentrated so much on traded interdependencies, input–output relations, we have little systematic knowledge of the geography of untraded interdependencies and its relationship to economic development and especially, to organizational and technological learning and competition.
2. Untraded interdependencies, whether territorially concentrated or not, are not static. Again, we know very little about such geographical dynamics.
3. The geography of an industry is not determined by either its input–output relations or its untraded interdependencies. I am not speaking here of political forces and all that, which obviously do affect economic geography in important ways. Perhaps there are degrees of freedom of these spheres. Schwartz and Romo recently have proposed that the American automobile industry pursued a logic of taking advantage of the possibility of geographically extending its input–output relations – due to the technological stabilization of the mass production paradigm – and so dispersed much of its production throughout North America and the world (Schwartz and Romo, 1993). But in so doing, managers may have weakened some of the crucial untraded interdependencies – what Schwartz and Romo call the 'regional production culture' – which would have been critical to a successful response by Detroit to the Japanese challenge and to the exhaustion of their own mass production model's productivity gains. This suggests that we need to know much more about when territorialized untraded interdependencies constitute real constraints on geographical behaviour, and when they are necessary to innovative behaviour but do not constrain locational behaviour. In the latter case, the failure to 'respect' such interdependencies could lead to the situation theorized by Schwartz and Romo (1993) for the North American car industry: loss of innovative capacity.
4. Untraded interdependencies, especially in the form of conventions, not only potentiate collective action, adjustment and

learning, but may impede it. Neo-institutionalist scholars call this the problem of institutional 'sclerosis' (North, 1981). Much more needs to be known about the kinds of conventions which construct efficient and dynamic 'real worlds of production' and those which favour technological or institutional lock-in.

CONCLUSION: IMPLICATIONS FOR THEORY

The purpose of this exercise has been to identify, in theoretical terms, why regions keep emerging as centres for new rounds of growth. The logic identified here is intended to apply to the recent re-emergence of regional economies as centres of technological and economic spaces, in spite of the largely unequivocal historical trend to transcend the limitations of raw distance in the transport and communication of goods, people, capital, and information. We have made a proposition that the region has a central theoretical status in the process of capitalist development which must be located in its untraded interdependencies.

Th[is] assertion means, simply, that these interdependencies are necessary to capitalist development and that they are, under certain conditions, necessarily regionalized. But this does not mean that there are not other reasons for regional economies to exist or to grow; politics, for example, may decide which regions grow. Established regions, especially those of a certain size, have strong auto-reproducing Keynesian dynamics which have little to do with the logic elaborated here, for example. And the dynamics of a global division of labour and the national or global shifting of productive activity continue, and require analytical tools other than those outlined in this paper. Thus, the proposition made here is not intended to stand in for a complete theory of regional development. But these other reasons, however empirically and politically interesting, do not tell us whether there is a necessary role for regions in the economic process in capitalism.

The task of researching untraded interdependencies as the basis of the ongoing resurgence of regional economies, patterns of regional growth, regional differentiation in development, trade, and technology accumulation, is an enormous and exciting multidisciplinary project.

Section 3.4 New Spaces of Production

Kuniko Fujita and Richard Child Hill

'Global Toyotaism and Local Development'

International Journal of Urban and Regional Research (1995)

Current complexities in the international division of labour reflect the interaction and interpenetration of two quite different transnational production systems: global fordism, most closely approximated by American postwar practice, and global toyotaism, typified by the post-1960s expansion of Japanese manufacturing.[1]

GLOBAL FORDISM

American practice inspired world manufacturing after the second world war. US companies hired legions of semiskilled workers to mass produce standardized goods with product-specific machines. Corporate marketing strategies created mass demand. With a high and predictable rate of profit, US firms invested in labour-saving technology and built a network of large assembly and parts factories exploiting economies of scale (Piore and Sabel, 1984).

As the American corporation extended its territorial reach, it extended the division between those who conceptualize and those who execute production, and between skilled and unskilled work, first within national boundaries, and then, with advances in transportation and communication, across different workforces in different countries. Conceptualization and skilled labour were concentrated in the advanced industrial countries, a mix of skilled and unskilled in the NIEs, and mainly unskilled in the least industrialized nations (Lipietz, 1987). [At the urban scale] corporations dispersed manufacturing to branch plant cities in response to the pull of labour power, markets and raw materials. General corporate offices were lodged in national and world cities (Hymer, 1971).

US TNCs initially set up production behind third world tariff walls to protect future market share. They could do this despite higher costs in less developed countries because they were protected from external competition. When world economic growth slowed in the 1970s, many TNCs extended production relations abroad to cut labour costs. In theory, the technological advantages US firms held over foreign competitors should have protected the jobs of American workers. In fact, however, the movement of US branch plants to less developed nations boomeranged and inflicted heavy economic damage on America's industrial cities (Bluestone and Harrison, 1982).

THE NEW GLOBAL FORDISM

The 1980s severely taxed the fordist mass production model. Technological discoveries in microelectronics and information processing altered the way work was performed. The fragmentation of world markets challenged the utility of scale economies, long product life-cycles and global sourcing. US TNCs confronted competitors who were innovating and reducing the time required to bring new products to the market.

American companies learned from the competition and tried to combine global integration with local responsiveness to more flexibly manage worldwide activities (Bartlett and Ghoshal, 1989; Ohmae, 1985). They organized multinational design teams and entered joint ventures to master the idiosyncracies of local markets. But American TNCs continued to stress economies of scale, to seek the lowest price for quality suppliers wherever in the world they happened to be, and to locate final assembly close to customers and parts manufacturing in outlying, lower-cost sites (McGrath and Hoole, 1992).

GLOBAL TOYOTAISM

The volatility of world markets in the 1980s favoured a manufacturing strategy – flexible specialization – that in many ways was the obverse of mass production (Reich, 1983; Piore and Sabel, 1984). In a flexibly specialized production system, skilled and adaptable workers produce a diverse and changing array of semi-customized goods with general-purpose equipment. Managers and workers, parent companies and suppliers, must cooperate to master the learning and investment challenges of rapidly shifting markets. Competitive advantage is sought less in lowest-cost production than in superior product design, continuous improvement in methods, products and labour processes, and shortened product cycles (Sabel, 1989).

Most analysts use the term 'flexible specialization' to refer to network systems of production among small producers (Hatch, 1988). Flexible specialization in Japan, however, is more often found in collaborative relations among large parent companies and smaller supplier firms as typified in the Toyota production system. Fordist mass production methods lowered costs by minimizing product diversity and maximizing economies of scale. Toyotaism, on the other hand, achieves efficiencies through flexible specialization, spatial agglomeration and just-in-time (JIT) delivery logistics. Combining the advantages of craft flexibility with the most advanced information technology requires considerable cooperation among managers, workers and suppliers. Toyotaism achieves cooperation through enterprise groups (Gerlach, 1989), lifetime employment for core employees, promotion of union officials to managerial positions, [and] worker participation schemes (Cusumano, 1985: Chap. 5).

The higher productivity, product diversity and quality generated by Japanese flexible manufacturing methods gradually translated into global competitive power. Computer-based machines reduced the need for manufacturing scale economies, allowing firms to shorten product life-cycles. Consumers experienced, then came to demand, more product variety. Markets fragmented, diminishing the effectiveness of mass production. By the early 1980s, Japanese market penetration had thoroughly shaken western assumptions about production organization.

Toyotaism integrates conception and execution in space as well as in the workplace, thereby fostering industrial districts rather than differentiated administrative and branch plant economies.[2] A premium is placed upon close working relations among people at all stages of the production process. Companies try to place their research and development and state-of-the-art production facilities as close as they can to their central offices (Obayashi, 1993).

Product diversity, short product cycles and JIT delivery in small batches also encourages close, cooperative contact among parent firms and subcontractors (Shimogawa, 1986). Suppliers must be expert in design and production technology if they are to deliver sophisticated parts for rapidly changing products just-in-time to the assembly line. This privileges tight social and spatial relations among assemblers and parts firms. Instead of bargaining with suppliers and customers at arm's length, parent companies continuously consult with them, facilitating the flow of advice, suggestions and all kinds of information. Companies organized into Toyota-style industrial districts possess the capacity to respond to economic problems and market uncertainties by continuously reshaping productive processes through the rearrangement of component activities (Womack *et al*, 1990).

Japanese industrialization depended upon government intervention to augment supply and demand. Industrial policy, the promotion of industrial structures that would enhance Japan's international competitive power, took priority. Japan's developmental state exists at the local as well as the national level. For example, Toyota City and Aichi Prefecture played a key part in the social and spatial development of Toyota's production system by providing transportation networks and other infrastructure, annexing territory, writing down land costs, abating taxes, building industrial parks, providing industrial and employment services, researching manufacturing trends and formulating regional plans (Fujita and Hill, 1993).

Japanese firms are reluctant multinationals due to the character of their production systems. From the Japanese perspective, overseas manufacture reduces competitiveness, efficiency and quality while raising costs. Close links among parent makers, a skilled and flexible workforce and high quality parts firms are difficult to reproduce outside of Japan.

During the 1970s, Japanese firms' global strategy was based on three principles: high productivity, domestically developed technology and expanded production through exports.

But as Japan's market strength grew during the 1980s, so did her trade surplus. Japan's trading partners reacted with protective measures, [while] the yen doubled in value between 1985 and 1988. Japanese production and labour costs shot up commensurately. Avoiding trade restrictions abroad, reducing costs at home, and taking advantage of a powerfully enhanced yen combined into an irresistible rationale for direct foreign investment [DFI].

The surge in Japanese foreign investment during the late 1980s indicated a shift in Japanese TNC response from cost reduction to a broader strategy of global localization. Global localization means that manufacturing is governed by a global outlook, but attention is also paid to growth in the host economy (Morita, 1992).

Global localization is a response, in the first instance, to political pressures from Japan's trading partners. As local assembly expanded, host governments demanded local production for domestic markets, technology transfer, research and development sharing, and improved coordination between Japanese manufacturers and indigenous firms. The Japanese accelerated direct investment in North America, western Europe and Southeast Asia to maintain market share and blunt trade frictions.

Japanese industrial corporations are now engaged in two global–local investment strategies: transplants in North America and western Europe and a complementary, regional division of labour in Southeast Asia.

TRANSPLANTS

Japanese transplants are transferring the organizational framework for flexible specialization to host localities. All things being equal, the same production system principles that made Japanese firms reluctant multinationals – capital-intensive production, collaboration among firms in product design, quality control, flexible production and JIT delivery, and spatial concentration in production complexes – predispose Japanese companies to transplant their spatially integrated production systems abroad once exports give way to DFI (Emmott, 1992).[3]

As Japanese TNCs transplanted manufacturing abroad during the 1980s, host-nation interest in Japan's developmental state grew. Japanese transplants in automobiles, electronics, steel, rubber and machine tools were often linked to revitalization strategies mounted by localities in declining industrial regions (Kenney and Florida, 1993; Garrahan and Stewart, 1992; Jones and North, 1991; Morris, 1991).

Japanese TNCs are in fact transferring the technological, social and organizational infrastructure of flexible specialization, forming industrial networks with domestic suppliers, and engaging in new relationships with American businesses and trade unions, city and state governments, universities and vocational schools, civic and community organizations (Hill *et al*, 1989).[4] American suppliers, unions and politicians are pressuring the Japanese to locate research and development centres in the United States. In the automobile industry, Toyota and Honda now have design centres in Japan, Germany and the United States. Toyota's new Avalon model, to be assembled at its Kentucky manufacturing complex, will be the first instance of a Japanese automobile maker carrying out the entire process from developing to marketing a car in the USA (Kato, 1993).

A COMPLEMENTARY REGIONAL DIVISION OF LABOUR IN SOUTHEAST ASIA

Japanese TNCs are also localizing production in Southeast Asia. The flow of Japanese capital to east Asian production sites conformed to the global fordism, NIDL model for a time. Japanese TNCs fashioned a vertical division of labour to reduce production costs by tapping overseas reserves of cheap labour (Fujita and Hill, 1989). But they are now constructing a horizontal as well as a vertical division of labour between Japan and less developed east Asian nations (Doner, 1991; Hill and Lee, 1994).

Asian governments demanded local production while the late-industrializing Japanese do appear to be more willing than western capitalists to transfer appropriate technology and operate within the framework of developing country industrial policies. The Japanese government has also been encouraging firms to manufacture in Southeast Asia.

As Japanese firms develop new technology, they are transferring mature technologies to the Asian NIEs and the ASEAN countries.[5] As Japanese TNCs turn over market entry production niches to Asian NIE companies, ASEAN producers can supply parts for the more standardized products being made in the Asian NIEs. This division of labour is promoting industrialization in the region.

Small national markets and weak technological capabilities make the Southeast Asia region as a whole the attractive site for TNCs, not any one individual country. Japanese TNC affiliates, many in joint ventures with local firms, are producing parts in each ASEAN country and then exporting them to other association countries for local assembly. This strategy allows for plant specialization and regional economies of scale, and furthers economic integration among the ASEAN nations. Taken together, Japanese state and TNC strategies and the industrial policies of developing countries are promoting regional integration in east Asia involving a multilayered division of labour based upon vertical and horizontal specialization and deepening mutual interdependence.

We think differences in industrial development between east Asian nations and those in other regions [in part] derive from global toyotaism; that is, a different pattern of TNC and state relations among Japan, the Asian NIEs and the ASEAN Four in contrast to those characterizing past US and European relations with Latin America and Africa. Japanese operations in Southeast Asia depart from the hierarchical axial communication pattern associated with global fordism and conform more closely to a polycentric system, with many local development centres each communicating with every other.

THE IMPACT OF GLOBAL TOYOTAISM ON JAPAN

In contrast to western capitalist nations, Japanese manufacturing investment abroad has been accompanied by a 'conscious and concerted policy of upgrading the industrial and technological structure of Japan's economy' (Morris-Suzuki, 1991, p. 146).

State policies encourage the transfer of more routine production to Southeast Asia and the retention of higher-value manufacturing in Japan. Japanese industrialists also consciously link their foreign investments abroad to the production of higher-value, more specialized products at home (Kojima, 1977). Because new employment in expanding sectors has steadily replaced job loss in contracting ones, Japan's high-technology supplier relationship with the Asian NIEs and ASEAN Four has garnered labour's support.[6]

Adjustment policies notwithstanding, the Japanese are now questioning some paradigmatic toyotaist principles. The short product cycle is criticized for engendering too intense a workplace and the bankruptcy of suppliers. The JIT delivery system may damage the environment through traffic congestion and air pollution. A domestic labour shortage is leading Japanese companies to experiment with new labour practices [and] to locate new assembly plants far outside their home districts[7] to secure a new workforce.

These product cycle, workplace, logistical and geographical experiments all indicate modifications in Japan's flexible production system, but none seem so fundamentally at odds with the principles of toyotaism as to call the system itself into question.

CONCLUSION

The current internationalization of the Japanese production system (global toyotaism) carries different implications for urban and regional development from the past internationalization of the American system (global fordism). Three trajectories of global toyotaism – transplants in North America and western Europe, a complementary regional division of labour in Southeast Asia, and flexibly specialized industrial districts in Japan – all indicate that global toyotaism, in comparison to global fordism, localizes more of the production process and therefore seems more conducive to local development in host nations.

Japanese transplants in North America and western Europe are designing and engineering new product models for local markets as well as transferring advanced process technologies. Japanese subsidiaries in Southeast Asia are engaging in regionally integrated production in conjunction with host-nation industrial strategies and Japanese government assistance. Manufacturers in Japan are networking automation equipment in industrial districts to produce high-value-added products in small lots and expending vast sums on research to strengthen technological capabilities (Buderi, 1992).

NOTES

1. In regulation theory, fordism refers to a regime of capital accumulation and its reproduction through the stable allocation of resources between consumption and accumulation. A regime of accumulation requires a mode of regulation – institutions, organizations and value commitments – that leads individuals to behave in accord with the scheme of reproduction (Lipietz, 1986). Transnational production systems are not governed by full fledged, global regimes of accumulation, but they are subject to regulatory mechanisms since their organization is conditioned by the state (e.g. foreign economic policy) and social (e.g. management and labour relations) practices of home and host societies. Economic, political and social relations are all implicated in global competition and interpenetration among production systems. This is the sense in which we use the terms 'global fordism' and 'global toyotaism' in this paper.
2. Researchers have identified three types of industrial district in Japan: (1) the classical marshallian type in which a myriad of small firms intertwine in locally based manufacturing networks; (2) corporate castle towns consisting of hierarchical, closely integrated, industrial networks of large parent companies and numerous medium- and small-sized supplier firms; and (3) technopolises in which R&D-oriented 'mother' plants and small pilot firms

agglomerate to incubate new industries (Hill and Fujita, 1993, pp. 287–8).

3. Please note that we are referring to a system governed not by the pure working out of a logic but by tendencies and compromises. For example, all things being equal, Toyota Motor Corporation would prefer to organize in concentrated production complexes as it has in Toyota City, and as it is now doing in Kentucky. But the company must adapt its production system to the social circumstances of each production location. Toyota's new assembly plant in Derbyshire, England, for example, is drawing upon suppliers located in western Europe to counteract western European fears that Toyota is using England as a Trojan Horse to penetrate the EC market. Toyota can organize its production system in this way so long as its competitors in the EC market are similarly dispersed; that is, so long as Toyota does not face competition from a Toyota-style production system in the European market.
4. For example, Japanese automobile transplants and their suppliers have concentrated along the interstate freeway, I-75, which runs from Ontario to Florida. Some now call I-75 the 'Sushi Line'. The Japanese have followed a path already taken by America's Big Three automobile manufacturers and parts suppliers who located new production facilities in lower-wage southeastern 'right-to-work' states. Research suggests that Japanese transplants are forging new social relations with local and state governments, universities and schools, civic and community institutions all along the Sushi Line (Chapman and McCullough, 1992).
5. The Asian NIEs include South Korea, Taiwan, Hong Kong and Singapore. The ASEAN Four comprises Malaysia, Thailand, the Philippines and Indonesia.
6. MITI also manages redeployment in sunset industries. To facilitate domestic structural adjustment the Japanese state is making enormous investments in new communication and transportation systems, including optical fibre telecommunication networks, magnetic levitated trains and high speed cargo transport (Rimmer, 1993).
7. Such as Nagoya, Osaka and Tokyo.

Richard Florida

'Regional Creative Destruction: Production Organization, Globalization, and the Economic Transformation of the Midwest'

Economic Geography (1996)

> Welcome to the new Midwest. Hammered by foreign competition during the 1980s, and left for dead only five years ago, America's heartland is booming. (America's Heartland: The Midwest's Role in the Global Economy, 1994)

Recent contributions to economic geography have focused attention on the changing organization of production in advanced industrial economies (Piore and Sabel, 1984; Kenney and Florida, 1993) and the spatial implications of such transformations (Storper and Walker, 1989; **Sayer and Walker**, 1992). According to an increasingly influential line of thought, particularly in economic geography, such changes have registered themselves in a new economic landscape, as traditional mass-production industries (Hounshell, 1984) and mass-production regions are supplanted by *new industrial spaces* composed mainly of flexible small-firm networks (Piore and Sabel, 1984; Scott, 1988b; Saxenian, 1994). These new industrial spaces are located at considerable geographic remove from the traditional regional manufacturing cores, which are seen as geographic repositories of an outmoded *Fordist* model of mass production. This perspective assumes that new forms of production organization are the province of newly emerging regions, while traditional manufacturing regions are essentially trapped in older, outmoded forms of production organization. The notion that economic change occurs through and is reflected in regional shifts is deeply embedded in economic geography, from notions of the product cycle (Vernon, 1966; Markusen, 1985) to the spatial division of labor (Froebel *et al*, 1980; Massey, 1984), as firms seek out less costly and less restrictive locations.

In this article, I question this view and suggest that the process of economic transformation need not be confined to new regions. Older regions can also experience a process of *regional economic transformation*. There is thus a geographic element to the strong transformative forces which Schumpeter identified as *gales of creative destruction*, which 'incessantly revolutionizes the economic structure *from within*, incessantly destroying the old one, incessantly creating a new one' (Schumpeter, 1942, p. 83).

Creative destruction involves both the creation of new industries and, just as importantly, the underlying transformation of existing industries by deep changes in technology and production organization. Creative destruction extends to regions as well as new technologies and new forms of production organization not only register themselves in new regions, but inform and shape the reconstitution and revitalization of existing regions. This emphasis on transformations in the underlying production systems and organizational fabric of regions is of special relevance today, when a large body of contemporary theorizing emphasizes the institutional rigidities (Olson, 1982) and so-called *lock-in* effects (Arthur, 1988, 1990) that constrain and limit the process of regional change.

To explore this issue, this article examines the economic transformation of the Industrial Midwest. Stretching from Buffalo, New York, and Pittsburgh, Pennsylvania, through Ohio, Indiana, Illinois, Michigan, and into Minnesota and Wisconsin, this great industrial belt developed as an integrated industrial complex producing huge quantities of steel, automobiles, machine tools, and later consumer electronic products during the late nineteenth and early twentieth centuries (Meyer, 1989; Page and Walker, 1991). Furthermore, the Midwest is typically portrayed as the paradigmatic example of a declining mass-production industrial region (see Scott, 1988b). [Various writers have argued] that the Industrial Midwest would face long-term, secular, and chronic disinvestment and deindustrialization (Bluestone and Harrison, 1982) as a result of its own internal organizational rigidities and *institutional sclerosis* (Olson, 1982), a shift of traditional industries to low-wage locations (Crandall, 1993), the rise of a postindustrial service economy (Bell, 1973), and the emergence of new industrial spaces elsewhere (Scott, 1988b).

In contrast to this view, the central argument advanced here is informed by three key points. First, [notwithstanding] the decline of manufacturing in the Midwest during the 1970s and 1980s, the industrial economy of the Midwest has recently enjoyed a pronounced revitalization, with substantial improvement in key indicators of economic performance, output, manufacturing investment, and productivity since the mid-1980s (Council of Great Lakes Governors, 1994).

Second, in contrast to the prevailing view that new forms of production organization tend to arise in new regions, new forms of production organization are being adopted in and are thus transforming the traditional Midwestern core. The Midwest industrial base is engaged in a shift from mass production to a new model of production organization characterized by a cluster of organizational techniques (e.g. the use of work teams, continuous improvement, the integration of suppliers into the product development process, and other organizational factors), which function collectively to harness intellectual and physical resources at all levels of the firm as well as the broader production system.

Third, the adoption of new forms of production organization has been accelerated by the global integration of the Midwest economy, particularly through increased foreign direct manufacturing investment. Mounting foreign competition has encouraged domestic manufacturers, particularly larger firms, to pursue new forms of production organization to increase their performance and competitiveness in global markets. The establishment of transplant manufacturing facilities by leading foreign manufacturing companies has resulted in the transfer of new manufacturing technology and production organization to key Midwest locations. The process of economic transformation is underpinned by a strong relationship between globalization (via foreign direct investment) and new production organization (see Figure 1). Foreign direct investment increases competition, forcing all producers to improve their performance, and facilitates the transfer of new production organization. Knowledge of new forms of production organization spreads through the regional economy via imitation as regional manufacturers become suppliers to transplants, through joint ventures between transplants and local firms, through the regular flow of information between transplants and their local suppliers and clients, and through the normal rotation of personnel. The diffusion of new forms of production organization, spurred by transplants, then conditions productivity improvement across the regional manufacturing base, setting in motion a virtuous cycle of imitation, adaptation, and improvement.

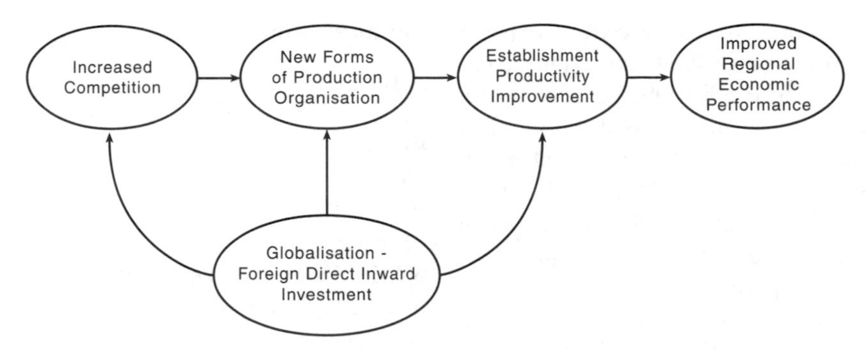

Figure 1 Relationship between new production organization, foreign direct investment, and economic performance

This article is concerned primarily with the underlying process of economic transformation, or regional creative destruction, as an organizational and spatial process. I seek to shed light on the processes by which older regions, in this case the Industrial Midwest, adopt new forms of production organization and thus overcome the institutional rigidities and lock-in effects that the literature suggests are almost completely binding and contribute to long-term economic decay. Thus, the primary focus is on the adoption and diffusion of new forms of industrial and production organization as an organizational and spatial process. I am secondarily concerned with the broader process of regional economic recovery, for two reasons. First, the literature suggests that the tendency for older regions to become locked into older forms of production organization is a key factor in the economic decay of those regions. Second, the Industrial Midwest has been inaccurately portrayed as a region undergoing long-term, secular, and irreversible economic decline, based to a large degree on its outmoded Fordist organizational configuration and inability to inculcate new and more advanced modes of production organization.

RESEARCH DESIGN AND METHODOLOGY

The research effort combined an analysis of existing published secondary source data with that of primary data from field research, personal interviews, and survey research. Site visits and extensive interviews were conducted with government, business, labor, and academic experts in selected Midwest cities and metropolitan areas. The research then collected information on the role of larger, *hub* manufacturing establishments in the process of economic transformation. Access was obtained to 12 US-owned manufacturing establishments as sites for intensive field research, involving more than 100 interviews with plant managers, human resources managers, purchasing officials, factory workers, and union representatives. A mailed survey was used to collect data on the adoption and diffusion of new forms of production organization by small and medium size manufacturers [using] a stratified random sample. Surveys were mailed to 1,933 firms and generated 193 useable responses, for a response rate of 10 percent.

ECONOMIC PERFORMANCE OF THE MIDWEST

Over the past decade or so in contrast to widespread predictions of decline, economic conditions in the Midwest have improved considerably. In 1989, the industrial Midwest produced more than $235 billion in manufacturing output, roughly one-quarter of the national total, and more than $800 billion in total output, roughly one-fifth of the national total. The region posted a 15 percent increase in output between 1987 and 1989, and expanded at a 4.9 percent annual rate in 1993 compared to a 3 percent rate for the nation as a whole. The region's manufacturing output increased by 16 percent between 1982 and 1987, after declining by more than 25 percent in real terms between 1977 and 1982. In 1982, the region's unemployment rate was 12 percent, significantly higher than the national rate of 9.7 percent. A decade later, in 1992, the region's unemployment rate was 6.6 percent, better than the national rate of 7.4 percent.

The Industrial Midwest has [also] made considerable gains in productivity. Between 1986 and 1988, productivity in the eight Great Lakes states increased by roughly 15 percent – 6 percent faster than Japan (9 percent), and considerably better than the United States (–1 percent) and Germany (–2 percent).

A central argument of the deindustrialization thesis was that older industrial regions were experiencing 'widespread and systematic disinvestment' and a shift in capital away from productive investment in plant and equipment (Bluestone and Harrison, 1982). The recovery of Midwest industry has been driven in part, however, by substantial real increases in investment in plant and equipment. Capital expenditures in the Midwest grew at a rate of nearly 12 percent in real terms from 1982 to 1987, outpacing the national trend, which registered a 10.4 percent decline.

EXPLANATIONS FOR THE MIDWEST'S INDUSTRIAL RECOVERY

There are a number of potential explanations for the economic recovery of the Industrial Midwest: (1) a shift from traditional manufacturing sectors to new high-technology industries and services, (2) productivity improvement from employment reductions and declining wage rates, (3) increased trade and exports to other nations, and (4) the transformation of the region's production system.

Sectoral Shift

Although the Midwest has certainly generated both service and high-technology employment, the region's industrial base remains firmly anchored in manufacturing. Manufacturing continues to comprise a greater share of the Midwest economy (25 percent) than for the nation as a whole (18.7 percent). Furthermore, the Midwest has the highest level of manufacturing output per capita of any region in the country. Between 1977 and 1989, the Midwest's share of the nation's manufacturing jobs declined only slightly, from 27 to 24 percent. More than one-third of the Midwest's manufacturing employment remains concentrated in heavy manufacturing sectors such as primary and fabricated metals, industrial machinery, and transportation equipment.

Downsizing and Wage Reduction

Although many Midwestern manufacturers have reduced the size of their operations and manufacturing employment has declined, the combination of downsizing and layoffs does not provide a full explanation for the region's economic transformation. The Midwest experienced a 15 percent decline in manufacturing employment between 1977 and 1982. Employment losses related to manufacturing have subsided since then, however, with manufacturing employment registering 2 percent growth from 1982 to 1987. Furthermore, in 1990, the average annual manufacturing wage in the eight Great Lakes states was $30,671, roughly 6 percent higher than the national average. The Midwest also retained relatively high levels of unionization in manufacturing. While clearly a part of the picture, the downsizing argument fails to account for the full scope of the region's substantial productivity gains and economic improvement.

International Trade Performance

Although the Midwest has certainly reversed its long history of producing almost exclusively for domestic markets, it is doubtful that exports and trade alone account for the region's considerably improved economic performance. In 1991, the six core Midwest states exported more than $65 billion in goods, 19 percent of the national total. From 1989 to 1990, the value of export trade for the Industrial Midwest grew by 13.4 percent, twice as fast as the national rate of 6.7 percent. The Midwest's level of exports as a share of gross regional product is slightly better than that for the United States, but significantly less than Japan, Germany, France, and Canada [while] the majority of the region's small and medium-size firms continue to produce mainly for regional and domestic markets. Overall, 40 percent of survey respondents reported that the Midwest region was their key market. The region's industrial base is heavily comprised of suppliers. In short, the region's manufacturing base is just beginning to make the transition from being regionally and domestically focused to a greater level of integration into the global economy.

The Industrial Midwest has thus experienced significant economic recovery. Furthermore, while existing explanations that emphasize the role of changes in industrial structure, corporate downsizing, and wage reductions, or increased trade and export performance, capture elements of the region's turnaround, they fail to provide a complete explanation for the region's industrial revival. What other factors may help to complete the explanation for the economic transformation and recovery of the Industrial Midwest?

SHIFT TO NEW FORMS OF PRODUCTION ORGANIZATION

The literature tends to view older industrial regions as having considerable organizational and institutional rigidity and as essentially being locked into older, outmoded forms of production organization, and hence virtually unable to make the transition to new forms of production organization. In contrast to this view, the argument advanced here is that the Midwest is undergoing the regional equivalent of what Schumpeter referred to as creative destruction, as new forms of production organization work to transform its historic industrial base.

The data from the Midwest Manufacturers' Survey provide considerable evidence of adoption and diffusion of new forms of production organization across the regional manufacturing base. Roughly one-third of respondents report the use of self-directed work teams, more than two-thirds report the use of quality teams with production workers, and more than half compensate workers for extra time spent working in quality teams. Thirty-seven percent of respondents report that they rotate workers across functional assignments, and 30 percent utilize a formal incentive system for workers engaged in continuous improvement activities. More than half (54 percent) of the respondents report that they had established a total quality management (TQM) program. In addition, more than half use statistical process control, and almost half use a just-in-time system for inventory control and production scheduling. Survey respondents report a 69 percent rate of capacity utilization, and roughly half operate on a round-the-clock basis. Furthermore, the survey respondents report an R&D-to-sales ratio of 11.5 percent, indicating that small and medium-size manufacturers are increasingly a source of innovation and technology development in the Industrial Midwest. These results thus indicate a clear trend toward the adoption and diffusion of new forms of production organization among respondent manufacturers.

Although measurement is difficult, a significant number of survey respondents are realizing substantial performance paybacks from the adoption of high-performance production organization while larger manufacturing establishments also report that they are realizing considerable performance payoffs from high-performance production systems. Xerox, for example, noted significant plant-level productivity gains associated with its companywide efforts at quality production and at deploying new systems of work and production organization.

New forms of production organization in some cases were a key factor in their very survival. A Big Three automotive transmission plant, for example, was able to avoid imminent closure by shifting to new production organization. The plant restructured its production process, reducing the number of job classifications, introducing self-directed teams, and decentralizing the decision-making process to harness the knowledge and full capabilities of factory workers. The plant instituted an extensive training program and also transformed its relationships with its principal suppliers, emphasizing quality and delivery as well as price. As a result of these efforts, productivity increased by nearly 60 percent and management of the parent company reversed its decision to close the plant (site visit and personal interview by Richard Florida and the Midwest research team, April 1993).

ROLE OF HUB FIRMS

The role of larger manufacturing establishments frequently goes beyond their individual experiences with new production systems. Indeed, larger manufacturing establishments typically act as *hubs* in broader production complexes. Close, interactive, and codependent relationships between these hubs and their suppliers play an important role in the transfer and diffusion of new manufacturing technologies and organizational practices. Furthermore, larger hub establishments encourage and assist in the adoption of high-performance practices by their suppliers as a vehicle for productivity improvement and continuous cost reduction.

The Midwest Manufacturers' Survey provide[s] clear evidence of a shift toward more codependent and interactive supplier relations. A large share of survey respondents report that they deliver according to a just-in-time schedule: 80 percent produce on customer order, and fully 40 percent make daily deliveries to their main customers. Eighty-two percent of survey respondents report that they interact with their customers in the early stages of product design, and 62 percent report that their customers evaluate them for certification. Furthermore, the survey suggests a high degree of concentration and integration between survey respondents and their suppliers, with roughly 30 percent of their production inputs coming from the same state and another 35 percent coming from the Midwest states. Fifty-one percent of survey respondents involve their suppliers in the design and development of new products, and 50 percent evaluate their suppliers at least once every two years. The survey data thus suggest the development of close and interactive supplier relationships in the Midwest, which is both an indication of a shift to new forms of production organization in its own right and represents an important mechanism for enhancing the diffusion of such practices through the region's manufacturing base.

GLOBALIZATION

The adoption of new production organization has been accelerated by the integration of the Midwest into the global economy, and especially heightened by foreign competition and rising foreign direct investment in manufacturing. Automotive and electronics manufacturers noted the competitive effect of highly efficient, high-quality Asian producers. Xerox specifically noted both the effect of competition in the low-end copier market from Canon and Ricoh and also the importance of its Japanese sister company, Fuji–Xerox, in helping to inform and structure its early quality efforts (site visits and personal interviews by Richard Florida 1992–94).

Moreover, the key to understanding the connection between globalization and the shift to new forms of production organization lies in the phenomenon of transplant manufacturing facilities. Transplant companies and foreign direct manufacturing investment in general are important sources of economic transformation and productivity improvement and economic growth across the advanced industrial nations (Graham and Krugman, 1991). An OECD study (1994) of 15 advanced industrial nations found that foreign-owned companies are typically more efficient than domestic firms in both absolute levels and rates of productivity growth, and that productivity gains result from more advanced technology than domestic industries, or from adding capacity, while productivity increases at domestically owned companies more often result from downsizing and layoffs. Furthermore, the McKinsey Global Institute (1993) found that foreign direct investment affects productivity by accelerating the transfer of world-class technology and production organization.

The Midwest is home to a large and growing body of foreign direct manufacturing investment. Between 1981 and 1989, the *gross book value* of foreign investment in the Midwest increased from $22.1 to $71.0 billion, an increase of 135 percent in real terms. Foreign investment in the Midwest is heavily concentrated in high value-added manufacturing sectors, particularly industrial machinery, chemicals, automotive assembly and automotive component parts, and steel. Furthermore, the Midwest has a particularly large concentration of Japanese transplant manufacturing facilities, especially in the automotive-related industries. The Midwest is home to 498 Japanese-affiliated plants, 40 percent of the national total. More than half of all Japanese foreign direct investment in automotive-related industries is concentrated in four Midwest states: Ohio, Indiana, Michigan, and Illinois (Florida and Kenney, 1991). Japanese investment has provided more than $7 billion to modernize the region's steel industry (Florida and Kenney, 1992), resulting in the establishment of technologically advanced steel finishing and galvanizing capabilities in the region.

Transplant factories of leading foreign manufacturing companies have played a key role in the transfer of state-of-the-art manufacturing technology and production organization to the Midwest locations, accelerated the diffusion of these practices through supplier complexes, and created powerful demonstration and learning effects for local companies. A 1994 cross-industry survey of Japanese transplants (Florida and Jenkins, 1996) indicate[s] that the automotive-related industries that are mainly concentrated in and around the Midwest show the highest rate of adoption of high-performance production organization. Transplants have brought new production organization to even extremely traditional industries like steel, which have experienced great resistance to new work and production systems (Florida and Kenney, 1992). Transplants have also stimulated the diffusion of state-of-the-art management practices into the US industrial base by working with their suppliers to help them adopt new forms of production organization. A recent survey of US suppliers to Japanese automotive transplants indicates an extremely high level of adoption of new forms of production organization by those US suppliers (Florida and Jenkins, 1996).

INSTITUTIONAL BARRIERS TO NEW PRODUCTION ORGANIZATION

Although the shift to new forms of production organization is clearly occurring, the research identified three factors that have impeded adoption and diffusion of new forms of production organization. First, a significant number of larger, hub manufacturing establishments remain locked into traditional production systems. In particular, a number of hubs continue to organize their supply chains along traditional mass-production lines focusing mainly on cost reduction and using erratic production scheduling, which requires suppliers to hold large inventories.

Second, existing banking and financial practices constitute another impediment to the adoption and diffusion of new forms of production organization. The financial system is a key element of the broader economic environment, sending strong signals to manufacturers via its loan requirements. Roughly 28 percent of respondents to the Midwest Manufacturers' Survey report that banks require inventory to be held as collateral on loans, thus creating a sizeable barrier to the just-in-time inventory and supply practices.

Third, the existing, largely Fordist public policy regime comprises an additional barrier to the adoption and diffusion of new forms of production organization. The Midwest Manufacturers' Survey data indicate that fully half of all respondents are unsatisfied with current government efforts to improve the regional manufacturing base. In the field research and interviews, manufacturing establishments, particularly those that have implemented new forms of production organization, were highly critical of government policies that seek to transfer new manufacturing technology or engage in industrial modernization assistance. Manufacturers also noted a contradiction in the existing environmental policy regime that favors clean-up technology and end-of-the-pipe solutions over efforts to introduce new production processes that could simultaneously improve productivity and prevent pollution. In other words, elements of the existing Fordist policy regime are in effect creating market failures, comprising an unnecessary barrier to the adoption and diffusion of new forms of production organization. This stems from the simple fact that the existing policy regime and the broader business climate it helps to inform grew up over the past century in conjunction with and to support the needs of mass-production organization. Simply put, this Fordist policy regime has not yet adjusted to the demands of new forms of production organization. This is suggestive of a lag between the rise of new forms of production organization and changes in government policy regimes and the broader regional business climate.

These factors notwithstanding, the research provides considerable quantitative and qualitative evidence of the adoption and diffusion of new forms of production organization among manufacturers in the Industrial Midwest. [This] evidence clearly confirms the central hypothesis: that new forms of production organization are not limited to new regions, that such practices can take root and flourish in older regions, and that they can play an important role in the reconstitution and revival of those regions as well.

CONCLUSIONS

The findings suggest two main conclusions. First, the Midwest is going through a deep and fundamental process of economic transformation, or regional creative destruction. The Midwest manufacturing base is shifting from traditional Fordist modes of production organization to new and more advanced or high-performance modes of production organization, characterized by a relatively high rate of adoption and diffusion of new forms of production organization, such as work teams, continuous improvement, and codependent supplier relations. This process of regional economic transformation is spatially uneven, occurring at higher rates in distinct geographic pockets within the region.

Second, the adoption of new forms of production organization across the region's manufacturing base has been accelerated by the global integration of the Midwest economy, particularly by increased levels of foreign direct investment in manufacturing. Escalating foreign competition has encouraged hub establishments to pursue new forms of production organization to increase their competitiveness and to push new and innovative production practices through their supply chains. The influx of a large number of transplant manufacturing facilities has stimulated the transfer of new forms of production organization, accelerated the diffusion of these practices through supplier relationships, and created demonstration and learning effects for regional manufacturing establishments.

At bottom, these findings raise questions for economic geography and regional theory. The research presented here clearly indicates that older regions can become focal points for new production systems. This contradicts the widely accepted belief, advanced in a huge body of literature, that older regions are inexorably doomed to disinvestment, deindustrialization, and decline (Bluestone and Harrison, 1982; Massey, 1984). Furthermore, it also contradicts the more recent, but nonetheless widely held, conviction that new forms of production organization are the province of new regions (Scott, 1988b; Saxenian, 1994). The overall thrust of this research thus calls for some rethinking and revision of existing theories of regional growth and decline. The idea that old regions decline because they are locked into old institutional practices, while new regions grow because they are home to that which is new, fails to capture the full breadth of regional economic dynamics. Simply put, simplistic metaphors of regional growth and decline, which served theory so well in the past, can no longer account for the full richness of regional economic transformation – an ongoing, evolutionary process in which many, varied outcomes are possible.

Nick Henry, Steven Pinch and Stephen Russell

'In Pole Position? Untraded Interdependencies, New Industrial Spaces and the British Motor Sport Industry'

Area (1996)

INTRODUCTION

Although British geographers have made many contributions to debates concerning new industrial spaces, Britain itself has not figured prominently in these debates. Hence, the few references to the City of London (**Amin and Thrift**, 1992; Pryke, 1991; Thrift, 1994) and various corridors, crescents or phenomenon in southern England (Hall *et al*, 1987; Henry, 1992; Keeble, 1988, 1989; Massey and Henry, 1992; Segal, Quince and Wicksteed, 1985) pale into insignificance compared with the vast amount of literature devoted to the Third Italy, Baden–Wurttemburg or Silicon Valley. Indeed, not only has Britain's economic performance after the Second World War been described as an example of 'flawed Fordism' (Jessop, 1989), but Scott (1991) argued that Britain's experience in the era of post-Fordism was as relevant as the experience of Spain in the previous Fordist era. Similarly, in Porter's (1990) examination of the factors leading to international competitiveness, Britain provides few examples of world-leading competitive industries. Instead, Britain is used primarily as an example of a nation in which Porter's 'diamond' of factors are locked into a downward cycle of relative economic decline.

Given this lamentable economic performance, it is therefore somewhat remarkable to find an activity in which Britain maintains – without any question of doubt – world dominance. In addition, this industry displays all the descriptive hallmarks of a 'New Industrial Space' (Scott, 1988a, b). We refer to the British motor sport industry (BMSI).

The aims of this paper are twofold: first, we demonstrate the regional success story of the BMSI; second, we use the BMSI to illustrate how concepts of culture and economy, specifically untraded interdependencies, provide further avenues in our attempts to explain localised growth.

THE BRITISH MOTOR SPORT INDUSTRY

Motor sport is 'a disarmingly simple term for what has become, particularly through the 1980s, a multi-faceted, international industry' (Griffiths, 1990). It may be broadly defined as competitive racing by similar machines on a frequent basis on designated tracks and circuits. These machines range from go-karts through Formula Ford, touring cars and rallying, to IndyCar racing and the sophisticated pinnacle of them all, Formula One. A concentration on Formula One reveals the astounding dominance of this international industry by British firms (Russell, 1994).

For example, the Formula One Constructors Championship has been won by a

Table 1 The sources of major parts for Formula One Racing Teams in 1994

Team	Base	Engine	Chassis	Designer
Benetton–Ford	UK	UK	UK	UK
Ferrari	Italy	Italy	Italy	UK
Footwork–Ford	UK	UK	UK	UK
Ligier Gitances	France	France	France	France
Lotus	UK	Japan	UK	UK
McLaren Peugeot	UK	France	UK	UK
Minardi	Italy	UK	Italy	France
Pacific Grand Prix	UK	UK	UK	UK
Sasol Jordan	UK	UK	UK	UK
Sauber Mercedes	Swiss	UK	Swiss	Swiss
Simtek Ford	UK	UK	UK	UK
Tourtel Larrousse	France	UK	France	UK
Tyrrell	UK	Japan/UK	UK	UK
Williams–Renault	UK	France	UK	UK

Source: *Sunday Times*, 7 August 1994.

British-based team in 29 of the last 38 years. Indeed, since Ferrari won the championship in 1983, the following 11 years have seen the title shared between the British-based teams of McLaren and Williams. Since the Constructors Championship began in 1958, 15 of the 24 teams who have won this coveted award have been British-based with McLaren topping the table as the most successful constructor ever in terms of Grand Prix wins (higher even than Ferrari). In fact, by the end of the 1994 season, a non-British based team had not won a grand prix since 1990, representing over 60 races.

Of course, the design and assembly of a car is only part of the story. Table 1, which lists the source of the major components which go to make up a Formula One racing car, reinforces the dominance of British-based firms, especially in engine manufacture. Indeed, the most successful engine in grand prix history was developed by Northampton-based Cosworth which is the largest manufacturer of racing machines in the world.

The US-based IndyCar Racing circuit and other less prominent forms of racing are also dominated by British-based constructors. All of the teams on the IndyCar circuit in 1994 used a car designed and built in Britain. In total, over 75% of single seater racing cars used in more than 80 countries across the world are British.

After the Second World War, Ferrari dominated Formula One. Yet, by 1989, almost complete British dominance of the industry finally led Ferrari to employ a British designer and, in 1992, relocate its design office from Italy to Surrey, England. On doing so, Luca di Montezemolo, President of Ferrari, made the following comment:

> In Italy we are cut away from the Silicon Valley of Formula One that has sprung up in England. (*The Sunday Times*, 6 September 1992)

Similar sentiments were expressed by Flavio Briatore the Managing Director of the Benetton Formula One racing team:

> If you like prosciutto, you come to Italy. If you like champagne, you come to France. For Formula One, you come to England. I don't like the English weather, but the best engineering is here. (*The Independent*, 4 March 1994)

GEOGRAPHY AND INDUSTRIAL ORGANISATION: TELL-TALE HALLMARKS

The BMSI displays many of the hallmarks of a new industrial space. The industry's production organisation revolves around the design and construction of high speed, high performance 'one-off' cars. It is a technologically advanced industry geared to small batch production at short notice such as between the roughly fortnightly race meetings of the grand prix circuit. Both the product (e.g. new materials, electronic systems) and the production process (e.g. CAD/CAM, aeronautical engineering, wind tunnels) are 'high-tech'.

The industry is based upon the activities of many small and medium-sized firms with only a handful such as Cosworth, Williams, Benetton and McLaren employing more than 200 workers. The majority produce specialist parts such as gearboxes, brakes and suspensions, yet these new technology-based SMEs (small to medium-sized manufacturing enterprises) are part of a dynamic process of growth, spin-off and new firm formation. For example, the highly successful engine manufacturer Ilmor, set up in 1984, was a spin-off from Cosworth, itself an outstandingly successful engine manufacturer, and itself also a spin-off from Lotus. Russell (1994) has studied 10 firms founded between 1958–1986. In total, employment on start-up amounted to 81 employees, in 1988 1063, and by 1993 1663 with the largest firm employing over 750 workers. In 1992, an RAC Motorsports Association report estimated that 50,000 people were employed in the UK motor sport industry (A. Henry, 1994).

The employment growth of the industry is attributable to its international success. The industry is an export earner contributing an estimated £600m annually to the balance of payments (Freeman, 1992). All bar one of the 10 firms in Russell's (1994) survey revealed that international sales accounted for 75% or more of their total sales.

Finally, the British motor sports industry is characterised by its regional concentration (one might more accurately refer to an *English* motor sports industry). An initial mapping of *all* the UK Formula One and IndyCar producers reveals a distinct, yet familiar, pattern of location which follows the work by Hall *et al* (1987) on the Western Crescent, the UK's premier high technology region stretching in an arc around West London from Cambridge to Surrey (see Figure 1). Mapping suppliers to the industry reinforces the message of regional concentration but through a pattern of sub-clusters. In particular, it highlights Oxfordshire (the home of Williams, Benetton, Reynard and Simtek) as the centre of this 'Silicon Valley of motor sport' (*FT Survey*, 2 March 1992) such that 'Virtually all of the industry – a web of small technologically advanced workshops – is located within a 50 mile radius of mid Oxfordshire and the Silverstone racetrack in Northamptonshire' (Freeman, 1992).

To sum up, the British motor sports industry consists of a set of geographically concentrated, technologically-advanced SMEs undergoing rapid growth on the back of success in international markets. Moreover, the vast majority of these firms are UK-owned, as well as UK-based. These operations represent further evidence of the importance of small firms in regional development (for example, Keeble, 1990; Keeble *et al*, 1991) and given the debate over the importance of networking amongst SMEs, one might even speculate that where the BMSI is concerned, the 'Third Italy' has come to Britain.

THE RESURGENT REGION: INDUSTRIAL DISTRICTS AND THE NEW INDUSTRIAL SPACES

Initial work on industrial districts by Italian economists Piore and Sabel (1984) represented them as the geographical manifestation of a re-emergent production system of flexible specialisation growing in opposition to a crisis-ridden system of mass production. Scott's (1988a, b) treatise on the New Industrial Spaces concentrated on the flexibility, transaction costs and geography of the critical input–output structures of the constituent firms that comprised the districts. However, critics argued that the linkages identified were not present in sufficient numbers or satisfactorily specified in terms of causal powers to account for the size and dynamism of the agglomerations (Gordon, 1992; Henry, 1992; Storper, 1995). To be fair

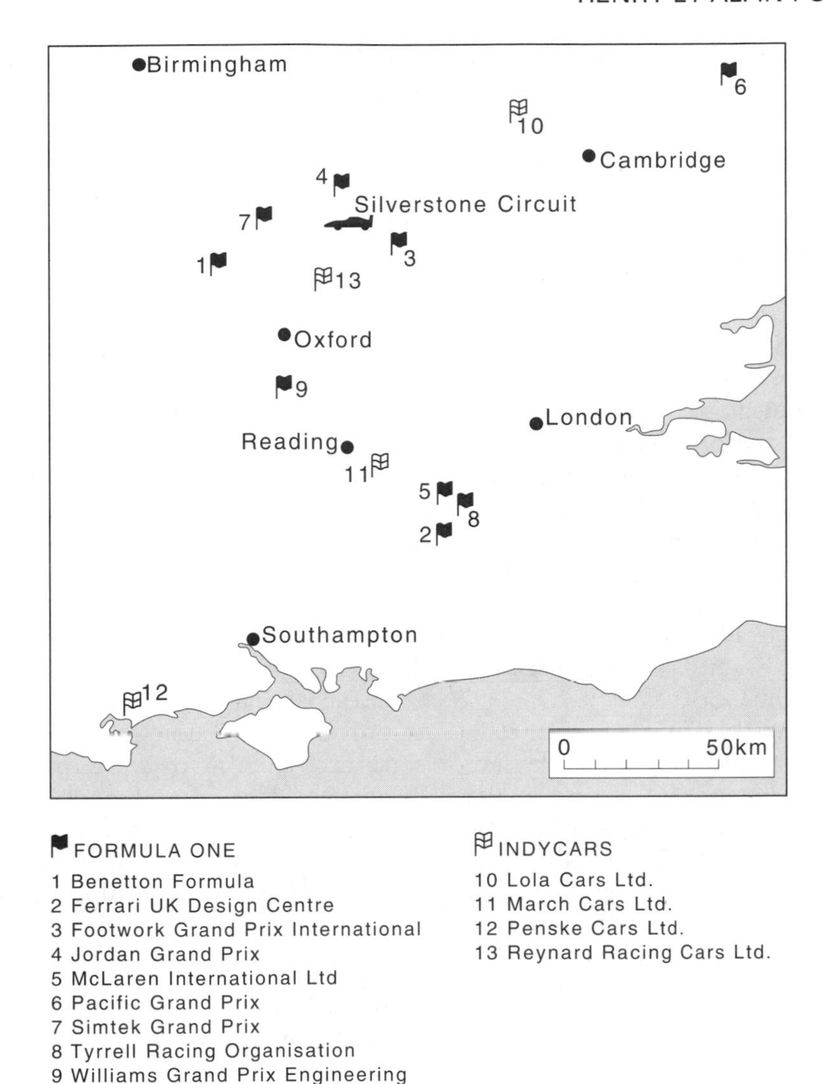

Figure 1 The location of British-based Formula One and IndyCar teams

Scott identified other agglomeration factors such as local labour markets and the slippery concept of Alfred Marshall's (1920) intangible 'the industry is in the air' as critical elements of self-sustaining growth. Yet, attempting to identify the 'transaction costs' of such intangibility highlighted the overall failings of this approach. Thus, other groups, such as GREMI (providing the label of the 'milieu' but not necessarily greater explanation), have looked elsewhere for answers (Aydalot, 1986; Aydalot and Keeble, 1988). However, the recent work of Storper (1993, 1995) and his concept of 'untraded interdependencies' may provide some of the answers to this 'intangibility' as an additional/alternative mechanism of regional growth.

UNTRADED INTERDEPENDENCIES: A CONCEPT OF ECONOMY AND CULTURE

Storper starts from the premise encapsulated in New Institutional Analysis that the economic is 'embedded' within social institutions (e.g.

Granovetter, 1985; Hodgson, 1988, 1993a, b). In breaking down such embeddedness, he goes on to argue how, especially in the case of advanced technology, there are behavioural–institutional sources of learning such as *untraded* interdependencies that are critical to the construction of any production system (and its technological trajectories). These untraded interdependencies comprise conventions, rules, practices and institutions which combine to produce both possible and real 'worlds of production' which present the action trajectories for firms within a world of uncertainty. They provide the capacity for firms to gain knowledge of, for example, technology and markets which enables them to sustain competitive advantage. In addition, however, such interdependencies may be *territorialised*, they may comprise 'local worlds of production' and thus local production complexes may be studied as the nexuses of untraded interdependencies.

Storper's work is one of a series of recent attempts at further unravelling the non-economic social relations and structures that 'embed' growth. The economic is, of course, always embedded in particular cultural practices but these attempts have been given impetus by the recognition of the increased knowledge-intensiveness and aestheticisation of contemporary economies or, put another way, the extent to which 'culture' penetrates the economy (**Lash and Urry**, 1994, see Part Two, Section 2.2). Some of the most recent of this work highlights that 'culture' is encapsulated not only in the consumption of the product but also in its production; the construction of meaning is an integral element of the production process across a range of industries (**Allen and du Gay**, 1994, see Part Five, Section 5.2; **Lash and Urry**, 1994, see Part Two, Section 2.2; Leidner, 1991; McDowell and Court, 1994).

Much of this work has concentrated on interactive service employment. However, such an analysis may hold for financial and business services where 'prosumption' (*simultaneous* production and consumption by an interacting employee and customer) does not take place. Recent work on merchant banking highlights how the global financial markets are essentially social networks in which 'relationship management' holds the key to profitable success. The streams of information on prices, rates, etc. which *are* the markets open to individual interpretation and one of the key skills in international finance is to construct a network of contacts, in effect an interpretative structure or network, so as to be part of the interpretation of what is actually happening, of what it all means (**Amin and Thrift**, 1992; Thrift, 1994).

Similarly, **Lash and Urry** (1994) argue that such 'knowledge structures' (Strange, 1988) or 'expert systems' (Giddens, 1990) are relevant too for our understanding of the production systems (and industrial districts) of high technology. Indeed, they go so far as to rename these high technology districts as *cultural districts* in as much as production is based on (cultural) expert-systems in which what matters is the exchange (communication, interpretation, co-ordination) of ideas and people. Thus we come back once more to Storper's concept of 'local worlds of production (knowledge)' (including untraded interdependencies) as the latest formulation for understanding some of the dynamics of localised growth.

THE BMSI AND UNTRADED INTERDEPENDENCIES

Whether the BMSI constitutes a 'local world of productive knowledge' requires further detailed analysis but there are certainly many strong indications that this is the case. For example, there is a very clear technological trajectory in the evolution of racing car design. The Darwinian-like struggle for survival amongst racing teams means that innovations which offer advantages are rapidly adopted by other teams. The history of Formula One is full of radical innovations that spread throughout the sport such as the aerodynamic skirts pioneered by Lotus in the early 1970s. One reason for the rapid diffusion of ideas is the fact that the drivers, designers and engineers move from team to team, taking with them considerable knowledge of how things are done in rival teams.

Another factor encouraging the spread of ideas is that there are only a limited number of race tracks available for testing. Furthermore, much of the technology – the aerodynamics and suspension geometry – is fully exposed on racing cars. The teams do their best to keep their cars secret but there are many stories of espionage and 'spying'. Ideas are also diffused by the fact that, each season the latest innovations are subject to intense rumour, gossip and speculation in the trade, technical and popular press. It might seem advantageous to be hidden away from this ferment but all the evidence indicates the contrary; it seems there is more knowledge to be gained about the technological trajectory by being close to 'Motor Sport Valley'. The local world of production in motor sport therefore has a strong territorial expression in Britain.

There is also evidence within the motor sport industry of conventions and the actors who produce them. For example, on the one hand, the companies in the industry are intensely competitive and hence highly secretive about their latest innovations. Yet, on the other hand, they are engaged in a *collective* endeavour, for these companies are reliant upon other teams to produce successful products to ensure exciting races. One of the most remarkable illustrations of this stems from the 1960s when the Cosworth-powered Lotus was all conquering. The predictable nature of wins by Lotus was threatening to kill off Formula One. Thus, Colin Chapman, then head of Lotus, was persuaded to make this engine available to other teams to increase rivalry in the sport. In the 1990s, the motor sport industry survives less as part of the R&D process of mass car production than on its delivery of 'spectacle'. Teams are funded to the tune of several tens of million of pounds a year by sponsors looking to catch the eyes of the 26 billion people across 103 countries who watched Formula One in 1992 (Forsythe, 1993).

Indeed, the tension between motor sport as technological development or advertising jamboree recently came to a head. Continued technological development and the dominance of the sport by Williams–Renault, it was argued, was making the driver obsolete such that the 'spectacle' of the industry was in danger. This threatened audience figures and sponsorship money. Thus, under pressure from the FIA, the leading teams met in July 1993 and agreed to 'abandon, from the start of next season, much of the technology which has long been claimed as a large part of the sport's raison d'être' (*Financial Times*, 31 July 1993). In essence, agreement was reached on the *management of competition* in a highly competitive industry to guarantee the spectacle.

Moreover, this historic agreement to limit *some* technological development in Formula One highlights the elaborate and continually changing regulations of the governing bodies of motor sport. Prior knowledge of such regulatory moves is clearly of paramount importance if a team is to stay competitive. Critically, however, the teams themselves are involved in setting the regulations. In Formula One, technical regulations are recommended by a working group of team engineers (5 out of 6 of whom are British). Regular meetings are held to gain a *consensus* viewpoint on particular technological changes thus making clear the cooperative nature of the technological trajectory of this intensely competitive world of production. Innovative solutions are devised (rapidly) but within a constantly *negotiated* framework. Untraded interdependencies determine the framework of production for this agglomeration of motor sport firms.

CONCLUSION

In this paper we have described an undoubted British success story, the motor sport industry. Moreover, we have revealed 'Motor Sport Valley' as an example of dynamic, localised small-firm growth. Much more work is required on the British motor sport industry before it may be labelled a New Industrial Space but we have shown that Storper's (1993, 1995) concept of 'untraded interdependencies' provides a further analytical avenue for understanding the success of Motor Sport Valley. Significantly, the concept of untraded interdependencies goes far beyond the narrow confines of neo-classical economic models such as transaction costs. It is an attempt to theorise the social and cultural determinants of the economic (more specifically, the cultural construction of competitive success) and is symbolic of a significant wider 'cultural turn' amongst economic geographers. The use of such concepts in the identification and analysis of economic success stories such as the BMSI remains of considerable import.

John Bryson, Peter Wood and David Keeble

'Business Networks, Small Firm Flexibility and Regional Development in UK Business Services'

***Entrepreneurship and Regional Development* (1993)**

INTRODUCTION

During the 1980s, turnover, employment and the stock of firms in a range of business service activities in the UK have all grown spectacularly (Keeble *et al*, 1991). Overall employment in 'other business services' (SIC 8395, which includes management consultants, market research, public relations consultants, employment agencies, translators, document copying services, etc.) more than doubled between 1981 and 1990, growing by 354,000 or +122%. This compares with a growth rate of +3.5% for all industries and services, and a decline in manufacturing employment of –17.5%. Both large and small firms are involved.

Much of this growth in the stock of business service firms has been dominated by small enterprises and has been most dramatic in information-intensive business services such as computer services and management consultancy. The growth of business services must be seen as part of a general phenomenon of small firm expansion in UK services and the economy generally during the 1980s (Keeble, 1990). Between 1985 and 1990, the number of small firms expanded rapidly, with increases in the total stock of businesses of 114% for management consultancy, 106% for personnel recruitment and employment agencies, and 37% for advertising and market research.

The growth of small firms in professional business services thus stands out as the single most important component of the rapid growth of small service sector businesses in the 1980s (Bryson and Daniels, 1998). Equally rapid employment growth reflects in part at least the intrinsically high labour-intensity of information-based business services, 'professional expertise' being a commodity whose production and delivery cannot easily be automated. This paper examines the characteristics of small business service firms which help explain their competitive success, focusing in particular on the nature of the 'networks' entrepreneurs utilize to acquire clients and to create and maintain organizational flexibility.

FLEXIBILITY, NETWORKS AND NETWORKING

The small firm sector is inherently flexible in being able to respond rapidly and flexibly to changing client requirements. An inherent part of this flexibility is a consequence of the relatively intangible and personal nature of the services these firms provide. Small business service firms are also able to satisfy client requirements by employing the complementary services of other small or large consultancy firms or even sole practitioners. Small business service firms which specialize in particular industries, techniques or regions may also be employed as subcontractors by large organizations. The central argument of this paper is that small business service firms, or even individuals, are able to compete successfully – by price and by expertise – with large firms through the use of a variety of personal contacts, associates and business contacts. This type of organizational flexibility, frequently termed 'networking', or a 'network' of contacts/associates, enables small firms to offer a wide range of services without employing a substantial full-time professional or support staff.

The term 'network' or 'networking' was originally developed in relation to formal exchanges of information within large organizations, especially under the influence of modern information technologies. Given this background two different and conflicting definitions of 'networking' have been developed in the academic literature. First, 'networking' and 'networks' have become increasingly fashionable conceptual devices for theorizing the internal organization of large businesses (see Bressand *et al*, 1989). For example, a recent paper by Leo and Philippe (1991) examines the geographical networks, or branch plants, of service companies and attempts to understand the localization of business services within France. Second, entrepreneurship has been conceptualized as a dynamic process which requires for its successful development linkages or networks between key components of the process (Aldrich and Zimmer, 1986, p. 3). In this approach entrepreneurship is viewed as 'embedded in a social context, channelled and facilitated or constrained and inhibited by people's positions in social networks' (Aldrich and Zimmer, 1986, p. 3). It is the second approach which is examined in this paper.

In recent papers Curran *et al* (1991) argue that networks and networking 'have emerged as fashionable conceptual devices for theorizing and researching a number of important aspects' of small businesses, but that much recent work is 'conceptually and methodologically poorly realized' (Curran *et al*, 1991, p. 1). To overcome this theoretical confusion they argue that 'networks' and 'networking' can be usefully divided into two types: compulsory and voluntary networks. Compulsory networks are those which an organization must belong to in order to survive and operate successfully, for example banks or accountants, while participation in the local chamber of commerce or golf club would be classified as voluntary networking. Many of these 'networks' are support networks (for example banks, enterprise agencies and business advisers), which function to provide advice, information and capital (Curran *et al*, 1992, pp. 3–4).

One difficulty with this model of networks rests on the identification of voluntary and compulsory networking. A particular firm's membership of the local chamber of commerce or professional association may be considered by its managing director as a compulsory activity vital for the company's success or its public image. For another company, membership of these organizations might be considered as voluntary and even unnecessary networking. The two types of network identified by Curran *et al* overlap with those identified by Birley (1985) who divides 'networks' into two types: formal and informal. Formal networks include banks, accountants, lawyers and the local chamber of commerce, while informal networks include family, friends, previous colleagues and previous employers (Birley, 1985, p. 109).

For small firms the definition of 'networking' and 'networks' must be founded on the relationship they have with the external environment. As far as this paper is concerned the informal networks, including those embodied within formal relationships (for example, with clients or banks), identified by Birley appear to be the most important and can be usefully further divided into three types depending on the nature of the relationship the firm has with the external environment. The first comprises 'networks' associated with clients, obtaining new business and the maintenance or establishment of contacts with clients, in other words *demand-related networks*. The second type covers 'networks' associated with the co-operative supply of a service or

Table 1 Markets and clients of small business service firms

	Mean % of business firms						
	Financial services	Business services	Other services	Manu-facturing	Local govt.	Central govt.	Other
Total firms	12.7	20.6	15.2	39.6	3.4	5.5	3.4

	Client size (%)		
	Small firms	Medium firms	Large firms
Total firms	12.7	15.7	71.6

product – *supply-related networks*. Finally, there are networks which can loosely be defined as *support functions*: for example banks, business advisers, the founder's family and friends. This paper examines the demand- and supply-related networks of small business service firms.

DEMAND-RELATED NETWORKS

The growth of small business service firms must be related to the part they play in wider changes in business organization, and especially in satisfying the needs of clients. The significant role played by small information-intensive business service enterprises in the wider economy is indicated in Table 1, which shows that clients of the surveyed firms are drawn from a wide range of sectors, but with manufacturing and business services accounting for three-fifths of sales. These sectors, along with financial services (13%), are especially important for national and regional economic growth. Moreover, the clients of small business service firms are predominantly large firms (with over £5 million turnover). In fact, the lists of clients served by management consultancy and market research firms are dominated by major 'blue-chip' clients. Thirty-eight (63%) of the small management consultancy firms surveyed depended on large firms for over half of their turnover, and only five (8%) had no large clients. Conversely, over half have no small clients (with less than £1 million turnover) (Wood *et al*, 1991, p. 10). Market research firms are even more dependent on large clients, with no less than 50 (83%) of the sample firms deriving over half their turnover from them.

The intangible nature of the products supplied by business service firms makes it difficult to acquire clients simply by advertisements. A client's selection of a management consultancy or market research firm is more of 'an art rather than a skill' (Aucamp, 1978). Factors which a potential client must take into consideration when selecting a business service firm are its reliability, its experience in a particular area, its potential ability to work productively with the client's personnel, the competition, the quality of its work and its cost. These factors are impossible to measure in advance, since business service firms sell skills, expertise and experience which result only in future, and maybe unclearly defined gains. The large number of management consultancy and market research firms, and the specialized intangible nature of their expertise create an imperfect market-place for their services. This is a direct result of clients' lack of information about the services, and quality of any business service firm (Holmstrom, 1985, p. 187). Consultancy and research firms also usually have very little knowledge of their competitors, and client choice of external advisers is dependent on personal recommendation and on an informal selection procedure.

The growth of small information-intensive business service firms reflects the often strategic nature of their advice and its relatively intangible nature, embodied in experienced individuals. These individuals provide a personal or 'customized' service to clients, based on their own reputations, experience, track records and contact client networks. Business service firms thus employ highly educated and qualified individuals whose professional expertise is not tied strongly to the firm in which it is developed. Such expertise is highly mobile and coupled with low barriers to entry, requiring little capital or equipment, enables individuals to leave established firms to set up new firms or to act as sole practitioners. These characteristics encourage the continuing fragmentation of the market research and management consultancy industries, enhancing organizational flexibility. The most important attributes possessed by the entrepreneur who establishes a small research/consultancy firm are professional expertise, an existing reputation and a network of client contacts. These essential requirements for competitive success explain the concentration of founders' previous employment in same-sector or client firms, and into large rather than small companies (Table 2). Thus 90% of the 184 founders involved were previously employed in either market research (28%), management consultancy (21%) or client companies (41%), while 63% were employed in large firms (over £5 million turnover), with only 20% coming from other small firms (less than £1 million turnover).

Regional differences are important here with a large proportion of inner London and southern firms spinning off from established consultancy firms and market research firms (Wood *et al*, 1991). In northern Britain new management consultancy and market research firms more often spin off from client companies which are the obvious training grounds in areas which do not possess a large base of established business service firms. Small business services firm founders thus acquire their reputation and initial client contacts whilst working for either a large supply or client company.

An important measure of the relationship between a business service firm and its clients is the level of repeat business. Repeat business reflects a client's satisfaction with an earlier assignment. In this case the only measure of the success of an earlier assignment is the client's perception of its success or failure. A 'successful' project will probably lead to further assignments. As far as all 120 firms are concerned, repeat business provided on average just under two-thirds (61%) of small firm assignments by value (proportion of annual turnover), although this rate was lower for management consultancy firms. For management consultancy firms based in the rest of the South East outside London repeat business provides a significant majority (73%) of their business by value. In contrast repeat business for consultancy firms based in northern Britain represents only 40% of their turnover. Market research firms are even more dependent on repeat business with 70% of the turnover of firms based in the rest of the South East and northern Britain coming from former clients.

Table 2 The process of new business service firm formation

Type	No. of firms					
	A	B	C	D	E	F
Market research	28	7	6	3	6	3
Management consultancy	21	17	8	4	0	1
Total firms	49	24	14	7	6	4

Notes:
A = Worked for an existing MR/MC firm and decided they could do it for themselves.
B = Worked for an existing client firm and decided they could do it for themselves.
C = Redundancy.
D = Spin-off, of 2–3 founders jointly from MR/MC firm.
E = Housewife with children re-entering profession/work.
F = Externalization, of 2–3 founders jointly from client company's in-house service department.

The importance of repeat business for small business service firms is further emphasized in an analysis of how firms obtained their last three clients (Table 3). Some 67% of the last three assignments of market research firms and 40% of those of management consultancy firms were obtained from former clients. Table 3 also provides evidence for another measure of the importance of demand-related networking for small business service firms, that of referrals between client companies. A significant minority of management consultancy firms (22%)

Table 3 Could you perhaps describe how your firm obtained its last three assignments?

Source	No. of mentions							
	Management consultancy				Market research			
	UK	IL	OSE	N	UK	IL	OSE	N
Referrals from clients	38	11	16	11	21	7	6	8
Referrals from other supply firms	11	5	4	2	12	7	1	4
Repeat business	70	29	23	18	116	37	45	34
Directories/advertisements	16	4	6	6	5	0	1	4
Cold calls	11	5	0	6	13	2	5	5
DTI	21	4	2	15	2	0	1	1
New business	9	2	5	2	5	3	1	1
Total	176	60	56	60	174	57	60	57

Notes:
UK = United Kingdom.
IL = Inner London.
OSE = Outer South East, excluding London.
N = North West and Yorkshire and Humberside.

obtained their last three projects from referrals between clients. This was particularly important for firms based in the rest of the South East, excluding London. Thus personal contacts which have developed either between a business service firm and its clients or between clients account for 61% of consultancy firms' and 76% of market research firms' last three assignments. Given these figures, the importance of personal contacts and an established reputation for business service firms cannot be overestimated.

The final measure of the importance of demand-related networks is the symmetry of the relationship. In serving client needs, the expertise offered by small business service firms must complement the client's own, in-house capabilities (Wood, 1991b). These will have been developed through staff training and career development programmes and will generally respond slowly to changes in organizational structure and in the commercial market-place. Specialized outside assistance is increasingly sought from business service firms to help companies respond to economic and political turbulence effectively. For our sample of small firms, taking on work which could have been carried out by a client's own in-house staff occurred in only a minority of cases (Wood *et al*, 1991, p. 14). Firms were asked to describe in detail their last three projects. Most clients required advice which either did not exist inside the firm or an independent view of a particular issue or problem. In the majority of cases the types of projects undertaken by consultancy and market research firms were specialist, and in many cases strategic, in their focus, whose success depended to a great extent on the relationship that existed between the client and the business service firm. The strategic nature of these projects is perhaps best demonstrated by the most frequently used type of management consultancy expertise, human resource management (Wood *et al*, 1991). The management of human resources is one of the most important aspects of corporate planning, but it is also one of the most difficult tasks for internal appraisal. A particular need arises for authoritative and experienced outside experts, able to take an independent view separate from the personal and political involvement of most 'insiders'. If successful, such projects would lead clients to seek further advice, when required, from the business service firm and to recommend the company to other potential clients.

SUPPLY-RELATED NETWORKS

Small business service firms are able to satisfy specialized client demand by co-operating with

Table 4 What role do associates play in your business?

Role	Management consultancy				Market research			
	UK	IL	OSE	N	UK	IL	OSE	N
Extend expertise	50	22	20	8	38	18	12	8
Work overload	26	9	9	8	13	7	3	3
Reduce costs/increase flexibility	15	5	4	6	4	0	2	2
Marketing	6	4	2	0	1	0	1	0
Data collection	2	0	1	1	3	0	0	3
Extend geographic range	6	3	3	0	3	1	1	1
Total	105	43	39	23	62	26	19	17

other business service firms or even with individuals. By using a network of contacts and associates, small business service firms are able to compete, by expertise and price, with large firms by being able to offer a wide range of services without employing a substantial full-time professional staff. These supply-related networks enable small business service firms to be extremely flexible in their responses to the demands of clients and to offer many of the services and facilities provided by large firms.

Supply-related networks can be divided into two types. First, small business service firms network with other small firms, or even individuals, to increase the range of services and advice they can provide. Many of these are national networks, but increasingly small business service firms are developing formal links with complementary firms based in other European countries. Second, sole practitioners combine together into a network to provide a formal vehicle for their activities. This type of network may be either an informal relationship between a small number of individuals, or a formal relationship involving many individuals governed by a series of rules and regulations.

A significant proportion of surveyed firms (74%) regularly use outside researchers or consultants to enhance the skills of their in-house staff. Management consultancy firms use associates (85%) more than research companies (64%). In both sectors more firms based in inner London or in the outer South East use external associates than northern-based companies. This is probably due to the low level of specialization and the limited number of research and consultancy companies based in the north (Wood *et al*, 1991). The number of associates used regularly by consultancy or market research companies is fairly small, the majority of companies (77%) having networks of 10 or less associates. Those supporting northern-based management consultancy companies are much smaller than for inner London or outer south-eastern firms. Just under one-third of northern-based management consultancy companies do not use associates, while 30% regularly use only between one and five associates. Exactly half of the firms based in inner London regularly use between six and 10 associates.

Table 4 records the six most important reasons for firms using associates. Responses were unprompted and classified after interview. The most common reason identified by 48% of management consultancy firms and 61% of market research firms was to extend their existing in-house expertise (Table 4). The second reason for using associates was during a period of work overload. Market research and management consultancy companies typically exhibit a cyclical work pattern related to economic cycles. This cyclical pattern may be accentuated by the limited time available for marketing the company and extending networks of potential client contacts during busy periods. The third reason, related to the first two factors, was to reduce costs and enhance flexibility. Perhaps the most important aspect of Table 4 is that it suggests that business service firms use associates and other firms primarily to obtain high-order professional expertise, specialist advice and assistance rather than to buy in low-order services, for example data processing. This suggests that networks are a powerful component in the assembly of expertise to manage business change.

We have argued that reputation and knowledge which reside in individuals rather than in companies is the basis of the market research and management consultancy industries. These characteristics encourage individuals to leave companies and establish themselves as sole practitioners. Both the management consultancy and market research industries have high proportions of sole practitioners. Sole practitioners can undertake work directly for clients, but sometimes operate in a network with other sole practitioners or as subcontractors. This latter type of individual-centred supply-related network is well illustrated by the example of a sole practitioner management consultant based in Berkshire. This consultant left a large management consultancy firm to establish his own client base to obtain greater control over his working life. Thirty per cent of the projects this consultant is offered by clients, however, are too large to be undertaken successfully by an individual. This difficulty has been overcome by establishing a limited management consultancy company, with a network of sole practitioners. While still predominantly operating as independent sole practitioners this company provides the 'resources' and formal company image to enable the individual professionals involved to undertake large projects. The sole practitioners use the company vehicle only when they are in danger of losing a project because the client considers it to be too big for an individual. This method of organization enables sole practitioners to compete, by open tender, with small and even large consultancy companies. It gives sole practitioners flexibility working together either as a formal group or as individuals.

The final measure of supply-related networking is that of referrals between business service firms. This, in fact, is not very important, accounting for only 6% of the last three projects obtained by management consultancy firms, and 7% of market research projects (Table 3). This low level suggests a reluctance to co-operate with potential competitors, perhaps affected by the current recession. This type of network also depends on the availability of associates to undertake specialist parts of projects which cannot be performed by the firm's professional staff. On the other hand six management consultancy firms argued that associates are extremely useful contacts, occasionally hearing about projects which they cannot undertake themselves. In these circumstances they may advise the potential client to seek the advice of the consultancy firm which has given them the most employment opportunities.

CONCLUSION

This paper has shown that supply- and demand-related 'networks' of informal personally based contacts are an inherent and vital component of the relationship which a small business service firm has with its clients and other business service firms. They are also extremely important for specialist companies attempting to fill particular market niches. Without their ability to 'network' with other firms many small business service firms would be unable to offer a wide range of services at a competitive price. To survive and compete successfully with large companies small business service firms must occupy a web of fairly well-developed demand- and supply-related 'networks'. Networking thus appears to have been an important element in the recent growth of small companies in information-intensive business services.

The imperfect market which characterizes the demand and supply of business services implies that 'networking' and 'networks' are especially useful operational tools for small management consultancy and market research companies. Consultancy and research firms have very little knowledge of their competitors while client choice of external advisers is dependent on personal recommendation and on an informal selection procedure (Table 4). 'Networking' is a useful approach as it allows the role and function of small firms to be judged in the context of their overall activities and in relation to their external environment. It emphasizes the interactive nature of the activities of business service firms, interactive with clients and with other business service firms, and highlights the importance of an individual's reputation and expertise.

Roger Mark Selya

'Taiwan as a Service Economy'

***Geoforum* (1994)**

INTRODUCTION

The economic development of Taiwan (Republic of China) has been extensively studied. For the most part, the literature on the economic development of Taiwan has focused on several narrow questions including description and explanation of change, and the role of the government in fostering change (e.g. Kuo *et al*, 1981). Discussions on specific sectors of the economy have been concerned mainly with the successful expansion of manufacturing and trade, although both the positive and negative impacts of modernization on regional development and agricultural change have appeared (Bello and Rosenfeld, 1990a, b). The published research of geographers parallels these general topical trends (Selya, 1975, 1982, 1993; Shaw and Williams, 1991). Overall the impression one gets from the voluminous literature is that Taiwan is basically an industrial economy with increasing per capita income and a weakening agricultural base.

The data on employment and origins of GDP, however, suggest that Taiwan should be considered a service dominated economy. In this article this argument is presented and analyzed.

EVIDENCE THAT TAIWAN IS A SERVICE ECONOMY

In terms of employment and origins of GDP it is clear that not only are services the largest single contributor to the economy but also by far the major employer as well. In the 38 years of economic development of Taiwan, growth in services has outperformed growth in manufacturing around half the time. In general both the absolute and relative (percentage) growth of services have been less volatile, less affected by recession, and quicker to recover from recession than the manufacturing sector. These growth patterns are found in other service dominated economies (Price and Blair, 1989). In terms of absolute changes in employment, some 56.82% of all the new jobs added to the economy since 1952 have been in the service sector; jobs in services have been added on average at a rate 2.62 times that of manufacturing positions. Since 1987 services have added more jobs and more GDP than manufacturing in both absolute and percentage terms. Growth in services has not been uniform however. Although total service employment in the period 1971 to 1986 increased some 120.69%, some services, such as commerce and finance, grew more rapidly. Within the lagging social and personal services sector some types, such as real estate and recreation, actually contracted (Table 1).

As the economy of Taiwan has grown in size and sophistication, so too the number and type of services has expanded. For example, totally new business services enumerated in the 1986 census include market place management, consulting services, data processing and

Table 1 Growth in service employment

Sector	Percentage growth 1971–1986
Construction	87.29
Commerce	169.54
Transportation	86.93
Finance	150.12
Social of which:	
Hotels, restaurants	203.97
Real estate	–54.55
Recreation	–10.71
Personal	27.88

information services, advertising services, product and packaging services, and machinery and equipment renting and leasing. One new social service category, mass media, was also added. In contrast, some 29 categories of manufacturing were either eliminated or consolidated. As with the growth in service employment, so too in the growth in the number and sophistication of Taiwan's experience mimics that found elsewhere (Price and Blair, 1989). However, Taiwan's experience can also be seen as a mix of the recent experience with services and manufacturing in both developed and underdeveloped countries. In terms of changes in the origins of GDP and employment, Taiwan resembles the developed world; in terms of retaining a sizeable manufacturing sector it resembles many developing countries (Britton, 1990; Daniels, 1985, 1993; Dicken, 1992).

The service sector shares one important trait with Taiwanese manufacturing: it is dominated by small establishments. The mean number of employees for all services except insurance are well below that of manufacturing. Small size is an additional trait which Taiwanese service industries also share with services elsewhere (Daniels, 1983). It has been suggested that most Taiwanese service firms are in fact too small to operate efficiently.

SPATIAL DISTRIBUTION OF SERVICES

Clearly the major cities and *hsien* (county) capitals dominate the pattern, a distribution found elsewhere (Marshall *et al*, 1988; Price and Blair, 1989). This urban orientation is confirmed by location quotients of services comparing the distribution of services to that of population (Figure 1). Only 28 townships and cities out of 318 in Taiwan had location quotients of one or higher. All of these areas are either cities or *hsien* capitals. There are several reasons to expect that services should be highly concentrated. On the one hand services may have located in urban centers since these areas traditionally have contained larger pools of highly skilled labor, and because urban areas are the location of both competing and complementary economic activities. Similarly, an urban location may be preferred as it may reduce the cost of delivering the service to the market. It may also be the case that service industries prefer rich information environments (Daniels, 1982), and obviously cities have an advantage in this regard over more remote areas. Part of that rich information environment is improved telecommunications (Daniels, 1983); in Taiwan improved telecommunications are found in major cities. Furthermore, the business environment of major cities tend to be more complex, and therefore attract more services (Kirn, 1987). The concentration in and about Taipei City is traceable to the fact that Taipei is at the lower end of the World City Hierarchy (Friedman and Wolff, 1982), and displays many of the functions and characteristics of a world orientated service economy (Selya, 1994).

It is possible to argue that service industries are more equally distributed than manufacturing, although the growth of services in absolute terms favored urban and suburban areas and not remote rural areas. It might be argued that the apparent disparities in the measuring of service location are inherent in the ambiguity of services in regard to what constitutes demand and market (Allen, 1988c). As such the growth, and distribution, of different types of services are the function of different causal processes which will be reflected in different measures of location. Finally the possible impact of government policies regarding access to public services and restructuring of

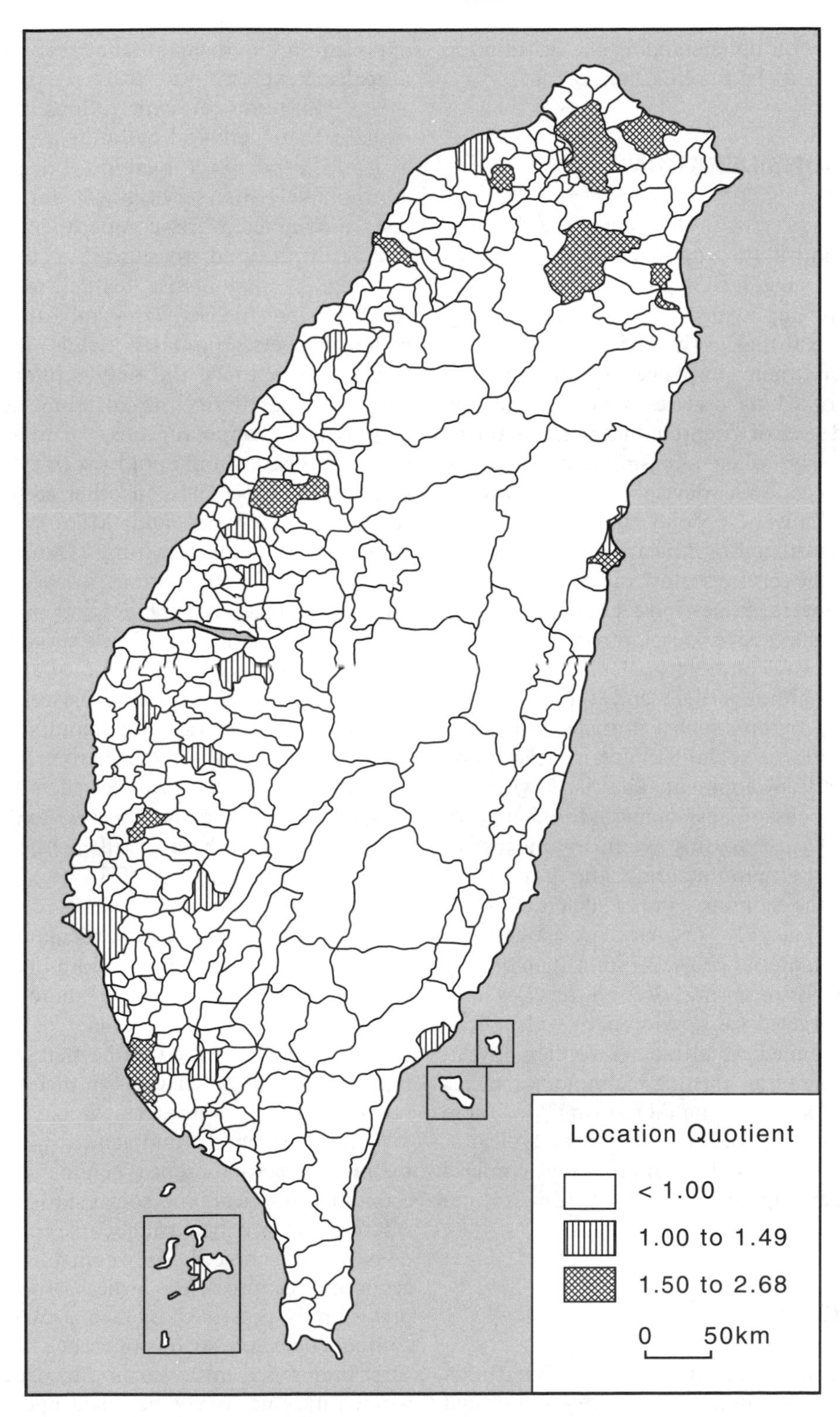

Figure 1 Location quotient of employment in services, 1986

the economy on understanding the distribution of services must be raised (Daniels, 1982).

POLICY CONSIDERATIONS

If services do truly dominate the Taiwan economy and if they do have locational attributes which suggest they provide greater spatial equality in employment than manufacturing, why, then, is it that in the various development plans, and their implementation, for the Republic of China services have not received the same degree of attention that manufacturing has? Similarly, when assessing the origins of successful economic development in Taiwan, why have Taiwanese politicians, planners and government officials not mentioned or analyzed the role of the service sector? In answering these questions several issues must be kept in mind. First, the successive development plans did include macro-economic goals and objectives for services, although they tend to be subsumed in sectoral programs such as transportation and communications, social welfare, tourism, and research and development. That is, services are never the focus of development for their own sake. They are mainly seen as means to improving the manufacturing and trade components of the economy, and as such investment in services usually involves investment in physical or material projects rather than quality of life or skills programs. Second, services have not been targeted for development with special legislation aimed at attracting foreign investment or upgrading existing technologies as has been the case for manufacturing. In these regards Taiwan appears to have been typical of most countries in its lack of conscious comprehensive development planning for services (Riddle, 1986).

CONCLUSIONS

In arguing that Taiwan is best analyzed as a service economy and that more attention should be given to services in the planning process, it is important not to overstate the case and to avoid unrealistic expectations.

In terms of expectations it must be realized that there will be limits as to what an expanded service sector can do. For example, it is most likely that services will not be able to restore completely the comparative advantage that Taiwan used to enjoy in textiles and clothing. It may assist textile and clothing manufacturers in discovering and capturing new niche markets. Similarly, while in general, services seem to be the only economic sector with the possibility of offering large-scale employment opportunities, growth of the service industry should not be seen as an overall panacea for joblessness in other sectors of the economy (Gershuny and Miles, 1983), and especially in manufacturing (Daniels, 1983). This may be less important in the short run, given Taiwan's low unemployment rate, and because there is no significant rural surplus of workers to be absorbed into the urban economy. In the long run, the growth of services may come to compete for workers in manufacturing for two reasons. First, as the Taiwan population continues to age, new services such as adult or senior day-care, nursing homes, and physical and occupational therapy will emerge as growth areas. Second, residents of Taiwan, like those elsewhere, seem to prefer working in services (Riddle, 1986). Women in particular do not see careers in manufacturing as more than temporary situations before marriage, and/or opening a small shop. Such preferences may be interpreted in many ways. It may well be that women say that they would rather work in their own shops because of the long term impact of wider systems of patriarchy and gender inequality. Or the preferences may reflect genuine shifts in the economic, political, and social status of women wherein they are increasingly able to make their own career choices and control their own economic resources. Regardless of which interpretation is preferred there are two practical consequences of such preference statements. First they represent a warning to planners that women may no longer be relied upon to automatically become a part of the manufacturing sector. Second they represent a major shift in the perception of services as a socially acceptable career area. There is a long time Chinese aversion to serving other people since such activities appear to be undignified. Training programs and the lure of higher salaries than available in manufacturing should help to overcome any lingering inhibitions to employment in the service economy. Finally, while expansion of services may foster regional economic development by creating new businesses, by increasing demands on existing ones, or expanding the export base of an area, in fact any such expansion may also lead to a reaffirmation of existing regional economic disparities, or may even exacerbate them. This may be true for both services and industry (Daniels, 1983). Service industries then may be limited in their ability to radically change or regenerate regional and local economies, while at the same time the lack of services may be a crucial factor in the stagnation of manufacturing. Growth of services may then require the same type of policies already in place to redress regional imbalances in Taiwan industry (Selya, 1993). Regardless of these limitations, the development of services via the processes of deregulation and internationalization appears to be a necessary condition if Taiwan is to restructure and thereby insure the long term health of its economy (Dicken, 1992).

Saskia Sassen

'Place and Production in the Global Economy'

from ***Cities in a World Economy* (1994)**

As the end of the twentieth century approaches, massive developments in telecommunications and the ascendance of information industries have led analysts and politicians to proclaim the end of cities. Cities, they tell us, should now be obsolete as economic entities. With large-scale relocations of offices and factories to less congested and lower cost areas than central cities, the computerized workplace can be located anywhere: in a clerical 'factory' in the Bahamas or in a home in the suburbs. The growth of information industries has made it possible for outputs to be transmitted around the globe instantaneously. And the globalization of economic activity suggests that place – particularly the type of place represented by cities – no longer matters.

This is but a partial account, however. These trends are indeed all taking place, but they represent only half of what is happening. Alongside the well-documented spatial dispersal of economic activities, new forms of territorial centralization of top-level management and control operations have appeared. National and global markets as well as globally integrated operations, require central places where the work of globalization gets done. Furthermore, information industries require a vast physical infrastructure containing strategic nodes with a hyperconcentration of facilities. Finally, even the most advanced information industries have a production process.

Once this process is brought into the analysis, funny things happen; secretaries are part of it, and so are the cleaners of the buildings where the professionals do their work. An economic configuration very different from that suggested by the concept of information economy emerges, whereby we recover the material conditions, production sites, and place-boundedness that are also part of globalization and the information economy. A detailed examination of the activities, firms, markets, and physical infrastructure that are involved in globalization and concentrated in cities allows us to see the actual role played by cities in a global economy. Thus when telecommunications were introduced on a large scale in all advanced industries in the 1980s, we saw the central business districts of the leading cities and international business centers of the world – New York, Los Angeles, London, Tokyo, Frankfurt, São Paulo, Hong Kong, and Sydney, among others – reach their highest densities ever. This explosion in the numbers of firms locating in the downtowns of major cities during that decade goes against what should have been expected according to models emphasizing territorial dispersal; this is especially true given the high cost of locating in a major downtown area.

If telecommunications has not made cities obsolete, has it at least altered the economic function of cities in a global economy? And if this is so, what does it tell us about the importance of place, of the locale, in an era dominated by the imagery and the language of economic globalization and information flows? Is there a new and strategic role for major cities, a role linked to the formation of a truly global economic system, a role not sufficiently recognized by analysts and policymakers? And could it be that the reason this new and strategic role has not been sufficiently recognized is that economic globalization – what it actually takes to implement global markets and processes – is misunderstood?

The notion of a global economy has become deeply entrenched in political and media circles all over the world. Yet its dominant images – the instantaneous transmission of money around the globe, the information economy, the neutralization of distance through telematics – are partial and hence profoundly inadequate representations of what globalization and the rise of information economies actually entail for cities. Missing from this abstract model are the actual material processes, activities, and infrastructures that are central to the implementation of globalization. Both overlooking the spatial dimension of economic globalization and overemphasizing the information dimensions have served to distort the role played by major cities in the current phase of economic globalization.

The last 20 years have seen pronounced changes in the geography, composition, and institutional framework of economic globalization. A world economy has been in existence for several centuries, but it has been reconstituted repeatedly over time. A key starting point for this book is the fact that, in each historical period, the world economy has consisted of a distinct combination of geographic areas, industries, and institutional arrangements. One of the important changes over the last 20 years has been the increase in the mobility of capital at both the national and especially the transnational level. The transnational mobility of capital brings about specific forms of articulation among different geographic areas and transformations in the role played by these areas in the world economy.

This trend in turn produces several types of locations for international transactions, the most familiar of which are export processing zones and offshore banking centers. One question for us, then, is the extent to which major cities are yet another type of location for international transactions, though clearly one at a very high level of complexity.

Increased capital mobility does not only bring about changes in the geographic organization of manufacturing production and in the network of financial markets. Increased capital mobility also generates a demand for types of production needed to ensure the management, control, and servicing of this new organization of manufacturing and finance. These new types of production range from the development of telecommunications to specialized services that are key inputs for the management of a global network of factories, offices, and financial markets. The mobility of capital also includes the production of a broad array of innovations in these sectors. These types of production have their own vocational patterns; they tend toward high levels of agglomeration. We will want to ask whether a focus on the production of these service inputs illuminates the question of place in processes of economic globalization, particularly the kind of place represented by cities.

Specialized services for firms and financial transactions, as well as the complex markets both entail, are a layer of activity that has been central to the organization of major global processes in the 1980s. To what extent is it useful to think in terms of the broader category of cities as key locations for such activities – in addition to the more narrowly defined locations represented by headquarters of transnational corporations or offshore banking centers – to further our understanding of major aspects of the organization and management of the world economy?

Much of the scholarly literature on cities has focused on internal aspects of the urban social, economic, and political systems, and it has considered cities to be part of national urban systems. International aspects typically have been considered the preserve of nation-states, not of cities. The literature on international economic activities, moreover, has traditionally focused on the activities of multinational corporations and banks and has seen the key to globalization in the power of multinational firms. Again, this conceptualization has the effect of leaving no room for a possible role by cities.

Including cities in the analysis adds two important dimensions to the study of economic internationalization. First, it breaks down the nation-state into a variety of components that may be significant in understanding international economic activity. Second, it displaces the focus from the power of large corporations over governments and economies to the range of activities and organizational arrangements necessary for the implementation and maintenance of a global network of factories, service operations, and markets; these are all processes only partly encompassed by the activities of transnational corporations and banks. Third, it contributes to a focus on place and on the urban social and political order associated with these activities. Processes of economic globalization are thereby reconstituted as concrete production complexes situated in specific places containing a multiplicity of activities and interests, many unconnected to global processes. Focusing on cities allows us to specify a geography of strategic places on a global scale, as well as the microgeographies and politics unfolding within these places.

The transformation during the last two decades in the composition of the world economy accompanying the shift to services and finance brings about a renewed importance of major cities as sites for certain types of activities and functions. In the current phase of the world economy, it is precisely the combination of the global dispersal of economic activities and global integration – under conditions of continued concentration of economic ownership and control – that has contributed to a strategic role for certain major cities that I call global cities (Sassen, 1991). Some have been centers for world trade and banking for centuries, but beyond these long-standing functions, today's global cities are (1) command points in the organization of the world economy; (2) key locations and marketplaces for the leading industries of the current period, which are finance and specialized services for firms; and (3) major sites of production for these industries, including the production of innovations. Several cities also fulfill equivalent functions on the smaller geographic scales of both trans- and subnational regions.

Alongside these new global and regional hierarchies of cities is a vast territory that has become increasingly peripheral, increasingly excluded from the major economic processes that fuel economic growth in the new global economy. A multiplicity of formerly important manufacturing centers and port cities have lost functions and are in decline, not only in the less developed countries but also in the most advanced economies. This is yet another meaning of economic globalization. We can think of these developments as constituting new geographies of centrality (that cut across the old divide of poor/rich countries) and of marginality that have become increasingly evident in the less developed world and in highly developed countries as well.

The most powerful of these new geographies of centrality binds the major international financial and business centers: New York, London, Tokyo, Paris, Frankfurt, Zurich, Amsterdam, Sydney, Hong Kong, among others. But this geography now also includes cities such as São Paulo and Mexico City. The intensity of transactions among these cities, particularly through the financial markets, flows of services, and investment has increased sharply, and so have the orders of magnitude involved. At the same time, there has been a sharpening inequality in the concentration of strategic resources and activities between each of these cities and others in the same country. For instance, Paris now concentrates a larger share of leading economic sectors and wealth in France than it did 20 years ago, whereas Marseilles, once a major economic center, has lost its share and is suffering severe decline. Some national capitals, for example, have lost central economic functions and power to the new global cities, which have taken over some of the coordination functions, markets, and production processes once concentrated in national capitals or in major regional centers. São Paulo has gained immense strength as a business and financial center in Brazil over Rio de Janeiro – once the capital and most important city in the country – and over the once powerful axis represented by Rio and Brasilia, the current capital. This is one of the meanings, or consequences, of the formation of a globally integrated economic system.

What is the impact of this type of economic growth on the broader social and economic order of these cities? A vast literature on the impact of a dynamic, high-growth manufacturing sector in highly developed countries shows that it raises wages, reduces economic inequality, and contributes to the formation of a middle class. There is much less literature on the impact of the service economy, especially the rapidly growing specialized services.

Specialized services, which have become a key component of all developed economies, are not usually analyzed in terms of a production or work process. Such services are usually seen as a type of output – that is, high-level technical expertise. Thus insufficient attention has been paid to the actual array of jobs, from high paying to low paying, involved in the production of these services. A focus on production displaces the emphasis from expertise to work. Services need to be produced, and the buildings that hold the workers need to be built and cleaned. The rapid growth of the financial industry and of highly specialized services generates not only high-level technical and administrative jobs but also low wage unskilled jobs. Together with the new inter-urban inequalities mentioned above, we are also seeing new economic inequalities within cities, especially within global cities and their regional counterparts.

The new urban economy is in many ways highly problematic. This is perhaps particularly evident in global cities and their regional counterparts. The new growth sectors of specialized services and finance contain capabilities for profit making that are vastly superior to those of more traditional economic sectors. The latter are essential to the operation of the urban economy and the daily needs of residents, but their survival is threatened in a situation where finance and specialized services can earn superprofits. This sharp polarization in the profit-making capabilities of different sectors of the economy has always existed. But what we see happening today takes place on a higher order of magnitude, and it is engendering massive distortions in the operations of various markets, from housing to labor. We can see this effect, for example, in the unusually sharp increase in the beginning salaries of MBAs and lawyers and in the precipitous fall in the wages of low-skilled manual workers and clerical workers. We can see the same effect in the retreat of many real estate developers from the low- and medium-income housing market who are attracted to the rapidly expanding housing demand by the new highly paid professionals and the possibility for vast over-pricing of this housing supply.

The rapid development of an international property market has made this disparity even worse. It means that real estate prices at the center of New York City are more connected to prices in London or Frankfurt than to the overall real estate market in the city. Powerful institutional investors from Japan, for instance, find it profitable to buy and sell property in Manhattan or central London. They force prices up because of the competition and raise them even further to sell at a profit. How can a small commercial operation in New York compete with such investors and the prices they can command?

The high profit-making capability of the new growth sectors rests partly on speculative activity. The extent of this dependence on speculation can be seen in the crisis of the 1990s that followed the unusually high profits in finance and real estate in the 1980s. The real estate and financial crisis, however, seems to have left the basic dynamic of the sector untouched. The crisis can thus be seen as an adjustment to more reasonable (that is, less speculative) profit levels. The overall dynamic of polarization in profit levels in the urban economy remains in place, as do the distortions in many markets.

The typical informed view of the global economy, cities, and the new growth sectors does not incorporate these multiple dimensions. Elsewhere I have argued that we could think of the dominant narrative or mainstream account of economic globalization as a narrative of eviction (Sassen, 1993). In the dominant account, the key concepts of globalization, information economy, and telematics all suggest that place no longer matters and that the only type of worker that matters is the highly educated professional. This account favors the capability for global transmission over the concentrations of established infrastructure that make transmission possible; favors information outputs over the workers producing those outputs, from specialists to secretaries; and favors the new transnational corporate culture over the multiplicity of cultural environments, including reterritorialized immigrant cultures within which many of the 'other' jobs of the global information economy take place. In brief, the dominant narrative concerns itself with the upper circuits of capital, not the lower ones.

This narrow focus has the effect of excluding from the account the place-boundedness of significant components of the global information economy; it thereby also excludes a whole array of activities and types of workers from the story of globalization that are as vital to it as international finance and global telecommunications are. By failing to include these activities and workers, it ignores the variety of cultural contexts within which they exist, a diversity as present in processes of globalization as is the new international corporate culture. When we focus on place and production, we can see that globalization is a process involving not only the corporate economy and the new transnational corporate culture but also, for example, the immigrant economies and work cultures evident in our large cities.

The new empirical trends and the new theoretical developments have made cities prominent once again in most of the social sciences. Cities have reemerged not only as objects of study but also as strategic sites for the theorization of a broad array of social, economic, and political processes central to the current era: economic globalization and international migration; the emergence of specialized services and finance as the leading growth sector in advanced economies; and new types of inequality.

Susan M. Roberts

'Small Place, Big Money: The Cayman Islands and the International Financial System'

***Economic Geography* (1995)**

In *The Condition of Postmodernity*, Harvey presented his 'basic (though tentative) conclusion' that 'if we are to look for anything truly distinctive (as opposed to 'capitalism as usual') in the present situation, then it is upon the financial aspects of capitalist organization and on the role of credit that we should concentrate our gaze' (Harvey, 1989, p. 196). A small but growing geographic literature now deals with aspects of money and finance. Certainly, the rapid developments in the international financial system during the past 20 years and the associated shifts in the spatiality of capitalism have been difficult to ignore. Prompted by the increased salience of finance, geographers have undertaken empirical work at a variety of scales.

The 1970s and 1980s saw massive reorganization and growth in the international financial system. The development of entirely new markets – fuelled by deregulation and technological advances – was a rapid and enormously complex process. Marxian and non-Marxian analysts agree that these financial innovations enabled capital to be much more flexible. In a Marxian framework this enhanced flexibility is seen as allowing capital to overcome (albeit temporarily) its own deep contradictions that tend toward crisis – even as new types of potential crisis are generated. It is through the dynamic financial markets that crisis may be both deferred (temporally) and displaced (spatially). Harvey (e.g. 1985) has shown how temporal and spatial 'mismatches' between the demand for and supply of surplus capital can be arbitraged through the financial markets. As this happens, new social understandings and experiences of time and space develop. Moreover, new types of material spaces, such as new financial centers, emerge. Contemporary capitalism, through its hallmark component, the dynamic global financial system, is, to use Lefebvre's (1991) term, secreting distinct new spatial forms.

Indeed, in the 1970s and 1980s, at the same time as the top financial centers such as London and Tokyo were undergoing far-reaching change in the way they functioned and in the spatiality of their operations, a series of new financial centers that came to account for a significant portion of the world's cross-border bank claims appeared on the world map. From the Bahamas to Luxembourg to Vanuatu, small and often marginal places became offshore financial centers (OFCs): new and distinctive spaces corresponding to nodes in the circuits of offshore financial markets that developed in the 1970s (Roberts, 1994). These new spaces correspond to the uneven global topography of financial regulation, composed of different national spaces of regulation. This paper examines one of these new regulatory spaces characteristic of contemporary capitalism's financial turn: the Cayman Islands.

FROM TAX HAVEN TO OFFSHORE FINANCIAL CENTER

'Imagine three tiny gems dotted in the azure waters of the Caribbean palm fringed white sand beaches, lazy days, balmy nights, you're almost there – The Cayman Islands.' So begins a tourist brochure elaborating the tempting pleasures of the Cayman Islands. In 1992, more than 241,000 vacationers stayed in the Caymans and more than 613,000 cruise ship passengers disembarked for a day ashore Grand Cayman (Cayman Islands Government, 1993). When these visitors first explore George Town, the Caymanian capital, they may be surprised to find a bustling town with several shiny new office buildings displaying the names of banks from around the globe. They would be perhaps even more surprised if they stepped into the air-conditioned marble lobby of such a building and saw rows of brass or plastic plates bearing the names of the hundreds of companies for which the building is the registered office. These wayward tourists would have stumbled upon the second pillar of the Caymanian economy: offshore finance. With no significant natural resources and little prospect for agriculture, the Caymans have opened themselves up to two kinds of foreign visitors – those lying on the beach and those inscribed on the nameplates in the lobbies of George Town office buildings.

The Cayman Islands, like many other established and aspiring OFCs, built upon their history as a tax haven. A tax haven is a jurisdiction

> (a) where there are no relevant taxes; (b) where taxes are levied only on internal taxable events but not at all, or at low rates, on profits from foreign sources; or (c) where special tax privileges are granted to certain types of taxable persons or events. Such special tax privileges may be accorded by the domestic internal tax system or may derive from a combination of domestic and treaty, provisions. (Spitz, 1983, p. 1)

Thus the tax haven designation is a relative one. A place or jurisdiction takes on haven attributes only in relation to another place or jurisdiction. The Cayman Islands are one of a number of places in the world that are 'no-tax jurisdictions'. While duties are levied on imports and there are sales taxes, the Islands have no income, capital gains, or inheritance taxes. The no-tax status of the Cayman Islands has a long history, but it was not until the 1960s that serious attention was given to developing the Islands' tax haven role. The two major obstacles to attracting tourists had been tackled in the previous decade: an airstrip had been opened in 1953, and the notoriously fierce mosquito population was being brought under control. But despite offering some chances for employment, the nascent vacation industry was unable to absorb all the Caymanian men who found themselves displaced from their jobs on merchant vessels as supertankers, containerized cargo, and cheaper, less-skilled labor changed the merchant shipping industry and labor market. In the 1960s, faced with this situation, Caymanian business leaders and politicians embarked on a deliberate strategy to develop the Islands' economy based dually on tourism and, by augmenting the tax haven role, offshore finance. Despite attempts at diversification, the Cayman Islands' economy remains almost entirely dependent on these two industries.

Some international banking was already present on Grand Cayman in the 1960s. Barclays Bank had arrived in 1953 and was followed ten years later by The Royal Bank of Canada and the Canadian Imperial Bank of Commerce. The 1960s were a time of laying the political and legislative foundations for an offshore financial sector. In 1962, after heated political battles, the Cayman Islands decided to become a British Crown Colony in their own right and not join Jamaica (of which they had been a dependency) in independence. The new Crown Colony took the first steps toward establishing the framework for a financial sector when, four years later, it refused to continue being a party to Jamaican double taxation treaties. The keystone Banks and Trust Companies Regulations Law and the

Exchange Control Regulations Law were also enacted in 1966. In addition, the Caymans began to emphasize several advantageous factors in promoting itself as an OFC: the Islands' location in the same time zone as Miami and New York (see Figure 1); the ease of communication, via frequent flights to the United States, in particular, and a reliable telecommunications service; and, above all, the stable political and social environment of a British Crown Colony. These basic factors are still listed in Caymanian promotional material today.

By 1972, more than 3,000 registered companies and more than 300 trust companies had been established in the Caymans. The Bank of Nova Scotia, First National City Bank (now Citibank), and three private banks were also added to the register (Great Britain, 1973, pp. 15–16). The Cayman Islands thus began as an offshore center focused on offshore company formation and trusts. Offshore banking soon developed, however. The Bahamas had been the regional leader in offshore banking, but the financial community there became fearful of restrictive measures when the Bahamas gained independence; the Cayman Islands were waiting with the necessary legal framework (much of it based on Bahamian law) and some basic infrastructure in place. Several companies, some banks and trust companies, and a number of personnel moved to the Caymans from the Bahamas, giving the Caymanian offshore financial sector a timely boost.

The Cayman Islands today rank as a well-established OFC in the Caribbean with a relatively diversified offshore sector. Offshore companies, offshore banking services, trusts, captive insurance, offshore mutual funds, and a variety of complex financial deals are to be found on Grand Cayman. The main offshore industries in the Cayman Islands and their salient characteristics are presented below.

Banks

Although a number of Canadian and British banks had located in the Cayman Islands by the early 1970s, the number of banks, especially from the United States, increased dramatically on Grand Cayman during the 1970s. Indeed, from 1972 to 1982 the average annual growth rate in the number of category 'B' (offshore) bank licenses was 23.4 percent. From 1982 to 1989 the average annual growth rate was just 3.1 percent (Figure 2). The explanation for the sudden rise in offshore banking in the Caymans between 1972 and 1982 lies in the development of the Euromarkets (see Yannopoulos, 1983, pp. 245–6). The Caymans' experience was not atypical. In fact, the history of the world's OFCs and the Euromarkets are parallel. The Euromarkets are the quintessential offshore market. They arose 'off the shore' of the United States as an exogenous market in US dollars.

Although there is debate as to the relative importance of the factors leading to the Eurodollar and eventually the Euromarkets (as other national currencies and bonds were traded outside their home countries), it is clear that in general terms they were a response to the uneven geography of financial regulation in the 'developed world'. First, in 1957, as a result of the 'sterling crisis', the British government restricted the availability of loans in sterling for trade financing outside the sterling area. In order to meet their customers' demands for finance, the banks turned to the unrestricted US dollar. The London-based market for US dollars continued to grow as banks sought ways around US regulations. The Interest Equalization Tax (1963–74) was designed to stem the outflow of dollars, but instead acted to herd investors toward the cheaper Eurodollar market (Versluysen, 1981, p. 25). In addition, the so-called Regulation Q gave the Federal Reserve the authority to set a ceiling on interest rates. Before the inflationary 1970s the maximum rates set under Regulation Q remained above the market rates. In the 1970s, however, the market rates rose higher than those the US banks were permitted to offer. Partly to circumvent these regulatory rigidities, US banks established branches in London. London became, and remains, the hub of the Euromarkets.

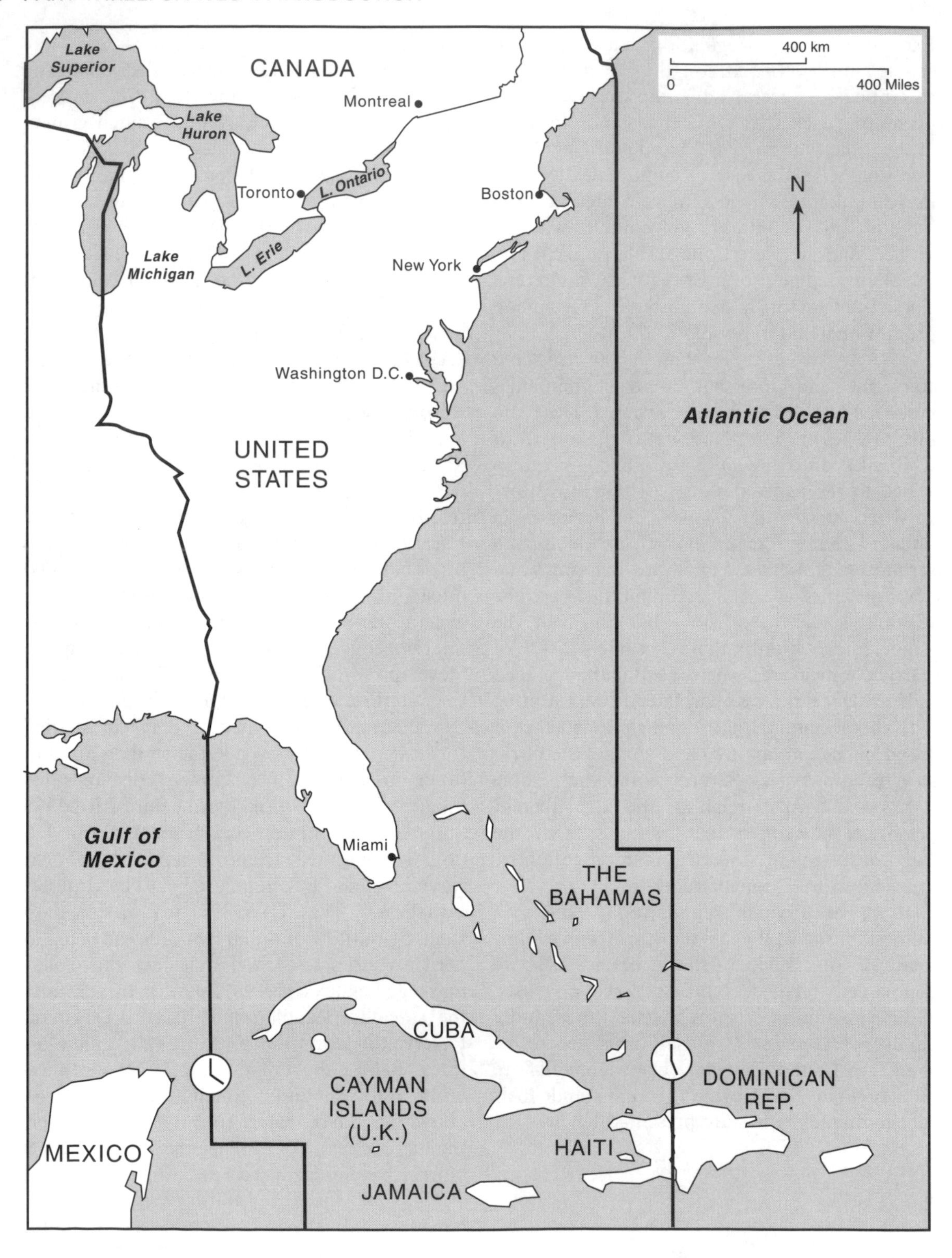

Figure 1 Location of the Cayman Islands

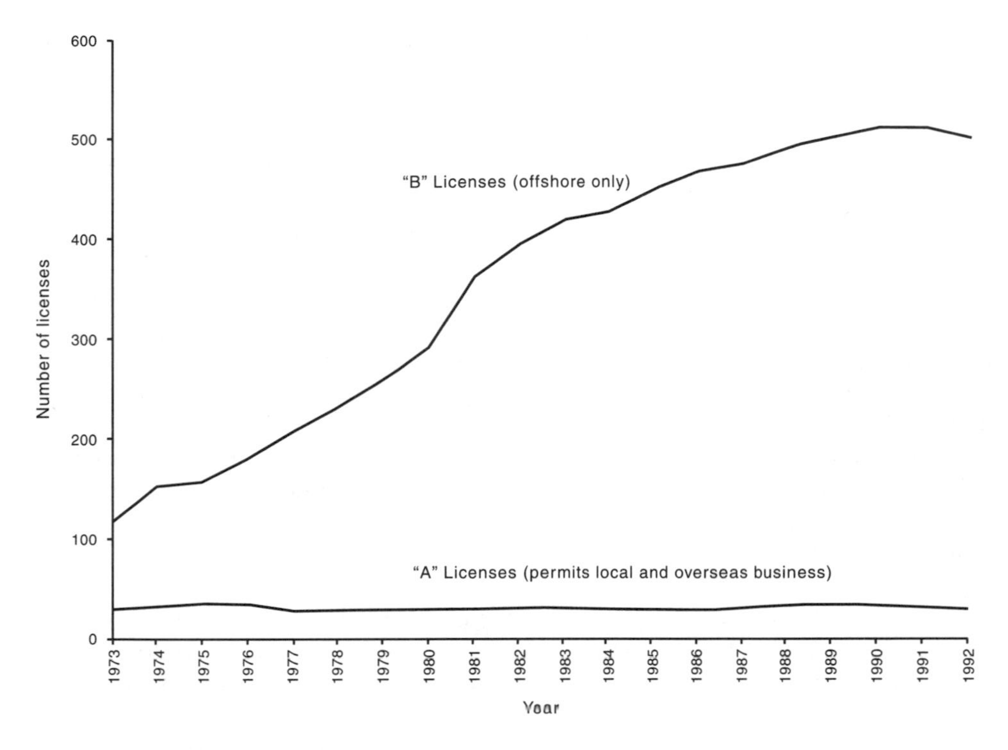

Figure 2 Cayman Islands, banking and trust licenses, 1972–92

Since its establishment, the Eurocurrency market has undergone significant widening. Currencies other than the US dollar are now important. The Japanese yen and some other currencies have grown in share of the Eurocurrency market, and the European Currency Unit (ECU) is also used. In addition to the currency market, there is also a Eurobond market that was originally a market for fixed-interest, longterm borrowing by corporations, governments, and municipalities. Now it includes floating-rate notes and has been the site of tremendous innovation. A host of ingenious financial instruments have been created as part of the global trend toward greater and greater securitization.

The Eurocurrency markets grew spectacularly in the 1970s. The Bank for International Settlements (BIS) estimates that the size of the Eurocurrency market grew from $57 billion in 1970 to a staggering $661 billion in 1981, reflecting the deposits of Organization of Petroleum Exporting Countries' surpluses, or 'petrodollars'. It was during this period that offshore banking in places such as the Cayman Islands took off, as internationalizing banks sought unregulated, 'tax-efficient' bases from which to conduct (or at least to which to 'book') their Eurocurrency business. Much of the 'recycling' of petrodollars undertaken by banks in the 1970s was done in the Euromarket. Syndicated 'Euroloans' (such as those to Third World countries) were often routed via an offshore center such as the Caymans.

A bank does not need to set up a staffed office on Grand Cayman to use the Caymans in this way. Indeed, more than 80 percent of the 532 banks registered do not maintain any physical presence in the Islands. These are sometimes called 'shell' branches, 'booking centers', or 'brass plate' banks and are found only as a brass or plastic nameplate in the lobby of another bank or law firm. The Eurocurrencies that are 'booked' to Cayman shell branches do not actually come to the Islands but are merely entered into a ledger or a computer file by the parent bank's staff onshore. Although London remains the center of the Eurocurrency market, the Bahamas and Caymans together account for about the same amount of US banks' branches abroad in terms of assets booked there. (For example, in October 1988 31.5 percent of assets were in London, while the Bahamas and Cayman together accounted for 31.45 percent.) The leading country of origin is the United States, followed by Switzerland (Inspector of Banks and Trust Companies). Six of the 532 banks are full-service clearing banks (Barclays, Royal Bank of Canada, Bank of Nova Scotia, Canadian Imperial Bank of Commerce, Cayman National Bank, and Butterfields). Another 24 banks hold licenses to operate onshore, but the vast majority of registered banks (502) are purely offshore operations. Cayman banks had some $400 billion in external assets in 1992 (Cayman Islands Government, 1993). By this measure, the Caymans rank as one of the world's largest financial centers.

Insurance

The Caymans rank second in the world (by numbers of companies) as a center for offshore insurance. Bermuda is the world's largest offshore insurance center, and Guernsey is in third place. Offshore insurance is the business of setting up 'captives'. A 'pure' captive is 'a wholly owned or controlled company that only insures or reinsures the risks of its non-insurance parent or affiliated companies' (KPMG, 1988, p. 5). Other kinds of captive insurance companies in the Caymans include 'association' or 'industry' captives, which are 'insurance companies owned by a group of companies or members of a professional association solely to insure or reinsure the risks of shareholders', as well as 'agency' and 'open-market' captives (KPMG, 1988, p. 5).

Self-provided risk management or insurance coverage cuts out another company, thus reducing costs. Forming a captive offshore eliminates the uncertainty of cost and availability of coverage in the notoriously cyclical North American insurance market (KPMG, 1988, pp. 7–8). Like offshore companies in general, a captive may also be used as a profit center or as a vehicle for minimizing or deferring taxes.

The first Cayman Islands captive, a medical malpractice captive formed by a group of Boston-area hospitals affiliated with Harvard University, was set up in 1976. At the time Bermuda, the premier site for captive insurance, was not welcoming medical malpractice captives. The fledgling insurance industry on Cayman received a welcome boost when social unrest was feared in Bermuda. As in the case of banking, where the Caymans benefited from fears about an independent Bahamas, they also profited from apprehension over the riots that followed the trial and execution of the murderers of Bermuda's governor. Several Bermuda-based captives set up bolt-hole companies in Cayman, which they later found they could operate successfully. Recognizing the need to regulate (at least minimally) the growing insurance industry, the Caymans government passed the Insurance Law in 1979. It came into effect in 1980, and after an initial drop in the number of companies (some did not meet the requirements) the number of captives continued to grow. Under this law captives are required to be managed by a licensed insurance manager, of which there are 28 in the Caymans. The majority of captives are medical malpractice captives of US origin. Other kinds of risks that Cayman captives cover include a range of professional liabilities and workers' compensation. In 1992 there were 372 offshore insurance companies licensed in the Cayman Islands (Cayman Islands Government, 1993).

EFFECTS OF THE OFFSHORE SECTOR IN CAYMAN

The overall contribution of the financial industry to the Caymanian economy is hard to assess. In 1992 the government received some $25 million directly in fees and licenses for registration and operation (Cayman Islands

Government, 1993). Estimating for fees paid by companies and banks that are onshore rather than offshore operations would still make that figure over $20 million, which is more than 15 percent of total government revenue for 1992. Of course the government also has expenditures for the Banking Inspectorate and so forth. These figures are not disaggregated, however. The benefits that the private sector derives from the offshore financial sector are likewise difficult to quantify. There are 25 accountancy firms, including the big transnationals, in the Caymans. There are 18 law firms and approximately 80 lawyers, with an estimated 40 to 50 specializing in offshore finance. Nineteen firms specialize in offshore company management, and 28 are licensed insurance managers. Available 1992 data show that, together, banking and insurance (offshore and onshore) account for 10 percent of total employment in the Caymans (Cayman Islands Government, 1993). The insurance industry as a whole (including onshore insurance) employs 226 people (of whom 200 are Caymanians) (Cayman Islands Government, 1993). About 70 banks and trust companies have a physical presence on Grand Cayman, and insurance managers and company managers all collect fees for their services. There are 1,363 people employed in banking alone, the vast majority (1,062) of whom are Caymanian (Cayman Islands Government, 1993). The majority of professionals in the financial sector (such as lawyers and accountants) are expatriates, usually British and usually male.

The large number of non-Caymanians is a highly sensitive political issue. Caymanians now make up only 67 percent of the total resident population of the Caymans, which totals 29,700 (Cayman Islands Government, 1993). The tourism industry, which relies on large amounts of imported (unskilled and semi-skilled) labor, rather than the financial sector, is seen as the chief culprit. The financial industry is, however, acutely aware of the importance of social and political stability in attracting finance to the Islands, and there are attempts under way to train more Caymanians for some of the better jobs in the financial sector.

THE CAYMAN ISLANDS' DYNAMIC ECONOMIC AND POLITICAL CONTEXTS

Banking: Offshore and Off-Balance Sheet

On the demand side the highly volatile world financial system and the increasingly interconnected global networks of investment, trade, and manufacturing have meant that internationalizing capital requires ways to deal with or manage risk. The first customers for global risk management services offered by transnational banks were US institutional investors such as pension funds and insurance companies that sought to manage international portfolios. Now banks compete aggressively for such clients, offering them a package of corporate services under the label 'global custody'.

> At the heart of global custody are the basic services of settlement, asset safe-keeping, tax reclamation, collecting dividends and reporting information to the customer on anything relevant to his [*sic*] stocks – rights issues, takeover bids and the like. But there are a lot of other services that have developed – stock lending, investor accounting, performance measurement, master trust services, risk analysis. (Sowton, 1989, p. 24)

The Cayman Islands and other OFCs are used as nodes in a network by global custodians, with banks often 'cross-selling' jurisdictions to customers. In many cases the trading and cash management operations entailed in the management of very large international portfolios may not be executed in the Caymans (they may be done at another offshore center, such as Bermuda) but are booked to the Cayman Islands. Global custody business allows banks to pair the older tax haven attributes of the Caymans (particularly the trust concept) with new 'financial' instruments. Such business is necessarily technology-dependent, and global custody requires massive investments in telecommunications and data management technologies. That the Caymans have a reliable power supply, good quality telecommunications links, and available professional and technical skills means that participation in global custody networks is feasible.

Another newly developed package of services that banks are offering is 'international private banking' (IPB). While this sort of service has a long history in banking, it is only recently that banks have reorganized their internal structure in such a way as to be able to offer a cluster of services marketed as IPB. Through IPB banks are 'providing a convenient, confidential, and comprehensive program of personal financial counselling, structured to meet the investment, banking, and estate planning needs of wealthy individuals' (Wacker, 1989, p. 71). For example, Citibank, a leader in the promotion of IPB, reorganized internally to tap the resources of targeted so-called 'high net-worth' populations. They targeted people who were self-made, specifically seeking out professional athletes and entertainers, in addition to their traditional targets: those with inherited wealth. Through their incorporation into IPB networks, the Caymans have benefited from two key developments: first, the huge increase in the wealth of the rich in many countries (including the United States – one aspect of a wider trend Lipietz (1992b, p. 35) calls 'Brazilianization') in the 1980s; and second, the internationalization of these private monies as transnational banks compete to sniff out and set them in motion. In an interview, a Cayman Islands banker commented that

> People are beginning to realize that there is a fantastic amount of capital in the hands of private individuals. In the past it tended to be static – a historic chain of family money – and they were always advised by the same people and the money stayed in the same place. For example, they had a merchant bank in London look after their affairs and the money went into bonds, or perhaps blue chip stocks, and it stayed in Sterling. Now with internationalization and communications, people can have assets in several jurisdictions and in several currencies. The mobility of capital is far greater and they need more services: this is where private banking comes in. People have all sorts of assets all over the place. They might have a stud farm in North Florida, a house in Monte Carlo, a place on Mustique, and so on. It is very diverse.

The legally guaranteed confidentiality (albeit with exceptions under the Mutual Legal Assistance Treaty (1989) negotiated with the United States) is an important factor attracting IPB to the Caymans. Bankers acknowledge that the biggest risk they face with IPB is 'criminal involvement' (Wacker, 1989, p. 79), because there is considerable overlap between funds in IPB and international 'hot' money flows. While the exact connections between the gains of drugs trading or securities fraud and IPB are notoriously difficult to trace, the general link between IPB and capital flight is easier to see. Capital flight is difficult to define, but Gulati offers a working definition: 'speculative short-term capital outflows based on economic or political apprehensions in the home country' (Gulati, 1988, p. 166). In his review of attempts to measure capital flight, Gulati notes that a frequently used gauge is the 'errors and omissions line of the balance of payments as calculated by the IMF' (1988, p. 171). In addressing the problem of assessing the potential market size of IPB, Wacker writes: 'One consultant advises that the best way to determine the size of the global private banking market is to analyse international capital flows, particularly the errors and omissions account. Sadly most of the funds in the international private banking market come from countries experiencing political and economic instability' (Wacker, 1989, p. 73). Through capital flight the Cayman Islands have a double involvement in the Latin American debt crisis: as a booking center for the original Euroloans, and now as a haven for the flight capital, which is probably itself a cause as well as an effect of the debt crisis. Of course IPB is not all flight capital, and not all flight capital is IPB. Capital flight may take many forms, from smuggling to sophisticated financial deals spread out over a number of centers. The Caymans are a center for flight capital in the form of reinvoicing for corporations, especially from Latin America, worried about the sudden imposition of exchange controls.

By examining the changes in transnational banking it may be seen that the Caymans, and OFCs in general, are being used as 'platforms' from which banks can enter specialized financial markets. These new markets have exploded in size and variety in the past 20 years or so as circuits of capital, enabled by technological developments and deregulation, have sped up and internationalized. In addition to the new markets, the institutional form of financial intermediation has also been restructured. Multinational corporations (including banks) have undergone substantial changes in their structure and operations and have become more interlinked. Thus, as new spaces in the global political economy, OFCs are components in major shifts in the institutional forms of capital (corporate restructuring) and in the development of new financial markets. In other words, both capital's (particularly financial capital's) restructuring and the creation of new markets have been, in significant ways, achieved through offshore financial centers.

CONCLUSIONS

Contemporary capitalism is characterized by the ongoing development of an enormous and unprecedentedly complex international financial system. This system has become the site of rapid innovation as tremendous risks are generated and managed for profit on a firm-by-firm basis. There is much to suggest, however, that risks are cumulating and compounding at the systemic level, and this is a major problem facing regulatory authorities. However, capital has, through international financial markets, met increased volatility with flexibility in an effort to 'deal with' this risk. The changing nature of the practice of banking and the proliferation of risk are symptomatic of the ways in which deeper contradictions in contemporary capitalism are being displaced through time (which is the purpose of credit) and space. As the international financial system has grown in sheer size and sophistication, it has 'secreted' new and dynamic geographies. Mapping these new geographies shows that a new type of financial center has arisen: the offshore financial center. These are new regulatory spaces through which the new financial markets and institutions operate. The Cayman Islands are only a small place, but they have positioned themselves at the nexus of a series of emergent (and ever-changing) relations between capital and risk.

The Cayman Islands, and OFCs more generally, are new distinct spaces within a changing global political economy. OFCs are spaces that exemplify the contradictory movements towards global integration (as the flows of capital around the globe move faster and faster, for example) and local variation and fragmentation (as places act entrepreneurially to exploit niches in world markets and as capital hones its ability to arbitrage difference). OFCs, however, are not merely the result of changes in the financial system. Rather, major shifts in the way in which capital operates internationally simply cannot be understood without reference to the spaces *through which* they are achieved.

PART FOUR

Spaces of Consumption

PART FOUR SPACES OF CONSUMPTION

Section 4.1 Introduction: Consumption, Geography and Identity

INTRODUCTION

Economic geography has been largely concerned with understanding the location of economic activity and, more recently, with describing the shift from manufacturing to service employment, exploring the transition from Fordist to more flexible modes of production, and with trying to understand the dynamics of the increasingly global economy. Over the last decade research on consumption has expanded dramatically (Jackson and Thrift, 1995; Miller *et al*, 1998). Much of the geographical research on consumption, however, draws heavily upon work undertaken by sociologists (Veblen, 1970) and anthropologists (see Miller, 1995). An indication of this is that the first paper in this section is written by a sociologist (**Bocock**). A fine line also exists between geographical and anthropological research on consumption (see Douglas and Isherwood, 1978; Miller *et al*, 1998). Our approach in this introduction and in this section of the Reader is to highlight and develop key themes, concepts and ideas that enable economic geographers to explore the relationship between production, consumption and the spaces of consumption.

Consumption provides the link between production, need and the creation of social identities (**Bocock**). The relationship between people, consumption and commodities is one about 'language'. Consumption is about the projection of an image which can be read by others. Commodities like food, houses, clothing and cars have practical uses, but they are also all a non-verbal, but visual, mode of communication. Ownership of a Rolls-Royce car as against a Lada places the owner in a symbolic framework which is readily understood by an audience. Ownership of a commodity can never be divorced from this process of social construction. To understand these linkages in greater depth requires an understanding of the complexity of economic and social activities.

CAPITALIZING ON CONSUMPTION

'Who now reads Marx?' So begins **Lash and Urry**'s (see Part Two, Section 2.2) analysis of the new economy of signs and space. The answer to this question is that there is another Marx besides that of communism in 'eastern Europe'. Understanding why one should read Marx involves a brief exploration of his method of analysis. Central to this is a belief in identifying the processes that lie behind the superficial appearance of things. In the *Grundrisse*, Marx (1973) describes his method as one in which simple abstract concepts are developed, followed by the identification of their interrelationships and internal contradictions. In the 'introduction' to the *Grundrisse*, Marx makes two important points concerning consumption: the interconnectivity of the capitalist system and the circuit of capital.

The Interconnectivity of the Capitalist System

Production, distribution, exchange and consumption are the key features of a capitalist economy. These are not separate processes. To Marx 'production is also immediately consumption' (Marx, 1973, p. 90), and the term 'productive consumption' may be used to describe the relationships between these activities. Marx provides a number of examples to highlight his belief that the 'act of production is . . . in all of its moments also an act of consumption' (Marx, p. 90). Eating is a form of consumption in which the consumer produces his/her own body. This is also true of all types of consumption which in 'one way or another produces human beings in some particular aspect' (Marx, p. 91). Without production there is no consumption and without consumption no production. It is only through the process of consumption that commodities become real objects. Thus, a pair of jeans becomes a real pair of jeans only through the act of being worn. Consumption is the final part of the process of production.

Consumption also creates the need for new commodities, but this is not a simple one-way relationship. Production, especially through its link with the advertising industry (see the paper by **Daniels** in Part Three, Section 3.2; Mattelart, 1991), can create a demand for new products by linking them into the lifestyle of a target consumer audience. A good example of this type of lifestyle targeting is found in the American shampoo, Mane 'N' Tail. This product was developed as a horse shampoo. The product was rebranded as 'Mane 'N' Tail' when the manufacturer realized that stable girls were taking home bottles of the shampoo. The product is now sold in drug stores and retail chemists like Boots, but in the sophisticated world of the beauty salon it is branded under the name 'Equenne'. Mane 'N' Tail is part of the lifestyle associated with country pursuits, and its success is indicated by world-wide sales of $50 million (Buckley, 1995).

Advertising can also create lifestyles, for example the lifestyle promulgated by the American television presenter and journalist Martha Stewart, or by magazines such as *Wallpaper, Martha Stewart Living, Cosmopolitan* and *GQ*. Good examples of this process are car accessories that are produced incorporating a car manufacturer's logo, or clothing and novelty items linked to a sports team, for example Manchester United or the Chicago Bears. One of the best examples of this type of linkage is consumer goods linked to films. This process began during the 1920s when organizations such as Hollywood Fashion Associates and the Modern Merchandising Bureau emerged to co-ordinate the display of fashions in Hollywood film and fan magazines. In 1998, the link between Hollywood and consumption reached a new height with the movie remake of the cult, 1960s British television series *The Avengers*. Warner Brothers' publicity department began working on the movie a year prior to its release. In the first two weeks of August 1998 the British newspapers printed 15 separate features about *The Avengers*. There had also been 'roughly a dozen magazine covers, a limited edition PVC bra (called the Sex-A-Peel), a special line of clothes in Miss Selfridge and a top 10 single' (Gibson, 1998, p. 5). All of this three weeks before the movie was placed on general cinema release. This is part of the process of globalization with consumer demand in the UK being generated by an American publicity department. One of the best documented examples of the link between consumer demand and the global film industry is found in Kinder's (1991) analysis of the Nintendo Entertainment System and the *Teenage Mutant Ninja Turtles* television series. Between June 1988 and December 1989 $20 million worth of Turtle-related products had been sold (Kinder, 1991, p. 122). Films and their associated consumer goods may also come with a set of concealed values. Thus the work of Dorfman and Mattelart (1975) attempts to show that the Disney comic is a powerful ideological tool of American imperialism. Comics appear to be harmless, but to Dorfman and Mattelart, Disney projects a reading of society which naturalizes the social relations of Western capitalism.

Lifestyle consumption and the concept of productive consumption emphasizes the social relationships that bind production and consumption together. It is these social relationships – between employee and employer, between producer and consumer and between consumer and consumer – that are fundamental to the functioning of the capitalist system. Without the social relationships of consumption, lifestyle shopping would be meaningless. What is the point in keeping up-to-date with the latest fashion if no one is able to read the messages you are trying to project. It is not enough to wear jeans – they need to be the right colour, fit and especially brand name – Armani, Diesel, Levi 501s, but not C&As, Wal-Mart or J.C. Penney.

Circuits of Capital

To Marx capital must be defined as a *process* rather than a thing and it is through the circuit of capital that this process becomes visible. Money is transformed into commodities and back again into money plus profit: M–C–(M + Profit). Capital is constantly circulating through the economy seeking profit. Central to the circuit of capital is the notion of *turnover time*, or the time taken for the value of a given capital to be realized through production and exchange. Escalation in the turnover time of capital results in an increase in profitability.

One consequence of the relationship between profit and turnover time is the increasing importance of fashion (**Harvey**). It is to the producer's advantage to encourage consumers to replace last year's commodities with the latest fashionable version of the same product. Advertising, the development of brands and the increasing emphemerality of goods are key elements of this trend. A good example in the UK is the turnover time of clothes sold in supermarkets. Such supermarket clothes have short sell-by dates, with each range lasting only six weeks before it is replaced. The logic behind this rapid turnover is that the consumer will treat themselves to items from the clothing range each time they visit the supermarket.

One of the most important elements dictating the turnover time of clothes is colour. Why is it that each season comes with its own colour range? The answer to this question lies in the relationship between the activities of textile companies, the advertising industry, catwalk shows and the fashion press. In May each year, 25 people from different areas of the fashion industry, but especially from textile development organizations, meet in Mayfair, London. This is the twice-yearly meeting of the British Textile Colour Group (BTCG) which meets to begin the process of predicting the colours that will be fashionable in two years time. The fashionable colours identified during these meetings will determine the colour ranges available to all types of consumer products, from cars to coats. By the end of the day the British group will have identified a palette of colours. The next stage in the process is a meeting in Paris in which the British colour palette is presented to representatives from other countries. Colour palettes are thus another symptom of globalization. What was once just a decision for UK manufacturers and retailers has become increasingly a European and global decision.

The BTCG is one of the organizations involved in the forecasting of colour palettes that will determine High Street colours in two years time. Time is important for the fabric mills to plan and manufacture the textiles that will be required. In two years time the fashion designers will hold their catwalk shows and, not surprisingly, the discerning consumer will be able to identify easily a common colour palette. This is very much of a simplification of the process by which a season's colours are chosen. The important point is that seasonal colours are not about fashion, but about old-fashioned economics. It is economics that forces clothing manufacturers to change annually their colour palette. The fashion industry demands new colours for each season, and one element of the industry, the fashion media, operates to convince the consumer that beige, black, brown or grey, etc. is the current fashionable colour. What is occurring is a complex set of processes that connect designers, with manufacturers and consumers. The colour forecasters are not always right, and every season will see colours that are deemed by consumers to be decidedly unfashionable.

Circuits of Cultural Capital

Sharon **Zukin** develops the concept of a circuit of capital by exploring the relationships between capital and culture. She suggests that there are important linkages between economic and cultural values which may increase the economic value of capital. A good example of this process is found in television programmes and magazine articles which explore historic buildings and the work of named architects. Such publicity enhances the market value of historic properties as well as the value of architects and their buildings. The circuit of capital becomes transformed into a circuit of cultural capital when ideas are propounded by one element of the advertising industry and then incorporated into products, buildings and lifestyles. It is through Zukin's circuits of cultural capital that consumption patterns have a direct impact on geography, for example via processes of gentrification (Zukin, 1988).

In the advanced economies a merger of cultures and religions is increasingly occurring with each contributing to a circuit of capital. Christmas and Divali are a case in point in which religious festivals develop into advertising productive consumption spectacles with important ramifications for employment (part-time work), credit, retail sales, the popular music and book industries and television (Miller, 1993). A religious event becomes interwoven with a set of circuits of capital.

Circuits of cultural capital also operate to link the economies and cultures of the underdeveloped world with those of the advanced economies. Supermarkets and other retail chains attempt to attract and retain consumers by stocking items which are not available in other stores. The retailer has to identify a commodity and then create a demand by developing awareness of the product amongst potential consumers. Good examples of this type of product can be found in exotic fresh fruit, for example lychees, mangos and kumquats (**Cook**). The successful introduction of a new fruit amongst consumers involves a marriage between the activities of the retailer and of women's magazines or television programmes. Consumers need to become aware of the fruits and recipes that require them. This process also involves capturing the producers of these products located in underdeveloped countries and incorporating them into a circuit of capital. The growers have to be taught how to produce the fruit to the 'standards' required by the supermarkets. In some cases the knowledge required to grow the fruit is held by the retailer's contractors who acquire the fruit from growers. In this case the growers become effectively deskilled agricultural workers.

LANDSCAPES OF CONSUMPTION AND THE CONSUMPTION OF SPACES

During the nineteenth and early twentieth centuries new spaces (see **Knox**) and patterns of consumption began to develop amongst the middle and working classes. This was the age of the department store and arcade, places in which to see new mass-produced goods and to be seen. Consumption was no longer the preserve of the super wealthy and even the poorly paid working class could be seen to consume, albeit from street stalls and markets. The concept of 'conspicuous consumption' was developed by Thorstein Veblen (1970 [1899]; see **Bocock**) during this period to describe patterns of consumption which are designed to impress others. Part of this involves ostentation in the form of the giving of expensive gifts and feasts (**Clammer**). Such gift-giving makes the individual's wealth highly visible, as well as establishing the standing of the relationship between giver and receiver (Carrier, 1990).

The space of consumption is as important as the physical activity of consumption. Where one shops is endowed with meaning, for example Harrods versus C&As or Wal-Mart. All of these say something about the consumer. Thus, catalogue shopping provides a basic resource for social groups 'precluded from mainstream, leisure imbued formal shopping' (Clarke, 1998, p. 98). Shoppers with limited budgets frequently use catalogues to avoid the trauma of '"shopping around" with a restricted income' (Clarke, 1998, p. 92). Spaces of consumption are either exclusive or inclusive. Exclusive spaces are designed to attract the wealthy and repel the not so wealthy. Designer boutiques, Saville Row tailors and expensive jewellers try to maintain an air of exclusivity to ensure that they only attract those that can afford to consume in such spaces. Inadvertent browsers may occasionally wander into such stores only to be shocked by the price tags or absence of price tags.

Inclusive spaces attempt to attract all types of consumer except those too poor to consume (see **Knox**). Such spaces range from inner city department stores through to out-of town shopping malls, charity shops and car boot sales (**Wrigley and Lowe**). Each space of consumption is associated with a set of unwritten rules for the consumer. Thus the shopping mall is designed to encourage people to consume and to discourage window shopping, conversation and anything that distracts from consumption. The shopping mall is the ultimate in designed spaces; designed on the basis of psychology and economics (**Goss**). They are spaces of consumption, but also spaces to be consumed (Philio and Kearns, 1994).

Large out-of-town shopping centres are created and designed to be marketed as special places (see **Knox**). Places that contain a representative sample of all shops, and in some cases cultures and townscapes (for example, West Edmonton Mall (WEM), Canada (see the paper by **Goss**); or the Metrocentre, Gateshead, UK). Some malls have become tourist attractions in their own right, as tourist come to 'gaze upon or view a set of different scenes, of landscapes or townscapes' (Urry, 1990, p. 1). WEM is 'a world where Spanish galleons sail up Main Street past Marks & Spencer to put in at "New Orleans", where everything is tame and happy shoppers mingle with smiling dolphins' (Shields, 1989, p. 154). The Metrocentre (Gateshead, UK) contains 3 miles of shopping mall with over 350 shops, 50 restaurants, a 10-screen cinema, a bowling alley and a fantasyland of fairground rides and attractions. It also has four themed areas: a 'Mediterranean Village' with bubbling fountains and pavement bistros, a 'Roman Forum' with classical-styled Tavernetta, a 'Garden Court' with luscious greenery and waterfalls and an 'Antique Village' with a village pond and 'olde worlde tea shop' with water wheel (*Metrocentre Guide*). Within the confines of the Metrocentre the tourist can gaze and consume a variety of different landscapes, entertainments and shops. Malls are spaces in which to see, and to be seen. The same is also true for particular shopping streets, for example Bond Street in London, or Faneuil Square, Boston (**Zukin**; see **Knox**). This goes as far as involving the acquisition of the right set of shopping bags with the right brand images. It is these bags, as well as the costume of the consumer, that are read by others, including shop assistants.

The shopping mall is an extremely interesting type of space. The Metrocentre appears to be in no way dissimilar to the main shopping street of a large city, except that it is an enclosed heated space. The mall, however, is a privately owned and regulated space subject to high levels of surveillance. Only certain types of behaviour will be tolerated and the mall's 'police' force will ensure that only desirable people are permitted to consume its spaces. Thus, the homeless and unemployed are excluded. Shields (1989) suggests that people can enter the world of the mall and pretend that they have just shopped or just about to shop. They are able to gaze, stroll and be gazed upon (Urry, 1990; Shields, 1989) and to consume the space rather than relate to the mall as a space of consumption.

Selling WEM or the Metrocentre is similar to the process of selling cities (Philio and Kearns, 1994). Cities sell themselves to attract inward investment, out-of-town shoppers and increasingly exhibitions, fairs and trade shows (Rubalcaba-Bermejo and Cuadrado-Roura, 1995). The same place marketing processes are at work in the city and the shopping mall. The Metrocentre has to attract shoppers whilst cities like Birmingham have to develop and maintain their position in the European urban system. Such inter-urban competition is all about the development of a national and increasingly 'international presence'. The hosting of national and international political and trade events is one way for a city to sell itself to potential investors. Birmingham is an excellent example with its National Exhibition Centre (NEC) and International Convention Centre (ICC). In 1998, the city hosted a G8 summit meeting as well as the Eurovision Song Contest – both events raised Birmingham's profile in Europe. Such events come with a set of economic (consumption) multipliers: shopping, hotels and most importantly the projection of an image of Birmingham as a place suitable for inward investment.

PICK 'N' MIX: THE COMMODIFICATION OF IDENTITY

Like Crang's (1994) restaurant, shopping is also about display (**Clammer**) as well as the ability of the shop assistant to read the customer. A good popular example is Julia Roberts' inability to shop in the film *Pretty Woman*. The shop assistants did not classify her as a consumer of their clothes. Similarly, Chua (1992) highlights the relationship between the customer and sales assistant in a designer boutique located in Singapore. Here the sales process is classified as follows: customer as audience, salesperson as audience and shopping companion as audience. The important point is that clothes shopping is all about dressing the body for display. Stone (1962) labels the process of putting the components of appearance together as 'programming'. Each consumer 'programmes' his/her appearance according to the identity he/she desires to project. Such programming includes all forms of consumption from the car one drives, the house one lives in to the clothes one wears. An individual's identity, and the stage on which they project their image, can be manipulated to present the correct appearance (**Thrift and Leyshon**). Money can purchase the right education for one's children. The right education provides access to the right social networks and friends and will eventually lead to the right career. The relationship between consumption, lifestyle and success is thus extremely important (Miller *et al*, 1998). Image, and especially image articulated through consumption (education, accent, dress, cars, house, etc.), determines to an extent the ability of an individual to obtain and retain well-paid employment.

The programming of appearance can be directly related to success or failure in employment. Disney theme parks have stringent appearance criteria for staff – from a clean shave for men and 'the maintenance of an appropriate weight and size' (McDowell, 1995,

p. 77). At Disneyland, the self-proclaimed 'Happiest Place on Earth', the identities of new employees are 'not so much dismantled as . . . set aside as employees are schooled in the use of new identities' as they learn the Disney codes of conduct (Van Maanen, 1991, p. 73). The work of McDowell (1994) on the embodiment of financial workers in the City of London emphasizes the importance of appearance in the workplace. Dress can be used to fit into a social situation or appearance can be manipulated to achieve a desired result (**Clammer**). Women can play 'on their femininity to achieve visibility' (McDowell and Court, 1994, p. 380). One survey has shown that women are offered accountancy jobs with salaries in excess of £35,000 at a rate 25% faster than men (*Accountancy Magazine*, 1992, p. 21). This is explained by the appearance of men as anonymous and drab and dull in the traditional pinstripe suit, whereas women in their 'alternative' dress were easily remembered by interviewers. Similarly, McDowell shows the way in which women can become more or less female, depending on the circumstances and the location. Thus one female manager noted that her dress:

> Depends who I'm going to be seeing. Sometimes I'll choose the 'executive bimbo' look; at other times, like today when I've got to make a cold call, it's easiest if I'll blend into the background. I think this [a plain but very smart tailored blue dress] looks tremendously, you know, professional. No statement about me at all. 'Don't look at me, look at these papers I'm talking to you about.' But I wear high heels too, so I'm six feet tall when I stand up. And I think that commands some small sense of 'well, I'd probably better listen to her, at least for a little while'. I do dress quite consciously because you've got to have some fun in life, and sometimes wearing a leather skirt to work is just fun because you know they can't cope with it. (quoted in McDowell, 1994, p. 199)

CONCLUSION

One of the problems we faced in putting together this collection of readings is the difficulty of isolating one element of reality. Production can only really be understood in relation to consumption, and the world of work cannot be separated from production or consumption. This is especially important given the increasing importance of service work with its emphasis on the relationship between the customer and service worker (see **Allen and du Gay**'s paper in Part 5, Section 5.2).

One element needs further attention, and you should think about this during your 'reading' of these readings. This is the relationship between consumption and culture, especially in multicultural societies. The relationship between the production of a good and its consumption is always problematic, especially when culture operates to produce different readings of goods. It is important to remember that different cultures from within the same society may read an individual's 'programmed' lifestyle in different ways, and maybe not in the intended way (Thrift, 1988). Usually market research attempts to link the producer and consumer together (Nixon, 1997), but there is no such thing as the average consumer and consumers are notoriously unpredictable. A good example of the impact of culture on consumption is found in Clarks' decision in 1996 to call two designs of its trendy footwear, Vishnu and Krishna. In November 1997 this decision rebounded on the company when the 900 000 strong British Hindu community called for a boycott of Clarks shoes for 'blasphemy' over the use of names of Hindu gods as shoe labels. The footwear was considered to be doubly offensive – for Hindus the cow is sacred, and shoes are considered unclean and unfit to be worn in temples.

REFERENCES

Accountancy Magazine (1992) Fitter for the job, *Accountancy Magazine*, November: 21.

Buckley C (1995) A $50m winner straight from the horse's mane, *The Times*, Monday October 30: 40.

Carrier J (1990) reconciling commodities and personal relations in industrial society, *Theory and Society*, **19**: 579–98.

Chua B H (1992) Shopping for women's fashion in Singapore, in R Shields (ed) *Lifestyle Shopping: The Subject of Consumption*, London: Routledge, pp. 114–35.

Clarke A J (1998) Window shopping at home: Classifieds, catalogues and new consumer skills, in D Miller (ed) *Material Cultures: Why Some Things Matter*, London: UCL Press, pp. 73–99.

Crang P (1994) It's showtime: On the workplace geography of display in a restaurant in southeast England, *Environment and Planning D: Society and Space* **12**: 675–704.

Dorfman A and Mattelart A (1975) *How to Read Donald Duck: Imperialist Ideology in the Disney Comic*, New York: International General Editions.

Douglas M and Isherwood B (1978) *A World of Goods: Towards an Anthropology of Consumption*, Harmondsworth: Penguin.

Gibson J (1998) Avengers hype bypasses critics, *The Guardian*, Tuesday 11 August: 5.

Jackson P and Thrift N (1995) Geographies of consumption, in D Miller (ed) *Acknowledging Consumption*, London: Routledge.

Kinder M (1991) *Playing with Power in Movies, Television and Video Games*, Berkeley: University of California Press.

Marx K (1973) *Grundrisse*, Harmondsworth: Penguin.

Mattelart A (1991) *Advertising International: The Privatisation of Public Space*, London: Routledge.

McDowell, L (1994) *Capital Cultures: Gender at Work in the City*, Oxford: Blackwell.

McDowell L (1995) Body work: Heterosexual gender performances in city workplaces, in D Bell and G Valentine (eds) *Mapping Desire*, London: Routledge, pp. 75–95.

McDowell L and Court G (1994) Missing subjects: Gender, sexuality and power in merchant banks, *Economic Geography*, **70**: 229–51.

Miller D (ed) (1993) *Unwrapping Christmas*, Oxford: Oxford University Press.

Miller D (ed) (1995) *Acknowledging Consumption*, London: Routledge.

Miller D, Jackson P, Thrift N, Holbrook B and Rowlands M (1998) *Shopping, Place and Identity*, London: Routledge.

Nixon S (1997) Circulating culture, in P du Gay (ed) *Production of Culture/Cultures of Production*, London: Sage, pp. 177–234.

Philio C and Kearns G (1994) *Selling Places*, London: Paul Chapman.

Rubalcaba-Bermejo L and Cuadrado-Roura J R (1995) Urban hierarchies and territorial competition in Europe: Exploring the role of fairs and exhibitions, *Urban Studies* **32**, 2: 379–400.

Shields R (1989) Social spatialization and the built environment: The West Edmonton Mall, *Environment and Planning D: Society and Space* **7**: 147–64.

Stone G (1962) Appearance and the self, in A Rose (ed) *Human Behaviour and Social Process*, London: Routledge & Kegan Paul, pp. 86–118.

Thrift N (1988) Images of social change, in C Hamnett, L McDowell and P Sarre (eds) *The Changing Social Structure*, London: Sage.

Urry J (1990) *The Tourist Gaze: Leisure and Travel in Contemporary Societies*, London: Sage.

Van Maanen J (1991) The smile factory: Work at Disneyland, in P J Frost, L E Moore, M R Louis, C C Lundberg and J Martin (ed) *Reframing Organisational Culture*, London: Sage, pp. 58–76.

Veblen T (1970) *The Theory of the Leisure Class*, London: Unwin Books.

Zukin S (1988) *Loft Living: Culture and Capital in Urban Change*, London: Radius.

SELECTED RECOMMENDED READING

Bell D and Valentine G (1997) *Consuming Geographies: We Are Where We Eat*, London: Routledge.

Clammer J (1997) *Contemporary Urban Japan: A Sociology of Consumption*, Oxford: Blackwell.

Featherstone M (1994) City cultures and post-modern lifestyles, in A Amin (ed) *Post-Fordism: A Reader*, Oxford: Blackwell, pp. 387–408.

Jackson P and Thrift N (1995) Geographies of consumption, in D Miller (ed) *Acknowledging Consumption*, London: Routledge.

Miller D (1998) (ed) *Material Cultures: Why Some Things Matter*, London: UCL Press, pp. 73–99.

Urry J (ed) (1995) *Consuming Places*, London: Routledge.

Section 4.2 Capitalizing on Consumption

Robert Bocock

'Consumption and Lifestyles'

from *Social and Cultural Forms of Modernity* (1992)

> Years ago a person, he was unhappy, didn't know what to do with himself – he'd go to church, start a revolution – something. Today you're unhappy? Can't figure it out? What is the salvation? Go shopping! (Arthur Miller, *The Price,* 1985, Act 1)

It maybe somewhat surprising to find shopping prescribed as an antidote to unhappiness in this quotation from Arthur Miller's play. However, shopping has become a very popular activity in Britain and the United States, in the last decades of the twentieth century, second only to watching television as a favourite leisure time activity. 'Shopping' may not be quite the term to describe what many people do when visiting shopping centres – they eat, drink, walk around, purchase a few things, but they also look, or gaze, at the goods displayed in the shops and at one another. Gazing, viewing, watching – either on the small screen, or in shopping precincts or sports stadia – has become a major social activity.

This looking at objects, places, events and other people can be seen as part of a wider social process – the consumption of goods and services. Literally 'consumption' means the use of commodities for the satisfaction of needs and desires. It includes not only the purchase and rise of a range of material goods, from cars to television sets, but the consumption of services, such as travel and of a variety of social experiences. Modern societies have developed the process of consumption into a major social activity which uses large amounts of time, money, energy, creativity and technological innovation to sustain it. Everyone now consumes in modern societies – the old and the poor, men and women, as well as the rich, the young and those in the middle, though levels of consumption between these social groups differ.

The quality and degree of attention given by sociologists to consumption changed considerably in the late 1980s and early 1990s. In part, this reflects the growth of mass consumption in advanced industrial societies since World War II. But it is also because the concept 'consumption' marked an important theoretical move away from seeing the mode of production as the major, even sole, determinant of how modern societies have been shaped and of the ways in which they have operated. The formation and composition of social classes in capitalist societies has been seen by sociologists in the past as fundamental to understanding the structure of such societies and the ways in which they change. In some of the major sociological traditions 'class' was defined principally in terms of the relations of production (Marxist theories) and work roles or occupations in productive industry (stratification theories). However, in people's own subjective perceptions of 'class', it has often been patterns of consumption, not primarily occupation, which have been seen to be significant; and some sociologists have always highlighted consumption in their analysis of modern industrial capitalism. It has recently acquired even greater importance in sociological analysis as mass consumption has spread, and the number of people employed in mass production has declined and those in the service sector and in marketing have expanded.

I want to explore how the process of consumption has developed in western capitalist societies in the twentieth century. Consumption appears to be rooted in the satisfaction of purely natural, biological or physical needs. However, there is nothing natural about the ways in which millions of people now shop for and consume goods and services such as foods, cars, travel, or the media. The social processes of modern consumption have developed historically, the most significant expansion occurring in the twentieth century particularly. Patterns of consumption have changed and the scale and intensity of consumption in the developed societies seems to have reached a qualitatively new stage. The second section of this chapter explores some important aspects of this historical process through which consumption emerged in its modern form, and how this has been conceptualized by an earlier group of sociologists. Two key processes – (a) the relationship between patterns of consumption and major socio-economic groups ('classes') in early modern capitalist societies, and (b) the emergence of 'mass' consumption (as it has been called) in the middle of the twentieth century – are explored in the third section. The emergence of a *plurality* of patterns of consumption in the 1970s and 1980s is examined in the fourth section, a patterning which some claim is far removed from the earlier pattern of mass consumption.

THE EMERGENCE OF MODERN CONSUMPTION

Wants and needs could be seen as 'natural' because they are given by human biology – food, liquids, medicines, warm clothing, safe housing, means of transport: all these seem to be requirements which are biologically determined. Consumption may be seen as satisfying these basic, given needs.

The main difficulty with this type of argument is that such basic needs are not easy to specify in detail, and, in any case, they never arise outside a social, cultural, historical context. For example, the need for food seems basic, biologically given. Yet the kinds of food that groups eat vary enormously from one culture to another, from one historical period to another. People in Britain at the end of the twentieth century, do not eat snakes, for instance, even though some snakes are harmless and nutritious as sources of protein. On the other hand, many people do eat chocolates. So there is something of a puzzle here. Chocolate is not a necessary food for human beings to eat. It is a luxury item. Early manufacturers produced from the very beginning luxury consumption items, satisfying something beyond basic biological needs.

Who bought these items? Who consumed them? In the mid 1700s, a group of families existed with enough surplus income to form the basis for patterns of consumption which went well beyond the basic necessities of simple food, shelter and clothing. This was initially restricted to 'the landed gentry', but then spread to middle class commercial farmers and their families in the eighteenth century and to a new urban bourgeoisie based, not only in the professions (the law, the church, medicine), but also in the new manufacturing and commercial enterprises in the nineteenth century. The new working class, in the industrial towns, also began to enter into consumerism when not totally poverty stricken.

Among such social groups, as it always had been for the aristocracy, 'consumption' became detached from the satisfaction of biological needs and entered into the processes surrounding the construction of social identities. This process of identity construction around patterns of consumption involved a 'symbolic' dimension, distinct from purely biological survival. Activities such as novel-reading developed in the early nineteenth century, for example. Ideas of romantic love began to be articulated in popular literature, in songs, and in poetry. Alongside these rather inner-life, social psychological developments, ideas of fashion in clothing became more widespread in the nineteenth century, particularly. Through such processes 'modern consumption' was born.

The important shift here from biologically-driven, or economistic, 'common-sense' notions of consumption, towards a more social, symbolic and psychological concept of modern consumption is controversial. Not all social scientists accept this move because, they argue, it obscures and diverts attention from the study of physical malnourishment, bad housing, the effects of poverty upon millions of people, not only among the poorest sections of western societies, but most importantly, among the poor in the 'Third World', and in what was, until 1990, the Second World of Eastern Europe and the Soviet Union. However, even in these societies, Western consumer goods, such as jeans, radios, television sets, do take on a *symbolic* value for groups, once they are in a position to take basic food supplies for granted.

CONSPICUOUS CONSUMPTION

In the late nineteenth and early twentieth centuries new patterns of consumption began to develop among the urban middle and working classes. These patterns centred especially around the new department stores in city centres. The new department stores were sites for the purchase and display of a variety of commodities – groceries, furniture, clothing, crockery, kitchen utensils, and new electrical equipment as these were developed and mass produced in the course of the twentieth century – all under one roof. Such city centre shops offered more choice than local ones could do, although butchers, fishmongers, greengrocers and bakers remained in local high streets, and groceries in particular were obtainable in local corner shops. The city centre department stores developed as trams, trolley buses and railways emerged to carry people into the centre from the outlying suburban areas during the late nineteenth and early twentieth centuries.

Cities such as London, Paris, Glasgow, New York, Chicago, and Berlin expanded their transport networks and developed large city centre department stores from the 1890s up to the First World War in 1914. Cities grew around centres of government as well as around particular industries, from steel to lace-making. The shops and leisure facilities such as theatres, music halls and sports stadia, all grew up to satisfy the social and psychological desires of the new urban classes.

A new, distinctive urban culture, linked to consumption, thus emerged in these metropolises. The daily life of people who lived in a great metropolis, Simmel (1903) argued, [was] affected by the need to cultivate a 'blasé attitude' towards others.

Some became 'dedicated followers of fashion'; others walked around just looking – providing an urban audience for others to parade before. The stress was on the individual: everyone tried to remain socially detached from one another, or blasé.

Modern consumerism, therefore, in part results from this new way of life in the metropolis, the city and its suburbs. The processes involved in living in the city increased the awareness of style, the need to consume within a repertory or code which is both distinctive to a specific social group, and expressive of individual preference. The metropolitan individual is no longer the older type who would not spend 'foolishly' on relatively trivial items of clothing or adornment. Rather the person in the big city consumes in order to articulate a sense of identity, of who they wish to be taken to be.

The signs or symbols which a particular individual uses as a means of marking themselves from others have to be interpreted and understood by others. Someone can only mark themselves as being different from others if they also share some common cultural codes with others within which these signs of difference can be read and interpreted. This produces a ceaseless striving for the distinctive, with the higher social status groups continually having to change their patterns of consumption as the middle middle, lower middle and working class strata copy their habits. For example, drinking Champagne or malt whisky, once the preserve of the aristocracy, has moved down the social status ladder in this century, so that the upper echelons either cease to drink these drinks, or consume more exclusive and expensive vintages.

This aspect of the consumption process was observed by the sociologist Thorstein Veblen (1953) in the United States during the late nineteenth century. He was concerned, particularly, with one specific social class, the *nouveaux riches* – the 'new rich' – of the late nineteenth century. These groups, whose wealth was recently acquired, aped the European aristocracy, or tried to do so in order to win social acceptance. The middle classes and the working classes, black and white, were not yet caught up in this process, which Veblen termed '*conspicuous consumption*'.

Veblen uses the concept of 'patriarchy'. It is, he argues, as a consequence of our patriarchal past, that 'our social system makes it the woman's function in an especial degree to put in evidence her household's ability to pay'. Here is a specific example of what Veblen called 'conspicuous consumption': that is social display, based upon a high surplus income, enabling people to indulge patterns of consumption which are designed to impress others in some way. The aim, Veblen argued, was to show that the family possessed 'good taste', a good background, and an ability to pay for consumption beyond what most other people can afford. The social values of patriarchy produce the feature Veblen commented on here, namely that husbands display their wealth and high income by keeping a wife at home, who wears good clothes and is able to be a leisured woman.

THE ANALYSIS OF PATTERNS OF CONSUMPTION

Economic Class

Simmel and Veblen's work marked a shift of attention in sociology from economistic to social definitions of class. Classical Marxism operated within a broadly economistic set of assumptions, and emphasized the concepts of economic class, especially the bourgeoisie and proletariat. These were firmly located in terms of their respective relationship to the means of production, that is the factories, the mines, the machinery and energy sources, and (to a lesser extent) the means of distribution (the shops, department stores and transport systems) of modern capitalism. According to Marx, the bourgeoisie, which owned these means of production and distribution, had a set of economic interests which were directly opposed to those they employed, for they sought to maximize profits and to minimize costs, including wage and salary costs. This economistic model remained of great intellectual and political importance in the first half of the twentieth century; but it did present problems in understanding, explaining and conceptualizing the broader processes of social change taking place in the United States, Britain and Western Europe in this period. These included the changes in the class and status systems of such capitalist societies, which were, in turn, related to the process of consumption rather than to that of production. In peace time, the members of the proletariat were preoccupied with building up and preserving, not only their wages and salaries, relative to others, but also their social status, that is their own sense of who they were, of how socially worthy they were: worthy, that is, of prestige or esteem from others.

Social Status

The concept of social status has its roots in the sociological perspective of the German sociologist, Max Weber (1864–1920). Whilst Weber accepted the strengths and explanatory power of Marxism as an analytic model in the economic sphere, he held that it underestimated the crucial role of social status in capitalist societies. Far from being a left-over from an earlier, feudal epoch, as some Marxists argued, Weber articulated the view that social status had remained important, and that new bases of status had developed, in modern industrial capitalism. With a focus upon consumption patterns, this notion of social status has considerable significance.

For Weber, the basic condition of 'class' lay in the unequal distribution of economic power and hence the unequal distribution of opportunity. But this economic determination did not exhaust the conditions of group formation. He formulated the concept of status in such a way as to encompass the influence of ideas, beliefs and values upon the formation of groups without losing sight of economic conditions.

Weber focused upon 'class situation': that is the access to goods and external living conditions which [one] can obtain by selling [one's] skills in the labour market in return for wages and salaries. This economically based concept of class is contrasted with status situation, which is related to esteem, or honour given by others, and claimed by status group members from others. Some groups experience negative honour or esteem; they may be treated as outcasts, or marginal members of a society. Status is linked with a specific style of life, which involves consumption, that is the kinds of clothing, or house furnishings, foods and drinks, thought to be appropriate to a specific status group.

Social status groups use patterns of consumption as a means of establishing their rank or worth and demarcating themselves from others. It is not only Veblen's *nouveaux riches*, or the metropolitan individuals analysed by Simmel, but all status groups, which use some markers to differentiate themselves from others, in the Weberian view. The markers they use to do this include group values about consumption, although as Parkin (1982) points out any social or physical attribute may be used to effect social closure, to mark who belongs to a particular social group and who is excluded. The types of housing, furnishings and decorations; the types of music enjoyed; the kinds of clothing which are worn; the type of transport used; all these aspects of the process of consumption may be used as markers of difference between social status groups.

The existence, persistence and increasing growth of social status groups in the twentieth century has helped to generate a plethora of patterns of consumption. The growth of consumption in modern capitalist societies now affects, not only the upper and middle class, but the two-thirds of the population which made up what most sociologists called the 'working classes', i.e. those earning a wage for manual work of some kind. In other words, the twentieth century has witnessed the increasing growth of consumerism among most, if not all, major social status groups, including those who have the lowest incomes, from the state or from paid employment. Everyone can desire, or dream, about consuming, even if they cannot afford to purchase the objects or experiences.

Since the end of the 1940s, the capitalist societies of North America, Western Europe, Britain, Japan and Australia, have experienced a massive development of 'mass consumption'. This phrase encapsulates the processes whereby the majority of the working classes in these societies became 'consumers', not only, or even primarily, 'workers' in the production process.

The Affluent Worker

The changes associated with the mass consumption which developed in the years after the end of the Second World War in Britain were summed up in the phrase 'the affluent worker'. During the economic expansion of the 1950s, unemployment was comparatively low, and manufacturing industry enjoyed boom conditions. Wage levels rose, especially where skilled labour was in short supply. The phrase 'the affluent worker' was used in both political science and in sociology to describe the new, well-paid worker who emerged in this period. The ideal-typical affluent worker seemed to be the car worker, for car manufacturing was one of the leading, new, mass consumer-oriented, industries of the period. A major study was carried out by a team of sociologists into the lifestyle, voting habits and intentions, of this new type of worker (see Goldthorpe *et al*, 1968–9). The affluent worker was contrasted with the traditional workers who worked in heavy, basic industries such as coal-mining, iron and steel-making, and ship-building.

This study suggested that the affluent workers, who emerged in the 1950s and 1960s, epitomized by the car worker in the Midlands and the South East, were more privatized, home-centred, spent more time at home with wife, or husband, and children. They watched television, first in black and white, later in colour, in their well-furnished homes. They spent time and money on do-it-yourself decorating and redesigning their houses. They had at least one car in the family, which was used for pleasure trips at weekends, increasingly for shopping trips. The workers in heavier industries, on the other hand, were not home-centred; the men spent more time with other males in pubs or going to football matches and were less interested in home decorating, child-care, or even spending time in the home, than the newer style affluent male workers. Thus, there were said to be at least two distinct patterns of consumption and lifestyle in the 1950s and 1960s among industrial workers: the post-war, affluent worker in new industries such as car production, and the traditional worker in heavy industries.

As the older, heavy industries declined in the 1970s and 1980s so the older patterns of life based on men doing heavy manual work in large groups and women doing part-time lighter jobs in factories, or in cleaning offices and other large organizations such as schools, universities or hospitals, began to decline. Male unemployment rose in those towns and areas dependent on one of the traditional 'heavy' industries, such as coal-mining, ship-building and steel, thereby increasing the pressures on women to work, especially when there were young children to be fed and clothed. Consumption patterns were, of course, altered by the rise in male unemployment in such areas: basic items becoming more important than conspicuous consumption of clothes, cars, holidays, home furniture or exotic foods.

However, those employed in occupations which continued to provide well-paid employment could consume more conspicuously, spending more on what were considered to be less essential items in the household budget. Among the relatively well-paid sections of the manual workers, and among the clerical and service industries' workers, a concern with earning to provide enough income to support a relatively affluent lifestyle developed. This marked a move from the primary source of identity being based upon the paid work role a person performed to identities being constructed around lifestyles and patterns of consumption.

As the majority (between two-thirds and three-quarters) of the populations of western capitalist societies became more affluent, the mode of consumption changed from one concerned primarily with basic material provision (which many people still lack in the major capitalist societies as well as in the world as a whole) to a mode concerned more with the status value and symbolic meaning of the commodity purchased.

THE NEW CONSUMERS

The emerging affluent working class in western societies from 1950 onwards was seen as a new relatively undifferentiated 'mass' market by producers, department stores, advertisers and distributors of all types of what were termed 'consumer durables', such as televisions, washing machines, cars, transistor radios and record players. However, by the end of the 1980s, many market researchers and advertisers were becoming dissatisfied with the older, social class, or social status group, categories which had been in use for thirty years or more. New categories of consumer began to be developed.

During the 1950s, 1960s and 1970s, the patterns of consumption, in Britain particularly, but in other western societies too, tended to follow well established social status group and economic class categories. Until the 1980s, most marketing researchers conducting research into which groups would buy what kinds of consumer goods, and advertisers in designing advertising campaigns for selling products, saw the population as divided into several categories of social class, distinguished by a combination of income level, occupation, and an associated pattern of expenditure.

The standard version of the stratification system used by many market researchers from the 1950s to the early 1980s was as in Table 1.

Table 1 Social classes

Social Class A	Higher managerial, administrative, or professional
Social Class B	Intermediate managerial, administrative, or professional
Social Class C1	Supervisory or clerical, and junior managerial, administrative, or professional
Social Class C2	Skilled manual workers
Social Class D	Semi and unskilled manual workers
Social Class E	State pensioners or widows (no other earners), casual or lowest grade workers, or long-term unemployed

On the whole, occupational categories correlated highly with income levels, with social class A having higher disposable income than social class B, and so on.

In 1987, for example, the proportion of the British population in each of these categories was as in Table 2.

Table 2 Percentage of UK population by social class, 1987

Social Class A	3%
Social Class B	15%
Social Class C1	23%
Social Class C2	28%
Social Class D	18%
Social Class E	13%

The social class categories in Tables 1 and 2 have been used, and found to 'work' by market researchers when trying to predict patterns of consumption. Such categories might be used to analyse alcohol consumption for both men and women, in each social class, for example. Wine drinking was found to be greater among A and B groups than other groups in Britain. Expensive drinks such as whisky, gin, or brandy were also more likely to be purchased by A and B groups. C2 and D groups would consume beers and lagers more than other kinds of alcohol.

However, some market researchers, advertisers and companies designing new products such as clothing for men and women, began to detect significant changes in consumer patterns in the 1980s. The old social class categories seemed less good at predicting who would consume what, whereas age-grades – young adult, middle-aged, or older people – became increasingly significant. The notion of 'lifestyle' emerged as part of the attempt to capture this set of changes in consumers' patterns of purchasing. 'Lifestyles' may differ, not only between social classes, as in the past, but also within social classes. Mike Featherstone has defined 'lifestyle' as follows:

> The term 'lifestyle' is currently in vogue. While the term has a more restricted sociological meaning in reference to the distinctive style of life of specific status groups, within contemporary consumer culture it connotes individuality, self-expression, and a stylistic selfconsciousness. One's body, clothes, speech, leisure pastimes, eating and drinking preferences, home, car, choice of holidays, etc. are to be regarded as indicators of the individuality of taste and sense of style of the owner/consumer. In contrast to the designation of the 1950s as an era of grey conformism, a time of mass consumption, changes in production techniques, market segmentation and consumer demand for

> a wider range of products, are often regarded as making possible greater choice (the management of which itself becomes an art form) not only for youth of the post 1960s generation, but increasingly for the middle aged and the elderly. . . . We are moving towards a society without fixed status groups in which the adoption of styles of life (manifest in choice of clothes, leisure activities, consumer goods, bodily dispositions) which are fixed to specific groups have been surpassed. (Featherstone, 1987, p. 55)

One example of changing patterns of consumption is the emergence of a distinctive youth culture [and youth 'sub-cultures', e.g. Teddy Boys and Mods] in the 1950s. The young emerged as a new major market in the 1950s in Europe, a little earlier in the United States. Some young people found relatively well paid jobs in the new industries manufacturing and selling consumer goods. The increase in disposable income enabled them to buy consumer goods such as motorbikes, scooters, clothes, records, radios, television sets, and nonalcoholic drinks in coffee bars – commodities closely associated with a distinctive youth lifestyle.

Another 'lifestyle' group related to distinctive consumer patterns can be identified at the other end of the age scale. The population of Britain, like that of most other western societies, excluding Japan and Australia, is growing older, unlike third world societies where young people predominate. Consumption patterns among the older segments of the population have become important since the explosion of youth culture(s) and the associated patterns of consumption during the 1950s, 1960s and 1970s. Market researchers have devised new categorizations of the population, based upon what they call life-stages. An example is given in Table 3 (notice that 'head of the household' is assumed to be male).

Table 3 Life-stages

Granny Power	People aged 55–70, living in households where neither the head of the household nor the housewife works full-time. They have no children and no young dependent adults, i.e. no non-working 16–24s live with them (14%).
Grey Power	People aged 45–60, living in households where either the head of the household or the housewife is working full-time. They have no children and no young dependent adults (12%).
Older Silver Power	Married people, with older children (5–15 years) but no under-fives (18%).
Young Silver Power	People who are married, with children aged 0–4 years.
Platinum Power	Married people aged 40 or under, but with no children (7%).
Golden Power	Single people, with no children, aged 40 or under (15%).

This categorization covers 82 per cent of the adult population, each group being described as a 'power' group *– power reflecting their spending power in terms of disposable income.*

Source: O'Brien and Ford (1988, pp. 293–4).

So far disposable income and stage in life have been mentioned as affecting consumption patterns. Gender has also been an important factor influencing consumption in the period 1950–90, and will continue to be so. Girls and women have been targeted as major consumers of perfume, jewellery, clothing, baby products, furniture, holidays, and foodstuffs for instance. In general, women do most of the household shopping in supermarkets and shopping centres and are thus thought of as the 'ideal consumers'. A great deal of advertising on television and in women's magazines is aimed at developing, eliciting, articulating and shaping the desires of women which will lead them to purchase particular products and to live according to a distinctive lifestyle.

In the 1980s yet another growing consumer market from manufacturers' and commercial sellers' points of view has emerged: that of men. Young men with reasonably well paid jobs, or with a disposable income as a result of living with parents, were the first major target group in the youth market in the 1960s. However, as patterns of consumption and relative affluence have spread in the 1980s, older groups of men have become targets for consumption by advertisers. It is important here not to conceptualize either women or men as target groups only, however. They are not only the passive targets of the advertisers. They desire to articulate and express their own sense of identity, their own sense of who they are, through what they wear, buy and consume. Such an articulation of identity is done through clothing, hairstyles, body decoration (from perfume to earrings), as well as through house-style, cars, travel, music and sport, hi-fi's, video cassette recorders, personal stereos, and other electronic consumer goods.

Frank Mort identified an increase in individuality between the 1950s and 1980s, articulated through clothes, hair, body decoration, and body movements among young men in the UK. This has been followed rather than simply created by advertisers and marketing people. The advertisements, together with photographs attached to fashion spreads and feature articles in the new men's style magazines, have contributed to a sexualization of the male body in ways which would have been unthinkable in the 1950s. The male imagery used is no longer the old macho image but has become more openly erotic, even narcissistic. Men no longer simply consume, but must be seen to consume: consumption is a badge of social and sexual identification. Street culture is a matter of glances, looks, making a quick impression: it is a highly visual culture (note the echo here of the analysis that Simmel made of the city of Berlin and its blasé attitude).

CONCEPTUALIZING CONSUMPTION

Traditionally, consumption has been seen as either a material process, rooted in human biological needs, or as an ideal practice, rooted in symbols, signs, codes. For example, the influential social theorist Herbert Marcuse (1964) used the notion of 'needs', based in human biology, in his critical theoretical analysis of modern capitalism.

Marcuse distinguished between 'true' needs and 'false' needs. 'True' needs are seen as based in human and social interaction, uninfluenced by modern consumer capitalism. 'False' needs are induced, or produced, by modern capitalism, by advertising and marketing strategies; they have no basis in genuine human social interaction.

In recent years, the approach Marcuse took has been subject to a fundamental critique. For example, Barthes (1973) argued that there was always a dual aspect to consumption – that it fulfilled a need, as with food or clothing, but also conveyed, and was embedded within, social, cultural symbols and structures. A sweater, for example, could both keep you warm and signify an image, like a romantic walk in the woods. The function of consumer goods in satisfying material human needs could not be separated from the symbolic meaning of commodities, or what Barthes call their 'significations'. Consumption is embedded within systems of signification, of making and maintaining distinctions, always establishing boundaries between groups.

Bourdieu on 'Distinction'

Bourdieu (1984) analysed the ways in which status and class groups differentiate themselves one from another by patterns of consumption which help to distinguish one status group's way of life from another.

Bourdieu examined distinctions between groups, especially in the top sectors of French society: similar processes, he argued, operate in Britain, West Germany and in some ways in the United States too. A major distinction Bourdieu made, was that between groups with access to two different types of capital. The business, entrepreneurial, management, commercial and financial groups emphasize economic capital. Such groups seek to amass money capital, real estate, factories, shops, shares and bonds. Their way of life is akin in some ways to the conspicuous consumption Veblen analysed in his analysis of the American *nouveaux riches* in the late nineteenth century.

The second meaning Bourdieu gives to the concept of 'capital' extends it into the realm of culture and education. He argued that there are forms of intellectual capital which are distinct from economic forms. The educational systems in modern capitalist societies generate another structure of capital, based upon being able to talk about, or to create new cultural products, from major philosophical or social scientific texts, to novels, paintings, buildings, films, television programmes, clothes, furniture and interior decor. In the universities the highest social prestige continues to be attached to non-utilitarian studies, especially in Europe, but also in the United States (in spite of the high prestige of institutions like the Harvard Business School, for example). In the period during which Bourdieu was writing, philosophy and literary studies held the highest status in western universities, in the eyes of many elite groups, followed by pure, rather than applied, mathematics and natural sciences.

This approach is a powerful one, which does not abandon the fundamental notion of there being structures (class structures, status group structures, structures affecting ethnicity and gender) which have real effects on people, independently of their own subjective consciousness. On the other hand, these structures may constrain but do not determine agents' actions, beliefs, values or desires. Poor people may desire to be rich, but remain poor because of their structural position, to take the most obvious example. However, the desires of the poor, their belief and value systems, are not produced, nor directly determined, by their structural position in the economic system. In this sense, Bourdieu's approach differs from more economistic versions of Marxism. Such desires, beliefs and values, he argues, have a high degree of autonomy from a group's position in the structure of capital, without being completely detached from it.

Bourdieu is anxious to emphasize that the positions in a structure do not produce unified groups who will act politically in a concerted way together in order to preserve or protect their way of life. (This was another false assumption of economistic Marxism.) They may do so; but if, and when this occurs, it is a result of a separate activity, such as political mobilization, through which they become constituted as agents of social action. The structural position is just that: a position which may be occupied by any specific individual as a result of upward or downward social mobility. Such positions in a structure do not generate ways of life, or symbolic meanings, of themselves. Symbolic activity, including consumption, is a relatively autonomous practice. It is not directly produced or determined by a position in the social structure of a social formation.

Bourdieu sought to combine the importance that Veblen, and Weber, gave to social status and to patterns of consumption as a way of marking one way of life from another, with the idea that consumption involves signs, symbols, ideas, not only the satisfaction of a biologically rooted set of needs. In this, Bourdieu may be seen as having attempted to combine the well-established approach to consumption through notions of social status groups (especially the upper, middle and working classes), with the newer analytical approach to signs, symbols, ideas and the cultural.

Bourdieu's analysis of the role of education in these processes is especially important: the sons and daughters of the wealthy industrial and commercial groups who consumed material objects, as in Veblen's leisured class, enter universities. They thus add cultural capital to their economic capital: if they are successful in absorbing the right kind of cultural capital at university: not business studies but philosophy or art criticism; not engineering but pure physics. 'Cultural capital' is not defined by the industrial and commercial classes but by intellectuals and artists whose tastes and definitions of what matters in culture typically differ from those of the industrial classes, both bourgeois, petit bourgeois and working class. The social sciences, incidentally, hover uneasily in this schema: they are neither 'pure' enough for some people, unlike philosophy or art criticism, nor practical enough for others.

Baudrillard on Consumption

An important critique both of the earlier conception of biologically ordered 'need' as a basis for an approach to consumption and of an approach such as that of Bourdieu which used a notion of social *structure*, may be found in the work of the French sociologist, Jean Baudrillard. He argued that we cannot operate within a theoretical framework based upon 'needs', nor one based upon economic class or social status groups. This is so because there is no way of fixing the categories of 'fundamental needs' versus 'media induced consumption' or of distinguishing between Marcuse's 'true' and 'false' needs.

All consumption is always in part the consumption of symbolic signs in Baudrillard's view. These signs, or symbols, do not express an already preexisting set of meanings for a person or a group such as a social class. The meanings are generated within the system of signs/symbols which engage the attention of a consumer. So, far from consumption being conceptualized as a process in which a purchaser of an item is either trying to satisfy a basic, pre-given human need, a need rooted in biology, or responding to a prompt, or a message they have received from advertising media, the consumer is always actively creating a sense of identity. 'Consumption' for Baudrillard is no longer seen as an action induced by advertising upon a passive audience which belongs to a specific social class or life-stage, but as an active process involving the symbolic construction of a sense of both collective and individual identity. Because this sense of identity is no longer seen as given to us by our membership of a specific economic class or social status group, consumption becomes an absolutely necessary process to the construction or articulation of a sense of identity. Baudrillard suggests that we do not purchase items of clothing, food, body decoration, furniture or entertainment to express a pre-given sense of who we are. Rather, we become that which what we buy makes us. In other words, he argues, the sphere of the symbolic has become primary in modern capitalism; the 'image' is more important than the satisfaction of material needs. In the words of a famous contemporary poster, 'I shop, therefore I am'.

The important element in this shift is the role of desire. We are not already constituted as an attractive woman, or handsome man. Rather, we try to become the beings we desire to be, by purchasing the clothes, foods, perfumes, cars, and experiences which will signify that we are x or y to ourselves and to others who share the same code of signifiers, the same system of signs/symbols and their socially produced meanings.

Baudrillard argues that 'in order to become object of consumption, the object must become sign'. It is the relation between signs which enables difference to be established. It is difference from others, which at the level of everyday experience, we can see is frequently one of the main 'uses' of consumption. People seek to establish that they have more 'taste' than others, that they support a better team, their team, by wearing special clothing to football matches; that they are English or French; well educated or 'all right as I am, take it or leave it'. The point Baudrillard is concerned to make here is that it is through being caught up in systems of signs that difference can be established, and the essence of the modern consumer society is this construction of difference.

CONCLUSION

Consumption has emerged in the course of the twentieth century as a core social process in western societies. From the work of Simmel, Veblen and Weber, there has emerged the notion of social status groups using consumption patterns as a major means of marking out and symbolizing or signifying their group identity, their lifestyle, from that of others, and this idea has been developed by sociologists writing after the 1950s period. The emphasis some feminists gave to the process of reproduction of the labour force, played an important role in effecting a move away from the analysis of modern societies in terms of the mode of production alone, towards an awareness of the importance of consumption. Later theorists have extended this shift to the sphere of the symbolic.

Sociologists continue to take different views about the significance of economic factors, such as income levels, in the analysis of patterns of consumption. Some argue that these are still of major importance in affecting who buys what. Others would argue that in recent decades, since the 1950s or 1960s, income levels are only one part of the explanation of the new patterns of consumption which emerged in the 1970s and 1980s. On this second view, consumer items from clothing to motor-cycles or cars, from food and drinks to tourism, have become involved in processes which have a high degree of autonomy from economic class or even traditional social status groups. Young men and women from different social status groups, as measured by the occupations of their fathers, or the male head of household, consume in ways which articulate to themselves and others a sense of identity which may be autonomous from such traditional social status groups.

Identities can be constructed through the desire for consumer goods as much as through actually purchasing them. Indeed, for many people, consumption is now their major means of establishing, and creating, who they wish to be. As Arthur Miller observed in the opening quotation, shopping has taken over from politics and religion as a means of warding off unhappiness – unhappiness produced by a loss of a sense of self, of identity.

David Harvey

'Time–Space Compression and the Postmodern Condition'

from *The Condition of Postmodernity* (1989)

Accelerating turnover time in production entails parallel accelerations in exchange and consumption. Improved systems of communication and information flow, coupled with rationalizations in techniques of distribution (packaging, inventory control, containerization, market feedback, etc.), made it possible to circulate commodities through the market system with greater speed. Electronic banking and plastic money were some of the innovations that improved the speed of the inverse flow of money. Financial services and markets (aided by computerized trading) likewise speeded up, so as to make, as the saying has it, 'twenty-four hours a very long time' in global stock markets.

Of the many developments in the arena of consumption, two stand out as being of particular importance. The mobilization of fashion in mass (as opposed to elite) markets provided a means to accelerate the pace of consumption not only in clothing, ornament, and decoration but also across a wide swathe of life-styles and recreational activities (leisure and sporting habits, pop music styles, video and children's games, and the like). A second trend was a shift away from the consumption of goods and into the consumption of services – not only personal, business, educational, and health services, but also into entertainments, spectacles, happenings, and distractions. The 'lifetime' of such services (a visit to a museum, going to a rock concert or movie, attending lectures or health clubs), though hard to estimate, is far shorter than that of an automobile or washing machine. If there are limits to the accumulation and turnover of physical goods (even counting the famous six thousand pairs of shoes of Imelda Marcos), then it makes sense for capitalists to turn to the provision of very ephemeral services in consumption. This quest may lie at the root of the rapid capitalist penetration, noted by Mandel and Jameson, of many sectors of cultural production from the mid-1960s onwards.

Of the innumerable consequences that have flowed from this general speed-up in the turnover times of capital, I shall focus on those that have particular bearing on postmodern ways of thinking, feeling and doing.

The first major consequence has been to accentuate volatility and ephemerality of fashions, products, production techniques, labour processes, ideas and ideologies, values and established practices. The sense that 'all that is solid melts into air' has rarely been more pervasive. My interest here is to look at the more general society-wide effects.

In the realm of commodity production, the primary effect has been to emphasize the values and virtues of instantaneity (instant and fast foods, meals, and other satisfactions) and of disposability (cups, plates, cutlery, packaging, napkins, clothing, etc.). The dynamics of a 'throwaway' society, as writers like Alvin Toffler (1970) dubbed it, began to become evident during the 1960s. It meant more than just throwing away produced goods (creating a monumental waste disposal problem), but also being able to throw away values, lifestyles, stable relationships, and attachments to things, buildings, places, people, and received ways of doing and being. These were the immediate and tangible ways in which the 'accelerative thrust in the larger society' crashed up against 'the ordinary daily experience of the individual' (Toffler, p. 40). Through such mechanisms (which proved highly effective from the standpoint of accelerating the turnover of goods in consumption) individuals were forced to cope with disposability, novelty, and the prospects for instant obsolescence. 'Compared to the life in a less rapidly changing society, more situations now flow through the channel in any given interval of time – and this implies profound changes in human psychology.' This transcience, Toffler goes on to suggest, creates 'a temporariness in the structure of both public and personal value systems' which in turn provides a context for the 'crack-up of consensus' and the diversification of values within a fragmenting society. The bombardment of stimuli, simply on the commodity front, creates problems of sensory overload that makes Simmel's dissection of the problems of modernist urban living at the turn of the century seem to pale into insignificance by comparison. Yet, precisely because of the relative qualities of the shift, the psychological responses exist roughly within the range of those which Simmel identified – the blocking out of sensory stimuli, denial, and cultivation of the blasé attitude, myopic specialization, reversion to images of a lost past (hence the importance of mementoes, museums, ruins), and excessive simplification (either in the presentation of self or in the interpretation of events). In this regard, it is instructive to see how Toffler (pp. 326–9), at a much later moment of time–space compression, echoes the thinking of Simmel, whose ideas were shaped at a moment of similar trauma more than seventy years before.

The volatility, of course, makes it extremely difficult to engage in any long-term planning. Indeed, learning to play the volatility right is now just as important as accelerating turnover time. This means either being highly adaptable and fast-moving in response to market shifts, or masterminding the volatility. The first strategy points mainly towards short-term rather than long-term planning, and cultivating the art of taking short-term gains wherever they are to be had. This has been a notorious feature of US management in recent times. The average tenure of company executive officers has come down to five years, and companies nominally involved in production frequently seek short-term gains through mergers, acquisitions, or operations in financial and currency markets. The tension of managerial performance in such an environment is considerable, producing all kinds of side-effects, such as the so-called 'yuppie flu' (a psychological stress condition that paralyses the performance of talented people and produces long-lasting flu-like symptoms) or the frenzied lifestyle of financial operators whose addiction to work, long hours, and the rush of power makes them excellent candidates for the kind of schizophrenic mentality that Jameson depicts.

Mastering or intervening actively in the production of volatility, on the other hand, entails manipulation of taste and opinion, either through being a fashion leader or by so saturating the market with images as to shape the volatility to particular ends. This means, in either case, the construction of new sign systems and imagery, which is itself an important aspect of the postmodern condition – one that needs to be considered from several different angles. To begin with, advertising and media images have come to play a very much more integrative role in cultural practices and now assume a much greater importance in the growth dynamics of capitalism. Advertising, moreover, is no longer built around the idea of informing or promoting in the ordinary sense, but is increasingly geared to manipulating desires and tastes through images that may or may not have anything to do with the product to be sold. If we stripped modern advertising of direct reference to the three themes of money, sex, and power there would be very little left. Furthermore, images have, in a sense, themselves become commodities. This phenomenon has led Baudrillard (1981) to argue that Marx's analysis of commodity production is outdated because capitalism is now predominantly concerned with the production of signs, images, and sign systems rather than with commodities themselves. The transition he points to is important, though there are in fact no serious difficulties in extending Marx's theory of commodity production to cope with it. To be sure, the systems of production and marketing of images (like markets for land, public goods, or labour power) do exhibit some special features that need to be taken into account. The consumer turnover time of certain images can be very short indeed (close to that ideal of the 'twinkling of an eye' that Marx saw as optimal from the standpoint of capital circulation). Many images can also be mass-marketed instantaneously over space. Given the pressures to accelerate turnover time (and to overcome spatial barriers), the commodification of images of the most ephemeral sort would seem to be a godsend from the standpoint of capital accumulation, particularly when other paths to relieve overaccumulation seem blocked. Ephemerality and instantaneous communicability over space then become virtues to be explored and appropriated by capitalists for their own purposes.

But images have to perform other functions. Corporations, governments, political and intellectual leaders, all value a stable (though dynamic) image as part of their aura of authority and power. The mediatization of politics has now become all pervasive. This becomes, in effect, the fleeting, superficial, and illusory means whereby an individualistic society of transients sets forth its nostalgia for common values. The production and marketing of such images of permanence and power require considerable sophistication, because the continuity and stability of the image have to be retained while stressing the adaptability, flexibility, and dynamism of whoever or whatever is being imaged. Moreover, image becomes all-important in competition, not only through name-brand recognition but also because of various associations of 'respectability', 'quality', 'prestige', 'reliability', and 'innovation'. Competition in the image-building trade becomes a vital aspect of inter-firm competition. Success is so plainly profitable that investment in image-building (sponsoring the arts, exhibitions, television productions, new buildings, as well as direct marketing) becomes as important as investment in new plant and machinery. The image serves to establish an identity in the market place. This is also true in labour markets. The acquisition of an image (by the purchase of a sign system such as designer clothes and the right car) becomes a singularly important element in the presentation of self in labour markets and, by extension, becomes integral to the quest for individual identity, self-realization, and meaning. Amusing yet sad signals of this sort of quest abound. A California firm manufactures imitation car telephones, indistinguishable from the real ones, and they sell like hot cakes to a populace desperate to acquire such a symbol of importance. Personal image consultants have become big business in New York City, the *International Herald Tribune* has reported, as a million or so people a year in the city region sign up for courses with firms called Image Assemblers, Image Builders, Image Crafters, and Image Creators. 'People make up their minds about you in around one tenth of a second these days', says one image consultant. 'Fake it till you make it', is the slogan of another.

It has always been the case, of course, that symbols of wealth, status, fame, and power as well as of class have been important in bourgeois society, but probably nowhere near as widely in the past as now. The increasing material affluence generated during the postwar Fordist boom posed the problem of converting rising incomes into an effective demand that satisfied the rising aspirations of youth, women, and the working class. Given the ability to produce images as commodities more or less at will, it becomes feasible for accumulation to proceed at least in part on the basis of pure image production and marketing. The ephemerality of such images can then be interpreted in part as a struggle on the part of the oppressed groups of whatever sort to establish their own identity (in terms of street culture, musical styles, fads and fashions made up for themselves) and the rush to convert those innovations to commercial advantage (Carnaby Street in the late 1960s proved an excellent pioneer). The effect is to make it seem as if we are living in a world of ephemeral created images. The psychological impacts of sensory overload, of the sort that Simmel and Toffler identify, are thereby put to work with a redoubled effect.

The materials to produce and reproduce such images, if they were not readily to hand, have themselves been the focus for innovation – the better the replication of the image, the greater the mass market for image making could become. This is in itself an important issue and it brings us more explicitly to consider the role of the 'simulacrum' in postmodernism. By 'simulacrum' is meant a state of such near perfect replication that the difference between the original and the copy becomes almost impossible to spot. The production of images as simulacra is relatively easy, given modern techniques. Insofar as identity is increasingly dependent upon images, this means that the serial and recursive replications of identities (individual, corporate, institutional, and political) becomes a very real possibility and problem. We can certainly see it at work in the realm of politics as the image makers and the media assume a more powerful role in the shaping of political identities. But there are many more tangible realms where the simulacrum has a heightened role. With modern building materials it is possible to replicate ancient buildings with such exactitude that authenticity or origins can be put into doubt. The manufacture of antiques and other art objects becomes entirely possible, making the high-class forgery a serious problem in the art collection business. We not only possess, therefore, the capacity to pile images from the past or from other places eclectically and simultaneously upon the television screen, but even to transform those images into material simulacra in the form of built environments, events and spectacles, and the like, which become in many respects indistinguishable from the originals. What happens to cultural forms when the imitations become real, and the real takes on many of the qualities of an imitation?

Sharon Zukin

'Real Cultural Capital'

from *Landscapes of Power: From Detroit to Disney World* (1991)

Cultural strategies of visual consumption permit the selective consumption of space and time. While gentrification and Disney World are fine examples, we can also look at the 'new' tourism that juxtaposes nature and culture in the form of home-style inns with French cuisine in rustic surroundings. From our perspective, the most interesting point about these flourishing strategies of cultural consumption is how they articulate with the service economy. Although they certainly manipulate and capitalize on symbols – hence their association with 'symbolic capital' – they also produce real economic value. Continuing to analyze cultural capital in only symbolic terms misses its relevance to structural transformation. For this reason I have turned around Fredric Jameson's assertion that 'architecture is the symbol of capitalism' and suggested that in an advanced service economy, architecture is the capital of symbolism.

Strategies of cultural consumption rely on effective demand among new demographic and social actors. But just as they are embedded in reflexive – or highly mediated and intellectualized – consumption, so they reinforce self-conscious production. On the supply side, cultural consumption creates employment for a self-conscious critical infrastructure (and lower-level service personnel), and is in turn created by its labor. Cultural consumption contributes to capital accumulation, moreover, by enhancing profits on entrepreneurial investment in production and distribution (e.g. gourmet cheese stores, Cuisinart plants). And as we see at both Faneuil Hall (Boston) and Disneyland, cultural consumption has a positive effect on capital accumulation in real estate development. Cultural goods and services truly constitute real capital – so long as they are integrated as commodities in the market-based circulation of capital.

We can see the economic significance of cultural goods and services by sketching their role in *interacting circuits of economic and cultural capital.* Drawing on David Harvey's illustrative use of Marx's circuits of capital, it is clear that the continuous circulation of capital in a capitalist economy increases the value of commodities produced. These increases in turn expand the value of capital. As it circulates, moreover, capital periodically crosses sectoral and institutional boundaries. Depending on rates of return, investment capital may in the aggregate swing from manufacturing to finance or services, from domestic ventures to offshore projects, from property in the inner city to suburban real estate development. As values fluctuate, capital shifts back, or circulates elsewhere.

It is interesting to speculate, as Harvey (1982) has recently done, on the conditions under which changing cultural values reflect change in economic values. But it is even more interesting to ask whether the continuous production of cultural commodities, moving between 'economic' and 'cultural' circuits, continually increases the economic value of investment capital. 'Circuits of cultural capital' may offer us a key to understanding the structural linkage between cultural and economic values today.

As the discussion of urban form suggests, architects' designs become more useful economic tools to speculative real estate developers when they are published, theorized about, and disseminated within the architectural profession. Plenty of magazines and reviews have been founded in the last few years for this sort of intra-professional publicity, and some architects have also published picture books featuring their projects. These are more than just coffee-table books. Publicity expands the cultural value of designers' ideas, and in the process enhances the market value of both the architects and their buildings. Similarly, downtown property values rise with the development of new cultural practices like nouvelle cuisine. In this context, nouvelle cuisine is not only the product or social practice of some expensive restaurants, but also a published, theorized, and professionally disseminated culinary discipline. The interacting circuits of economic and cultural capital that underlie gentrification can be charted in a fairly simple way (see Figure 1).

In a physical infrastructure of old townhouses and lofts in the heart of the city, architectural restorations provide a new element in market culture that both comments on existing modes of new construction and initiates a new mode of its own. At the same time, in the service economy, a set of interrelated amenities caters to the local market of townhouse and loft residents. The concentration of these markets, and the relative autonomy of cultural producers, creates a downtown 'scene'.

This scene in turn attracts more investment in local real estate. As more people move into the area, adopting the cultural values of the innovators, they create more demand for replicas of old crafts products. The increase in demand provides more of a mass market for Victoriana and chintz, which eventually encourages the establishment of new magazines and more publicity for this kind of cultural strategy. When the proponents of this strategy succeed in appropriating central spaces, they seek to protect their claims (in terms of both economic and cultural values) by establishing a historic landmark district. Once the local government designates such a district, changes that do not conform to these cultural values are declared 'out of character' and not allowed.

Landmarking has great economic value, for it spurs more publications about historic preservation and architectural restoration, and also attracts tourists and shoppers. These new economic values now encourage large-scale developers to undertake the expansion of the central business district by means of new office and apartment construction. The central district is transformed into a new marketplace for relatively upscale shopping and residence as well as business services. The new buildings and the international circulation of ideas and creative, professional, and financial personnel who work in them expand the marketplace, and the landscape of nouvelle cuisine. This in turn enhances the economic value of the built environment downtown, increasing investment pressure on old townhouses and lofts.

Disney World, for its part, is a complex of service-sector and entertainment functions that enhance both economic and cultural values. Mickey and Minnie are mass products in all their guises, and the numbers of people who stay in Disney World hotels, or buy houses or timeshares in Orlando or near mixed-use developments inspired by the Disney World model, are far greater than the numbers of people involved in gentrification. From the point of view of circuits of capital, however, Disney World is a set of cultural goods and services that articulate with the mass production system in a service economy (see Figure 2).

These circuits of economic and cultural capital begin with the actual stage sets and movie studios that had moved to Hollywood from New York – to decrease production costs

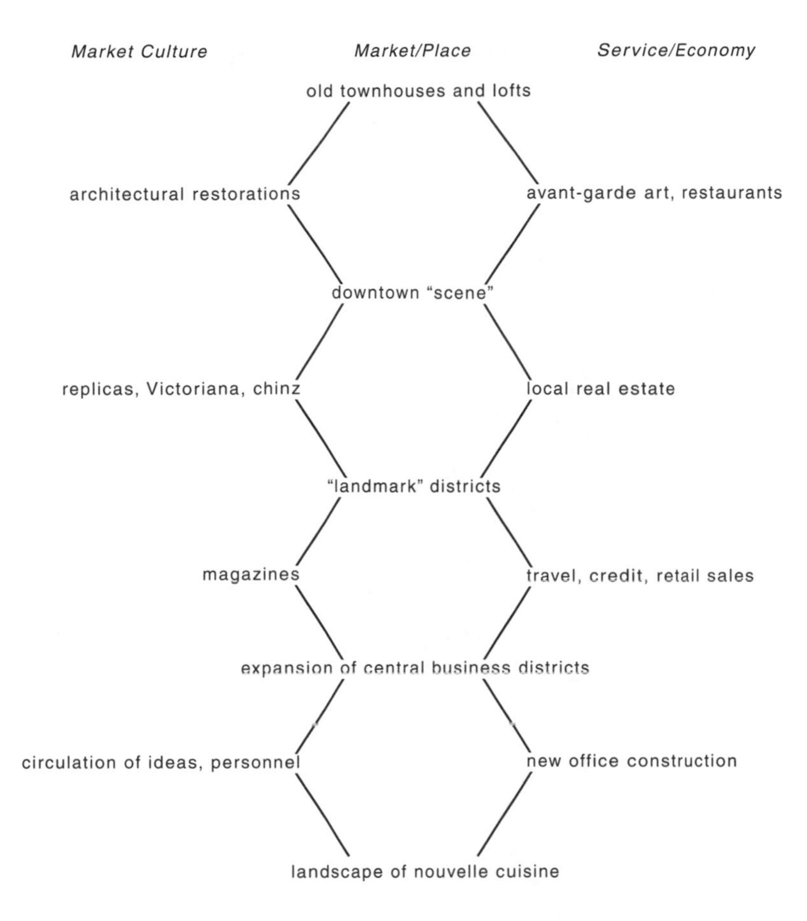

Figure 1 Circuits of cultural capital: gentrification

– by 1920. This double fantasy land of dream products in a dream climate inspired its own sort of vernacular architecture in and around Los Angeles. The vernacular grew into a market culture, both commenting on and reproducing the norms of fantasy production in films. Just as Hollywood films became an important US export, so the film industry located in Hollywood generated real estate development.

As one of Hollywood's products, animated cartoons, developed bigger markets, it inspired the postwar construction of a new sort of amusement park in the orange groves of Anaheim. Disneyland created a marketplace in Orange County for new service occupations, new housing, and other kinds of development. At the same time, it generated still more products in the image of Mickey, Donald *et al.* These developments inspired new architectural designs, which were incorporated into the fantasy ecology of Los Angeles. Publicized in magazines and seen in television series, this in turn encouraged tourism and retail sales. By the same token, when the buildings of Arquitectonica were shown on *Miami Vice*, and the firm's principals appeared on *The Today Show*, as they did in 1989, they enhanced the value of Miami as a marketplace for the work of creative personnel.

The imagineers and others created television shows featuring Disney characters and more product spin-offs. These fruits of their labor constituted a market culture that both commented on Disney creations and responded to demand for Mickey and Minnie facsimiles. At the same time, the flow of visitors who

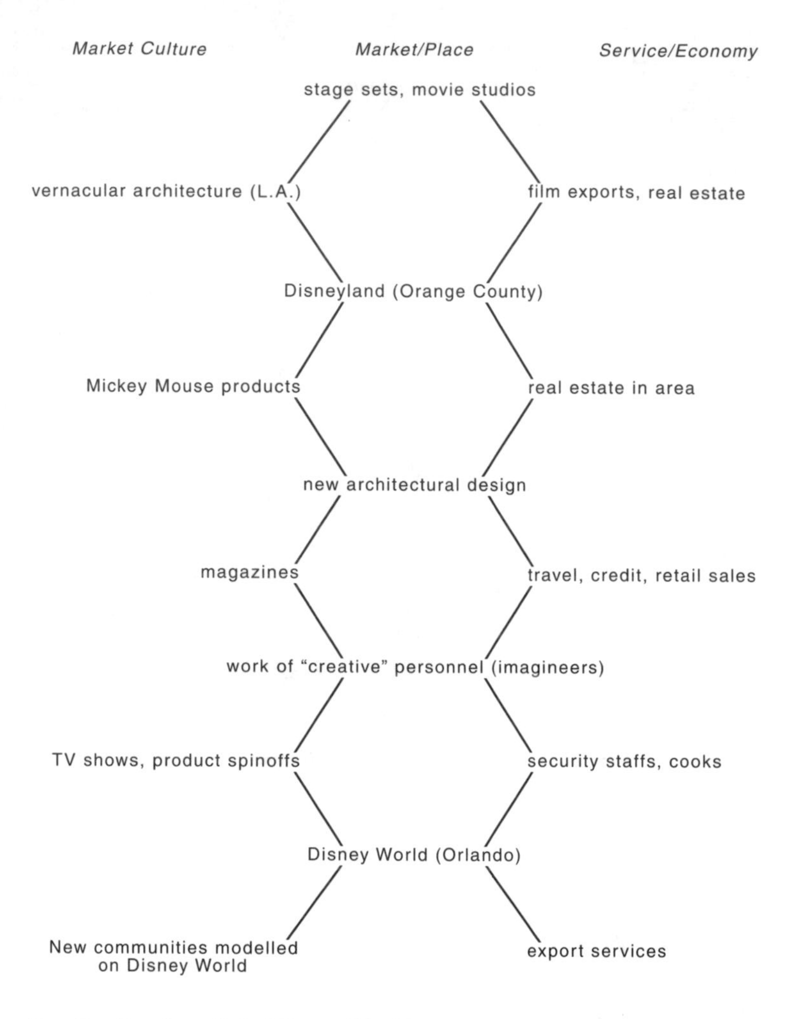

Figure 2 Circuits of cultural capital: Disney World

sought a more immediate experience in Disneyland generated more jobs in the service economy for security guards, hotel staff, and cooks.

These markets in turn inspired the development of an even bigger, more complex Disney World in an underutilized area near Orlando, Florida. As a multi-use exurban complex of new construction, Disney World represented a considerable marketplace. But it had even greater potential. On the one hand, its socially harmonious elaboration of Main Street created a market culture that commented on the fragmentation of American (and by extension, modern) society, while it quickly became a model for the development of new residential communities from Seaside, Florida, to Mashpee Commons on Cape Cod, both of which were designed by the historicist New Urban architects Andres Duany and Elizabeth Plater-Zyberk. On the other hand, Disney World's tremendous commercial success created demand for more export services, notably by 'franchising' Disney theme parks in France and Japan. (Tokyo Disneyland is operated under license from Walt Disney Company; Euro Disneyland is 49 percent owned by Disney.) These have in turn initiated their own regional circuits of economic and cultural capital: developing land, training a service labor force, supporting film, video, and television production.

In the United States, the creation of place by means of Disney markets was confirmed by Disney World's addition of an MGM movie set and Hollywood theme park in 1989. In 1990, Universal Studios opened a competing theme park nearby. These reproduced the initial circuit of cultural capital.

From our viewpoint, it is irrelevant whether cultural capital constitutes a primary, secondary, or tertiary circuit of capital, in line with the ideas of Harvey, Henri Lefebvre, and other urban political economists. But it seems only logical that in the contemporary market economy, investment in cultural capital would offset cyclical devaluation in other parts of the same circuit – for example, in local industries and the built environment. Downriver near Detroit, for example, has drawn investment in leisure and entertainment facilities, although not to an extent that would compensate for structural disinvestment in autos and steel. Countercyclical investment also reflects relative rates of return. In this sense, investment in the mass production or distribution of cultural commodities grows out of the entertainment industry. Finally, investment in cultural capital becomes more profitable because of the inelasticity of demand for certain cultural goods and services that are now deemed essential, at least by that richest stratum of the population with an increasing share of income. While the demand for socially accepted cultural symbols like Mickey and Minnie Mouse remains relatively high, their supply is expanded by licensing agreements, product replicas, and reproduction through different cultural forms (TV, films, toys, advertisements). Yet the economic efficacy of cultural capital may also be a matter of image-creation. Cultural investment 'screens' investment in the services. As cultural capital, Disney World suggests both the power of façade and a façade of power.

Because circuits of cultural capital are formed in real spaces, they suggest how space in an advanced service economy is really formed. Neither solely 'productionist' nor merely 'local', space is structured by, and structures, circuits of capital that incorporate real estate development, amenities and services, and visual consumption. On the one hand, the linkage between cultural capital and real estate development enables new economic structures (e.g. the service economy, global financial markets) to be *localized.* On the other hand, the choice of specific sites as landscapes of cultural consumption (e.g. Times Square, Orange County) represents the geographical *location* of these economic structures. Moreover, building theme parks, theme towns, and other artificial complexes (such as an Amazon rain forest) is now a favored strategy of economic renewal. Unable to attract front or back offices, some cities use abandoned manufacturing sites as their doorway to a service economy.

Although magazines may seem a trivial part of this scenario, they play an increasingly important role, along with other means of mass communication, in developing new circuits of capital based on cultural consumption. The American Express Company, for example, during the last years of the 1980s bought a number of 'lifestyle' magazines produced for specific local markets, including *L.A. Style* and *New York Woman,* as well as the more panoramic *Food and Wine.* American Express based its publishing strategy on the fact that 70 percent of its twenty million credit card holders live in the top twenty cities of the country. 'We want magazines in cities where they live,' the president of American Express Publishing says. 'And the advertising side can build its relationships with our customers.' By the same token, a small number of rich real estate developers in New York City have recently bought or founded newspapers in the city. As individuals they are expressing a whim, a desire, eccentricity. But they are also active agents in a circuit of cultural capital, linking mass communications, the local service economy, and real estate development.

The spatial mediation of cultural consumption affects the redistribution of benefits among social classes. In suburbs of big cities, actions of landed, political, and corporate elites use a suburb's visual homogeneity as a useful base for encouraging social homogeneity among the upper middle class. Gentrifiers' and developers' ability to appropriate central spaces by means of visual consumption has the real effect of displacing lower-income residents. Building the Disney-like complex of Opryland outside Nashville, rather than renewing and expanding it on the Grand Ol' Opry's original site in the center of the city, prolongs the economic crisis of the downtown area. While such an expansion might well create the same sort of displacement as gentrification and publicly sponsored urban renewal, it nonetheless deprives the inner city of a significant redevelopment project linked to the new service economy. But this suggests an important constraint imposed on cultural capital by political economy. Strategies of cultural consumption may only complement, rather than contradict, strategies of capital accumulation.

Ian Cook

'New Fruits and Vanity: Symbolic Production in the Global Food Industry'

from *From Columbus to Conagra: The Global Station of Agriculture and Food Order* (1994)

In the penultimate chapter of *The Sociology of Agriculture*, Buttel *et al* (1990) identify nine trends and six gaps in the contemporary sociology of agriculture. Among these are the trend of commodity systems analysis and the gap to be filled by critical ethnographic fieldwork. A number of writers have attempted to apply the latter in their work on the former to assess how the distinctive industrial structures of the global food economy are produced, reproduced, and transformed by the day-to-day actions of those people who work within them.

Like these writers, I believe that a deeper understanding of agricultural change can come through ethnographic fieldwork and the development of theory, that engages local with global, agency with structure, and the symbolic with the material. In this paper I begin to explore these engagements through an analysis of interviews with a number of executives working for J. Sainsbury, Tesco, and Safeway, the three largest food retailing chains in the United Kingdom. An examination of the work that goes into the introduction of new – or 'exotic' – fruits to customers suggests that the meanings that companies attempt to ascribe to these fruits play a crucial role in the articulation of commodity systems. Put differently, just because they are produced and packed in one place and shipped, ripened, and delivered fresh to a store in another, it does not necessarily follow that anyone will buy them. In short, there is a symbiotic relationship between the 'material' production of a fruit or vegetable and the 'symbolic' production of its meaning(s).

The major link between a fruit's material and symbolic production are its specifications. These include its acceptable shape, size, weight, internal pressure, blemishes, seasonal sourcing, and price as defined by the retailers. Day by day – albeit through highly asymmetrical power relations – these are negotiated through both symbolic and material labor processes. Here we will explore this link between the symbolic and the material.

CONTRACT FARMING

The main reason that I have chosen to discuss contract farming – as opposed to other forms of agricultural production – is that, here, explicit attention has been paid to the day-to-day negotiation of specifications by growers.

The Nature of Contracts

In the underdeveloped world, where many exotic or new (to Westerners) fruits are grown, contract farming has emerged for two major reasons. First, in many places, agribusiness corporations have been forced to give up their rights to land as a result of nationalist pressures and the attendant threats of expropriation and local regulation (Watts, 1992). Second, independent peasant households have been institutionally captured by contract farming with its promises of modernization and credit (Clapp, 1988).

Perhaps the most important features of contract farming are that its labor force consists of peasants who work their own land and that it is usually promoted as a 'dynamic partnership' between the rural poor and private capital. This partnership, to quote Watts (1992, p. 15) 'promises rapid market integration, economic growth and technical innovation while protecting the rights and autonomy of the grower via the contract'. It legitimates the continued activities of agribusiness corporations, organizations such as the World Bank and the US Agency for International Development (USAID), and ruling classes in the underdeveloped world. Through contracts, smallholding peasants become tied to local merchants, the state, agribusiness corporations, or joint enterprises that buy, process, and/or export their produce. However, whatever the nature of these enterprises, they demand produce that satisfies the retailers' specifications. This, in turn, means that peasants must follow a detailed checklist of instructions concerning the methods and dates of soil preparation, pesticide and fertilizer application, planting and harvesting established in the contract. Furthermore, even when this has been done, the contractor's inspectors can reject the crop on the grounds of 'quality' standards, which they also specify (Clapp, 1988). Only when all of these conditions have been satisfied can the crop enter the processing and transportation networks that make up its commodity system and the grower be paid for his or her work.

Meaning for Growers

From the contractor's perspective, this form of production is extremely efficient: It allows the programming of produce in advance at a predictable price, quality, and schedule of supply. At the same time, it insulates contractors against the risks of weather, insect attack, disease, and labor bottlenecks because they pay based on results, not by the hour. These risks are further minimized through multiple sourcing.

Though the contractor cannot avoid acts of God, the suppression of labor unrest is a key strategy of risk avoidance. Contract farming is extremely labor intensive, as great care has to be taken to produce fruit that is the right color, texture, shape, size, and ripeness on the supermarket shelf. So, for instance, one study of snow pea cultivation in Central America found that it was six times more labor intensive than the local staple, maize (Watts, 1992). To provide this labor, growers draw on family and other social networks in times of need. At the same time, the specialized knowledge required to successfully produce contracted fruit is usually in the hands of the contractor's outreach workers. Thus, while they own their own land, growers are effectively deskilled agricultural laborers working for a piecework wage for the contractor.

Behind the image of a dynamic partnership between the parties concerned, then, lies a mode of production in which contractors can obtain a continuous supply of high-quality produce by placing the risks of production almost entirely in the hands of growers and their families. Having entered into a legal agreement with growers, companies effectively absolve themselves of any welfare responsibilities. A newspaper article about Margaret Thatcher's visit to, and praise of, the South African farming sector in mid-1991, for instance, pointed out that its 1.5 million black laborers 'enjoy no statutory protection on minimum wages, conditions of service, health and safety standards, unemployment insurance, working hours, overtime pay, maternity leave, [or] holidays' (Beresford 1991, p. 26). Although South Africa may be an extreme example, it is fair to say that contracts do not promote the 'rights and autonomy' of growers but exploit their poverty and procure the grower's 'self-exploitation' (Clapp, 1988, p. 10). Thus, struggles over the production process are often fought not against the contractor but within the household. As a result these struggles become conjugal and/or generational and are often over 'the customary rights, responsibilities and obligations linking [the] labor claims and property relations' of household members (Watts, 1989, p. 18; cf. Thomas, 1985).

SYMBOLIC PRODUCTION

At the other end of the chain in the UK, there have been two major developments in the sale of fresh fruit in recent years. First, the market has grown quite considerably: Between 1983 and 1989 imports of fresh fruit into the UK grew by about 25 percent. In the retail industry, this has been explained largely in terms of an increased consumer interest in healthy diets and a breakdown of traditional meal patterns. Consequently, more fruit is being consumed both in regular meals and as snacks (Marketing Strategies for Industry, 1988).

Second, during this same period, unusual and exotic produce has become increasingly important. This is illustrated by the fact that between 1983 and 1989, although the percentage of UK fruit imports by weight changed very little, by value the 'other' fruit category (in which the exotics have been placed) registered a significant increase (Commonwealth Secretariat, 1990). This pattern can be explained largely as the result of changes made by the major supermarket chains (or multiples) both at the point of sale and in terms of their buying strategies. At the point of sale, self-selection is the key issue, as it allows customers to handle the produce, select the particular pieces of fruit they want, and thereby to 'discover' new varieties rather than by having to ask for them by name. Moreover, the chains have organized their suppliers to provide more and more produce year round, allowing formerly exotic fruits to become a regular part of their customers' diets.

In this arena, the supermarket chains have been able to carve out an increasingly large slice of fresh fruit sales. For instance, their share of the market grew by 2 percent per year between 1985 and 1987, when it stood at 46 percent. Furthermore, this growth happened at a time when the largest of them were undergoing a long-term expansion in terms of selling area, turnover, and profits and has continued even through the current recession.

From the perspective of the supermarket chains, the symbolic production of exotic produce can be seen as rooted in the daily lives of a small number of trading managers and technologists working in their headquarters. At Tesco, for example, the exotic fruit team established in December of 1990 consists of just one technologist and one trading manager. For the other supermarket chains, exotic produce usually falls within the purview of much larger soft and hard fruit teams.

Trading managers and technologists work very closely with the representatives of a small number of suppliers (e.g. Geest, Del Monte, Brooks Brothers, and Worldfresh) to set the specifications for their chosen range of fruit. They do this with two major corporate obligations in mind: to increase turnover by getting the best quality fruit year round; and to set their prices to meet the budgetary targets set by their senior management. Here, struggles ensue as to which new fruits to sell and how to promote them. Interestingly, however, immediate profits are not the sole determinant of whether the introduction of an exotic fruit can be termed 'successful'. These struggles and successes can both be seen to take place in a circuit of cultural forms.

Struggles

Struggles over the introduction of new fruits involve a blend of meanings and money, neither of which can be entirely separated from the other. A key problem, as a senior manager at Sainsbury's told me, is that trading managers have to deal with the fact that 'one of the problems with exotic fruits is they're a bit like armadillos – you can't get into the damn things! You've got to have an instruction manual on how to eat it and what to do with it.' But, given their budgets, the trading teams often cannot ordinarily provide these for customers because, as a Safeway trading manager explained:

> They are very low volume, they are very high price and, to try to get things off the ground, you've got to spend a lot of money on them – and there isn't much money available. We've just launched a new exotic called the Pitahaya. . . . On average, we're looking for 35, 36 percent gross profit somewhere round there – on the exotics. And, for the Pitahaya, we're talking about less than 10 percent. So, by the time the store, inevitably, has thrown one out of the box of ten away, you haven't got any profit – and they're more likely to throw four or five out of the box of ten away. So, there's no profit in it, but it's just trying to establish that product. Now, because the sales are so low, you can't go out and get a glossy brochure.

Faced with this situation, the stores have experimented with a wide variety of manuals to get their message across. All appear to be designed to draw new fruits into consumers' 'maps of meaning' (Burgess, 1990; Hall, 1980; Jackson, 1989) through associating them with popular notions of class, place, healthy living, and sensual experience. The latter includes, but is by no means limited to, that involved in tasting other more mainstream fruits. To illustrate these associations, consider three of the many types of manuals that were available on supermarket shelves in the UK in 1990 and 1991.

First, supermarket chains have produced wide-ranging manuals for sale. Take, for instance, the exotic fruit and vegetable manual available in early 1991 at the checkouts of Asda – the fourth largest chain – for twenty pence. For instance, on the page that promotes limes, mangoes, lychees, and papayas, all are described in terms of their origins, properties and brief suggestions for their preparation. Finally, the back of the manual contains a number of recipes including 'kumquats in honey butter' and 'fluffy mango fool'.

Another type to be found on supermarket shelves is the single fruit manual put together by its suppliers. Consider, for instance, the one provided free by Morris, sole producers of the Kiwano melon. The rather outlandish approach to the marketing of this exotic fruit can be better understood by taking into account the motives of its producer. In November 1990, in an article in *The Sunday Times* color supplement, John Morris described why he had paid $25,000 to register 'Kiwano' as a trademark: 'There is an advantage in brand names because we are in an age when people buy labels. What we're after for the horned melon is a Pierre Cardin image' (Martin, 1990, p. 27).

My final example is a brochure that was available free in Safeway produce departments in the fall of 1990. It begins by describing how they are able to provide 'the widest possible choice of fresh fruit and vegetables – throughout the year' and, among the pages devoted to their grapes, apples, aubergines, beansprouts, tomatoes, potatoes, and oranges is one on carambolas. Interestingly, unlike the other manuals I have seen, this paints an image of their material production.

Rather than being the sole source of information on new fruits, trading managers see the production and distribution of these manuals as part of a wider circuit of cultural forms in which, as one at Safeway put it,

> The people who buy exotics . . . are looking for them basically. You get a sort of gourmet person – they don't necessarily have to be on a high income, but . . . they'll go out and buy the Pitahaya. What we've got to do is then convince some of the people who don't like trying things, but might have a lot of money. . . . What we've got to try and do is break some of them into eating that product. It's a slow process. Gradually, . . . they may see recipes in . . . women's magazines or television, or recipe leaflets that we produce as well. It's a very slow process.

He continued by explaining that the information available in these circuits was not always complimentary and referred specifically to the media:

> They will phone our public relations department and say, 'Were going to do a programme', or 'We're going to do an article on so-and-so. Can we have some?' and 'Can we have your comments?' . . . So then we will supply them – with some wariness because they don't always give particularly good comments about it. We could do a lot of work – there are one or two in particular where we do a lot of background work – and then all . . . we get is just slagged off about how bad it is. Obviously they've got to be truthful, but . . . we can't bend over backwards to help someone who's just got an axe out.

Success

After having listened to my contacts talking about the difficulties of introducing new fruits to the British public and the long periods of making little or no profit on them, I wondered what purpose is served by these exotic fruits. I put this question to the man at Safeway and he provided an explanation: 'First of all, it's a point of difference for Safeway. There's a lot of products that are exactly the same, any supermarket you go in. People won't come to you because you sell Heinz Beans, because everybody sells Heinz Beans. But, people may just do their 75, £100s worth of shopping a week with you because you sell the Pitahaya and . . . Safeway's the only place you can get it. So, you could actually attract customers by making the shopping interesting.' In the UK, as in the United States, supermarkets are commonly designed so that customers enter directly into the fresh produce department. With their colors and smells, and, through a European Community (EC) directive, the place of origin of all produce is clearly displayed on the shelf alongside the price. Sainsbury's marketing director described entering the store as a geography lesson or a trip around the world.

Moving back to Safeway, my contact gave a second reason for selling exotic fruit: 'For any new product to take off, you've got to work at it for ten years. Avocados came in, first, into Safeway and . . . they were an exotic item for ten years and suddenly they're a major item for us now.' As well as avocados, he could have mentioned kiwifruit and mangoes – both of which have shed their exotic image to become everyday fruits available, like apples and oranges, throughout the year. In a relatively short time, both have grown into approximately £20 million markets in the UK. Therefore, trading managers, technologists, and suppliers are all searching for the kiwi or mango of the future.

CRITICAL ETHNOGRAPHIES IN THE MODERN WORLD SYSTEM

Having argued that symbolic production is a key process in the articulation of commodity systems, I would like to conclude with some notes on the prospect of a broader critical ethnography of the modern world system. Perhaps the most important insights that ethnographic research can provide for us are found in the 'maps of meaning' drawn upon and contested within and between groups of people working at different stages in a commodity system. Looked at critically, these are not 'merely subjective'. Rather, growers, outreach workers, trading managers, technologists, and consumers – as well as other related groups – conceptualize the horizons of their power and future fortunes with these maps. Thus, it is interesting to note Clapp's observation that in many rural areas in Latin America,

> The Coca-Cola sign on the local *bodega* may be ten years old and its glow filtered by a coat of dust, but the Dow poster in the extension agent's office has the aura of the new and successful in a town where not much does. These advertisements may not persuade a farmer to buy one brand rather than another, but the cluster of them together, reinforced by the urging of state and company promoters, may define what it is to be a modern and 'progressive' farmer. It should not surprise us that [in one Guatemalan study] . . . personal crop sprayers had become a sign of status and that farmers with access to credit through the contract had gone deeply into debt to buy them. (Clapp, 1988, pp. 13–14)

This imaging of the Other is also a key component drawn on by supermarket executives.

My main goal in talking with them was to trace connections between their work and that of the people who grow the fruit that they sell. In their companies' annual reports, claims are often made that they 'care' about the production process. Consider, for example, one page from Tesco's 1990 report. It begins: 'Food retailing still offers enormous growth potential – potential which can be realized by developing attractive new products of the highest quality, and by satisfying new consumer values such as health, the environment, and the *source and methods of manufacture*'. It ends by noting that, 'in terms of variety, absolute quality, value for money, *product integrity* and new product development, we set and achieve exceptionally high standards' (emphases added).

Yet, when challenged about sourcing and methods, they responded with rather limited conceptions of 'product integrity' that had little or nothing to do with labor processes. The trading director of Sainsbury's explained:

> I think we would assume that anybody we dealt with . . . would be fulfilling any criteria that we set down as a bona fide supplier – which would mean having clean factories, and standards, and all the rest of it. And the factories would be vetted, and they're vetted . . . often on unannounced audits. Now, all of our Kenya dwarf beans packed in Kenya – you will note that they all point the same way. We had a customer letter about it two weeks ago asking, 'Why do they all point the same way?' The simple reason is that there's . . . lots of black ladies out there who put them in all facing the same way. Now, that's not forced labor. They just choose to do it.

This, of course, is by no means a free or simple 'choice'. Yet, thinking so allows those involved in symbolic production to quite sincerely peddle their caring approach in an exotic, rather than systematically exploited, world. At a general level, much of this would seem to be embroiled in British culture with its colonial legacy and, more generally, a disassociation of exotic or holiday locations from the widespread poverty and hunger of many of those who live there. Moreover, those executives who do trot the globe visiting sites of material production do not spend any great length of time sharing experiences with their growers. They are unlikely even to speak the same language (Hannerz, 1990).

In sum, the articulation of commodity systems in space and time is not just a technical matter. It necessarily involves the production and consumption of the commodity's meaning(s) within, as well as between, all stages of its existence (Burgess, 1990; Johnson, 1986). As such, the 'struggles' shaping the course(s) of agricultural change are by no means confined to growers and their families negotiating new relations of production that have resulted from their incorporation into the capitalist world system. Rather, as Marcus and Fischer (1986, p. 91) argue, we should see this system as being produced, reproduced, and transformed more broadly by 'the activities of dispersed groups of individuals whose actions have mutual, often unintended, consequences for each other, as they are connected by markets and other major institutions that make the world a system'.

Thus, a key component of critical ethnography – particularly with respect to the exploration of power relations – is that its fieldwork should take place in more than one locale (Marcus, 1986) within a given commodity system in order that the local and the global, agency and structure, and the symbolic and the material can be more directly engaged.

Section 4.3 Landscapes of Consumption

Paul L. Knox

'The Restless Urban Landscape: Economic and Sociocultural Change and the Transformation of Metropolitan Washington, DC'

***Annals of the Association of American Geographers* (1991)**

NEW LANDSCAPE ELEMENTS

As in nearly every central city, the built environment in Washington, DC has been strikingly altered by the accelerated spread of 'designer neighborhoods' through rehabilitation and gentrification – a kind of do-it-yourself post-modern design that has been the product of the confluence of land economics and the desire on the part of the 'new' middle classes to establish the physical framework for a distinctive 'habitus' (Jager, 1986; Smith and Williams, 1986).

Consumers who want style and distinctiveness in more sequestered settings are the target market for private, master-planned suburban and exurban communities. The essential features of these communities are 'a definable boundary; a consistent, but not necessarily uniform, character; overall control during the development process by a single development entity; private ownership of recreational amenities; and enforcement of covenants, conditions and restrictions by a master community association' (Suchman, 1990, p. 35). They typically exploit cluster zoning to provide comprehensively designed environments in which the amenities (tennis, golf, swimming, etc.), architectural styles (colonial revival, Tudor, larger-than-life Cotswold, etc.), landscaping (lavish, meticulously trimmed, and brimming with special features like cast iron-and-brass signwork), layout (maximizing rustic views and amenity frontage) and security (symbolized by gatehouses or imposing gateways and operationalized by card-key systems) are carefully targeted to appeal to the lifestyles and status identities of particular fractions of the middle income group. Such communities have a distinct marketing advantage over other kinds of residential development, and individual homes in such communities have been shown to maintain their value better than similar homes elsewhere (Schwanke, 1990).

The net result is a collage of intensely private worlds, each entered through brick or timber portals in the manner of an English landed estate and each announcing itself with a name that draws heavily on a repertoire of historic and distinctive associations: King's Forest, The Chase, Woodlea Manor, Wellington ('The Dawn of an Old Era'), and Sully

Station ('Sometimes It's Better to Live in the Past'). Here are Boyer's 'artful fragments' in serial production, providing the foundation settings for the aestheticization of everyday life described by Berman and others. They also provide the opportunity for placemaking through consumption and an alternative physical framework for the development of a distinctive habitus for the new bourgeoisie. It is in the suburban periphery, however, that the master-planned community comes into its own as an upper-middle class vivarium and a setting for commodification. Avenel, for example, provides a wooded setting of 970 acres that contains some 700 single-family detached dwellings (all of them, reassuringly, 'priced well into six figures') and about 150 townhouses and courtyard homes, faced with bike paths, jogging trails and bridle paths and featuring an equestrian center and an 18-hole PGA golf course. At Sully Station, a 1200 acre, 3300-home development adjacent to the Dulles corridor, the centerpiece is a $1 million brick clubhouse/sales center in what is claimed to be the style of a 1930s small-town English railway station, complete with wooden waiting room benches and a stationmaster's clock. Perhaps the most ambitious of Washington's master-planned communities is Kentlands, currently being developed on a $40 million parcel of farmland near Gaithersburg, Maryland. Kentlands is planned to include 1600 homes (a mix of single-family detached houses, townhouses, apartments, senior housing, condominiums, and artist/craft studios), a 1.2 million square-foot shopping mall and more than 1 million square feet of office space. Special features include: a nineteenth-century style street grid punctuated by a 'Main Street' town center and public squares; the preservation of an 1852 farmhouse and its outbuildings; the construction of a brick barn, mill, firestation, guest house and carriage house as accommodation for cultural activities; a wetlands environment – a twelve-acre lake with a boathouse and surrounding parkland; along with churches, an elementary school, a post office, and a recreation center. The kitchen garden of the farmhouse will be maintained by a gardener who will also operate a holding nursery to sell shrubs and plants deemed appropriate to Kentlands' residents.

Such amenities are central to the marketing strategies of the large developers who have identified Washington's rapidly-expanding new bourgeoisie and new petite bourgeoisie as the most desirable segment of the residential construction market. It is not simply the amenity packages that are thought to be necessary for marketing success. *Design and style* are recognized as crucial ingredients and in this respect the dominant theme is neotraditional planning, emphasizing not only period architecture and landscaping, but also a return to pedestrian-oriented settings with a comprehensive mix of land uses and amenities (Lecesse, 1988; Forum, 1988).

In larger metropolitan areas are to be found the *festival settings* produced by the 'Rousification' of downtown districts (Hall, 1988; Martin, 1985; Sawicki, 1989). These large-scale redevelopments, usually the product of public–private partnerships, create focal settings (waterfronts, market areas) for integrated packages of upscale offices, tourist shops, 'impulse' retailing, restaurants, hotels, offices, concert halls and galleries. They are distinctive as new landscape elements merely because of their scale and their consequent ability to stage – or merely to be – the spectacular. They are settings for events such as concerts, outdoor exhibits, street entertainment and New Year's Eve festivities, often subsidized by the developer's budget for promotion. They are described by Harvey (1989, p. 92) as the 'carnival mask' of contemporary urbanization, their spectacular spaces being a means 'to attract capital and people (of the right sort) in a period (since 1973) of intensified inter-urban competition and urban entrepreneurialism.' Well-known examples include Harbor Place in Baltimore, Riverwalk in San Antonio, Fisherman's Wharf (including The Cannery and Ghirardelli Square) in San Francisco, South Street Seaport in New York, and Faneuil Hall in Boston.

Washington has two major focal settings of 'spectacle and display' in Alexandria (Old Town King Street, the Torpedo Factory) and Georgetown (Georgetown Waterfront, the Georgetown Park Mall, the Paper Mill, the Flour Mill, Dodge Warehouse, the refurbished Chesapeake and Ohio canal, and Canal Square) that provide settings for concentrations of tourist shops and restaurants, hotels, galleries, and exclusive condominiums. These are both 'organic' festival settings (like the Fisherman's Wharf area) that have emerged through piecemeal development.

Finally, the expansion of the suburban space-economy has produced new commercial landscapes: *high-tech corridors* along limited-access highways. Development in these corridors has typically taken a form very different from the infilling, multinucleation, commercial strips, bypass strips and mixed-use suburban freeway corridors of the modern metropolis (Baerwald, 1978; Hart, 1983, Erickson, 1983, McDonald, 1985). The dominant components are set on a framework of large lots, usually several acres, and include large-scale structures, extensive onsite parking, generous amounts of landscaped parkland (including waterfalls, lagoons, terraces, gazebos, and sculptures), and a variety of services and amenities such as fitness centers, cycling trails, coffee shops, flower shops and day care facilities. Office blocks, R&D labs, clean industries and hotels (not shops, showrooms, outlets or restaurants) dominate these corridors, while a new component of increasing importance is 'flex space': sprawling, single-story structures that look like offices in front but have loading docks in the rear, giving access to R&D labs or small manufacturing or warehousing operations. The Washington metropolitan area experienced a gain of more than 36 million square feet of flex/R&D/warehouse space between 1978–88, an increase of 174 percent (Birch and Jain, 1989). There are concentrations of flex space within all of the recent speculative development that has taken place near major routeways that pass through the metropolitan fringe, but the most marked concentration is at the western end of the Dulles corridor, in Loudon County, Virginia.

CONCLUSIONS: NEW URBAN LANDSCAPES AND A NEW URBAN GEOGRAPHY

I have argued that a series of distinctive new urban landscapes is emerging from a socio-spatial dialectic dominated by the effects of the reconfiguration of economic and cultural life. I have also attempted to show that the emergence of these new urban landscapes provides a rich source of conceptual and empirical issues that link urban geography with economic, social and cultural geography. The particular approach adopted here clearly lends itself to a theoretical framework rooted in historical materialism, in which the built environment is seen as part of the superstructure that is not only produced by but also helps to sustain the dominant relations of production. Postmodern design, along with postmodern culture and philosophy, thus becomes the latest, incipient, dialectical expression of the transformation from rationalist/Modernist/Fordist capitalism to an emergent, globalizing, advanced capitalism. According to this perspective, the new settings described here can be interpreted as sustaining the transformation of capitalism because of their capacity to legitimize new social, professional, commercial and financial elites and their capacity to enhance the circulation and accumulation of capital within an emergent global political economy. More specifically, they can be held to reflect the ideology and values of elements of the new social formations that have emerged with this transformation and be implicated in some of the consequent tensions and conflicts between them. Similarly, they can be interpreted as reflecting and reinforcing the broader fragmentation and polarization of urban space. As Boyer puts it,

> we can take the current recycling of inner city neighborboods, the rise of designed environments from restaurants to city structures, the pleasures of sophisticated food taste – in short the whole realm of aesthetic consumption from town houses to sun dried tomatoes – as cultural reinforcement of urban

> spatial structures: those of fragmentation and hierarchicalization at the local level, as well as homogenization at the transurban level. (1987, p. 21)

The examination of new urban landscapes in the Washington metropolitan area lends support to the notion of a new urban geography, with a radically different form and ecology from that of the classic American city depicted by factorial ecologies and explained, with varying degrees of success, by bid-rent theory, theories of residential mobility, Weberian theory, and neomarxist theories. The spatial patterns associated with the landscape elements described in this essay do not fit comfortably within the sectors, zones or mosaic patterns that have been the focus of academic debate surrounding the classic American city. Fragmentation, multinodality, fluidity, plurality, and diffusion are more in evidence than homogeneity, nodality and hierarchy (Gottdiener and Gephart, 1991). Peirce Lewis coined the term 'galactic metropolis' to capture a new urban geography in which 'The residential subdivision, the shopping centers, the industrial parks seem to float in space; seen together, they resemble a galaxy of stars and planets, held together by mutual gravitational attraction, but with large empty areas between clusters' (Lewis, 1983, p. 35). This new urban geography, as some geographers have pointed out (Fonseca, 1976; Wood, 1988), is characterized by landscapes that are semiurban: landscapes of mixed densities and unexpected juxtapositions of forms and functions. All this should not imply, of course, that the 'old' urban geography has been completely overwritten or that the restlessness of the built environment is somehow linked to an attenuation of polarization and segregation within the city. Rather, we have to contend with new economic and sociocultural trends, a changing framework for metropolitan growth, changing urban landscapes and, probably, changing patterns of economic, social, and cultural differentiation. Most of this, clearly, has yet to be mapped out and set within appropriate theoretical frameworks. It remains to be seen whether the new urban landscapes described in this essay will become paradigmatic of the new urban geography: the symbolic landscapes of the postmodern, fin de siecle American metropolis.

Neil Wrigley and Michelle Lowe

'New Landscapes of Consumption'

from ***Retailing, Consumption and Capital: Towards the New Retail Geography* (1996)**

NEW LANDSCAPES OF CONSUMPTION

[T]he consumption sites and spaces (and associated consumption experiences) under consideration here are increasingly ubiquitous aspects of contemporary (western and often non-western) societies. Indeed the omnipresence of such new consumption landscapes has led **Knox** (1991) to the view that western societies are being transformed into a landscape which is like a supermarket writ large in which the whole of the landscape is geared towards consumption. We have much sympathy with this viewpoint and would like here to consider the various ways in which this is the case. Here then we shall examine these 'new landscapes of consumption' paying specific attention to the ways in which consideration of such landscapes may direct our 'new cultural geographies of retailing' in the future. We do this under two main headings. First, we reconsider consumption sites and spaces that have been the subjects of academic attention in the past. Second, we examine new consumption sites and spaces which are, in our estimation at least, likely to prove important in the future. We confine our discussion to the examples we know best. Hence we focus on examples of these new landscapes in the UK. Readers will doubtless recognize, however, that the themes that emerge here have much broader applications.

RECONSIDERING CONSUMPTION SITES AND SPACES

[T]he primary consumption sites and spaces that have thus far been considered by geographers have been 'the department store' and 'the mall' with some secondary (largely historical) research work being directed at 'the street'. Here we would like to reconsider these consumption sites and spaces via an examination of their contemporary importance.

The Department Store

In the decade of the 1990s 'the department store' appears to be enjoying something of a revival. Such stores are no longer associated purely and simply with major retail groups such as John Lewis or House of Fraser. Rather the contemporary department store has appeared in an altogether different guise – the so-called 'flagship store'. Incorporating the 'own label' as their sole product identity, stores such as the new 'Dr Marten's Department Store' in Covent Garden, London, or the myriad of new flagship stores in New Bond Street, London, takes us full circle back to the department stores of the nineteenth century. Like their nineteenth century equivalents the contemporary department stores combine within their 'spaces', designer interiors, rituals of display and leisure, sexuality and food. At Donna Karen in Bond Street, for example, DKNY mineral water, New York bagels and New England cheesecake allow the contemporary consumer to lose themselves in the ultimate 'own label' experience. At Dr Marten's Department Store the 'Doctor's Orders' Canteen/Cafe offers Wollaston Shepherd's Pie or Northamptonshire Bangers and Mash and even includes a children's hairdressing department, while the interior of the Dolce and Gabbana 'flagship store' in Sloane Street has been designed on a palazzo style and combines both classical and modern elements. Furniture includes gilt picture frames and baroque divans. The association in these new consumption landscapes of product and place identity is profound and offers much in the way of research potential in the future.

The Mall

Crewe and Lowe (1995) suggest that it is important to shift 'the research focus away from dominant accounts of the malling of retail space and of global homogeneity and [to] . . . focus instead on the complexity and differentiation of retail spaces'. From our perspective it is also essential to move beyond a homogeneous view of shopping malls and to focus instead on the various ways in which such 'centres often strive to establish their unique identity' (Jackson, 1994). While some such matters have been considered we suggest here that the differentiation of shopping malls is far more advanced than a reading of the literature to date would suggest. Virtually every local shopping centre has within it some version of the 1980s-built mall (or often more accurately the 1960s-built and 1980s-renovated mall). Such in-town shopping precincts stand in marked contrast to their out-of-town mega-mall equivalents, while in-town centres vary in scale and type from that at Neal's Yard in Covent Garden to the Broad Street Mall in Reading. Then there are the malls constructed – once again in the 1980s or 1990s – as part and parcel of the tourist strategies of particular towns or cities – Canute's Pavilion at Ocean Village in Southampton is a case in point. Centred mainly around fast food outlets and selling primarily Belgian chocolates, fold-up cards, candles and massage oils, such malls are an essential means by which particular localities are being reconstructed and 'sold' no longer as centres of production but of consumption. Finally, there are the 'discount malls' such as that at the 'Clarks Village' at Street in Somerset. Here the differentiation of the mall experience can be viewed at its most extreme with whole coachloads of consumers being bussed in for that extra special 'bargain'. The latest phase of the Clarks Village includes the redevelopment of the former engineering workshops and medical departments of the original Clarks footwear factory.

The Street

We noted above that 'the street' as a consumption landscape has been largely neglected except in a few historical studies. However, Crewe and Lowe (1995) have begun to address this neglect via their consideration of 'pioneering retailers and their creation of differentiated spaces of consumption'. Paralleling suggestions regarding the 'remaking of the mall' discussed above, certain streets are seen to acquire particular consumption identities. Hence, Little Clarendon Street in Oxford with its mixture of coffee shops, tapas bars and charity stores, intermingled with up-market clothing retailers, becomes a central place to 'hang out' for the young (and often not so young) and trendy of Oxford. Carnaby Street in London – famed as a centre for fashion in the 1960s – still maintains a certain 'aura'. An account of such differentiated 'streets of style' leads us to a further important theme. The evolution of 'style' in public places – especially 'the street' has become the subject of much popular attention. More specifically, the reciprocal relationship between international (high) designer fashion and 'streetstyle' is an inherently geographical project.

NEW CONSUMPTION SITES AND SPACES

During the 1990s a number of new landscapes of consumption have emerged and continue to emerge in the UK. Given the rapidity of developments it is almost impossible to mention them all. Here then we have chosen to structure our discussion of such landscapes under three main headings. First, 'the captured market' whereby individuals confined in places for other very different reasons are induced to consume. Second, 'taking consumption to the consumer' specifically in 'the home'. Finally, via an examination of consumption opportunities associated with more conventional leisure pursuits, 'the leisured consumer'. All of the above are centred on the driving mechanism of intensifying consumption and are engaged in a calculated reworking of space to ensure that such intensification takes place.

The Captured Market

The essential conditions for a captured market are a guaranteed throughflow of large numbers of individuals who are likely to consume, some enforced wait time, and/or a lack of locally available alternatives. At its most advanced at the airport (where security measures and air traffic controllers ensure a minimum wait time), at the railway station, or on cruise ships or cross channel ferries, such consumption sites are increasingly being targeted as essential new opportunities by retailers. 'Gatwick Village' at Gatwick Airport's South Terminal includes a vast array of High Street chains. Here the ultimate blurring of leisure and retail takes place. With vast numbers of potential 'customers' on holiday, extended 'seasons' for particular product lines can be achieved, and customer socio-economic profiles compare favourably with those in more traditional shopping centres. Moreover, anxious about their impending flights and unknown holiday locations 'travellers' are likely to spend significantly more than more conventional High Street customers. In a similar vein petrol/service stations have become 'mini markets'. Indeed, 'filling station snacking, as well as full scale shopping is the fastest growing but least predicted area in retailing today. . . . Out of hours diners . . . include late night clubbers, police patrols and lorry drivers' (Margolis, 1995).

In considering the 'captured consumer', two further examples come to mind. The first is the office development – often in a strategic new location – which includes in its 'bowels' some combination of retail facilities. The much vaunted Canary Wharf in London is a case in point, but new business centres and allied retail outlets are an increasingly common form of 'consumption landscape'. At Merry Hill in the West Midlands, for example, the latest phase of development comprises primarily offices at Waterfront East and West and – despite their locations adjacent to the Merry Hill Centre – additional retail facilities have been developed on the ground floors of the office blocks. The final example of the 'captured consumer' is the case of the hospital. In the wake of health service privatization what was once the remit of the Hospital League of Friends – selling fruit and flowers at the hospital entrance – has been broadened to include national chainstores such as John Menzies.

Taking Consumption to the Consumer

The 'home' as the traditional site of reproduction/consumption has been re-cast as a 'new consumption landscape' of considerable potential. This expansion of 'home shopping' has taken several forms. First, and most obviously, via the dramatic growth of mail order catalogues which allow someone who is not close to a shopping centre or prefers not to venture out to shop from home. The household names of home shopping in the UK – like Littlewoods, Kays and Freemans – have been joined by a wealth of 'niche' targeted shopping catalogues such as Next, Racing Green and Lands End. Interestingly here there is often cross-fertilization between the catalogue and the High Street/mall with Racing Green, for example, opening stores following the success of its mail order operations. Second, the advent of 'lifestyle magazines' such as those produced by Sainsbury, Marks & Spencer or Harvey Nichols have become an essential means of marketing new season product lines at the very heart of the consumption landscape – the living rooms and kitchens of the middle classes. Finally, the Internet is being investigated as a new means of taking consumption to the consumer. IBM, for example, has begun a pilot shopping service for PC Gifts and Flowers which will initially sell flowers, balloons, teddy bears and gourmet food. If successful the company proposes to add video games, watches, cameras and coffee to its product range.

The Leisured Consumer

The Walt Disney and Warner Brothers empires were the first to take a successful leisure product – film – and roll it out into a complete consumption experience (via theme park and store). The Disney Store and the Warner Brothers Studio Store can be viewed as the ultimate shopping and entertainment mix and have become increasingly popular in UK shopping centres. A less well known (but significantly growing) example of this leisure/retail blurring are the more traditional leisure pursuits of the British – football clubs. As football has become increasingly big business – and often itself moved 'out of town' – the old club shops have become superstores 'open six days a week and crammed with everything from club duvets, to gnomes, aftershave, jeans, teddy bears and mountain bikes' (Longmore, 1994). But the commercialization of this particular leisure experience does not end here. Manchester United, for example, 'plans to open stores in Tokyo and Sydney to add to its branches in Plymouth, Dublin, Belfast and the centre of Manchester'. These stores 'boast a loyalty that other retail brands would die for' (Longmore, 1994).

John Goss

'The "Magic of the Mall": An Analysis of Form, Function, and Meaning in the Contemporary Retail Built Environment'

***Annals of the Association of American Geographers* (1993)**

Shopping is the second most important leisure activity in North America, and although watching television is indisputably the first, much of its programming actually promotes shopping, both through advertising and the depiction of model consumer lifestyles. The existential significance of shopping is proclaimed in popular slogans such as: 'Born to Shop', 'Shop 'Til You Drop', and 'I Shop Therefore I am'. An advertisement for Tyson's Corner, Virginia, asks: 'The joy of cooking? The joy of sex? What's left?' and the answer provided is, of course, 'The joy of shopping'! As Tyson's obviously knows, recent market research shows that many Americans prefer shopping to sex (Levine, 1990, p. 187).

Nevertheless, there persists a high-cultural disdain for conspicuous mass-consumption resulting from the legacy of a puritanical fear of the moral corruption inherent in commercialism and materialism, and sustained by a modern intellectual contempt for consumer society. This latter critique condemns the system of correspondences between material possessions and social worth (Veblen, 1953), the homogenization of culture and alienation of the individual (Adorno and Horkheimer, 1969; Marcuse, 1964) and the distortion of human needs through the manipulation of desire (Haug, 1986). The contemporary shopper, while taking pleasure in consumption, cannot but be aware of this authoritative censure, and is therefore, like the tourist (Frow, 1991, p. 127), driven by a simultaneous desire and self-contempt, constantly alternating between assertion and denial of identity. This ambivalence is, I think, precisely expressed in the play of the slogans cited above, which cock a snook at the dominant order of values, but in so doing also acknowledge its inevitable authority.

This paper argues that developers have sought to assuage this collective guilt over conspicuous consumption by designing into the retail built environment the means for a fantasized dissociation from the act of shopping. That is, in recognition of the culturally perceived emptiness of the activity for which they provide the main social space, designers manufacture the illusion that something else other than mere shopping is going on, while also mediating the materialist relations of mass consumption and disguising the identity and rootedness of the shopping center in the contemporary capitalist social order. The product is effectively a pseudoplace which works through spatial strategies of dissemblance and duplicity.

THE MAKING OF THE MALL

The developer's profit accrues from the construction and sale of shopping centers, lease rent, and deductions from retail revenues. Unlike other forms of real estate, where markets have been rapidly saturated and are dependent upon urban and regional economic fortunes, shopping center construction has been a relatively secure investment, whether in the suburbs, always provided a big name department store could be enticed to sign an agreement (Frieden and Sagalyn, 1989, p. 79), or downtown, provided subsidies could be negotiated from cooperative municipal governments. Recently however, there has been a marked slowdown in the speculative development observed in the 1970s and early 1980s. This trend is attributed to a variety of factors: the combination of a shortage of suitable greenfield sites; escalating costs of land assembly, construction, and operation; tightened developmental controls; declining federal government programs that provide infrastructure and capital incentives; organized resistance from local communities; the financial vulnerability of highly leveraged retail chains; changing market demographics; and the segmentation of the retail industry (Goss, 1992, p. 168). As a result, many regions are effectively saturated and intercenter competition is intense. An extreme example is Dallas, where three megacenters (Galleria, Prestonwood and Valley View) are within two miles of each other. Profit increasingly depends, therefore, upon image making and the creative management of shopping centers.

The costs of initial development, maintenance, and overhead are typically covered by fixed charges, including lease of floor space, common-area maintenance charges and promotional expenses, levied upon retailers. Profit derives from average, or a proportion (typically 6 percent) of store turnover above an agreed base for each retailer, requiring open-book accounting agreements, and leading to management pressure for high value and volume trade. The measure of success of the center is 'operating balance per square foot of Gross Leasable Area [CIA]' (*Retail Uses*, 1991) and in the professional literature the figure for 'sales per square foot of CIA' is ascribed a special mystique.

The shopping centers profit from an internalization of externalities; that is, by ensuring strict complementarity of retail and service functions through an appropriate tenant mix (Goss, 1992, p. 167). Leasing agents plan the mix of tenants and their locations within the center, inevitably excluding repair shops, laundromats, or thrift stores that might remind the consumer of the materiality of the commodity and attract those whose presence might challenge the normality of consumption. Where resale shops are found, they conventionally indicate difficulty in attracting more desirable tenants (Ricks, 1991, p. 56). Similarly, vacant stores are hidden behind gaily painted hoardings, and we are assured that a store will be 'opening soon', in case we might suspect that this, like downtown, is not the thriving place where everyone wants to be. Detailed lease agreements create the appropriate atmosphere by insuring uniform store opening hours; regulating signage, sightlines, lighting, store front design, and window display; and stipulating advertising minima for each store (see Frieden and Sagalyn, 1989, p. 66).

While individual retailers may pursue their own strategies for profit within limited bounds, the center operates as a whole to maximize 'foot traffic' by attracting the target consumers and keeping them on the premises for as long as possible. The logic is apparently simple:

> Our surveys show [that] the amount of spending is related *directly* to the amount of time spent at centers. . . . *Anything* that can prolong shoppers' visits are [*sic*] in our best interests overall. (A senior vice-president of leasing and marketing cited in Reynolds, 1990, p. 52, emphasis added)

The task begins with the manufacture and marketing of an appropriate sense of place (Richards, 1990, p. 24), an attractive place image that will entice people from their suburban homes and downtown offices, keep them contentedly on the premises, and encourage them to return. This occurs in an increasingly competitive retail market resulting from the 'overmalling of America' and in response to consumer loyalties shifting from name-retailers to specific shopping centers, the personality of the center is critical.

IMAG(IN)ING THE MALL

In constructing an attractive place image for the shopping center, developers have, with remarkable persistence, exploited a modernist nostalgia for authentic community, perceived to exist only in past and distant places, and have promoted the concept of the shopping center as an alternative focus for modern community life. Shopping districts of the early years of this century, for example, were based on traditional market towns and villages, and a strong sense of place was evoked using stylized historical architecture and landscaping (typically evoking the village green). They were built on a modest scale, functionally and spatially integrated into local communities, in order to provide an idyllic context for consumption by the new gentry (Rowe, 1991, p. 141). The picturesque Country Club Plaza in Kansas City, Missouri, built in 1922, is a prototypic example. With the contemporary postmodernist penchant for the vernacular, this original form is undergoing a renaissance in the specialty center, a collection of high-end outlets that pursue a particular retail and architectural theme. Typically these are also idealizations of villages and small towns, chock-full of historical and regional details to convince the consumer of their authenticity (Goss, 1992, p. 172).

In contrast, the modern regional shopping center was built on a large scale with regular, unified architecture. Its harsh exterior modernism and automobile-focused landscaping refused any compromise with the rustic aesthetic. As Relph (1987, p. 215) notes, however, 'modernism . . . never wholly succeeded in the landscape of retailing', and the interior contained pedestrian walkways, courts, fountains and statuary that referred reassuringly to the traditional urbanism of southern Europe (Gruen, 1973; Rowe, 1991, p. 126), Victorian Britain or New England. According to Victor Gruen, the acknowledged pioneer of the modern mall, his 'shopping towns' would be not only pleasant places to shop, but also centers of cultural enrichment, education, and relaxation, a suburban alternative to the decaying downtown (Gruen and Smith, 1960).

Gruen's shopping centers proved phenomenally successful, and he later argued that by applying the lessons of environmental design learned in the suburbs to downtown, 'we can restore the lost sense of commitment and belonging; we can counteract the phenomenon of alienation, isolation and loneliness and achieve a sense of identity' (Gruen, 1973, p. 11). James Rouse, effectively heir to Gruen and heralded as 'the savior of downtown America' (Sawicki, 1989, p. 347), similarly argued that shopping centers 'will help dignify and uplift the families who use them, . . . promote friendly contact among the people of the community, . . . [and] expose the community to art, music, crafts and culture' (Rouse, 1962, p. 105). Thus, if the developers could create the illusion of urban community in the suburbs, they could also create this illusion in the city itself. The key, Rouse argues, is not so much the design features of the shopping mall, but centralized retail management (CRM) and leasing strategies (cited in Stokvis and Cloar, 1991, p. 7), which would include levels of security and maintenance well beyond that provided by municipal authorities, market research, cooperative advertising, common business hours, common covenants, and a regulated tenant mix (Cloar, 1990). Downtown is now 'learning from the mall': as the director of the National Mainstreet Center, an organization established by the National Trust for Historic Preservation, argues, 'shopping centers . . . are well-planned, well-funded, and well-organized. Main streets need management like that' (Huffman, 1989, p. 95).

The new downtown retail built environment has taken two essential forms, which in practice may be mixed. First is the commercial gentrification of decaying historical business and waterfront districts, pioneered by James Rouse with Quincy Market in Boston. Its opening in 1979 supposedly marked 'the day the urban renaissance began' (Rouse, cited in Teaford, 1990, p. 253) and subsequently no self-respecting city seems complete without its own festival marketplace, replicating more or less the original formula. Historical landmarks and 'water exposure' (Scott, 1989, p. 185) are critical features, as this retail environment is consciously reminiscent of the commercial world city, with its quaysides and urban produce markets replete with open stalls, colorful awnings, costermonger barrows, and nautical paraphernalia liberally scattered around.

A second form is the galleria, the historic referent of which is the Victorian shopping arcade and especially the famous Calleria Vittorio Emanuele in Milan. After Cesar Pelli pioneered the galleried arcade in the early 1970s (at The Commons in Columbus, Ohio and the Winter Gardens in Niagara Falls, New York), glazed gallery and atria became standard feature in downtown mixed-use developments, their huge vaulted spaces suggesting a sacred liturgical or secular-civic function. They have since been retrofitted to suburban malls and natural daylight has enabled support of softscapes – interiorized palms, trees, and shrubs – reminiscent of the street in the model garden city, the courts of Babylon, and most especially, the tropical vacation setting. Enclosed streetscapes refer to the idealized, historic middle-American Main Street or to exotic streets of faraway cities, including Parisian boulevards, Mexican paseos, and Arabic souks or casbahs, if only because the contemporary North American street invokes fear and loathing in the middle classes. They reclaim, for the middle-class imagination, 'The Street' – an idealized social space free, by virtue of private property, planning, and strict control, from the inconvenience of the weather and the danger and pollution of the automobile, but most important from the terror of crime associated with today's urban environment.

The malling of downtown could not work, however, without the legislative and financial support of the local state. These developments exploit historic preservation laws and federal and municipal funds to subsidize commercial development. Newport Center in Jersey City, for example, is the recipient of the largest-ever Urban Development Action Grant (Osborne, 1988). Frieden and Sagalyn (1989) provide a particularly incisive analysis of the coalitions of private capital and municipal government necessary to the successful development of the new urban retail built environment.

In creating these spaces, developers and public officials articulate an ideology of nostalgia, a reactionary modernism that expresses the 'disease' of the present (see Stewart, 1984, p. 23), a lament on the perceived loss of the moral conviction, authenticity, spontaneity, and community of the past; a profound disillusionment with contemporary society and fear of the future. More specifically, we collectively miss a public space organized on a pedestrian scale, that is, a setting for free personal expression and association, for collective cultural expression and transgression, and for unencumbered human interaction and material transaction. Such spaces no longer exist in the city, where open spaces are windswept tunnels between towering buildings, abandoned in fear to marginal populations; nor were they found after all in the suburb, which is subdivided and segregated, dominated by the automobile, and repressively predictable and safe. Such spaces only exist intact in our *musees imaginaire*, but their forms can now be expertly reproduced for us in the retail built environment. Below, I discuss the form and the contradictions inherent in the reproduction of such spaces as conceived in their idealized civic, liminal and transactional forms.

The Shopping Center as Civic Space

By virtue of their scale, design, and function, shopping centers appear to be public spaces, more or less open to anyone and relatively sanitary and safe. This appearance is important to their success for they aim to offer to middle-Americans a third place beyond home and work/school, a venue where people, old and young, can congregate, commune, and 'see and be seen' (Oldenburg, 1989, p. 17). Several strategies enhance the appearance of vital public space, and foremost is the metaphor of the urban street sustained by street signs, street lamps, benches, shrubbery, and statuary – all well-kept and protected from vandalism. Also like the ideal, benign civic government, shopping centers are extremely sensitive to the needs of the shopper, providing a range of 'inconspicuous artifacts of consideration' (Tuan, 1988, p. 316), such as rest areas and special facilities for the handicapped, elderly, and shoppers with young children (recently including diaper changing stations). For a fee they may provide other conveniences such as gift wrapping and shipping, coat checking, valet parking, strollers, electric shopping carts, lockers, customer service centers, and videotext information kiosks. They may house post offices, satellite municipal halls, automated government services, and public libraries; space is sometimes provided for public meetings or religious services. They stage events not only to directly promote consumption (fashion and car shows), but also for public edification (educational exhibits and musical recitals). Many open their doors early to provide a safe, sheltered space for morning constitutionals – mall-walking – and some have public exercise stations with health and fitness programs sponsored by the American Heart Association and YMCAs (Jacobs, 1988, p. 12). Some even offer adult literacy classes and university courses. Examples of the former include Middlesboro Mall in Middlesboro, Kentucky and Sunland Park Mall in El Paso, Texas; an example of the latter is Governors State University, University Park, Illinois, which offers 28 classes at Oriand Park Place Mall in Chicago.

Such services obviously address the needs of the public and attest to the responsiveness of management. Many facilities, however, are not so much civic gestures as political maneuvers to persuade local government to permit construction on the desired scale. This is particularly the case with day care facilities now featured in many shopping centers. It is also clear from the professional literature that many concessions are made in order to enhance the atmosphere of public concern precisely because it significantly increases retail traffic (McCloud, 1991, p. 25). Public services not consistent with the context of consumption are omitted or only reluctantly provided, often inadequate to actual needs and relegated to the periphery. This includes [*sic*], for example: drinking fountains, which would reduce soft drink sales; restrooms, which are costly to maintain and which attract activities such as drug dealing and sex that are offensive to the legitimate patrons of the mall (Hazel, 1992, p. 28); and public telephones, which may be monopolized by teenagers or drug dealers. As a result, telephones in some malls only allow outgoing calls (Hazel, 1992, p. 29).

The idealized public street is a relatively democratic space with all citizens enjoying access, with participatory entertainment and opportunities for social mixing, and the shopping center represents a similarly liberal vision of consumption, in which credit-card citizenship allows all to buy an identity and vicariously experience preferred lifestyles, without principles of exclusion based on accumulated wealth or cultural capital (Zukin, 1990, p. 41). It is, however, a strongly bounded or purified social space (Sibley, 1988, p. 409) that excludes a significant minority of the population and so protects patrons from the moral confusion that a confrontation with social difference might provoke. Suburban malls, in particular, are essentially spaces for *white* middle classes. There have been several court cases claiming that shopping centers actively discriminate against potential minority tenants, employees, and mall users. Copley Place in Boston, for example, has been charged with excluding minority tenants; a Columbia, South Carolina mall was accused of discriminatory hiring practices; and security personnel have been widely suspected of harassing minority teenagers. Security personnel target those who, despite implicit signs and posted notices that this is not the place for them, seek to hang-out, to take shelter or to solicit alms. Rowdy teenagers may spill out of the amusement arcades designed purposefully to keep them on the periphery, or use the parking lot for cruising, disrupting the comfortable shopping process of adults and particularly the elderly. Consequently some managers have even tried to regulate hours during which teenagers can shop without adult supervision, and passed ordinances and erected barricades in parking lots to prevent unnecessary and repetitive driving. 'Street people' are harassed because their appearance, panhandling, and inappropriate use of bathrooms (Pawlak *et al*, 1985) offend the sensibility of shoppers, their presence subverting the normality of conspicuous consumption and perverting the pleasure of consumption by challenging our righteous possession of commodities. Even the Salvation Army may be excluded from making its traditional Christmas collections, perhaps because they remind the consumer of the existence of less-privileged populations and so diminish the joy of buying.

Developers must of course protect their property and guard themselves against liability (Hazel, 1992, p. 29), but the key to successful security apparently lies more in an overt security presence that reassures preferred customers that the unseemly and seamy side of the real public world will be excluded from the mall. It is argued that the image of security is more important than its substance:

> Perception is perhaps even more important than reality. In a business that is as dependent as film or theater on appearances, the illusion of safety 'is as vital, or even more so, than its reality'. (Hazel, 1992, p. 28)

In extreme cases, however, overt and pervasive security may itself be part of the attraction, and this applies particularly to the defensible commercial zones which reclaim part of the decaying inner city for the display of cultural capital and lifestyles of the middle classes. For example, the trademark of Alexander Haagen Development Co., a pioneering inner city developer much celebrated in the professional literature, is an 8ft ornamental security fence with remote controlled gates. Haagen's centers in Los Angeles (Kenneth Hahn Plaza, Baldwin Hills Crenshaw Plaza, Martin Luther King, Jr. Center, and Vermont-Slauson Shopping Center), include perimeters patrolled by infra-red motion detectors, manual observation decks, armed security personnel, and closed-circuit TV monitoring, all coordinated through 24-hour command posts with state-of-the-art 'alarm processing technology.' At Crenshaw Plaza, for example, 'a study in state-of-the-art security and stylish consumerism,' the 'Omni 1000 Security Management System provides around-the-clock surveillance [to] deter crime and attract customers' (Bond, 1989, p. 181). Such pan-optical presence has been enhanced in some cases by donating mall space for local police.

The Shopping Center as Liminal Space

The market, standing between the sacred and secular, the mundane and exotic, and the local and global, has always been a place of liminality; that is, according to Turner (1982), a state between social stations, a transitional moment in which established rules and norms are temporarily suspended (see also Zukin, 1991, and Shields, 1989). The marketplace is a liminoid zone, a place where potentiality and transgression is engendered by the exciting diversity of humanity, the mystique of exotic objects, the intoxicating energy of the crowd channeled within the confined public space, the prospects of fortunes to be made and lost in trade, the possibility of unplanned meetings and spontaneous adventures, and the continuous assertion of collective rights and freedoms or *communitas* (Bakhtin, 1984, pp. 8–9). The market thrives on the possibility of 'letting yourself go', 'treating yourself', and of 'trying it on' without risk of moral censure, and free from institutional surveillance.

Places traditionally associated with liminoid experiences are liberally quoted in the contemporary retail built environment, including most notably seaports and exotic tropical tourist destinations, and Greek agora, Italian piazzas, and other traditional marketplaces. Colorful banners, balloons and flags, clowns and street theater, games and fun rides, are evocative of a permanent carnival or festival. Lavish expenditure on state-of-the-art entertainment and historic reconstruction, and the explosion of apparent liminality is perfectly consistent with the logic of the shopping center, for it is designed explicitly to attract shoppers and keep them on the premises for as long as possible:

> The entertainment at Franklin Mills keeps shoppers at the center for 3-4 hours, or twice as long as a regular mall and the more you give shoppers to do, the longer they stay and the more they buy. (Marketing executive, cited in Entertainment Anchors, 1989, p. 54)

This strategy reaches its contemporary apotheosis in the monster malls that contrive to combine with retailing the experiences of carnival, festival, and tourism in a single, total environment. This includes the famous West Edmonton Mall (WEM), Canada, which has already become a special concern of contemporary culture studies (Shields, 1989), and others inspired by its extravagant excess: Franklin Hills in Philadelphia, River Falls in Clarksville, Tennessee; the controversial new Mall of the Americas in Bloomington, Minnesota; Meadowhall in Sheffield and Metrocentre in Gateshead in England; and Lotte World in Seoul, South Korea. The shopping center has become hedonopolis (Sommer, 1975). Shopping centers have become tourist resorts in their own right, recreating the archetypical modern liminal zone by providing the multiple attractions, accommodations, guided tours, and souvenirs essential to the mass touristic experience, all under a single roof. WEM which receives 15 million visitors a year claims that:

> Tourists will no longer have to travel to Disneyland, Miami Beach, the Epcot Park New Orleans, California Sea World, the San Diego Zoo, the Grand Canyon. It's all here at the WEM. Everything you've wanted in a lifetime and more. (Winter City Showcase cited in Hopkins, 1990, p. 13)

There are necessarily strict limits to any experience of liminality in these environments. Developers are well aware of the 'more unsavoury trappings of carnival life' (McCloud, 1989, p. 35), and order must be preserved. As a management consultant to Forest Fair (River Falls) says:

> You have to be very cautious. Everything has to be kept very high quality and maintain family appeal. You have to create a safe, secure feeling and make sure it's not intimidating to anyone. (Cited in McCloud, 1989, p. 35)

Liminality is thus experienced in the nostalgic mode, without the inherent danger of the real thing: the fairground is recreated without the threat to the social order that the itinerant, marginal population and the libidinal temptations that traveling shows might bring, while the revitalized waterfronts lack the itinerant sailors, the red lights, the threatening presence of foreign travellers and shiphands. The contrived retail carnival denies the potentiality for disorder and collective social transgression of the liminal zone at the same time that it celebrates its form. It is ironic, therefore, that WEM is struggling to cope with the liminality it has unintentionally unleashed, including accidental deaths on fairground rides, terroristic activity, drug trading, and prostitution (Hopkins, 1990, p. 14).

The Shopping Center is Instrumental Space

Most shoppers know that the shopping center is a contrived and highly controlled space, and we all probably complain about design features such as the escalators that alternate in order to prevent the shopper moving quickly between floors without maximum exposure to shopfronts, or the difficulty finding restrooms. Some of us are also disquieted by the constant reminders of surveillance in the sweep of the cameras and the patrols of security personnel. Yet those of us for whom it is designed are willing to suspend the privileges of public urban space to its relatively benevolent authority, for our desire is such that we will readily accept nostalgia as a substitute for experience, absence for presence, and representation for authenticity. We overlook the fact that the shopping center is a contrived, dominated space that seeks only to resemble a spontaneous, social space. Perhaps also, we are simply ignorant of the extent to which there is a will to deceive us. The professional literature is revealing. *Urban Land*, for example, congratulates the Paseo Nuevo project in Santa Barbara for its deception: it '*appears* to be a longstanding part of downtown' (when it isn't) and is a '*seemingly* random arrangement of shops, tree-shaded courtyards, splashing fountains, and sunny terraces' when it is a carefully designed stage for '*choreographing* pedestrian movement' ('Fitting a Shopping Center', 1991, p. 28, emphasis added). In this professional literature, the consumer is characterized as an object to be mechanistically manipulated – to be drawn, pulled, pushed, and led to flow magnets, anchors, generators, and attractions; or as a naive dupe to be deceived, persuaded, induced, tempted, and seduced by ploys, ruses, tricks, strategies, and games of the design. Adopting a relatively vulgar psychogeography, designers seek to environmentally condition emotional and behavioral response from those whom they see as their malleable customers.

The ultimate conceit of the developers, however, lies in their attempt to recapture the essence of tradition through modern technology, to harness abstract space and exchange value in order to retrieve the essence of use value of social space (Lefebvre, 1971). The original intention may have been more noble, but the contradiction soon became apparent, and the dream of community and public place was subordinated to the logic of private profit. Victor Gruen himself returned to his home city of Vienna disillusioned and disgusted at the greed of developers (Gillette, 1985), while James Rouse formed a nonprofit organization engaged in urban renewal. The contemporary generation of developers may still express the modernist faith in the capacity of environmental design to realize social goals, but one somehow doubts that Nader Ghermezian, one of the developers of the monstrous WEM, is genuine when he claims their goal is 'to serve as a community, social, entertainment, and recreation center' (cited in Davis, 1991, p. 4).

THE SHOPPING CENTER AS A SPATIAL SYSTEM

The built environment forms a spatial system in which, through principles of separation and containment, spatial practices are routinized and sedimented (Giddens, 1985, p. 272) and social relations are reproduced. First, the locale provides the context in which particular roles are habitually played and actions predictably occur, establishing spatiotemporal fields of absence and presence, and affecting the potentialities for social interaction. The association of regions with particular group membership, activities, and dispositions allows the individual to orient to the context and infer the appropriate social role to play – one literally comes to know one's place. The built environment is, therefore, socially and psychologically persuasive (Eco, 1986, p. 77). Second, the configuration of spatial forms determines the relative permeability of structures, physically limiting the possibility of movement and interaction. The relative connectivity, transitivity, and commutativity of spaces serves to segregate individuals and practices, and to (re)inforce the differential capacities of agents for social action. Social relations are realized in homologous geometrical relations. For example, the dialectic of inside–outside (Bachelard, 1964, p. 215) realizes principles of inclusion–exclusion, while that of open and closed realizes distinctions between public and private realms. The built environment is then also physically persuasive or coercive.

What role does the retail built environment play in the structuration of social class? While a full discussion is beyond the scope of this paper, some preliminary suggestions can be made. Market researchers develop stereotypical profiles of customers and apply a concept of social class, conflated with lifestyle categories into market segments. The center is then designed to explicitly meet the presumed environmental needs and desires of the segments dominating market areas; thus the 'look' of centers reflects and reinforces conceptions of social class (see, for example, Levine, 1990, p. 187). The professional literature is quite plain about the conscious social differentiation of the retail built environment and the use of class-loaded cues to effect the sociospatial segregation of consumption activity.

Within the shopping center itself, social segregation is reproduced through separation of specific functions and of class-based retail districts. Fiske *et al* (1987, p. 110) describe an example of the vertical structuring of mall space according to the social status of the targeted consumers, and while the exact homology (high–low level and upper–lower class) is seldom realized so neatly elsewhere, interior spaces are carefully structured to produce appropriate microcontexts for consumption. Bridgewater Commons in Bridgewater, New Jersey, for example, has three distinct leasing districts designed to appeal to specific market segments and, by implication, not to appeal to others: The Commons Collection contains upscale boutiques and includes marble floors, gold leaf signage, brass accents, individual wooden seating, and extensive foliage; The Promenade contains stores catering to home and family needs, storefronts have a more conservative look, and aluminum and steel features and seating are predominant; The Campus contains stores catering to a 'contemporary clientele' with dynamic window displays, plastic laminate, ceramic tiling, bright colors, and neon signage (see Rathbun, 1990, pp. 19–21). Almost every shopping center marks the distinction between high-end and low-end retail by such environmental cues.

The shopping center is designed to persuade the targeted users to move through the retail space and to adopt certain physical and social dispositions conducive to shopping. Let us begin with the entrance to the regional mall. The approved mode of approach is obviously the automobile, and the shopper proceeds across the bleak desert of the parking lot towards the beckoning entrance, usually the only break in the harsh, uniform exterior and typically announced with canopies, columns, and glass atria, surrounded by lush vegetation, all suggestive of an oasis or sanctuary inside. Formal entrances are increasingly dramatic (Rathbun, 1990), providing an appropriate sense of grand arrival and literally 'entrancing' the shopper into the fantasy world inside. Here external reality is immediately displaced: the temperature is kept at a scientifically determined optimum for human comfort, typically a pleasant 68 degrees in winter and a refreshing 72 degrees in summer. Shophouse-style storefronts are often reduced to 5/8 scale (as in Disney's theme parks) to give shoppers an exaggerated sense of importance, transporting them into a looking glass world.

Indoor lighting is soft to prevent glare on shopfronts and to highlight the natural colors of the commodities on display. Lights act as, 'silent salesmen [which] showcase the most pricey merchandise to stellar advantage and transform the most pedestrian goods into must-haves' (Connor, 1989, p. 191) and may be engineered 'according to the mood or emotion they are seeking to elicit within the shopper' (Connor, 1989, p. 193). Similarly, psychologically researched music covers the silence and soothes shoppers in 'an anesthetic or tonic aural fluid' (Boorstin, 1961, 176), although the traditional Muzak has been replaced by customized foreground music which research shows may increase retail sales by up to 40 percent. Mirrors and reflective glass add to the decorative multiplication of images and colors, double the space and the shopping crowd (Fiske *et al*, 1987, p. 101), and reflect shoppers, asking them to compare themselves with the manikins and magical commodities on display in the fantasy world of the shop window. Even in glasshouse malls, there are no windows that look out on the world except up at the sky; there are no means but the seasonal promotional activities to determine the time of year, no clock to tell the time of day, and no means but the identity of retail chains to determine regional location. The modern shopping center is literally a Utopia, an idealized nowhere (ou = no; *topos* = place), and thus on a Saturday afternoon at about 2pm, the terror of time and space evaporates for the millions of Americans at the mall.

This utopia is kept scrupulously clean and orderly, without any material contamination nor hint of the gradual obsolescence that characterizes material objects. It is kept perfect and ageless by personnel who may be employed to do nothing else but constantly polish or touch up the spotless shiny surfaces. At the Esplanade Center in New Orleans, for example, the walls of the telephone recesses are washed at least twice a day, and completely repainted every two weeks (Scott, 1989, p. 69). The backstage areas, where commodities are delivered, prepared and serviced, are concealed by landscaping, painted panels, and underground construction to protect the customers from knowledge of the activities that take place there, so preserving the myth of the pure, abstract commodity for sale. Access to these areas is impossible for those who do not know the plan.

The floorplan exerts strong centripetal tendencies, and the shopper is drawn further into the fantasy by tantalizing glimpses of attractive central features, past the relatively drab marginal tenants (mostly services) into the colorful and well-lit wonderland of consumption. In WEM, for example, 'from each of the 58 entrances an unusual sight pulls visitors toward an *illusive* vortex' (Davis, 1991, p. 13, emphasis added). Escalators sweep them up to galleries decorated with mobiles – typically birds, flags and balloons that dramatically evoke flight and colorful action – or take them down to underworld grottos, under arches and hanging gardens. This experience disorients the shopper and, just as in the fantasy worlds of popular literature and film, it is then notoriously difficult to find one's way out. According to one designer, 'a too direct and obvious a route between the entrance and exits must be avoided' (Beddington, 1982, p. 16), and exits must be carefully designed because 'if too prominent and inviting as seen from within they may sweep the unsuspected [*sic*] shopper from the centre' (Beddington, 1982, p. 27). Even fire exits are disguised as shopfronts or hidden behind mirrors almost to the point of invisibility. The mall is thus designed as a non-commutative space, and the goal is to trap the consumer in the world of consumption.

As Walter Benjamin saw the Victorian arcade as the spatial metaphor for the cultural experience of commercial capitalism, so Jameson sees the hyperspace as a metaphor for global capitalism. Individuals cannot orient themselves within a 'cognitive map' of this complex and confusing space, just as they cannot locate their immediate experience within the unimaginable totality of class relations and cultural institutions structured on a global scale. This sign-saturated place and its constant motion represent the spatial and temporal displacement characteristic of the postmodern world. We have, therefore, progressed from the shopping center as a modern rational Utopia to a postmodern Heterotopia – a disorder in which the fragments of a large number of possible orders glitter separately without law or geometry' (Foucault, 1970, p. xvii).

Nevertheless, the megastructure, like the conventional shopping center, expresses the will of the plan, effecting circulation of patrons in order to optimally expose them to commodities on display and offer them the opportunity to make impulse purchases. The shopping center is a machine for shopping: it employs crude, but very effective, behaviorist principles to move patrons efficiently through the retail built environment. The developer's first law of shopper behavior says that the American shopper will not willingly walk more than 600 feet (Garreau, 1991, pp. 117–18, 464). Mall length is conventionally limited to this distance lest shoppers be disinclined to walk to the next department store (or be tempted to get into their cars to drive to it!) There are a number of generic designs depending on the number of anchors: a wheel spoke layout draws customers to a single anchor from surrounding car parks; the classic dumbbell design (developed by Gruen) channels consumers along a corridor between two anchors; and a T or L shape is used for three and a cruciform for four anchors. If mall distances are longer, this fact must be concealed from the consumer, typically by breaking the space with strong focal points and attractions, or by obscuring the view with pop-out shopfronts.

Progress through the mall is encouraged by careful pacing of attractions and displays, and even the width of storefronts is regulated by covenant to create a sense of predictable rhythm. Mall widths are conventionally restricted to about 6 m in order to allow shoppers to take in shopfronts on both sides, and to maintain the sense of intimate, human scale. Wider malls allow for placement of seating, softscape, and kiosks in the center, obstacles that might draw shoppers along while also deflecting them towards intervening stores (Gottdiener, 1986). Pop-out displays and open storefronts are designed to coax shoppers into the interior to make the impulse purchase. An excellent example of this spatial manipulation is provided by San Diego's Horton Plaza, where, from the vantage of an overhead walkway or its cloisters, one can watch shoppers enter the center at the top of escalators, hesitate, and make forward to a goal, only having to immediately negotiate vendors' carts and sculpted plants which deflect them toward storefronts. The strategy does not always work, of course, but some shoppers dally in front of the window display and a few enter the store, perhaps to make purchases where their original path would not have taken them. More subtle is the use of floor patterns to suggest pathways through the mall and towards open storefronts, a strategy employed, for example, in Pearlridge Shopping Center in Pearl City, Hawaii.

In multi-storey shopping centers, the design must also encourage vertical movement so that pedestrian traffic is exposed to shop displays on all floors. Maitland (1990, pp. 49–50), in a design manual, suggests 'devices' to 'persuade' and 'invite' people to move upward; these include 'glass-bubble' elevators, stacked escalator banks (as in the Trump Tower), overhanging platforms and aerial walkways (as at Pier 39 and Horton Plaza respectively), towering waterfalls and fountains, and mobiles of birds, manikins, balloons and aircraft. Such design features celebrate the drama and aesthetics of motion, drawing the eye and the person to upper levels.

Shoppers cannot be kept moving all the time, of course; they must be allowed to rest from the arduous tasks of shopping, particularly as the average trip to the shopping center has reportedly increased from only 20 minutes in 1960 to nearly three hours today (Crawford, 1992, p. 14). However,

> Pause points for shoppers to rest, review their programmes and re-arrange their purchases etc. also need planning with care. Seating, while offering a convenient stopping point, must not be too luxurious or comfortable. Shoppers must move on and allow re-occupation of seating and the danger of attracting the 'down and outs' of various categories must be avoided. (Beddington, 1982, p. 36)

The need to rest for longer periods is recognized mainly in the food court, where, of course, shoppers will be consuming at the same time. Food courts have become an absolute necessity, in part because of the increased role of food as a marker of social taste, in part also because the presentation of diverse culinary experiences enhances the sense of elsewhere (food courts now typically present a range of 'ethnic' cuisines), and because it provides a vantage point for watching others display their commodified lifestyles (Goss, 1992, p. 174). Although development costs are greater than for other outlets, food courts are significant determinants of the shoppers' choice of shopping center, and are the main attraction for downtown office workers during lunch hour. Located in the interior or on upper floors, they can also, like department stores, draw customers past the speciality stores. Research finds that food courts can prolong a visitor's stay an average of 10–15 minutes (Reynolds, 1990, p. 51).

The space created by the developer – pedestrian malls and mock street cafes – and the activity it is designed to sustain – relaxed strolling, window-shopping, and people-watching – seem reminiscent of *flanerie*, the progress of the voyeuristic dandy who strolled the streets and arcades of Paris in the nineteenth century. Several authors have drawn on the work of Benjamin (1973) in making this observation, and, with appropriate gender neutralization of the term, have been predisposed to see in it a recovery of a lost form of public behavior and personal expression (see for example, Shields, 1989; Hopkins, 1990). But while the 'mallies' (Jacobs, 1988) seek pleasure in the display, the commercialization of the context has radically altered meaning, and what we witness, I suspect, is not the recovery of *flanerie*, but a nostalgia for its form which only marks its effective absence. Shopping centers would not function if shoppers were not asked to validate their presence by purchases, in questions posed both in environmental cues and, if necessary, by the security personnel. The contemporary flaneur cannot escape the imperative to consume: she or he cannot loiter in the mall unless implicitly invited to do so, and this generally only applies to the respectable elderly; those without shopping bags and other suspicious individuals (teenagers, single men, the unkempt, and social science researchers) will draw the attention of security, who use the charge of loitering as grounds for eviction. Moreover, shoppers do not independently pick their way like the leisurely flaneur, but follow the meticulously conceived plan which has plotted paths, set lures, and planted decoys for its purpose. There is little chance of taking a route or occupying a position unforeseen by this plan.

The shopping center is, therefore, a strategic space, owned and controlled by an institutional power, which, by its nature, depends upon the definition, appropriation and control of territory. Its designers seek to deny the possibility of tactics, an oppositional occupation by everyday practices, that is, activities which do not require a specific localization or spatiality but which may temporarily use, occupy or take possession of strategic space (De Certeau, 1984). There are no spaces that might be claimed by uninvited gestures or unprescribed 'pedestrian utterances' (De Certeau, 1985, p. 129) since potential microspaces are preemptively filled: whether dignified by static features (such as sculptures or potted plants), 'animated' by active, permanent features (such as mobiles, mechanical displays, or fountains), or 'programmed' with a performance by musicians, mimes, or street artist (see Garreau, 1991, pp. 443, 456). Performers are carefully screened and hired by the management, of course, and 'real' street performers should only be found outside. In case anyone should be inspired to spontaneous performance, the stages and gazebos provided for programs are inevitably roped off and signposted to discourage them. When activated, these installations nevertheless provide a sense of public space and help draw shoppers through the mall. Graphics and murals are also used to enliven routes, dramatize motion, and avoid 'the depressing effects of dead areas' (Beddington, 1982, p. 82), such as the hoardings obscuring vacant stores. Spaces and surfaces should be filled because, if everywhere in this environment there is a sign, the absence of a sign becomes a sign of absence: perhaps signifying a lack of anticipation and consideration on behalf of the developer, or more seriously, the perceived emptiness of consumption itself, but inevitably inviting a motion to fill the void.

On the other hand, designers may provide spaces precisely to contain any such gestures that individuals may be disposed to make. Small, intricate and irregular openings – what Relph (1987, p. 253) calls quaintspace – invite a personal claim, and the planned concession of such spaces for sanctioned private or interpersonal activities then facilitates surveillance. This is not to deny the possibility of tactics entirely, for 'mall rats' will claim public space, by sitting on the floor or 'mallingering' (Kowinski, 1985, p. 26); threaten quaintspace, by disturbing adults or engaging in unsanctioned activities; or programming their own spaces by performing 'The Robot' etc., until chased off by security guards.

Section 4.4 Pick 'N' Mix: The Commodification of Identity

John Clammer

'Aesthetics of the Self: Shopping and Social Being in Contemporary Urban Japan'

from ***Lifestyle Shopping*** **(1992)**

INTRODUCTION: SHOPPING, SELF AND SOCIETY

Shopping is not merely the acquisition of things: it is the buying of identity. This is true of all cultures where shopping takes place, and the consumption even of 'necessities' in situations where there is some choice, reflects decisions about self, taste, images of the body and social distinctions.

Shopping, although of course men also do it, is largely regarded as a female preserve, both shopping for daily necessities and for major consumer objects such as cars. There are several reasons for this: large numbers of Japanese women are housewives, expected after marriage, or certainly after the first baby, to devote themselves exclusively to the domestic well-being of their children and spouses, and possibly also of parents or elderly in-laws. Not only are women thus 'functionally' associated with shopping but they are also thought, because of the small sizes of Japanese homes and the generally absent characteristics of their husbands, to have both the leisure and the interest. Given also the fact that many Japanese housewives largely control the family budgets, the 'femininity' of shopping comes to be established. And women do often shop for their husbands as well as for their children and themselves. But there are deeper reasons than these operative. Married women are seen as being primarily responsible for the education of their children and for the physical presentation of their families (in clean, up-to-date and neatly ironed clothes). Accusations of scruffiness – *darashi ga nai* – against one's children or husband involve a serious loss of face for the mother/wife. Japanese houses are frequently crowded and untidy within, but the family will leave the home invariably impeccably dressed (even when in casual clothes). With husbands usually absent for long working days, and children at school (including Saturdays), networks of friendships come into being amongst women of similar age in a neighbourhood (in addition to kin networks), and shopping together is one of the activities that cements and promotes these friendships. Interestingly it is out of these female networks that the albeit embryonic Japanese consumer movement has arisen. Concern about price, quality, safety and durability has quite naturally become a major concern of women, who are the primary purchasers of the things that their families consume.

The presentation of self in a very self-conscious culture – which Japan is, and meaning here both concern with the image of the country itself as it is perceived by outsiders, and concern amounting often to anxiety with the 'correct' appearance of one's individual self – requires the acquisition of the emblems appropriate to both self-image and objective status (as ranked that is by the rest of the society, since Japan, while relatively classless, is nevertheless very status-conscious). While education, career, travel and cultural accomplishments are also important aspects of this, so is the array of things with which one adorns oneself, family and home. What one *does* and what one *is* are to a great extent the same, and it is thought very desirable to present a rounded or 'total' and consistent image of oneself. A very visible aspect of this is the 'uniforms' that almost everyone in Japan wears (all the more obvious in a society in which 'real' uniforms are rarely seen, except on policemen). Students wear tartan shirts, jeans and clumpy boot-like footwear, and, if the weather is cold, bomber-jackets; businessmen and bureaucrats wear suits (blue is the favoured colour) with ties and white shirts. 'Intellectuals', which in Japan means writers, artists, poets, well-known journalists and classical musicians, and university teachers, wear either rather tweedy clothes, possibly with an open-necked shirt, or the same uniform as the businessmen, but with the vital addition of a beret, the sure sign of intellectual status. Youth-culture persons wear youth culture uniforms, 'office ladies' (clerical workers) wear skirts with white blouses or business women's suits. The key is appropriateness: being not so much tidy as dressed for one's role. In Japan, all the world is indeed a stage.

CHOICE AND BEING

This project has four key elements – eclecticism, wrapping, choosing and discarding. The modern Japanese house often contains the most amazing collection of artifacts – Japanese dolls, bits of western furniture, a piano standing on the *tatami* (rush matting), a kitchen containing the latest electrical conveniences and an ancient and primitive stove, pictures and souvenirs representing the spoils of foreign travels or of trips within Japan by family or friends. This wild eclecticism is not just, or even, bad taste: it represents a quality of the Japanese character – the ability not so much to synthesize as to juxtapose the inharmonious, and to live happily with the result because it works. Function is beauty. The excellence of Japanese industrial design is an example of this phenomenon working at another level: it is in the design of the practical and mundane that the merging of the function and aesthetics is most satisfactorily carried out. The first thing that must strike a shopper from a Second or Third World society, or even a neophyte shopper from a First World one, is the sheer profusion of things to be found, a cornucopia and an orgy of overchoice in which practically everything exists in multiple varieties. A visit to the extraordinary 'Electric City' district of Akihabara illustrates this very well. Dozens of stores, many of them multi-storied, contain literally millions of items of electrical equipment of all conceivable kinds. Actually choosing one, even a simple item like an electric fan or clock, becomes a feat of decision making. This range of choice gives a sense of power to the shopper – there is so much to choose or reject – but also a sense of confusion: the thing bought may well not in fact be the best or the latest. The role of the salesperson is important here, and most of them are men, in some cases on loan from the companies that actually make the stuff. They are supposed to *know*: they have a guru-like quality in this existential quandary, and their advice is taken very seriously and rarely rejected. They are not just sales-clerks, but priests: mediators between the innocent, ill-informed and choice-fatigued would-be consumer and the plethora of things themselves. Overchoice itself promotes eclecticism, especially in the absence of a central scale of values to structure that choice, and it is for this reason that Japan has been proposed as the pre-eminently postmodern society – as one having no central core of values (say in the Judeo-Christian sense) and as never historically having had one, except for an aestheticism joined to strong pressures of group conformity. According to this argument, Japan has not just become postmodern, it has always been that way ('transmodern'), and contemporary eclecticism is simply an expression of this. But the eclecticism is empirically apparent whatever its origins, and has different expressions: it is not just the random collection of unrelated objects, but differs with class, age, sex and self-image. Seen from one angle a 'random' collection may be just that; seen from another it may be an indication of the cosmopolitan character of the individual collector.

One may regard oneself as a *kokusaijin*, an international person – for instance, a person of varied cultural persona. Eclecticism would also seem to be a reflection of the Japanese characteristic of simultaneous self-confidence and inferiority complex. For every Japanese who is proud of his/her country's enormous economic success, there is another who is convinced of the inferiority of Japanese culture and character and the superiority of things foreign. Often the two attitudes are found in the same person. The use of things, especially objects that are semiotically ambiguous, to mediate this, is an understandable reaction. With careful planning one can be cosmopolitan and indigenous at the same time. Snoopy (of 'Peanuts' fame) is an example of this possibility, and (as a result?) is something of a cult figure in Japan. Little shops everywhere sell trinkets bearing his likeness, and young girls carry his image on bags, key-rings, umbrellas, and T-shirts. He is vaguely American and therefore modern, but he has also been assimilated in the same way that baseball has – indeed both are widely thought to be Japanese ideas borrowed by the North Americans. Above all, however, he is *kawaii* (cute), a concept used with incredible frequency in modern spoken Japanese (especially as used by women). Objects, then, are not neutral, but can be exploited in different and even contradictory ways to illustrate different facets of one's shifting or evolving identity.

The thing bought, however, is not just 'itself' and nor is it just the cluster of symbolic meanings attaching to it. It is indeed all of these too, but it is also transformed by one additional and quintessentially Japanese procedure: its wrapping. This may sound trivial, but in Japanese culture it is not. A serious literature exists in the art (literally the art, for it is so regarded) of wrapping things in paper, straw, cloth and in packing them in wooden boxes and other kinds of containers. Shop assistants, assuming they have not already learnt the skill from their mothers, are taught how to wrap and tie expertly the merchandise that they are selling. The cheapest of the ubiquitous *obento* (lunch-boxes) bought at a railway station will be elegantly packaged and its contents laid out inside it in an aesthetically pleasing fashion. And a small pair of disposable wooden chopsticks will be included with the purchase, also neatly packaged in a paper wrapping often decorated with elegant calligraphy. A Japanese is as likely to give as much attention to the wrapping – the material, the way it is folded, the ribbons used to secure it – as to the contents of the package. To give a badly or inelegantly wrapped gift, or one not wrapped at all, is both rude and a negative reflection on one's own taste and sensibilities. Stationery stores have large sections for wrapping paper, ribbons, labels, cards and the special kinds of decorative envelopes used for giving gifts of money. The humblest purchase is carefully wrapped and one is usually asked if it is a present. If it is, it will be wrapped and tied in an even more elaborate way, at no extra cost to the purchaser, the buying and the packaging being intimately linked and both part of the same philosophy of service.

This emphasis on packaging is an important facet of the fact that Japan is, on a very large scale, a gift economy of a kind that would be instantly recognized by any anthropologist. Gift giving and receiving are ingredients of everyday culture. Gifts are given not only for birthdays and weddings, but also for funerals, when visiting someone's home, on the occasion of promotions or other similar events, and especially at mid-year and at New Year (*chugen* and *seibu*) when half-yearly bonuses are paid and when a nationwide boom in gift giving occurs – to those who have done you favours, to those from whom you hope to receive favours, to bosses and to one's children's schoolteachers. All the year round department stores stock appropriate gifts, but on these occasions they are full of them, advertising appears extolling certain products and parcel-delivery companies are inundated with work. Many families keep detailed record books of gifts sent and received, so that mistakes of reciprocity will not be made. The traditional gift-giving seasons are not the only occasions either. The Japanese (almost entirely non-Christian though they are) have discovered Christmas in a big way and even Easter to a degree, Mothers' Day, Fathers' day, Hallowe'en and Valentine's Day (on which occasion women give gifts of chocolate to men – husbands, boyfriends, lovers, bosses). Life in Japan sometimes seems to be an endless round of gift giving, reflecting the networks of close social relationships that abound. On moving into a house or apartment one gives gifts to the neighbours on both sides and opposite; on returning from holiday one always brings gifts (*omiage*) for family, friends, secretary, office mates and neighbours, usually a regional speciality of wherever one has been, cakes, cookies and sweets being common choices. A great deal of shopping activity is not for oneself, but for gifts for others, and shops cater for this huge volume of gift giving, and every hotel in a resort area has a little shop selling souvenirs and local foods. Even railway stations reflect the gift economy, and in a popular tourist destination like Kyoto there are gift shops not only around the station but within it and even on the platform, so that gifts for a suddenly remembered acquaintance can be bought at the very point of boarding the *Shinkansen* (bullet train) for the ride home. The actual shopping for gifts requires considerable cultural skill – for whom? How expensive? What kind of thing? And yet much of what is received is never consumed – it piles up or is given away again. This perpetual and enormous circulation of commodities – a gigantic kula-ring-like cycle of obligations and reciprocities – represents a key dimension of shopping behaviour in Japan, and a fascinating extension of the economics of the gift.

The art of choosing, as we have suggested, is a sophisticated one, whether for oneself or others. In buying a gift, for example, price is important, not only for one's own pocket book, but because to give a too expensive gift is to impose a heavy burden of reciprocity on the recipient, to give too cheap a one is an insult. In choosing for oneself, self-image comes into play. Here there are also some important contextual factors at work. Two groups are amongst the biggest consumers – youngish unmarried women who are between college and marriage, currently working and who are living at home and have few overheads; and youngish married or unmarried professionals who, because they cannot possibly ever afford a house in Tokyo, divert what would in other societies be mortgage savings into consumption. Both groups consider themselves sophisticated consumers, for whom consumption is indeed a way of life, and they are aided in this by the huge range of media, not only that carrying general advertising, but specialist consumer magazines and guides as well. The Japanese verb for shopping (*kaimono*) is written with the Chinese character for 'thing', and a leading consumer magazine, on sale at most news-kiosks, carries exactly this splendidly practical title – *Things*. It is, as a representative of its genre, a very interesting magazine – glossy, of nearly two hundred pages per issue, and containing nothing but advertisements and short articles on new products, including TVs and VCRs, cameras, watches, luggage, clothes, chairs, fashion accessories, toys, cars, pens, sunglasses, cigarettes, personal computers, lawnmowers, an endless succession of trivia – jigsaws, kits for making dinosaurs, tissue-holders, tiny first-aid kits, snuff, Wild Western-style embossed riding saddles, exotic clocks, art-nouveau decorations – new-product test reports on several items and a lead article on bourbon, with an illustrated guide to several dozen brands. Product guides of this kind (and there are many others, both rival general ones and ones specializing in cars, sporting goods, bridal wear, etc.) are themselves widely consumed – either for their aid in actually selecting an item or being alerted to the enormous range available, or because they are interesting in themselves, as attractive and fascinating guides to the inexhaustible range of possibilities – even if you do not actually want or need any of them. Such magazines are also an expression of the fine Japanese art of advertising. Adverts are everywhere – not only in magazines and newspapers and on six of the eight TV channels (two being state-run, non-commercial channels), but on walls, telephone poles, billboards, trains and subways, on every available surface, in neon all over commercial buildings and thrust in endless quantities into one's mail-box. As an incentive to consume, it is undoubtedly a powerful force, and like wrapping and the self, it is considered an art form.

A thing once chosen, however, will not always be retained. It will, if bought as a gift, be given away. But it is just as likely to be rapidly discarded simply because it is no longer new. A certain day each month is 'heavy rubbish day' when unwanted large objects can be put out on the sidewalk for collection by the municipal rubbish collectors or by private contractors. The most astonishing variety and volume of things are discarded – furniture, TVs, bicycles, golf-clubs, all kinds of electrical appliances and just about everything else that a modern household might possess. Students and poorer people often furnish their dwellings with cast-offs of this kind, which is not a bad idea since the objects are often in almost mint condition. It is not uncommon for middle- and uppermiddle-class households to change their furniture, appliances, curtains, even cutlery, every few years. New is good (and there may be a deep-seated cultural attitude here originating in Shinto ideas of purity). This high turnover means that constant shopping is necessary, and constant awareness of what is on the market and what is in fashion, which in turn requires a never-ceasing vigilance on the part of the consumer. This mindset produces huge quantities of perfectly serviceable junk, and this logic applies even, or especially, to cars. Very few old ones are to be seen on city roads, and those that are sometimes called *gaijin-kuruma*: 'foreigners' cars' as only foreigners, who have little or no 'face', are willing to drive around in them. Even the climate is roped in as a justification here. Japan is markedly seasonal (Japanese sometimes refer to themselves as 'ninety-day people' – fickle and rather changeable, like the seasons which only last three months each) and certain kinds of clothes are thought appropriate for each season, not only in terms of warmth or coolness but also of colour and style. The non-appropriate clothes are stored and/or discarded, and stores exploit this seasonality by introducing even finer distinctions. In 1990, there were not only autumn clothes, but 'early autumn clothes'. The logic of consumption expands here in culturally interesting ways. Overnight, stores are transformed from late summer to early autumn, and then instantly it is impossible to buy late-summer clothes unless one knows the discount stores and boutiques where out-of-season fashion is retailed at knock-down prices. And this transformation is literally overnight. The Christmas season ends on Christmas Eve when the big stores simultaneously remove Christmas trees, decorations, seasonal music and display themes and replace them by the following morning with decorations and gifts for the much more culturally significant New Year, which again, commercially speaking, ends as the shops close on 31 December.

The situation that we are analysing here, while clearly a recognizable 'culture of consumption' by western capitalist standards, is also one with its own distinctive characteristics. These include shopping to give away (the gift relationship), the predictability of the shopping cycle (its seasonality, constantly emphasized by the media, and especially on TV every night, where the characteristics of each season are dwelt on at length, and advice is given on where to go to view the best cherry blossoms, plum blossoms or maple leaves, as is appropriate), and in a sense its orderliness. The craziest object bought will still be wrapped in the most traditional manner and certain colours are still thought appropriate for each age group, season and even occupation. And there are yet other peculiarities. One of these is travelling abroad for the purpose, essentially, of shopping. Given the very high cost of living and inflated prices of Japanese commodities, it can be as cheap to vacation in Korea, South-East Asia, Taiwan or Hong Kong as it is to holiday in Japan, with the added incentive of being able to buy both international 'brand name' goods and Japanese-made products at far below their Tokyo prices (there being big price differentials between the cost of the same Japanese-made product at home and abroad). The desire for brand-name goods, especially French, Italian and British products – watches, luggage, shoes, leather products, perfumes and fashions – is a well-known characteristic of Japanese shoppers, and one thoroughly exploited by shopkeepers in places like Singapore, Hawaii and Hong Kong. The motive is partly to buy quality goods, something that does, for once, last; and since one's foreign travels are likely to be less frequent after one's brief 'office lady' phase, it makes sense to buy while one can and has the ready cash. But functionalism is not all: brand names possess a mystique, a cachet that creates the impression of sophistication, of internationalism, and of taste. The almost metaphysical levels that this can reach in Japan seem to transcend those found elsewhere, as revealed by the extraordinary success of the best-selling 'novel', *Nantonaku, kirisutaru* (*Somehow, crystal*), by Tanaka Yasuo, virtually plotless and consisting in large part of lists of brand-name goods and quasi-scholarly notes discussing these commodities, and which sold over three-quarters of a million copies and in doing so became something of a brand-name commodity itself. The big stores in the major cities often have either special promotions of foreign brand-name goods or stock them permanently, Harrods of London and Laura Ashley being two currently popular examples in large stores in the Ginza area of Tokyo.

Nigel Thrift and Andrew Leyshon

'In the Wake of Money: The City of London and the Accumulation of Value'

from ***Global Finance & Urban Living*** **(1992)**

It is clear that in the 1980s the City became more visible to the public gaze as a result of the interaction between changes in British society and changes in the City. In the case of the economy, the visibility of the City was made more acute by a general worldwide boom in the financial services industry and by specific increases in financial services employment in many parts of Britain, drawing new workers into the financial services labour force. In the case of the state, the City's visibility grew with the increasing scale and influence of large financial services conglomerates which periodically were able to outflank the various state controls based upon the Bank of England and assorted regulatory mechanisms. In the case of British society, the more dynamic sections of the City became part of a new 'disestablishment', a highly visible coalition of private business interests that was 'meritocratic rather than egalitarian, efficient rather than generous, individualistic rather than corporate' (Lloyd, 1988, p. 155). Finally, the City became imprinted on the national culture as an exemplar of a new Britain which was more conscious of wealth and more careless of egalitarian concerns. The rapidly growing cultural industries were able to serve up the City to audiences as a set of stereotypes. For example, young men and women working in the City were interpreted as Thatcher's stormtroopers in large numbers of plays, television series and films, and books, all the way from *Serious Money*, through *Capital City*, to *Nice Work*.

This chapter will take the measure of some of the economic, social and cultural aspects (the three are inseparable) of the City's influence on Britain in the 1980s. The argument is a straightforward one that starts with the payment of large salaries and bonuses to many of those working in the City in the 1980s. In turn, these salaries and bonuses were converted into personal wealth. One of the ways in which this conversion was accomplished was via the purchase of scarce assets and which therefore had a 'positional' value. Such positional assets have appreciated particularly rapidly in the 1980s. But these assets are not just a store of economic value. They also have a social and cultural value (indeed, this is one of the ways in which their positionality is defined). The purchase of these assets, therefore, became one of the ways by which the City's newly wealthy defined themselves socially and culturally and, over time, this definition (or set of definitions) spread to other parts of the business community, becoming more general in the process, both socially and geographically.

MAKING MONEY: THE GROWTH OF THE CITY OF LONDON THROUGH THE 1980S

The City's labour market expanded rapidly throughout the 1980s. The rate of expansion is in some dispute, but Rajan and Fryatt (1988) calculate that in the particularly frantic 'Big Bang' to 'Big Crash' period from 1984 to 1987, the average annual growth rate of employment was as much as 7.5 per cent. In the period from 1987 to 1992, Rajan and Fryatt calculated that the growth rate would still be in the order of 3 per cent per annum with the most sustained growth, over both periods, being found in banking, accountancy and management consultancy, and software services.

Of course, the City's labour market is not an undifferentiated mass. Over time, its composition has been changing as a result of changes in employer demand. Most importantly, the 1980s saw a general upgrading of the City's skills base which has meant that the proportion of employment in managerial and professional occupations increased, whilst the share of employment held by clerical staff decreased relatively (Rajan and Fryatt, 1988). The shortages of skilled labour induced by the City's need to upgrade its skills base also meant that through the 1980s more women and those from ethnic minorities have been hired. However, these groups remained under-represented and tended to be concentrated in the poorer paying clerical jobs which still account for nearly one half of the City's jobs.

PAY IN THE CITY IN THE 1980S

According to New Earnings Survey data, earnings in the City underwent a dramatic increase in the course of the 1980s. Average earnings were not just above average earnings in Britain, but also above average earnings in the rest of London, and in the south-east of England.

It is important to note here that earnings in the City do not consist only of salaries. In many City jobs, additions to basic salary are of crucial importance in boosting income. Amongst these, the most important is the use of bonuses. Another important addition to salary will consist of various perquisites, including mortgage subsidy, a company car or allowance for use of own car, help with costs of rail travel, life insurance, free medical assurance, low interest loans, subsidised meals and subsidised private telephone.

In summary, it is clear that the number of well-off people in the City increased quite rapidly in the 1980s as a result of generally improved levels of income, combined with a gradual increase in the number of professional and managerial workers who were the chief beneficiaries of these increases in income, both relatively and absolutely.

But the story cannot stop with income. It is important to point out that the high incomes of the kinds which could be found in the City can be rapidly converted into a stock of assets, into personal wealth. The use of stock options as bonuses is just the most direct conversion of income into assets that is offered to those with substantial incomes. More usually, they make the conversion themselves by buying stocks, and shares, or opening deposit accounts, which provide them with another stream of income. Alternatively, they can buy appreciating assets like houses, or antiques and other collectables.

SPENDING MONEY

The effects of the spending power provided by the boosted incomes and enhanced personal wealth formation of the City on the economy and society of London, the south-east of England and Britain as a whole remain a matter of fierce debate.

What seems certain is that spending generated by the incomes and personal wealth connected to the City of London has had identifiable effects on particular high-value markets. Of these markets, the most visible are those which have not only economic but also social and cultural resonance, that is, in which demand for goods and services is socially and culturally defined. These are markets in which money is deployed to buy goods and services which, when combined with other goods and services in particular practices, produce social and cultural advantage, what Bourdieu (1986, 1987) identifies as the accruing of 'social' and 'cultural' capital. Social capital refers to each person's insertion into, and accumulation of, a network of 'relationships of acquaintance and mutual recognition', that is a network of family, friends and business contacts. Cultural or symbolic capital refers to each person's accumulation of a stock of socially accepted competences all the way from a family name through ownership of 'appropriate' goods to formal educational qualifications. Roughly, the distinction Bourdieu draws is between knowing the right people and knowing the right things, although clearly the two are heavily interrelated, not least through their geography. Being in the right place is a crucial element of each type of capital. For example, 'to be seen at the right places, the right ski resorts in winter and the right watering places in summer, is important both for the symbolic capital involved and for the social and quasi-professional contacts that can be made or confirmed by simultaneous presence' (Marceau, 1989, p. 146).

For those in the City who, in the 1980s, found themselves in possession of substantial incomes and, in some cases, considerable personal wealth, the challenge was to spend their money in tasteful ways, that is, in ways which would accrue social and cultural capital for themselves and their children. The new City wealthy were not alone in this objective. It was one shared by the rising 'disestablishment' which needed to stamp its authority on Britain, not just in the economic realm, but in the social and cultural realms too. In other words, Marx's abstract community of money had to be transformed into a concrete community of the moneyed. That required seeking out goods and services that could become a part of practices that would allow this to happen. Examples of markets for such goods and services were legion in Britain in the 1980s, extending all the way from the boom in simple status goods that demonstrate what Bourdieu calls 'honourability' (like expensive automobiles or antiques), through the expansion of arenas for aiding the formation of social capital (such as the interrelated rise of the Season and the corporate hospitality business, or the rise of certain field sports), to the rise of private day school education (which allows a child to accrue both social and cultural capital).

THE COUNTRY HOUSE MARKET

The country house market has a number of common features which define it above and beyond the formal architectural parameters. Three of these stand out. First, the country house is 'historic'. Second, the country house should be vernacular in the broad sense; that is, it is built in materials and style typical of its local area and therefore blends harmoniously into the landscape. Third, and related, country houses should, as the term implies, be in rural locations.

The pages of *Country Life* attest to the fact that a country house market existed before the 1980s but, except for some areas in and around London and the Home Counties, it was not a particularly active market. In the 1980s all this changed: the market exploded into life. First of all, the wealth became available:

> More than anything else, the Big Bang in the City of London has altered people's attitudes to country houses. For one thing, it has produced overnight a generation of young people who can afford to buy the houses that their elders were struggling to keep up for years. (*Country Life*, 1987, p. 10)

In 1986, most new owners still came from the City. They were relatively young, with most men in their 40s and most women in their 30s. They were well off with 74 per cent earning more than £50,000 per annum and 41 per cent earning more than £100,000. Over half were purchasing their houses with cash. Those who had loans were willing to pay more than £2000 a month for a mortgage. As money from the newly wealthy started to pour into the country house market so substantial capital gains became possible, adding further to the market's attractions. Country houses are, after all, an important part of the British positional economy, made up of goods which are inherently scarce.

Second, the 1980s saw a truly national market for country houses gradually come into existence. Third, the 1980s saw tastes changing. The country house has been a potent symbol in Britain for a considerable period of time, whether at its most blatant as the patrician grand home, 'the abstraction of success, power and money . . . founded elsewhere' (Williams, 1973, p. 299), or in its more domestic manifestations as the solid patrician period farmhouse, complete with squire, or as the Helen Allingham-like period cottage. Most particularly, country houses have become both a part of and a way of reflecting back a set of larger cultural discourses in a way that advantages the person who owns them. Their value as cultural capital has been inflated. These discourses are many but chief amongst them we can count the revival of historical feeling in the 1980s, helped along by the growth of conservation movements and the use of the past as a resource by the retailing and heritage industries (Wright, 1985; Hewison, 1987; Samuel, 1989; Thrift, 1989a), and the increased interest in nature and the countryside, aided by the new devotion to pastoral versions of Englishness and the growth of the environmental movements (Samuel, 1989; Howkins, 1986; Thrift, 1989a). The country house has, in its different forms, amplified and extended these discourses, buoyed up by a wave of publications which have all helped to focus and fuel the fires of desire for ownership, ranging from stalwarts like *Country Life* to the vast flood of country house books which shows no signs of ebbing.

The origins of the revival of the country house market and its subsequent pattern of expansion can be traced through the market's geography. Undoubtedly, in the early years of the revival, proximity to London was a critical factor. Country houses had to be within commuting distance of London: 'the stockbroker who now has to be in front of his SEAQ computer at 7 am does not want to live too far away from London' (*Country Life*, 1987, p. 10). But the country house buying habit soon began to spread outwards from the immediate environs of London, for at least five reasons. First, there was the sheer expense of country houses nearer the capital. This expense pushed buyers without the requisite capital out of areas nearer to London. The effect was cumulative over time – as the frontier of high prices moved further and further out so those coming into the market were less likely to have the requisite capital and had to move further out again to find a house. Second, improved communications helped to open up certain areas to interested London buyers, especially the completion of the M25, and faster train times on some inter-city rail lines. The most famous example of this effect was the town of Grantham in Lincolnshire. With the introduction of high-speed trains, journey times to London were reduced to 75 minutes, and electrification subsequently reduced them to an hour. The result was an influx of more than 200 London commuters and a country house price increase of 45 per cent in 1987.

Third, the stock of country houses near London is finite. Thus the search had to be extended outwards. Fourth, there was the increasing spread of agents' offices out from London and of information about houses in the rest of the country back into London. Finally, there was the large rise in London house prices in the 1980s. This resulted in a considerable pool of people, who already had large salaries and were in a position to trade up, using their capital gains.

But, even in the mid-1980s, the country house market was clearly not restricted only to London (and especially City) buyers. The increasing wealth of the wealthier parts of the population in other parts of the country meant that the country house market was to some extent becoming independent of London. Thus by 1987 a number of major cities were forcing up country house prices within their own spheres of influence.

> Birmingham now affects Staffordshire almost as much as Warwickshire, while Hereford and Worcester are not far behind . . . Shropshire lags and continues to show plenty of scope for price improvement to match the other counties barely further in miles from Birmingham. Manchester's influence on Cheshire has produced in excess of a two-fold increase but Derbyshire and Lancashire have been less responsive. Bristol's interest in Gloucestershire, Avon and Somerset now collides cheerfully with that of London to bring about pronounced price increases. (Lilwall and Allcock, 1988, p. x)

CONCLUSIONS

This paper has attempted to trace some of the effects of the proliferation of high incomes in the City of London and consequent changes in patterns of demand and consumption on the economy, society and culture of Britain. This has been done by reference to one commodity market only, the market for country houses. In the process it has demonstrated the importance of positional goods in asset appreciation. Positionality is the result of a complex interaction between not only economic but also social and cultural processes of the definition of scarcity. The country house market thrived in the 1980s because the commodities that it had on offer held out not only the possibility of economic gain, but also social and cultural gains. Indeed, it was precisely the ability of the country house to act simultaneously as a store of economic, social and cultural value which was its attraction. The scarcity of the commodity on offer was clearly an essential part of that ability. In particular, country houses were seized on by the new private sector 'disestablishment' as a way of accumulating assets, and as a way of providing themselves with social and cultural credibility.

[First] what is certain is that expenditure by those with higher incomes has hardly been restricted to the country house market. A whole constellation of goods and services is involved. Paradoxically, perhaps, this point can be well illustrated by turning back to the example of the country house. Expenditure on country houses hardly stops with the purchase of the house. It involves a whole series of other commodities, all of them with their effects on the structure of demand for particular goods and services.

Second, it is important not to lose sight of the fact that commodities are part of the process of social relation. They are not something separate. The country house provides the example again. Country houses are simply one of the most visible parts of a wider process of social interaction and class formation. The dinner parties, the weekend visits by friends or acquaintances, the reciprocal visits to kin, all these social activities centred around the country house will help to build up the 'invisible resources' (Marceau, 1989) that will sustain the new disestablishment over future generations. The wake of money is long and sinuous.

PART FIVE

Work, Employment and Society

PART FIVE WORK, EMPLOYMENT AND SOCIETY

Section 5.1 Introduction

INTRODUCTION

The preceding parts of the Reader have covered, in diverse detail, the driving forces of change in 'global capitalism' and the consequent new economic geographies which have been, and are being, put in place. Dealing with such macro processes, abstract forces and patterns of change can lead us to forget *why* we are studying such things. The major part of the answer is, of course, that these processes are not happening somewhere out there, somewhere in the 'ether', but are very much part of *our* everyday lives. A particular message of theorizations of contemporary capitalism is that of its qualitative distinctiveness; for example, the impact of an Asian economic crisis is transmitted, at speed, through global processes such that the cost of credit used on the card which might have bought this book can be altered overnight (whatever 'overnight' means in an era of time–space compression). This is the latest 'scaled- and speeded-up' version of the impact of economic geography in our daily lives, yet it is only one small example of the myriad *impacts* on our lives which are the essence of studying economic geography.

The new economic landscapes strongly define whether or not we do (paid) work, what that work will entail (mental, manual, 'on-the-line', craft, service, knowledge, etc.) and who our work colleagues are likely to be – young, old, black, white, male, female to list just a few non-exclusive possibilities. These landscapes are the 'where' of work (home, telecottage, factory, office, city, suburb, Europe, Asia, Internet, etc.), they frame the wealth and/or welfare which may follow and, in turn, consumption possibilities or not (from the local store to the 'exotic' holiday in the increasingly crowded 'wilderness'). In short, the processes of economic geography help define in numerous ways the lives we lead and the societies within which we live (although the exact weight of economic determination of our lives is always inextricably bound up with other political, social or cultural processes; see, for example, the debate between Sayer, 1994, and Barnes, 1995). It is, first and foremost, the impacts of economic change on work, employment and society which are the focus of this part of the Reader.

In virtually all capitalist societies, paid work is viewed as the key to societal involvement, recognition and, often, an individual's sense of identity. If it is argued that consumption is increasingly taking over this role (see Part Four), then the ability to consume is still clearly linked to income. Moreover, if one is not engaged in paid work, the assumption is that one has done one's stint (retired), is being educated as future labour, or is supporting others to work (housewife and rarely househusband). Yet one of the strongest empirical realities of current economic change is the questioning of this basic expectation of paid work. This involves not only whether one can assume future paid employment but, more particularly, the form this may take. The present and future reality for many seems to hold out the prospect of periods of less secure, less uniform jobs such as part-time, temporary or contract work in contrast to a Fordist era characterized by protected, full-time, lifelong, wage employment (Allen and Henry, 1997). For many, the employment relations of Ulrich Beck's (1992) *Risk Society* have arrived. The possible repercussions for society are immense.

A starting point is our expectations of what work will entail; expectations that vary greatly depending on who we are. For example, the advanced economies have experienced the mass movement of married women and mothers into (predominantly part-time) jobs in the service industries. At the same time a hard core of long-term unemployed men, the victims of deindustrialization, provide a sobering sight to an increasingly disaffected tranche of young men, who themselves are viewed less as a labour source by society than an issue of law and order. Meanwhile, those in employment face a diversity of employment relations and terms and conditions, many of which raise serious problems about the ability to produce a 'living wage' and continuity of employment critical for quality of life throughout one's life course.

In the past, one could rely on the Keynesian welfare state (from sickness pay to a State pension, for example) but this too is under strain. Predicated on the belief that it serves a society of full, and stable, employment the 'safety net' is cracking under the increasing pressure of dealing with the 'risks' of new forms of employment. The result is both new, and old, forms of social exclusion: income polarization, gender inequality, racism, environmental risk, etc. As Part Two shows, all this is taking place in a 'globalizing' world simultaneously increasing the competition for the supposed 'passport to quality of life' of paid work between peoples and places whilst at the same time undermining the ability of the nation-state to manage this process and the 'fall-out' of failure.

The question this begs is: 'What future society?' Visions of the answer are no more prevalent than in the work on Los Angeles as future urban model by authors such as Soja (1989), Davis (1990) and Dear (1995) but also in the work of others like Bird (1993), Christopherson (1994), Jacobs (1996) and Robins (1995) amongst others.

THE TRANSFORMATION OF WORK

If there is one word that has dominated thoughts of economic transformation in the advanced economies in the last decade or so it is 'flexibility'. As the 'Golden Era' of Fordism has waned, so many commentators have put forward alternative futures (see **Harvey** and **Lash and Urry** in Part Two, Section 2.2, for example) which if not actually labelled flexible (e.g. flexible accumulation) have been quick to use the word on numerous occasions within their schemas of the future nature of capitalist organization. Invoked at a number of scales in the analysis (labour process and machinery through to regime of accumulation, for example), a key element of 'the flexibility debate' (Gertler, 1988, 1989, 1992; Schoenberger, 1989) has been its argument for the fundamental transformation of work and employment. This part of the Reader begins with **Pollert** and her critique of 'flexibility' and all that it entails for work and employment. Whilst Pollert is sceptical of 'flexibility' as more of a powerful ideology of capitalist management (how many times have you heard on the news in recent years of workers threatened with the need to become 'more flexible' or lose their jobs?) than an economic necessity, she does recognize that some things have changed.

One aspect of change, the 'identity of work', is the subject of the following two papers by **Beynon** and **Allen and du Gay**. Thinking through the 'shift to the service economy' (see the papers by **Sayer and Walker** and **Daniels** in Part Three, Section 3.2), Beynon asks whether or not this implies the 'end of the industrial worker' in society. He concludes that whilst the dominant identity of work may have changed, from coal miner to office worker for example, and partly because of who actually does these jobs, industrial labour and the industrial worker remain even if they do so within new (service) sectors of the economy or new regions of the world (less in Birmingham but more in Kuala Lumpur, for example). In contrast, Allen and duGay argue that there is something distinctive about certain forms of 'service work'; what they term a 'hybrid identity'. In essence, service work is about human's selling 'meaning' as well as, in some cases, a material product itself, whether this be the 'welcoming smile' of the air hostess or the agreement by the shop assistant of the 'excellent fit' of a piece of clothing (see Part Four Introduction; Crang, 1994; Chua, 1992). Selling 'meaning' as part of your work makes a difference to our analysis of work.

RETHINKING LABOUR

All the papers in Section 5.2, in recognizing the transformation of work, are acutely aware that this transformation is an issue also of who, what body/labour, is doing the work. For authors such as McDowell, '. . . if we are to understand the ways in which waged work is currently being restructured, gender is an essential part of our analysis' (McDowell, 1989, p. 161). McDowell is referring to the movement of women into paid work in the advanced economies but she is aware also that 'the feminization of the economy' has been a global trend. Section 5.3 begins with a paper by **Pearson** on 'Female Workers in the First and Third Worlds: The Greening of Women's Labour' which reviews this feminization of work across economic worlds. Pearson does so in recognition not only of gender but also race/ethnicity, and, of course, geography. Fr Pearson is explicit in her understanding of the industrial links between First and Third World and when **Beynon** talks of the geographical expansion of the 'industrial worker' he means precisely those women of the Third World who are the subject of Pearson's paper.

This, of course, is also a story of that particular spatial division of labour (Massey, 1995) known as the New International Division of Labour (Froebel *et al*, 1980; see the paper by **Lakha** in Part Three, Section 3.2) but Pearson takes issue with the passive view of 'labour' portrayed by this theory. As an understanding of economic location, the NIDL is very much about multinational capital (firms) 'finding' cheap labour, namely the 'green' labour in the title of Pearson's paper. Yet such 'labour' doesn't just exist. Pearson shows how 'green (i.e. female) labour' means different things in different parts of the world, both First and Third, and, most importantly, that it is actively produced as green labour, particularly through the role of the state. And this is the basis of **Herod**'s ground-breaking article 'From a Geography of Labor to a Labor Geography'. For if Pearson focuses our minds on the active construction of certain forms of 'labour' by the state, then Herod goes further by arguing for the active agency of labour itself in defining its own characteristics – flexible, technologically advanced, individualist, etc. – and working conditions. Too often in his eyes, and even within Marxist geography, workers are seen as passive recipients of the actions of capital. In this manner, economic geographies are made by capital, such as searching around the world for 'green' labour. Through empirical examples, Herod makes the fundamental point that labour, often through the traditional route of unions (Martin *et al*, 1996), is also able to shape the actions of capital (for example, where to invest in a new plant) and thus economic geographies. In so doing, one moves from a 'geography of labour' to a 'labour geography'.

This rethinking of labour as an active agent in economic geography is addressed in differing ways by the remaining two papers in Section 5.3 by **Pugliese** and **McDowell**. If the NIDL is about capital going to labour, then Pugliese's paper is about labour going to capital; patterns of economic migration from the Third World to the labour markets of Europe (see **Sassen**'s paper in Part Three). What **Pugliese** shows is how geographies of migration have changed with economic change (from industrial society to the post-industrial society) and that whilst these patterns are primarily economic in nature their particular geographies may be shaped by other political and social factors also; such as historical connections of colonialism and the role of the Catholic Church in the particular migratory route from the Philippines to Italy. Indeed, this point of the construction of an economic concept called '(migratory) labour' by other 'non-economic' factors, such as religion or the role of the state, brings us back to the starting point of this section and the quote from McDowell on the need to understand gender divisions of labour if we are to understand processes of economic restructuring. For in the final paper of this section, 'Life Without Father and Ford: The New Gender Order of Post-Fordism', **McDowell** highlights how the 'feminization' of the economy, has been enabled in part by wider changes in society. Crudely put, capital has been keen to incorporate this 'new' source of labour but the reason for it being a new source is as much to do with changing gender roles in society (social and cultural factors) which have seen large numbers of women 'come out of the home' and into the labour market. Precisely what gain this has afforded women is the final part of this paper's analysis, especially as the largest growth in female employment has been in part-time work and the term 'double burden' has become common parlance during this period. McDowell is quick to point out also that some men too have lost out: studying gender means much more than studying women (Women and Geography Study Group, 1997).

NEW (?) SOCIAL INEQUALITIES

Whereas a message of Section 5.3 is how 'the economic', such as the concept 'labour', is constructed by more than just economic factors (e.g. the role of the state, kinship networks of migrants, gender roles in society), Section 5.4 follows the rather more well-trodden path of analysing the social, cultural and political outcomes of economic change. Picking up on gender inequalities is the subject of the first paper of this section. **Massey**'s starting point is the feted 'Cambridge Phenomenon', the high-technology industrial agglomeration often symbolized as the future. She begins with an analysis of the labour process of the high-technology scientist/engineers who are the pinnacle of the Phenomenon (and in so doing provides a further example of the transformation of work covered in Section 5.2). Unsurprisingly, the concept of 'flexible working' is centre-stage and rather more surprisingly, given the changing identities of work and the movement of women into paid labour associated with current economic change, this elite workforce is almost entirely composed of men. As Massey unpacks the reasons for this, so gender roles in society come to the fore once again, including a technological version of 'what men should do' far removed from **Beynon**'s depiction of the yesteryear men of the coalfields. Massey labels this gender role as 'high-tech masculinity'. Yet the point Massey ends on very much resonates with the coalfield regions of old. For Massey's point is to argue that the ability of these men to do their jobs is very much dependent on the ability of others – principally women – to look after them and their children. And so something old joins something new, for whilst the coalfield regions have been criticized for their patriarchal gender orders so, too, it would seem can the rather newer, post-Fordist, Cambridge Phenomenon be criticized (see Walby, 1997, for a wider review).

The second paper in this section concentrates on an axis of social inequality which has been much neglected in the literature of economic geography, namely race/ethnicity. The earlier paper by **Pugliese** highlights that economic migration has a long history but one that takes different forms dependent on the economic times. **Waldinger and Lapp**'s paper is a particular, and contemporary, 'take' on this debate. Within the paper they take issue with the contemporary argument that not only have we seen labour migrating from Third to First World, but also forms of industrial organization, namely 'the informal economy' and, most dramatically, 'the return of the sweatshop' to the post-industrial West. Put another way, we have seen the migration of a form of 'cultural capitalism' rather less welcome than that described by **Mitchell** (Section 2.4).

A review of the empirical evidence leads the authors to conclude that the supposed 'return' is very much a myth, so begging the question: Why has the 'sweatshop' returned as a social problem for contemporary policymakers? For the authors, the argument lies in the social and political association of the informal sector and sweatshops with (illegal) immigration. In the increasingly hostile political environment faced by migrants, and as highlighted by **Pugliese**, the pseudo-problem of 'sweatshops' becomes one more avenue for putting pressure on migrants. In contrast, Waldinger and Lapp draw a rather different message from their investigation; namely, that those most able to profit from the informal economy tend to be in the higher occupational categories such as lawyers (see Pahl, 1984, for similar findings in the UK) and that the 'real scandal' is the indecently low wages and abhorrent conditions of the all too legal garment industry.

Waldinger and Lapp's conclusion provides one more piece of evidence for what has become a widely accepted fact of current economic times, namely that of increased social polarization. In what has become a confusing debate, **Pinch** picks his way through the theoretical and empirical arguments around social polarization. Arguing that there are two dominant versions of social polarization, a UK and a US version, he highlights the range of processes that have led to increased social polarization such that it is not just about income and wealth, although these are obviously major determining factors in levels of social inequality. Furthermore, reiterating the need to unpack our economic concepts, his work highlights how 'income' is a product of a number of intertwined economic, political and social processes. Nevertheless, the fact of almost obscene increasing income inequality, both within and between nations, and especially within the neo-liberal regimes of the US and UK, is for all to see in their daily lives (Esping-Anderson, 1990). This is no more evident than in the two papers in Part Four on the rise of conspicuous consumption (**Bocock**, Section 4.2) and by **Thrift and Leyshon** (Section 4.4) on the country house buying spree of 'bonus-bloated' City employees which could have been inserted just as easily at this juncture.

The final very short paper of this section by **Hutton** also draws on notions of polarization in the concept of the 30–30–40 Society. For Hutton, one of the legacies of the Fordist age is the close relationship between quality of life, throughout life, and full-time employment. The onset of an increasing array of employment relationships with paid work – including none, temporary, insecure and part-time – as part of economic transformation is splitting society three ways, and for many it will be an uncomfortable present and future. Whilst in the past the welfare state was there to provide the safety net (for example, the state pension) it too was constructed around a concept of a full-time, family wage and lifelong employment and it too is struggling to deal with the new realities of 'non-standard' employment histories.

BEYOND THE WELFARE STATE?

And thus the final section of Part Five deals with 'what next?' for the welfare state. Drawing on regulationist arguments (see the paper by **Tickell and Peck** in Part Two), **Pierson** reviews the argument that just as Fordism was accompanied by the Keynesian welfare state, so the 'crisis of Fordism' has been also a 'crisis of the Keynesian welfare state'. Indeed, some have argued that the strain of funding the Keynesian welfare state was a contributory factor in the crisis of Fordism. Furthermore, as the economic contours of the post-Fordist era become clearer so there is a need to build (restructure) a post-Fordist welfare state (see **Hutton** above). However, exactly what form this restructured welfare state will, and can, take, and exactly what point we are at in the process is the subject of much conjecture. **Standing**, too, recognizes that we are at something of a crossroads; indeed, he suggests that the nature of the crisis is such that we have lost a vision of what a future 'Good Society' might look like. The starting point for Standing, too, is the stark fact that regular wage labour will not be a realistic aspiration for a growing proportion of individuals. The slow but steady reality of this has been the inexorable spread through firms, sectors, regions and economies during the last decade of what he terms as the neo-liberalist agenda of 'subordinated flexibility'. The outcome has been labour force fragmentation, the erosion of economic entitlements (e.g. welfare), the growing problems of dealing with an increasingly excluded minority and a continued failure, despite changes, to raise productivity. The scenario he depicts is a continuing downward spiral for society. His alternative is 'cooperative flexibility' and he provides examples of small-scale, but fragmented, moves in this direction. Fundamental to this scenario, however, and a challenge to capital and labour alike, is his argument for the need to decouple employment (paid work) from income. For if regular wage labour is not the future for many (see **Hutton**, for example), then income security for everyone must be guaranteed through another mechanism. Standing argues for three forms of democracy – political democracy, industrial democracy and economic democracy – but most pointedly for a way of working with flexibility, an alternative path, that '. . . would effectively reverse the traditional social democratic and socialist agenda, since rather than nationalize the means of production, it would privatize the management and ownership functions while socializing the surplus' (p. 275). Many would see in this scenario aspects of the 'Third Way/sector' so dear to policymakers' hearts today, although Standing might not be so sure.

THE CHALLENGE AHEAD?

What is interesting about this final paper is how it returns us to some of the themes of Part Two of the Reader, The Economy in Transition: Globalization and Beyond, and two themes in particular. First, and the continued message of geographers, is that globalization is a set of processes of uneven economic development which need 'placing' (see **Amin**'s paper in Part Two). This placing highlights both the diverse geographical constructions and outcomes of the globalization process but also numerous points of leverage into 'the system'. At the heart of cooperative flexibility (**Standing**), and the current debates about the Third Way, is the recognition of myriad local and regional schemes and programmes that are putting in place alternative versions of 'the economy' (see Lipietz, 1994, also). Second, however, is the recognition that this 'bottom-up' approach to globalization is likely to be doomed without co-ordination at a global level (**Tickell and Peck**, Part Two, Section 2.5). For in a system driven by neo-liberalist competition, any attempt to 'upgrade social provision' as part of an alternative is likely to come face-to-face with a competitor region advertising the fact of its 'good business climate' by sticking to a path of subordinate flexibility (Peck and Tickell, 1994). The challenge of the after-Fordist fix is to regulate the new global capitalism to allow the 'bottom-up' to flourish. And if this challenge seems daunting (well, it is) and a challenge for others to take up (well, maybe or maybe not) then take heart, for as social anthropologist Margaret Mead once said:

> Never doubt that a small group of committed citizens can change the world, indeed it is the only thing that ever has.

REFERENCES

Allen J and Henry N (1997) Ulrich Beck's *Risk Society* at work: Labour and employment in the contract service industries, *Transactions of the Institute of British Geographers*, NS **22**: 180–96.

Barnes T (1995) Political economy 1: 'The culture stupid', *Progress in Human Geography* **19**: 423–31.

Beck U (1992) *Risk Society: Towards a New Modernity*, London: Sage.

Bird J (1993) Dystopia on the Thames, in J Bird, B Curtis, T Putnam, G Robertson and L Tickner (eds) *Mapping the Futures: Local Cultures, Global Change*, London: Routledge, pp. 120–35.

Christopherson S (1994) The Fortress City: Privatized spaces, consumer citizenship, in A Amin (ed) *Post-Fordism: A Reader*, Oxford: Blackwell, pp. 409–27.

Chua B H (1992) Shopping for women's fashion in Singapore, in R Shields (ed) *Lifestyle Shopping: The Subject of Consumption*, London: Routledge, pp. 114–35.

Crang P (1994) It's showtime: On the workplace geographies of display in a restaurant in southeast England, *Environment and Planning D: Society and Space* **12**: 675–704.

Davis M (1990) *City of Quartz*, London: Verso.

Dear M (1995) Prolegomena to a postmodern urbanism, in P Healey, S Cameron, S Davoudi, S Graham and A Madani-Pour (eds) *Managing Cities: The New Urban Context*, Chichester: Wiley, pp. 27–44.

Esping-Anderson G (1990) *The Three Worlds of Welfare Capitalism*, Princeton: Princeton University Press.

Frobel F, Heinrichs J and Kreye O (1980) *The New International Division of Labour*, Cambridge: Cambridge University Press.

Gertler M (1988) The limits to flexibility: Comments on the post-Fordist vision of production and its geography, *Transactions of the Institute of British Geographers*, NS **13**: 419–32.

Gertler M (1989) Restructuring flexibility? A reply to Schoenberger, *Transactions of the Institute of British Geographers*, NS **14**: 109–12.

Gertler M (1992) Flexibility revisited: districts, nation-states, and the forces of production, *Transactions of the Institute of British Geographers*, NS **17**: 259–78.

Jacobs J (1996) *Edge of Empire: Postcolonialism and the City*, London: Routledge.

Lipietz A (1994) Post-Fordism and democracy, in A Amin (ed) *Post-Fordism: A Reader*, Oxford: Blackwell, pp. 338–57.

Martin R, Sunley P and Wills J (1996) *Union Retreat and the Regions: The Shrinking Landscape of Organised Labour*, London: Jessica Kingsley.

Massey D (1995) *Spatial Divisions of Labour: Social Structures and the Geography of Production*, second edition, London: Macmillan.

McDowell L (1989) Gender divisions, in C Hamnett, L McDowell and P Sarre (eds) *The Changing Social Structure*, London: Sage, pp. 158–98.

Pahl R (1984) *Divisions of Labour*, Oxford: Blackwell.

Peck J and Tickell A (1994) Jungle law breaks out: Neoliberalism and global-local disorder, *Area* **26**: 317–26.

Robins K (1995) Collective emotion and urban culture, in P Healey, S Cameron, S Davoudi, S Graham and A Madani-Pour (eds) *Managing Cities: The New Urban Context*, Chichester: Wiley, pp. 45–63.

Sayer A (1994) Cultural studies and 'the economy, stupid', *Environment and Planning D: Society and Space* **7**: 635–7.

Schoenberger E (1989) Thinking about flexibility: A response to Gertler, *Transactions of the Institute of British Geographers*, NS **14**: 98–108.
Soja E (1989) *Postmodern Geographies*, New York: Verso.
Walby S (1997) *Gender Transformations*, London: Routledge.
Women and Geography Study Group (1997) *Feminist Geographies: Explorations in Diversity and Difference*, Harlow: Longman.

SELECTED FURTHER READING

Davis M (1990) *City of Quartz*, London: Verso.
Hanson S and Pratt G (1995) *Gender, Work, and Space*, London: Routledge.
Jessop B (1994) Post-Fordism and the state, in A Amin (ed) *Post-Fordism: A Reader*, Oxford: Blackwell, pp. 251–79.
Lipietz A (1994) Post-Fordism and democracy, in A Amin (ed) *Post-Fordism: A Reader*, Oxford: Blackwell, pp. 338–57.
Pahl R (1993) Review article: Rigid flexibilities? Work between men and women, *Work, Employment and Society* **7**, 4: 629–42.
Peck J (1996) *Work-Place: The Social Regulation of Labor Markets*, New York: Guilford Press.

Section 5.2 The Transformation of Work

Anna Pollert

'Dismantling Flexibility'

from ***Capital and Class*** **(1988)**

'Flexibility' has engulfed conceptions of employment structure and restructuring since the early 1980s. To call it 'the flexibility debate' is over-complimentary. For most of the flexibility literature tends to be either prescriptive, or assumes there is a 'new' trend, which it then proceeds to describe and generalise. There has been no discussion on the origins of the term, to unravel its many connotations, to question what is indeed 'new', and to set this against the ideological processes which have unleashed it. Such a situation has political dangers: it defines the agenda of debate, assumes a radical break with the past, conflates and obscures complex and contradictory processes within the organisation of work, and by asserting a sea-change of management strategy and employment structure, fuses description, prediction and prescription towards a self-fulfilling prophesy. A deconstruction of 'flexibility' as an extremely powerful term which legitimises an array of policy practices is long overdue.

On a wider level, the whole 'flexibility' ethos is notable for its futurological discourse. With the growth of insecure, or irregular forms of work, and unemployment, social scientists have 'discovered' what casual wage labourers and those outside paid employment – very largely women and youth – already knew, that not all 'work' is stable and secure. Libertarian futurologists have heralded work in the small business, self-employed, informal and domestic economy as a 'new' form of work, not only more competitive (because more 'flexible' in adjusting to markets), but offering autonomy, 'flexible' hours, and in general 'flexible working lives'. A body of futurological writing based on a postindustrial conception of a 'radical break' in 'industrial society' and a largely technically determinist prophecy of the 'collapse of work' and 'discovery of leisure' (Jenkins and Sherman, 1979) [celebrates] [i]n essence, a demise of regular wage labour, indeed, of wage labour itself [as] both inevitable and progressive.

FLEXIBILITY IN EMPLOYMENT AND IN WORK – WHAT'S NEW?

The 'flexibility' debate has concentrated on two aspects of labour flexibility: flexibility in *employment* and flexibility in *work*. The neo-classical revival has provided the cue for international concern with 'Eurosclerosis', the solutions for recession and unemployment sought in wages and labour market flexibility. Labour market deregulation in the form of the legislative offensive on labour organisation and employment rights and protection, the ideological and practical sponsorship of private enterprise and the small business sector, in the form of privatisation and competitive tendering, and the maintenance of a vast reserve army of the unemployed set the material agenda for using the competitive labour market as the solution to economic regeneration.

Linked to the concern with labour market flexibility, the implicit focus on the flexibility or rigidity of labour within *work*, perpetuates the concern with flexibility manifest in British productivity dealing in the 1960s and 1970s, albeit with a crucial shift in bargaining power away from labour.

Concern with both forms of flexibility is not new. Capital has always required flexibility of labour; the struggle over its control has structured management development, the capitalist labour process, and forms of labour organisation.

In the early days of industrial capitalism, the emphasis was on defining rules and demarcations, both in time and task. The tension between the discipline of wage labour and the rhythms of life outside this social relation has been a source of conflict throughout the development of capitalism. E. P. Thompson charts the harnessing to time discipline through the 18th and 19th centuries, when the employers, far from wanting flexibility, imposed rigidity on the 'irregularity of the working day and week . . . framed until the first decades of the 19th century, within the larger irregularity of the working year, punctuated by traditional holidays and fairs. Eventually, the rigid wall between work and non-work was built':

> The first generation of factory workers were taught by their masters the importance of time; the second . . . formed their short time committees . . . the third struck for overtime or time-and-a-half. They had accepted the categories of their employers and learned to fight back within them. They had learned their lesson, that time is money, only too well. (Thompson, 1963, pp. 76, 86)

The debate from this period about the benefits and costs of regular employment, organised or 'flexible' labour markets, was a perennial issue. Only when 'rigidity' of time and task become the basis of skill, or of occupational ownership and bargaining strength did management unequivocally stress 'flexibility'.

Having said this, however, there is something new about the current articulation of an imperative towards labour flexibility.

THE 'FLEXIBLE FIRM' MODEL: AN OUTLINE

The 'flexible firm' model asserts there is an 'emerging manpower system within the large primary sector firm; a "new" polarisation between a "core" and a "periphery" workforce, the one with functional and the other with numerical flexibility, to provide adaptability to changing product markets' (Atkinson, 1984). The multi-skilled 'core' group offers 'functional' flexibility in the labour process, by crossing occupational boundaries. It also offers flexibility by time, in terms of adjusting more closely to production demands. The 'periphery' provides 'numerical flexibility'; workers may be insecurely or irregularly employed, or not have a direct relationship with the firm at all, being, for example, sub-contracted or self-employed. Here, it is the precariousness of the employment relationship and competitiveness of the labour market which provides the employer with a numerically variable workforce.

FLEXIBLE SPECIALISATION: AN OUTLINE

The strategy of 'flexible specialisation' is accomplished by a new form of artisan production made easily adaptable with the aid of programmable technology. It supplies customised goods to satisfy a plethora of individual tastes and needs and revitalises markets in a way similar to Atkinson's 'flexible firm', the labour process resembling that of the 'functionally flexible' 'core' workers.

FLEXIBILITY AS DESCRIPTION, PRESCRIPTION AND PREDICTION

One of the distinctive traits of both the model of the 'flexible firm' and the FS thesis, is that the writing on flexibility is couched in a complex form of 'post-industrial' discourse. This asserts that the current period is a radical break from the past; the 'flexibility' managerial imperative is 'new'.

Why is this so important politically? The novelty of the polarisation between a privileged core and a marginalised periphery is one of the elements of the view that the working class as an active force is dead. It is irretrievably divided, and the 'core' can only consolidate itself by working *with* capital. This is the case for the new realism, the case for the broad alliance of the soft left with any 'progressive' group, and the relegation of a large part of the working class – the periphery – to an unorganisable underclass. It is a conception built on a new accommodation with 'flexible patterns of work' legitimising pliability, insecurity, unemployment, and 'getting by' with self-employment.

THE 'FLEXIBLE FIRM'

If the 'flexible firm' is indeed the 'new' trend in employment structure, it should be possible to show evidence both for an expanding 'periphery', and for a consolidating 'core'. On existing evidence, there is little support for either.

The Evidence for a Growing 'Periphery'

The 'core'–'periphery' dichotomy has been picked up and broadcast by some observers as a way to emphasise the importance of casual employment in the economy. Hakim has argued that, 'By the mid-1980s the labour force divided neatly into two-thirds "permanent" and one-third "flexible" . . . the importance of the "flexible" sector has clearly been underestimated; it is hardly a narrow and insignificant fringe on the edge of the labour market' (Hakim, 1987a, p. 93). Unfortunately, this conceptualisation has always been a muddy conflation of legally disadvantaged, but long-term work, such as part-time work, and temporary work, so that, from the start, 'numerical flexibility' has added to a confused understanding of employment. This has been further complicated by the ahistorical 'discovery' of insecurity, as though it were radically new, and the calling of this a 'flexibility' trend.

. . . [I]f 'peripheral' work is taken to mean 'non-standard' contractual status, deviating from a full-time 'norm', there is little evidence that there has been a recent increase.

Temporary work

The most recent study of temporary work (Casey, 1987) provides further evidence of the absence of significant change in the use of temporary work. Excluding the 12.5 per cent on special employment measures, there was 'no real growth in the period 1983 to 1985. In 1983 temporary workers not in special schemes made up 5.5 per cent of the labour force; in 1985 they made up 5.5 per cent'.

Ironically, the private sector focus of the 'flexible firm' model excludes the one area where the use of temporary work may be most significant – the *public sector*. In 1984, fixed-term contracts were more common in the public sector services than elsewhere and the education sector was responsible for half of these. The Civil Service, particularly the DHSS, and the Water Authorities, are two other services found to be introducing 'new' temporary workers.

There is thus little evidence that there has been an overall increase in temporary work, and its appearance may well be the cyclical response witnessed in previous periods. Sectoral variation gives little support to a view of overall restructuring, and also highlights what the generalisation obscures: *the importance of changes in the public service sector*.

Part-time work

Between 1981 and 1986 the proportion of part-timers in the workforce grew from 21 to 23 per cent. From 1983, the only employment growth was an increase in female employment in the service sector, which was almost all part-time work. Between 1979 and 1986, 62 per cent of the rise in women's service sector jobs was in part-time work.

The 1984 Census of Employment and 1985 Labour Force Survey cast further doubt on a simplistic view of increasing part-time work. It may well be that there are contradictory processes of growth and decline, as well as what may be more qualitatively significant, changes in the type of part-time work (hours, length and regularity of shifts), rather than a dramatic increase.

Finally, turning to the public sector, we find once more that it is highly relevant: 'Two fifths of the manual workforce in the public services sector consisted of part-time workers in 1984, compared with just over a quarter in the private services sector' (Millward and Stevens, 1986, p. 207). And its main growth between 1980 and 1984 appears to be 'amongst the largest workplaces in the public services sector' (*ibid.*, p. 209). These findings demonstrate that the 'flexible firm' conflates *changes in the public sector, and the services in general, with management strategy* in the private sector.

Outworking, freelancing and homeworking

This is a notoriously difficult area to research. Homeworkers, particularly ethnic minority women, are vulnerable to state or employer harassment and more likely to remain invisible in surveys, while the small firm sector, perhaps the most important for these forms of employment, is one of the most difficult areas to study. But in large firms (the subject of the 'flexible firm' model), Millward and Stevens (1986) found a decline in all these forms of employment between 1980 and 1984. This is explained mainly by the contraction in manufacturing. There is thus little evidence of the growth of a 'periphery' composed of these workers.

Subcontracting

The alleged increase in subcontracting is perhaps one of the most topical areas of the 'flexible firm', since it relates most directly to the interest in Japanisation, in the fragmentation of production and in the growth of small firms. But there is little evidence at an aggregate level of a dramatic increase in subcontracting in the private sector. The Warwick IRRU Company Level Survey found that 61 per cent of establishment respondents reported no change in the level of subcontracting over the previous five years (Marginson *et al*, 1988). Rather, subcontracting of services was *already* extremely widespread; 83 per cent of establishment managers reported that they subcontracted out at least one service, and 39 per cent of this group contracted out three or more services.

The blinkered concern with *increases* in subcontracting masks crucial changes at the qualitative level. The ESRC-funded Steel Project, which investigated the experience of redundant British Steel Corporation workers in Port Talbot between 1980 and 1985, found that while BSC expanded the range of work put out to subcontract, what was new was newcomer subcontractors' undercutting of established subcontract firms. This was achieved through breaking union membership, lowering wage rates, ignoring health and safety precautions, and organising work 'off the cards' (Febre, 1986, p. 23). An explanation of these varied and often contradictory trends would provide a more penetrating analysis than the imposition of a managerial imperative to 'peripheralise'.

Self-employment

Estimates of the increase in self-employment since 1979 also vary enormously. Most of this appears to be due to an increase in single-person businesses: between 1981 and 1984 the number of people who were self-employed in their main job grew by 442,000 to 2.6 million, or from 9.2 to 11.2 per cent of total employment (Creigh *et al*, 1986). While this may be associated with redundancy, it does not seem to be associated with flexibility strategies in large companies.

The Evidence for a 'Core'

To start with the problem of defining a 'core', the labelling of all permanent employees as the 'core' workforce is either tautologous or highly debatable if it overlooks segmentation in status, terms and conditions within the permanently employed.

Functional flexibility as the basis for a 'core'

Other research has looked in detail at the developments in 'functional flexibility' as chief ingredient of the 'core'. But the extent of multiskilling appears limited. IDS Study 360 of capital-intensive manufacturing industry, questions whether, in fact, multiskilling is as widely desired as a 'flexibility' trend suggests:

> Companies have widely different aims. Competitive pressures vary between sectors. Skill requirements are dependent on widely different technologies. More skill, different skill, less skill, can all be legitimate objectives. Generalisations about moving towards a small, highly skilled and highly flexible workforce may fit one company and be contradicted by the pressure towards a quite different workforce in another. (IDS, 1986, p. 1)

Similarly, a trade union based study of 'flexibility' concludes that, 'in spite of the impression created by a small number of widely publicised case studies . . . a full range of flexible working practices are not frequently implemented' (LRD, 1986, p. 5). Many 'flexibility' agreements were only 'enabling' not faits accomplis. Nor was there, apart from rare exceptions, evidence of a 'core' in terms of improved job security.

In Britain 'core' status thus has a hollow ring, especially when 'employment security boiled down in practice to a reduction in the threat of job loss rather than anything more positive' (NEDO, 1986, p. 79). This is hardly a convincing case for a consolidating 'core' workforce.

The Flexible Firm: Conclusions on the Evidence

From this review of the changes in the workforce and in employment practices, the 'flexible firm' model is left standing with few clothes. Where there has been most major restructuring, this has been led by the state as employer. But in the private sector, sectoral continuity is far more in evidence than change, with little evidence of polarisation between an (ill-defined) 'periphery' and a privileged 'core'. Despite this, the 'core'–'periphery' model has become a conventional wisdom of today's alleged 'flexibility' trend.

FLEXIBLE SPECIALISATION

Flexible specialisation is part of a much wider debate on early industrialisation which cannot be addressed here. For the present purposes, the assessment of the FS thesis will be judged on the validity of its key premises:

1. The concept of the 'technological paradigm' based on the equation of a 'system of industrial technology' and a dominant system of production.
2. The alleged decline of mass production.
3. The alleged decline of mass markets.
4. The alleged new opportunities of new technology for small firms.
5. The actual spread of skill-upgrading FS as the prevalent use of programmable technology in large firms.

1. Mass Production as a Technological Paradigm

The concept of Fordist mass production as a dominant system of 'industrial technology' is a crude and inaccurate delineation of the main developments of production systems in 20th-century capitalism. Types of technology, their organisation into different systems of production, and types of labour processes cannot be conflated into a single paradigm; the empirical evidence is far more complex.

To set up a false dichotomy between 'mass production' on the one hand, and 'craft production' on the other, and discover product diversity as proof of an alternative 'paradigm', is thus a distortion of a much more complex reality.

2. The Decline of Mass Production

To demonstrate a basic opposition between two types of production – mass production relying on mass markets, dedicated equipment and semi-skilled workers making standardised goods, versus flexible equipment and skilled workers producing specialist goods – poses a false problem. No such clear opposition exists.

3. The Decline of Mass Markets?

If there is no clear opposition between mass production and batch production, and if product markets have always been a mixture of standardised and differentiated goods, then what is the evidence for the break up of mass markets?

4. Production Flexibility: New Opportunities for Small Firms?

One of the lynchpins of FS as both the explanation of the persistence of artisan production and facilitator of decentralised restructuring is the suitability of the new, flexible technology for small firms. The evidence does not support this case.

5. Production Flexibility and the Rise of Flexible Specialisation?

The continuation of debate over the skill upgrading or deskilling utilisation of microprocessor technology, challenges any assertion of a dominant trajectory towards the revitalisation of craft skills in FS.

Elger's (1987) review of recent British studies of restructuring using new technology for more flexible production, reinforces a growing body of evidence that both management and labour can wrest gains and suffer costs in the negotiation of change. Indeed, the outcomes for labour of shifts to more flexible production vary between continuing dependence on traditional skills, deskilling, skill increases and skill polarisation. What does seem to emerge repeatedly is that there is no clear development towards an upgraded craft revival.

The critiques of both the managerial and the 'radical' applications of both labour and labour market flexibility highlight one common characteristic: both cases are based on 'ideal type' models of reality, which do not bear up well to empirical scrutiny. This leaves us with a fundamental question: is there a deeper conceptual and ideological affinity between these apparently distinct constructions, which serves to explain why, in spite of their shaky factual foundations, they are pursued so intently? We must turn then to a conceptual and ideological analysis.

Labour Flexibility and 'Core–Periphery' as a New State of Capitalist Integration

The policy implications of the 'flexible firm' model and of FS are neither identical in scope nor content. The first is a managerial strategy for controlling labour, the second, a strategy for restructuring both the labour process and capital as a whole. In the first, the concept of 'functional flexibility' at the 'core' allows for the multiskilled craft worker, but also the more pliable employee and the removal of demarcations. But in both models, centrality in the labour process strengthens labour not against capital as the most advanced, organised section of the working class, but to work with capital in a new partnership.

But there is also a much wider ideological message of social integration. The 'core' and 'periphery' model is one of organisational balance, labour process flexibility in the one supplemented by labour market flexibility in the other. It rejects an analysis of capitalism as a system based on contradictory class interests, and wholeheartedly supports market regeneration. Economic and social equilibrium is maintained by labour market division.

Core–Periphery and Flexible Specialisation as Social Division

In FS, the question of underprivilege fades into silence. But the stubborn persistence of inequality in the FS utopia will not go away.

Hints that women may not do very well under flexible specialisation begin to niggle at feminists while reading Piore and Sabel's fond references to Proudhon. Embedded at the heart of his understanding of the independent, craft based working class – which he celebrated – was the belief that women should only work in the home and that any self-respecting working man ought to be able to support his non-waged wife and children (Jenson, 1989).

Murray (1987) has shown the seamier side of decentralised production in the Third Italy: racial, gender, skill and age divisions are essential to its success. Competition between firms results in a survival of the fittest which appears in no way ameliorated by local state intervention. Geographical fragmentation between firms and phases of production creates 'maximum wage differentials between different groups of workers', undermining solidarity, and making union organisation 'an uphill task'. Meanwhile 'much work can in no way be described as prized and non-alienating craft labour' (Murray, 1987).

Sharp differences between workforces and firms extends to geographical fragmentation, so that privileged and successful industrial regions, in the Third Italy, for example, can export the insecurity, poor pay and working conditions needed to provide 'flexibility' with fluctuations in demand, to surrounding areas (Solinas, 1982, pp. 334, 350).

Piore and Sabel rightly feel the need to defend FS from being compared to 'the old Bourbon kingdom of Naples, where an island of craftsmen, producing luxury goods for the court, was surrounded by a subproletarian sea of misery' (1984, p. 279). Their answer rests on a voluntarist vision of co-ordination of market relations by local community structures and by national social welfare regulation. But the international perspectives are not quite so rosy, or if so, only on the basis of harmonious inequality. While it *may* be possible 'to modernise the burgeoning craft sector of third world countries along the lines of flexible specialisation – rather than urging these countries to initiate the mass production history of the advanced countries,' Piore and Sabel continue,

> It is conceivable that flexible specialisation and mass production could be combined in a unified *international economy*. In this system, the old mass-production industries might migrate to the underdeveloped world, leaving behind in the industrialised world the high-tech industries and the traditional dispersed conglomerates in machine tools, garments, footwear, textiles, and the like – all revitalised through the fusion of traditional skills and high technology. To the underdeveloped world, this hybrid system would provide industrialisation. To the developed world, it would provide a chance to moderate the decline of mass production and its de-facto emigration from its homelands. (1984, p. 279)

This blithely imperialist perspective is a further dimension of the unconsciously divisive and elitist complacency implicit in the FS school, for all its utopian collectivist aspirations. It further confirms the compatability of this programme of restructuring for *labour*, with the advocacy of labour flexibility as a panacea for *capital alone*, in the 'core–periphery' model as a managerial strategy.

CONCLUSIONS

From this analysis of 'flexibility' presented as a 'radical break', spearheading something 'new', it is worth reiterating that flexibility is far from new. It was argued earlier that capital's harnessing of the inherent flexibility of human labour is the defining characteristic of labour power. It is, and always has been, essential to capital accumulation. How this flexibility was organised has been, and is, part of class conflict. In the early days of the factory system, division and rigidification of time and task were the central management priorities, while flexibility on the workpeople's terms was attenuated. Today, flexibility on *labour's* terms remains an area of struggle as it was before; in spite of unemployment and the appropriation of 'flexibility' as a managerial and market concept, the organisation of work and employment continues to be socially negotiated (whether through formal industrial relations or informal means), and labour can, as in previous periods, recoup and transform managerial initiatives to its own ends. However, such outcomes depend, as 'participation' exercises, productivity deals and other managerial initiatives to gain worker co-operation show, on the groundwork of worker organisation and self-confidence. Whether the libertarian legitimations of casualisation and 'flexible' working lives can thus become re-appropriated by labour as positive improvements in terms of working hours and arrangements, paid for by capital, not labour, is an open question. But it is not a new terrain of struggle, just as the 'flexible workforce' is not the radically new phenomenon which it has been perceived to be.

The effect of 'flexibility' as both managerial strategy and ideology is complex. There can be no doubt that workplace and trade union organisation are both under attack. And yet, as the survey of the empirical evidence shows, the extent of change in the direction of casualisation is more uneven and complex than the 'radical break' perspective implies. In this sense, the 'discovery' of the 'flexible workforce' is part of an ideological offensive which celebrates pliability and casualisation, *and makes them seem inevitable.*

This paper has therefore attempted to analyse the interweaving of 'flexibility' both in its concrete applications, and as ideology. Seen in this way, the language of 'flexibility' reveals itself as the language of social integration of the 1980s: how to live with insecurity and unemployment, and learn to love it.

Huw Beynon

'The End of the Industrial Worker?'

from *Social Change in Britain* (1992)

The image of the industrial worker has been a powerful one which has sustained many popular and political ideologies over the past century. Nostrums about hard work not hurting anybody, and about not being afraid to get your hands dirty were fashioned into a central part of a code of ethics for the industrial working class in this country. It was both a method of judging your fellows, and an acerbic assessment of those who considered themselves to be your superiors. In many accounts, this condition of labour rendered heroic images within which the worker was seen to be both powerful and exploited; the antithesis of middle-class comfort and respectability yet the linchpin for it. Orwell's writings were of this kind. Lacking sympathy with the superficiality and condescension of middle-class life, he identified emotionally with manual workers. In the 1930s he went down a coal-mine, and he marvelled at the ability of his guide to work strenuously under conditions which, in themselves, were exhausting to the visitor.

Yet, as Orwell observed:

> You could easily drive a car right across the North of England and never once remember that hundreds of feet below the road you are on, the miners are hewing out coal. Yet in a sense it is the miners who are driving your car forward. Their lamp-lit world down there is necessary to the day-light world above, as the root is to the flower. (Orwell, 1957, p. 63)

This contrast (extended by reference to the comfort of sitting rooms heated by coal) was not lost on the miners either. In the first decade of this century, the Webbs observed that 10 per cent of the children born in Britain were brought up in the homes of coal miners. In places like Durham, miners formed 80 per cent of the working population. Yet the social conditions of their lives set them apart from the urban centres of power and [enabled the miners] to develop on the coalfield a distinctive culture associated with their trade.

In the 1960s, as the coal-mines closed down and miners and their families moved into different kinds of work, public attention focused upon the car industry. Here was a modern industry which was at the centre of what became known as the 'Fordist' period of capitalist expansion. In the 1930s – when the coal-mines were laid off – new car factories were established and places like Dagenham, Oxford, Birmingham and Coventry became deeply linked with the mass production of motorcars.

Here, in a changed industrial context, the concern expressed by Orwell had a continuing (if changed) relevance. The mass-assembly industries carried with them none of the cultural forms so closely associated with coal mining. The coal-mine and the pit village were replaced by the assembly plant and the city [w]hen Ford first moved to the UK in 1911, [and] the company established a plant on the Trafford Park estate to the west of Manchester.

The Trafford Park estate built upon this resource and (in spite of the loss of Ford to Dagenham) established an enormous production system which in 1945 employed 75,000 factory workers. In 1989, this number has declined to 23,000. The famed Dagenham and Longbridge estates have seen their work-forces decline in similar proportions. Coventry is no longer a centre of car production and there is much talk of the closure of the Cowley plant. The car workers, it seems, have gone the way of the miners; as have the shipyard workers, the steel workers and those men who in mechanical engineering factories supplied components for the consumer industries. The scale of the transformation involved here is perhaps best captured by the fact that employment in all of these industries (coal, steel, shipyard, cars, mechanical engineering) is now exceeded in the hotel and catering industries.

Today, the new industrial jobs are being concentrated away from the cities in the new electronic industries which are growing in East Anglia, along the M4 Corridor, in Silicon Glen in Scotland and along the coastal plain in South Wales. More and more people, however, are now employed in the 'service sector'.

Writing in *The Observer*, Ann Barr and Peter York (1987) asserted that:

> The key stereotypes of late nineteenth- and early twentieth-century work – people in productive industry, organised in factories – are now a dwindling minority sector (manufacturing jobs are 4.8 per cent of the workforce), and the perfecting of the 'lights out' (i.e. totally robotic) factory will hasten the decline in the 1990s.

Here, in what they termed 'the New Babylon', we were witnessing:

> the possibility of new careers of an astonishingly exotic – almost whimsical – variety . . . and these in turn, influenced the idea of what work could be in more everyday situations. The 'explosions' in media, financial services, property values and leisure of all kinds have opened up extra-ordinary opportunities for people who were lucky, (mostly) well educated, and well placed (Southern, etc.). There were openings not only in investment banking and the television industry but interior decoration and specialist retailing. (Barr and York, 1987)

In the 1980s these themes came together to provide a powerful and pervasive account of social change. On the one hand, the movement from manufacturing to service employment was explained as a historical shift from industrial to *post*industrial society. Equally, the changes in the organization of work and employment were variously interpreted as a move from Fordist to *post-Fordist* practices; or as the replacement of mass production with flexible-specialization. In this, unfortunately, critical traditions which once emphasized the damaging effect of work upon workers, gave way to those which celebrated the enriching aspects of employment. These were regularly contrasted with the debilitating experience of *un*employment. In this way, the major division within society became understood as that between those *with* and those *without* work. Here too sociologists developed radical accounts which suggested a historical change of a different order. To some it seemed that we were in the process of changing from one kind of society (a 'labour society' dominated by work and paid employment) to another (a more open society where people were freed from the excessive demands of work). These arguments were most strongly developed in Europe where, in Jahn's words, 'individuals are able to spend more time in non-work areas which became significant sources of their social identity' (Jahn, 1986). In France and Germany writers like Gorz and Offe warmed to this theme, postulating the emergence of societies where consumption was increasingly replacing work as both a 'social duty' and the 'central motivating force' in people's lives.

Ironically perhaps, in this new context, many of the 'new workers' began speaking in the language of the old. Teachers and university lecturers talked of working at 'the chalk face' and their trade unions developed this metaphor in their negotiations with employers, when they threatened 'industrial action'. More than this; in offices, people began to refer to the computers as their 'machines', and 'workaholism' became an acceptable illness amongst the new middle classes. While in Orwell's day these people looked down upon industrial workers and their work, in the 1980s they publicly embraced the ethic of work and endeavour. *The Times*, 7 June 1990, commented on how:

> In the 1980s a new type of working – 'macho working' – became increasingly prevalent. An urge to perform seemed to overtake people, perhaps most notably in the City . . . It became *de rigueur* to arrive at the office before everyone else and to be the last to leave. The 'power breakfast' was born, the better to get meetings out of the way before the rest of the day started. At the office, meetings were scheduled at 8 a.m. then, ludicrously, 7 a.m., when the people around the table might have said goodbye to each other only a few hours earlier. British Rail was asked to provide more early morning trains.

On TV-AM the Duchess of Kent insisted that: 'the busier you are, the more you get done. It takes about twenty five hours a day, but I just make sure there is time. The girls in the office insist that I'm a workaholic.' Mick Jagger who once sang 'Let's spend the night together' returned to the charts with 'Let's work'.

Undoubtedly, in the post-war period we have witnessed enormous changes in the organization of work, in its content and distribution between people and places in Britain. Whether these changes contribute so totally and uncomplicatedly to a general transformation in the nature of our society and 'the end of the industrial worker' is, however, open to question.

It is important to remember that there are still over five million people working in manufacturing industry – a considerably bigger grouping than some of the more enthusiastic accounts of economic change would allow for. Equally, the British figure is low in relation to other OECD countries. In Brazil, Mexico and Malaysia (not to mention Japan, South Korea and Hong Kong) manufacturing employment has expanded in the post-war period, and dramatically in the 1970s and 1980s. Arguably – within the 'global production system' – the industrial worker remains the largest single economic group, and in no sense in decline.

In the context of the UK and the OECD countries generally, however, the case for decline has some veracity. And if we believe, as Marx seemed to, that these countries paint the future for the rest, it needs to be addressed seriously. In so doing, several points of qualification and reservation are in order.

To begin with, it is hard to sustain the image of a manufacturing process that is built substantially upon the sophisticated operations of intelligent machines, with workers employed as creative adjuncts or onlookers to the process of production. In spite of the hyperbole, and the predictions of ten years ago, the 'lights out' factory is not operating in the UK, and the examples of complex manufacturing units are few and far between. The cautious assessment made of these changes by Tony Elger (1990) seems empirically sound. There *are* examples of industrial workplaces where the organization and content of the work has significantly improved the working lives of the workforce. But they are few and far between. The content of the work of most industrial workers still has little in common with the cybernetic cornucopia envisaged by many writers. A more adequate assessment of the 1980s would be one which stressed the increasing *pace* of work and the greater commitment *demanded* by employers.

So too, you might think, in some of the jobs in the service sector. For while many of these are creative and cerebral, others are distinctly manual, repetitive and unpleasant. Also, many of them in their content and organization resemble jobs previously done by industrial workers. At McDonald's for example, we are told that 'a quarter-pounder is cooked for exactly 107 seconds. Our fries are never more than 7 minutes old when sold.' In one of their restaurants they 'aim to serve any order within 60 seconds. At lunch-time in a busy restaurant, we serve 2,000 meals an hour'. In this advertisement, of course, the company insists that 'anyone who cooks hamburgers at McDonald's can join the management.' But this doesn't detract from the fact that most of the 27,000 workers in their 342 restaurants are involved in some form of manual work, and that in addition to providing a service ('Have a nice day' and all that) many of them are *producing* things. At this point the severe logic of the distinction between 'manufacture' and 'services' breaks down. In the past (and today) manufacturing industry was made up of a range of manufacturing and service jobs or occupations. The same applies to service industries, and this suggests a number of things.

To begin with, it seems likely that the rise of service employment has, to some extent, been associated with corporate realignment and specialization rather than a real increase in new kinds of jobs. Where once the car manufacturers employed line workers as well as janitors and cleaners, at Nissan these ancillary tasks are subcontracted to specialized cleaning firms.

Equally, and more centrally given the theme of this paper, it is clear that many of the jobs created in service industries are manual jobs, many requiring few skills, most of them done by women. Perhaps it was Nigel Lawson and not Tebbit and Walker who was most in tune with economic developments when he forecast 'not a high-tech but a no-tech' solution to the problem of employment growth. More important, perhaps, has been the significant presence of women in service sector jobs. This has contributed greatly to a general and profound increase in the proportion of women in the labour force, a change which, perhaps more than any other, provides a clue to the detailed ways in which work and employment have altered since the war. The manual industrial worker was archetypically male. It was this which changed in the 1970s and 1980s. In part this has been associated with the rise of industries based upon information technology, in which women play a central part: they manufacture the microchip in factories in the Far East and they assemble the computer boards in the lowlands of Scotland. It is also women who, for the most part, look at computer screens and hammer away at keyboards. Studies of audio-typists have noted how:

> Consequent upon their isolation, the typists' work appears doubly monotonous. Like manual workers in mechanised jobs, they are prevented from making conversation with their workmates by the noise of the machinery . . . In the building society, the audio-typists turned the machinery to their own advantage by listening to music cassettes on playback machines while copy-typing! While this might be seen as an attempt by the typist to gain control over the work environment, it might more accurately be seen as one of the ameliorating aspects of routine, alienating work long established in factories. (Webster, 1986, p. 128)

The parallels with factory work are not overdrawn, and this represents but one example of a general feminist ethnography which has illuminated the changing nature of work and employment in the 1970s and 1980s. These studies have drawn attention to the fact that manual employment is emerging in new and different ways and being constructed differently in relation to other kinds of work (housework and domestic work generally). They have helped draw attention to the fact that our previous understanding of industrial work and labour was deeply coloured by notions of sexual identity.

They also, less directly, serve to raise interesting questions about the idea of a *post*-industrial society. If we take industrialization to mean the production of commodities through the use of machinery aided with rational systems of organization, the postwar period can be seen as one in which areas of life hitherto unaffected by the onward march of capital were subjected to this process. I have mentioned how we – in the universities – now have 'machines' on our desks. Add to this the mechanization of banking, transportation and the home and we have the bones of an ongoing industrialization thesis and the *extended* rather than the *post*industrial society. Such a thesis would argue for the *continuity* of manual labour, maintained within different sets of relationships and contexts.

John Allen and Paul du Gay

'Industry and the Rest: The Economic Identity of Services'

***Work, Employment and Society* (1994)**

INTRODUCTION

The debate over post-industrialism initiated by Daniel Bell in the 1970s was centrally a disagreement over the interpretation of the sectoral shifts under way in the western economies and the implications that they may hold for the nature of work and employment. At the time, a variety of claims were made as to the coming shape of the western economies: that they were becoming service-based, driven by information, and characterised by 'think' work as manual jobs and skills gave way to white-collar and professional dexterities. In the UK, such claims were roundly rejected, principally by Kumar (1978) who stressed the persistence and continuity of industrial forms of work, and by Gershuny (1978) who argued that the growth of service employment was simply the result of an expanded and deepened division of labour brought about by an increased demand for manufactured goods. In laying considerable stress upon the ongoing dynamic of industrial society, however, the questions and lines of enquiry opened up by the post-industrial debate tended to become overshadowed by a celebration of the marked continuities of work and the economy. Drawing attention, correctly in our view, to a foreshortened history of industrialism, the critique of the post-industrial thesis, perhaps unwittingly, also foreclosed a debate on the nature of services and service work. Even though there have been a number of studies of white-collar work and also a scattering of studies on manual services, especially catering, service sector research has been unable to establish itself as a field of study. This is all the more surprising given the relative decline of manufacturing industry and employment in the western economies, especially in the UK, and the simple fact that people's work lives are shaped overwhelmingly by the experience of delivering a service. It is perhaps an appropriate moment, therefore, to think through what is and what is not distinctive about services and service work in an industrial economy.

IDENTIFYING SERVICE WORK

The slippage between the wastefulness and usefulness of service activity has its corollary in service work, namely between the representation of such work as, on the one hand, fundamentally 'unproductive and alienating' and, on the other, as capable of 'adding value' only when industrialised. 'Real work', of course, has long been associated with men making things and it is no accident that women's work, much of which is in service occupations, has been portrayed as less than productive. What we wish to consider is the status of manufacturing as 'real work', when the overwhelming bulk of employment in western societies is now located in services.

While there have been a variety of interpretations of this issue, two in particular are of considerable note. The first of these is explicitly nostalgic in focus. Deriving its force from the aesthetic romanticism of the 'alienation' hypothesis, this line of argument bemoans the dominance of service employment as the latest in a series of developments which have robbed modern 'humanity' of the possibility of dignity through work. Services, so the argument goes, are deemed to deprive 'humanity' of the possibility of creative labour even more than Fordist factories, because they remove from people the possibility of making anything at all, even the tiniest bit of a larger whole. Thus, services are the ultimate form of alienated labour. This line of argument has little time for services because they are viewed as antithetical to the essential 'desires', 'needs' and 'capacities' of human beings, which are uniformly defined in terms of the ability to 'make things'.

The second line of argument shares many of the characteristics of the first although with one crucial difference. In this version, services are represented as 'real work' simply because they are increasingly organised in industrial terms. In other words, services are attributed the status of 'real work' because the organisation and experience of work that people are exposed to is increasingly indistinguishable from that found in manufacturing. In effect, services represent the extension of industrial organisation – and hence the discourse of manufacturing – into new arenas and relationships. According to **Beynon** (1992, pp. 177–82), for example, the distinction between service and manufacturing work is often highly exaggerated, in that the content and organisation of most servicing jobs involves an extension rather than an erosion of manual industrial labour. In this reading, manufacturing metaphors are deployed to capture the 'truth' about work in services: the experience of working in fast-food catering, for instance, is described largely in terms of being 'on the line'. Similarly, the changing nature of work in banking, retailing and other service occupations is represented largely in terms of mechanisation, routinisation and rationalisation. Most contemporary servicing work may thus be labelled as 'real work' precisely because it entails the continuity of industrial – basically manufacturing derived – activities and processes 'maintained within different sets of relationships and contexts' (**Beynon**, 1992).

There is a slippage between, on the one hand, the representation of 'service work' as an essentially intangible and therefore 'unproductive' activity (i.e. it can never be 'real work' because it does not involve people making things) and, on the other, the representation of 'service work' as capable of adding value only when 'industrialised' (i.e. it can only don the mantle of 'real work' when it assumes the characteristics of industrial labour).

As we have had cause to note, however, while many forms of contemporary service work have been industrialised (that is, subject to processes of standardisation and routinisation), this does not imply that the identity of such work is solely explicable in terms of an industrial dynamic.

According to Urry (1990a, pp. 271–4), for example, an important feature of much service work is the more or less direct relationship it involves between one or more service provider and one or more service consumer. This leads him to observe that the traditional separation between the 'production' and 'consumption' which characterises manufacturing employment is not readily applicable to most services. As the consumption of a service is less easily divorced from its production, a crucial economic element of much service work is the manner in which that service is delivered. In other words, the quality of service delivery is, in many cases, an important part of the consumption of that service: 'part of what is consumed is the quality of the social interaction'. Thus what Urry refers to as the inherently 'social' nature of much servicing work may be seen to entail a distinct break with the characteristics of industrial manufacturing labour. If this is the case, then it follows that the economic identity of service work cannot be conceptualised simply in 'production' terms; that is, within the discourse of manufacturing industry. For if a worker's relationship with his or her work is modified by the shift to service employment, there is no longer the same identity – the industrial worker – in a new situation, but a qualitatively different identity.

SERVICE WORK: A HYBRID IDENTITY

If the growth of service sector employment establishes a qualitatively new identity for 'work', it implies that new approaches to understanding that identity should also be sought. For example, if the work-based skills associated with much service employment relate to issues of presentation, communication and display, then service work cannot be conceptualised solely as an economic phenomenon. It should also be understood in terms of what we may broadly refer to as 'cultural relations'. 'Culture', in this sense, refers to the production of meaning. As part of what is consumed by the customer of a service is the social interaction; service work necessarily involves the production of distinct meanings. In other words, a profitable service relation is one in which distinct meanings are produced for the customer. To suggest this, however, is not to deny that service work has an economic identity; it is rather to indicate that the economic identity is always already 'cultural'. Which is to say that what is properly 'economic' and what is properly 'cultural' about service work are inseparable, notably because the very act of servicing is both 'cultural' and 'economic' at one and the same time. Put another way, the identity of contemporary service work is irreducibly '*hybrid*'.

By the term 'hybrid identity', we wish to imply that the boundaries between the economic and the cultural are blurred in much contemporary service work. Take, for instance, the case of retail service work or, for that matter, many forms of tourist-related service work. As the quality of interactive service delivery is increasingly seen as 'prime determinant of service firms' competitive success or failure' (Fuller and Smith, 1991, p. 2) the links between workers and consumers (and between the 'inside' and 'outside' of the service organisation) have become tighter. Through the medium of a variety of 'soft' cultural or 'human' technologies of interpersonal and emotion management, employees engaged in service work have been encouraged to develop particular predispositions and capacities which are aimed at enabling them to win over the 'hearts and minds' of customers. Increasingly, direct service workers are encouraged to assemble and deploy their own experience and identity as consumers in their paid work of providing 'quality service'. These 'soft' technologies are not designed to suppress, but rather to incite and utilise the subjectivity of service workers. As such, they do not represent the simple extension of the various 'deskilling' impulses of industrial labour into a service context; rather they represent a medium of personal development and 'enskilling'. Servicing is regarded as a skill which may be learnt, and one which will stand its practitioner in good stead in many forms of social interaction in and outside of the particular work context.

While it is certainly true that the term 'industrialisation of services' provides a useful shorthand description of certain changes in the labour process of particular services – such as shop-floor retail activity and fast-food catering – even in these cases service work is never reduced to a purely 'economic' or 'productionist' phenomenon. In retailing, for example, the 'de-skilling' and 'flexibilising' effects of the utilisation of particular new technologies – EPOS (Electronic Point of Sale) and EFTPOS (Electronic Funds at Point of Sale) – have proceeded hand-in-hand with a growing focus on the improvement of labour effectiveness. While retailers have sought to improve their 'physical proximity' with customers through the use of EPOS technologies, the associated introduction of 'just-in-time' delivery systems and flexible employment practices designed to ensure a closer 'fit' between customer flow and staffing levels, they have simultaneously attempted to achieve a greater degree of 'emotional proximity' through improvements in customer service.

Techniques such as Transactional Analysis are designed to incite and channel, rather than repress, the subjectivity of service workers. These 'soft' technologies are fashioned to assist staff in managing the service relation; that is, to produce pleasurable meaning for the customer and a sale for the company. At the same time, however, they are also deemed to provide service staff with the practical means – or 'skill' – of empowering themselves in their work. In this way, service work in occupations such as retailing concerns the simultaneous 'production of meaning' and the 'production of profit'. The two goals are symbiotic. Moreover, the 'effective management' of the service relation is deemed to provide meaning and fulfilment for the producer of the service, the retail employee, and also for the consumer of that service, the customer.

Rather than representing a simple extension of industrial labour to a different context or the latest 'alienating' form of modern work suppressing people's innate need to make things, contemporary service work in occupations such as retailing has a complex economic identity. The latter is not susceptible to appropriation by the discourse of manufacturing, even when many of its constituent elements have been subject to standardisation and routinisation over time, simply because that discourse is unable to comprehend the subtle imbrication of economy and culture which service work increasingly entails.

This observation holds for financial and commercial service work as much as it does for the more direct servicing work found in parts of the retail, hotel, and tourist-related trades. In finance for example, particularly merchant banking, the elements of communication, display and presentation are not simply restricted to the culture of the 'deal'. Financial networks, especially global networks, are essentially social networks in which 'relationship management' holds the key to profitable success. Insofar as global financial centres are effectively characterised by streams of information, each of which is open to interpretation and each with its own contacts, one of the significant skills within international finance is the ability to make and to hold contacts, to construct relationships of trust, and to be part of the interpretation of what is actually happening (Thrift, 1994).

As Thrift points out, far from a reduction in the need for face-to-face contact in the global financial centres, even allowing for the profusion of 'hard' electronic technologies, there is now a greater emphasis on the 'presentation of self, face work, negotiating skills . . . because of the increasing requirement to be able to read people . . . because of the increasingly transactional nature of business relationships between firms and clients'. Trust in this context is something that has to be 'worked on', rather than simply assumed from one's social standing.

Even if we take an example of service work which contrasts sharply with the customer-oriented services or those of global finance, that of contract security work, the qualities of display, communication and presentation are an essential component of guarding work. Security work is reactive, whether on day or night shift. During the day, guards monitor arrivals and departures, often through a variety of 'hard' technologies, with a brief to control entry and exits. Social contact, however, is not a significant aspect of security work. The work of security is primarily performance, involving a standardised body and a range of ritualised and codified gestures. Security guards produce a controlled space simply through their uniformed presence. Being male and over six foot adds to the performance and communicates authority. There are few spoken lines to the performance, as it is the body itself which generates the controlled space rather than verbal interaction. In this instance, the skills involved are those of presentation management and they are as essential to the task as the forms of emotional management and relationship management referred to earlier in respect of quite different forms of service work.

None of these examples of service work can be explained solely in terms of an industrial dynamic. Indeed, that is basically our argument. It is possible to consider further forms of service work, although in our view they would – in various combinations of 'hard' and 'soft' technologies, work-based skills, degrees of social interaction, tangible and less tangible outputs – produce a 'hybrid' service identity.

Section 5.3 Rethinking Labour

Ruth Pearson

'Female Workers in the First and Third Worlds: The Greening of Women's Labour'

from ***On Work: Historical, Comparative and Theoretical Approaches*** **(1988)**

WOMEN WORKERS AND THE NEW INTERNATIONAL DIVISION OF LABOUR

Nimble-fingered young women working in serried ranks in a South East Asian electronics factory is by now a widespread image. More than that, it is the image of women industrial workers in the Third World. Virtually all of the analysis of women's work in the industrial sector in the Third World is based on the experience of export platform factories. From a number of often first-hand research and other reports about women working in a variety of sectors, regions and countries in the Third World an ideal and universal picture has emerged which has tended to coalesce the Third World into a single undifferentiated country where women factory workers are young, industrious, naive and passive. And it seems likely that the same kind of generalizations are being translated to the analysis of women working in the new technology industries of the industrialized world.

This simplistic analysis captures some aspects of the emergence of a specific demand for female labour for assembly operations in factories mainly owned by foreign capital and producing for the world market as part of what has been termed the new international division of labour. However, this framework of analysis, which refers to the trend during the 1960s and 1970s for foreign companies to relocate production to low-wage countries, belies the complexity of the wider process of capitalist restructuring at a global level. The analysis assumes the existence of an ubiquitous pool of 'suitable' female labour – a kind of global reserve army activated directly, and without contradictions, by international capital seeking low-paid workers with high productivity.

Partly responsible for this undifferentiated image of Third World women workers is the way in which the analysis of the new international division of labour has ignored the complexities and contradictions of producing the desired social relations of production involved in creating a new sector of waged labour. The analysis focuses primarily on the international mobility of capital which facilitates rational location decisions on the basis of comparative costs. Labour is cheap in Mexico but cheaper still in Sri Lanka and even cheaper in Malawi. Which location was chosen depended, of course, on a number of other considerations, including the scope of incentives provided by Third World governments eager to attract foreign investment in the industrial sector, for such investment offers foreign exchange earnings and industrial employment opportunities and at least the promise of escape from the Third World's traditional place in the international division of labour. Indeed the competition to attract international investment of this kind is so intense that many Third World countries have established special locations – free trade zones (FTZs) or export processing zones (EPZs) which provide international capital with relevant industrial infrastructure and services and effectively cede large areas of sovereignty over foreign companies operating within these areas in terms of trade and employment regulations.

It is clear, therefore, that location decisions are not just a matter of seeking the lowest cost environment; in addition the state, in the guise of the host government, has to intervene to deliver such environments to international capital in a variety of ways. They have to guarantee political and economic security, which could be provided by a strong military regime or could be delivered by enacting legislation about the control or absence of labour unions as well as provide the industrial inputs necessary to make the environment feasible for an internationally controlled and organized operation. This means providing telecommunications, air freight services, power and water supplies and basic infrastructural and internal transportation investments. The governments representing the multinational companies' home State have also been required to intervene to provide a feasible environment for this mode of international accumulation. The United States, for example, made special provision in its tariff schedule to allow US components assembled abroad to be allowed back into the United States free of import tax on the value of those components.

What has not been addressed is the availability or construction of cheap labour. It has not been acknowledged that either capital or the State might need to intervene to deliver the suitable labour required; it has been assumed that this was axiomatic on the existence of high levels of unemployment or underemployment in the Third World locations. Given that it has been female labour which was targeted to provide labour power for Third World export factories it was assumed that the absence of industrial employment for women in the immediate economic history of the country meant that there would be no problem in making this labour available in the quantities and qualities required.

THE IDEAL WOMAN FACTORY WORKER IN THE THIRD WORLD

The analysis of women's employment by multinationals involved in manufacturing for export in the Third World has established that women constitute an overwhelming proportion of the 'operator' (i.e. unskilled and manual worker) level of employment and that such employment constitutes up to 90 per cent of total employment generated by such investments. It is also clear that women are employed in both traditional and 'new technology industries' in spite of the unequivocal existence of unemployed male labour in the Third World. The reason why women's labour is the preferred 'cheap' labour in a situation of surplus labour of both sexes is complex. Women's wages are generally lower than those paid to male workers in comparable occupations, though this is not always the case. But also it has been demonstrated that women's productivity under the production conditions determined by specific production processes are higher than men working under the same conditions.

From this analysis of how and why women are the preferred labour force, a stereotypical picture of the average or ideal Third World woman factory worker has emerged, comprising four essential components:

1. that she is young – recruited from an age cohort ranging from fifteen to twenty-five, concentrated in the 18–21 age group;
2. that she is single and childless;

3. that she is 'unskilled' in the sense of having no recognized qualifications or training;
4. that she has no previous experience of formal wage employment in the industrial sector – 'virgins in terms of industrial employment that need not be retrained or untrained' (Konig, 1975), to quote one not untypical researcher writing about women's employment in the Mexican border industries.

BEYOND THE STEREOTYPE

In fact, when we come to examine in detail different case studies of women factory workers in different areas and regions of the Third World it becomes clear that there exists considerable variation in the characteristics of the workers recruited. For what the management of multinational companies consciously, if not explicitly, operated was indeed a strategy of providing for themselves a labour force which would incur minimum costs in terms of wages, fringe benefits, management control, discipline and militancy. And these are not necessarily supplied by recruiting a single age cohort from the vicinity of these factories and setting them to work in a standard context.

The mostly deeply held aspect of the stereotype of women workers in Third World factories is that they are young women; the age range varies in different accounts, but it is generally within the range of 15–25, bunching in the 18–21 age group. However, research has indicated that there is in fact a considerable variation in the age range. In some countries women as young as twelve are employed; in other countries (e.g. Barbados) the labour force in the electronics factories are considerably older, starting in the late twenties and going throughout the thirties age group.

This last example is interesting because it relates to another aspect of the stereotype, that women factory workers in the Third World are *single*. It is clear why this should form part of the employers' construction of their ideal labour force; single women, with few alternative industrial job opportunities, can deliver the highest level of compliance and loyalty to the firm. They are deemed not to have domestic and economic responsibilities to their own conjugal households and children – a kind of international teenager ready to exchange their hours of industrial activity for the monetary rewards and concomitant independence this brings. However, what single really means in the context of this analysis is *childless*, and in different social contexts the two characteristics do not necessarily go together.

In Barbados, where the age of first childbirth is earlier, and age at marriage later (which is the English speaking Caribbean pattern), and the provision of maternity leave and payments more historically integrated into the island's labour practice, older women are recruited who have passed through their intensive phase of childbearing.

In locations where there is a large supply of women applicants for jobs in export processing factories, the criterion of childlessness can be used as part of a complex recruitment mechanism for selecting 'ideal' applicants. In Mexico where there is an excessive supply of female labour and a range of export factories in different sectors, and where for social, economic and cultural reasons there is a high rate of illegitimate births, electronics factories include pregnancy tests as well as declarations of childlessness as a routine measure. In Malaysia where recruitment takes place within a less homogeneous and different social and cultural context, different strategies have been adopted. Amongst the social classes from whom electronics workers are recruited, there is a strong prejudice against married women working in factories, and a much more cohesive family structure, so recruitment of single high-school graduates will provide motivated childless women. But where the prejudice against factory work extends to daughters because of its implications in undermining forms of control of fathers and brothers over young women, capital may have to alter its strategy in order to 'release' the required labour power. One multinational company operating in Malaysia pursues a policy of reinforcing traditional forms of patriarchal power. Instead of undermining the father's authority over the daughter by encouraging modern, Western independent behaviour, it pursues a policy of reinforcement: 'the company has installed prayer rooms in the factory itself, does not have modern uniforms and lets the girls wear their traditional attire, and enforces a strict and rigid discipline in the workplace' (Lim, 1978, p. 37). In another case the firm has allowed traditional leaders onto the production line to talk to the women and check the modesty of the company uniform (*ibid.*, p. 36). Young women recruited from rural areas may be provided with supervised hostel accommodation and in some cases the wage is paid to the male kin rather than directly to the women workers.

What these variations in the composition of the labour force and in the employment conditions demonstrate is that women's potential labour power, as a commodity available for exploitation by capital, has to be negotiated for with forms of patriarchal control and with her childbearing and reproductive role; and at the same time these can be used to control the composition and characteristics of those employed.

Moreover the firms do not face an undifferentiated supply of homogeneous cheap female labour and can utilize differences within the potential female labour force to structure their recruitment strategy according to their own perceived requirements.

HOW WOMEN WORKERS BECOME CHEAP LABOUR

One such strategy is the practice of recruiting childless women for certain sectors of export manufacturing. This provides capital with a mechanism which serves a number of purposes in its attempt to release the ideal female labour power. Recruiting women with children, or who are liable to bear children during the course of their employment, potentially involves additional costs to the firm, such as maternity benefits, maternity leave, sick leave, absence from work, contributions to State health services for dependants. But it is not just the potential cost of employing women actively involved in these stages of reproduction which leads management to avoid them, for in many instances legal requirements for maternity payments are waived for this particular form of industrial production. It also provides a mechanism for ensuring whatever turnover rates are appropriate, for if the argument is that Third World women factory workers perform unskilled tasks, then several years of experience will not enhance a woman's productivity compared with a newly recruited worker. But the dispensability of the labour force in terms of the ease with which it can be retrenched in response to fluctuations in demand, or made redundant if a decision is taken to cease production or relocate in another country, is a further advantage of this carefully selected labour force. Avoiding recruitment of pregnant women and those with children and terminating the employment of those who become pregnant is one of the many mechanisms which is used to maintain an ideal structure of the labour force.

WORKING CONDITIONS

A further mechanism which increases the turnover of a given factory's labour force is the working conditions themselves. Occupation-related health hazards in the different sectors, ranging from myopia caused by close microscope work, nausea and cancer from contact with chemicals and solvents, bronchial and respiratory disorders from working with textiles, etc. are extensively detailed in the literature. In such situations health problems, for which the firm takes no formal responsibility, can be used to maintain and increase productivity in the plant; workers whose physical condition prevents them maintaining the required level of productivity will withdraw themselves from employment, without needing to be dismissed by management.

Another feature of working conditions in many multinational Third World operations is the absence of any effective structure for promotion, increased earnings and technical advancement for the women operators (Ong, 1984). While there may be minimal possibilities for advancement, e.g. to group heads (i.e. head of sections) in the electronics assembly plant, this only offers minimal improvements in hourly wage rates and has the disincentive of separating the woman from the companionship of her fellow workers, which is highly rated as being one of the main sources of job satisfaction for the women concerned.

The lack of promotional structure must be seen in the context of the rigidly enforced sexual hierarchy within the plants. This is a feature common to all manufacturing industry regardless of location and ownership, but one which is significantly intensified in Third World export processing plants in the electronics sector. Estimates vary between countries but the sexual stratification is so marked in Third World electronics that women workers often account for over 95 per cent of all unskilled jobs, themselves 80–90 per cent of total employment. The few supervisory, management and technical jobs are generally occupied by men.

A further aspect of working conditions refers to one of the characteristics of the multinational company's target female labour force frequently cited by commentators, that of docility. Many commentators, often quoting management and industrial promotion agencies in the Third World, have pointed to the docility of women operators in the face of boring, tedious and repetitive work. The explanations for why women should be so accepting of these conditions varies from the fact that women are naturally submissive, that domestic work suffers from the same disadvantages, to cultural factors concerning the pattern of behaviour of women in a given society. The rigid sexual hierarchy of production in the factories contributes to the promotion of such a response from the women workers as all authority and responsibility is firmly in the hands of male technical, managerial and supervisory personnel.

One final point about the variations in conditions of employment in the Third World must be made. Much of the literature assumes a single type of production situation and relations of production; that is, of young women recruited from the school leaver cohorts with no industrial experience, working in manufacturing plants organized along conventional factory lines. In reality, as we have seen, there is a range of production relations which vary according to the historically determined situation of women in any given situation. While the majority of Third World women industrial workers are employed as 'free' wage labour, this is not always the case. In Turkey, women weaving carpets for export in village-based workshops do not receive any payment from the subcontractor, who instead pays according to a piece-rate scale to the male head of the household. In Haiti, women employed in American-owned firms making toys and soft goods frequently take work off the plant to their homes to complete; in Puerto Rico, a large proportion of the production of garments for the export market takes place in illegal 'underground' domestic workshops whose output is then marketed by the multinational retail groups in the United States.

What this analysis demonstrates is that the female labour recruited for Third World market factories is not available in a prepackaged form. While it is clear that women workers offer capital labour power which can be low paid and highly productive, both the State and capital need to intervene to release this labour power in the particular form required by concrete production conditions. Nor can it be assumed that this labour is available in unlimited quantities from a given age cohort anywhere in the world; the characteristics of the female labour force – in terms of age, education, marital status, class and ethnic origin – and how their work process is organized will depend as much on the historically determined availability of female labour and the interaction of gender and class systems as on the demand for cheap labour from foreign capital.

MULTINATIONAL COMPANIES AND WOMEN WORKERS IN THE FIRST WORLD

We have argued that much of the analysis of women industrial workers in the Third World rests on the assumption that changes in the international division of labour are responsible for recruiting women for the first time into waged industrial work. It is, therefore, curious to find that the literature discussing the composition of the labour force recruited by multinational companies to work, particularly in the new technology sectors located within the industrial countries such as Britain, uses a parallel set of stereotypes to explain the location decisions of international firms and to describe (or dismiss) the composition of the female labour force. For it is clear that women workers in advanced industrial countries have also been targeted by management to provide, for given production tasks, a category of labour power which can be differentiated from the male labour force and utilized under specific conditions to provide a highly efficient and cost effective labour force.

THE BRITISH EXPERIENCE

As far as British experience is concerned, there is no doubt that firms starting up new production sites expected to recruit women workers for the labour intensive, unskilled or semi-skilled operations involved in the assembly of electronics-based consumer goods and electronic components.

CHANGES IN PRODUCTION TECHNOLOGY AND THE SEXUAL DIVISION OF LABOUR

Unskilled women operators still constitute the largest single category of employment in industry. As we argue below, the changing technical and managerial policies within industry are tending to utilize female unskilled workers in a different way from their use in assembly plants or conventional technology. However, in spite of the changing technical and sexual composition of the labour force, the undifferentiated notion of women's labour as being suitable and available for employment is still part of the accepted analysis of why firms locate production in depressed peripheral regions and why they recruit women's labour. This notion is described as 'green' labour and differs little in terms of the stereotypical assumptions and characteristics it carries from the notion of 'cheap labour' used in the analysis of the female labour force in world market factories in the Third World.

Technical change in the electronics industry in both product and process technology is also forcing a reformulation of the role of female labour. Some firms with conventional assembly plants still require women, as a flexible, efficient, docile and dispensable manual labour force. New automated silicon wafer fabrication plants require different skills and attributes from their female labour force. Rather than required manual dexterity for microscope assembly tasks, women are to become 'machine minders, monitoring the movement of wafer batches through highly complex equipment, reading the computerized performance data output and sounding the alarm if something goes wrong'. Workers will be required to work within a 'clean room environment' where meticulous adherence to antistatic procedures is required (Goldstein, 1984, p. 7).

While such jobs continue to be described as 'unskilled or semi-skilled' they are also crucial to the successful operation of these plants, and it is clear that women are recruited to do them because they can be relied on to adhere to the 'clean room' procedures, to remain within the environment for long stretches of time, and to learn how to monitor the computerized data for problems in the production line. The companies are, therefore, anxious to maintain a stable female workforce, seeing characteristics previously regarded as negative ones, such as docility, willingness to carry out monotonous tasks, as positive attributes of reliability and conscientiousness. Far from being dispensable, in a situation where there is a local shortage of skilled and trained technicians, management is concerned that the women workers who carry out their required jobs satisfactorily, should remain at the same plant because of their crucial role in the production process which cannot be instantly fulfilled by 'green' labour recruited from the reserve army of female school leavers and the unemployed.

CONCLUSION: THE GREENING OF WOMEN'S LABOUR

This chapter has argued that women employed by multinational companies producing consumer goods for the world market do not necessarily have identical characteristics, nor are one group of workers infinitely and instantly substitutable for another. But it also demonstrates that in both the First World and the Third World women are confined to relatively low paid jobs which are classified as unskilled or semi-skilled. This confinement is implemented through a number of complex mechanisms provided by capital through the recruitment preferences and production practices of management, the legislative and political actions (or inactions) of the State, and the ideology of gender roles which provides sex stereotyping of male and female jobs which are enforced by the community in general, the organized male skilled working class, and by women's own perceptions of what constitutes appropriate work for women.

We are, therefore, arguing two different but related propositions. Firstly, that women are sought out by capital for specific roles in new and emerging forms of production, as well as old and declining forms. And that in both old and new production processes, women continue to occupy the bottom layers of the occupational structure, reflecting the way in which women workers, doing women's work, are socially constructed as a subordinate group differentiated from the dominant labour force. But secondly, we have argued that the recruitment of women workers in new industrial situations – either new sectors and processes or parts of the world new to given kinds of industrial processes – does not of itself provide capital with suitable labour power. This labour has to be constituted, taking into account the preexisting sexual division of labour. It is constructed directly by the recruitment, selection, management and personnel policies of individual companies and indirectly by the intervention of the State, and negotiation within local and traditional modes of gender control. Women's labour, in an appropriate form, that is 'cheap' labour, or 'green' labour, does not exist in nature; it has to be directly cultivated, a process we have called the greening of women's labour.

Enrico Pugliese

'Restructuring of the Labour Market and the Role of Third World Migrations in Europe'[1]

***Environment and Planning D: Society and Space* (1993)**

PREMISE

In this paper I intend to point out some of the changes taking place in the structure of the labour market in Europe and to highlight the role played by recent waves of immigration of people from Third World countries. Contrary to the intra-European migrations of the 1950s and 1960s, present international migration flows are also directed towards southern European countries, such as Italy, Spain, and Greece. Not only is the number of final destinations greater than in the past, but also the nationalities and the ethnic groups which constitute present European immigration are many more than in previous decades.

'PUSH' FACTORS FROM THE SOUTH AND THE RESTRICTIVE IMMIGRATION POLICIES IN THE NORTH

Scholars of migrations devote a lot of attention to the question of 'push' and 'pull' effects.

In this section I shall first deal with the causes of expulsion (push) from Third World countries, and then I shall comment on the reasons why the flow which originates in the Third World ends up in Europe and then in a specific country. As far as the push effect is concerned, a little time should be spent on the reasons for its ever greater acceleration. First of all, there are the demographic factors which have been very much talked about. It is well known that the rate of population increase in these countries is such that local resources cannot satisfy the needs of an increasing population.

Increasing poverty is not the only cause strengthening the push effect. There are other causes, including the inability of the local economy to absorb a labour force with high levels of education (which explains the frequency of highly educated immigrants employed in menial jobs in Europe). And there are also cultural factors (which of course cannot be viewed independently from the pull effect) which contribute to the decisions to migrate, particularly in the less backward countries.

The present migratory flow is made up of people coming from almost every Third World country and it is directed to various countries of the First World at various levels of development in accordance with very peculiar trajectories which cannot be explained solely on the basis of categories of the labour market. In

[1] Pugliese, E (1993) 'Restructuring of the labour market and the role of Third World migrations in Europe', *Environment and Planning D: Society and Space* **11** 5: 513–22. Reprinted by permission of Pion Limited, London.

fact, in international migrations some main flows can be singled out that originate from a few main Third World areas and lead to one or more developed countries (Salt, 1989). The income (or wage) differentials are able to explain only the obvious fact that people move from a place with a given wage level to countries where wages are ten to twenty times higher. But they can by no means explain why, for example, migrants from the Philippines go in massive numbers to Italy. There are a series of historical, cultural, religious, and geographic factors which account for a given flow.

In some cases preexisting colonial relations explain why some nationalities and ethnic groups are present in a given country. In other cases geographical proximity is the most important factor. In addition to that, the role of some 'connecting agencies' which create specific trajectories should be considered. This is the case of some catholic organizations as far as the connection between the Phillipines and Italy is concerned.

Finally, an important reason which explains the presence of immigrants in some countries is the simple fact that they have not been allowed into others. This is certainly the case for a part of the first wave of immigration towards Italy. The migratory flow to other, richer, European destinations has been diverted to Italy because of the strict immigration policies of the other destinations. This introduces a further element of explanation for the relationships between pull and push factors: migratory policies. As is known, present-day Third World migrations are taking place in a context of restrictions. Restrictive policies are an old tradition in countries such as the USA, but they are relatively new for European countries. This is not to say that frontiers in Europe have always been open, but it is now extremely difficult – if not impossible – for people from the Third World to enter legally another country for work purposes.

The restrictive policies have a double effect: on the one hand they actually reduce the number of people who enter the country; on the other hand they tend to modify the character of the immigrant population and extend significantly the percentage of illegal workers.

In all cases a large quota of Third World immigrants, both to countries of traditional immigration (such as Germany, or Switzerland, or France) and to new immigration countries (such as Italy, or Spain, or Greece), do not have regular status, they are generally alegal or illegal. This is the result of the coexistence of a strong push effect from Third World countries and a restrictive immigration policy in the developed countries. There is no reason, therefore, to be surprised if the majority of these new immigrants are located in the informal sector or if they also have an irregular work situation when they work in the formal sector. We are now experiencing a paradoxical situation with an acceleration of the processes of internationalization of the labour market, and, at the same time, a tendency to restrict the mobility of the labour force, at least as far as the relations between the First World and the Third World are concerned.

IMMIGRATION IN INDUSTRIAL SOCIETY

There are many differences between present-day migratory flows and the intra-European migrations of the golden age of the welfare state. The most important aspect lies in the occupational destination of the immigrants, which in turn reflects the prevailing occupational structure in society. The migratory experience of the 1950s and 1960s has been extensively illustrated in studies which are now classic (Boehning, 1984; Castles and Kosack, 1973). A majority of the immigrants entered the working classes of the host countries and became part of them, although in different ways and with many contradictions.

As far as these migrations are concerned, industrial development in the core countries has been the motor of the labour demand and of the migratory inflows. The labour demand in the manufacturing sector (besides mining and the construction industry) was the factor activating population movements. This does not mean of course that all immigrants everywhere entered industrial employment. Italian ice cream parlours as well as Greek restaurants have a rather long history in France and Great Britain. But for the majority of intra-European migrants, industrial employment was certainly the main destination.

Of course there were differences in the specific conditions of immigrants in the various countries. Nevertheless, some common features in this industrial migratory flow could be singled out. In Europe it was a migration from poorer rural regions to richer industrial regions. It changed the occupational and the class position of millions of people; migrants, who had been farm-workers, peasants, or artisans, became part of the working class of the host countries. Immigrations have been a powerful permissive factor for the development of industrial capitalism in these countries (Castles and Kosack, 1973), but at the same time they have caused an enlargement and an internal modification of the working class. People of different nationalities, different cultures, and different habits were attracted to the urban-industrial areas. And at times, of course, divisions within the working class as a result of ethnic conflicts took place. But these differences – and the consequent divisions and contradictions – were, in a certain way, counteracted by a homogenizing element: that of having a similar position in the industrial relations of production. The Taylorist production model in industry fostered this characteristic. The mass worker, the international mass worker, became an important social actor in the industrial conflict in some countries. One example is the Turkish and Italian workers in the car industry in Germany in the 1970s and their active role in industrial conflict.

Of course this was only one dimension of the complex human and social experience of migration. Other aspects of the phenomenon led to different and contrasting results. First, only some of the immigrants in the industrial capitalist countries considered immigration as a definitive experience leading to a definitive settlement in the new country. The immigration policy of the various host countries influenced the temporary or definitive character of immigration. For example, the fact that Germany has always defined itself as 'a nonimmigration country' has not been irrelevant. It is not by chance that the number of Italian workers who had a migratory experience in Germany is several times higher than the number who chose (and were able) to remain in Germany. In this specific case the aspiration of the Italian workers to spend a period of time abroad just in order to accumulate some savings – an aspiration very common to first-generation migrants in general – has been functional in the tendency of the German immigration policy to favour return migrations. In other countries, at the time of the industrial expansion, the immigration policy was less restrictive and return migrations were less encouraged. In any case a noticeable section of the total migrants settled definitively in the host countries [and] a rather stable employment situation, and consequently a territorial stability, characterized the intra-European migrations of the postwar industrial development in the 'golden age' of the welfare state.

Things are quite different now. Even a superficial glance at present-day migration to Europe shows a noticeable change in the occupational destination of the majority of migrants, with the prevailing role of tertiary employment (commerce and services in general). As we shall see, other aspects of the work conditions are also different. Jobs are, in general, less stable and are short term. The lower work stability corresponds to a lower housing stability (particularly in a country of new immigration).

These new features and this new occupational composition of the migratory flows can be observed both in Europe and in the USA. And this is easily understandable if one keeps in mind the dramatic changes in the economy and employment structure of the advanced countries, corresponding to the decline of the Fordist–Taylorist age. It is impossible to understand migrations today, including the increasing difficulties in the process of integration of the immigrants, if the new trends in the occupational structure and in the labour market are not kept in mind.

CHANGES IN THE LABOUR DEMAND AND THE POSTINDUSTRIAL MIGRATIONS

The crisis of the Fordist model of production, with its consequences on the social and occupational structure, has been a widely debated issue in the last decade. In a schematic way the most relevant trends in the employment structure, which also cause the new occupational and social destinations of Third World migrants are:

- For the first time in the last decade, industrial employment has stopped growing in absolute terms and, in many countries, a decrease, both in absolute and in relative terms, has taken place.
- The renewed increase in the self-employed population (or at least a stop in its decline) within the private sector, not only in manufacturing but also in the tertiary sector.
- The decrease in the amount of regular, steady, year-round employment. Precarious employment tends to characterize many of the new jobs in the industrial and service sectors. Casualization of the labour force is one of the most relevant trends in the labour market.
- Casualization leads to a [further] aspect consisting of the increase in unemployment and its changing character in most advanced societies. Long-term unemployment affects the lives of people more now than in the past, and temporary unemployment, a characteristic feature of the secondary labour force, is becoming increasingly frequent.
- More generally, we are now witnessing a decline in the numbers of protected primary-market jobs.

This process of the enlargement of the secondary labour market can also be understood as – and actually it corresponds to – the process of development of what is generally referred to as the informal economy (Mingione and Redclift, 1986). More specifically, Mingione points out that three processes, strictly interwoven, cause changes in the labour demand, with strong implications for the demand for an immigrant labour force. These processes are: technological development, tertiarization of the occupational structure, and development of the informal economy. The first process aggravates the dualism between those sectors where technological innovations are possible – and therefore where productivity may increase so that higher labour costs (and labour rigidity) can be afforded – and those sectors where the degree of technological innovation is more modest. Needless to say, together with the industrial decline and the tertiarization of the economy, the less innovative sectors (including state-provided services) tend to expand. This expansion requires a cheaper and less demanding labour force; the process of informalization tends to satisfy this request. At the same time international migrations, and in particular from Third World countries, provide, on a massive scale, a labour force ready to accept low productivity and low-paid jobs in the informal sector or, more generally, in the secondary labour market. The illegal status of a sizeable part of the migrants favours their 'ghettoization' in the informal sector and, anyway, in the secondary labour market.

While analysing migrations in the age of industrial development I have mentioned only intra-European migratory flows. But immigration from southern Europe was only a component (although the predominant and almost exclusive one in some countries) of the postwar migratory flows. In the same period in Great Britain, for example, citizens from the colonies (or, better, from the Commonwealth countries) almost completely made up the migratory inflow. Gordon (1995) makes a rich and articulate survey of the prevailing aspects of European international migration, pointing out the similarities and differences between the various countries both at the time of the industrial development and today. He provides evidence on the declining employment of foreign and minority workers in manufacturing and industrial jobs, following the trend affecting the national labour force. The overrepresentation of immigrant workers in the secondary labour market is stronger now. In the past, immigrants in some countries also entered primary labour-market jobs (the ones requiring lower skills) in core industries. But the situation is now different because 'the overall balance of employment has shifted, with many of the primary jobs in heavy industry being casualties of recession and industrial restructuring' (Gordon, 1995).

The situation for these 'postindustrial' migrants is made more difficult by the changing and, in fact, more restrictive immigration legislation. 'Uncertainties about rights to continued residence or employment weaken the bargaining power of those involved, and this increases chances of exploitation, poor wages, working conditions and so on – and also diminish[es] the incentives to make investments in specific human or cultural capital' (Gordon, 1995).

CONCLUSIONS

These last aspects singled out by Gordon are very important for understanding the situations of the new immigrants coming from Third World countries. [In the past,] the growth of industrial employment gave occupational chances to immigrants who took the jobs made available by the social mobility of local workers. Although these went up the occupational ladder the newcomers took the less skilled and less preferred jobs. Occupational stability was in any case a general condition or at least a realistic aspiration.

This is not true any more for the local and, still more, for the immigrant workers. On the basis of the arguments used so far the basic differences between the two situations should be rather clear. The main differences, however, do not concern only the labour market and the labour demand. Two additional elements which characterize present-day immigration in the old and the new immigration countries are: (a) the increasing number of national and ethnic groups which make up the present flow; and (b) the much stricter immigration policies, at least as far as European countries are concerned.

Starting from the Anwerbwenstop in Germany in 1973, and with a generalization of the restriction during the 1980s, European frontiers can be considered completely closed for non-European (and more so for Third World) potential immigrants. The overrestrictive immigration policies justify the expression 'Fortress Europa', which more than reflects a condition of siege of the European countries by the overwhelming masses of Third World aspiring migrants. Both the migratory policies and the attitude of local populations seem to be very much conditioned by this 'siege syndrome'. It is not surprising that some political groups have tried to exploit the concern and anxiety of local populations by organizing xenophobic campaigns and asking for ever stricter immigration policies.

As an effect of that, migratory policies are becoming more restrictive and less benevolent. The progressive views and policies concerning social policies for immigrants have been contradicted by the other principle, which stops any new entrants. And, as is known, one of the main effects of these measures is the increase in illegal immigration, or, more precisely, in the presence of immigrants of illegal status.

The process of integration in the host society becomes particularly difficult for this specific (and growing) component of Third World migration. Their restriction to the informal economy or to the secondary labour force is strengthened by their legal situation. Their difficult situation in the labour market does not lead generally to unemployment, on the contrary it leads to marginal employment and acceptance of worsening work and pay conditions, and to social marginality. The informal work relations render their organization in the work place more difficult. As a general consequence traditional forms of horizontal solidarity, such as the ones expressed by the labour unions, lose their role. At the same time, other forms of solidarity – based more on ethnic or community ties – develop. Migrants look for new forms of organizations which are based less on their identity as workers. Also, the national, religious, and ethnic organizations become more important as a reaction against discriminations and anti-immigration movements. This fosters the specific cultural identity of the various groups of immigrants, but it is more likely to contribute to self-defence than to class solidarity.

Andrew Herod

'From a Geography of Labor to a Labor Geography: Labor's Spatial Fix and the Geography of Capitalism'

***Antipode* (1997)**

The notion of 'spatial praxis' is firmly on the intellectual agenda in human geography. The making of the economic and social landscape in particular ways is now recognized as being fundamental to the articulation of political power (cf. Harvey, 1982; Soja, 1989; Lefebvre, 1991). In this paper I argue, however, that whilst they recognize landscapes are socially constructed, many economic geographers and theorizers of the geography of the capitalist space-economy – both mainstream and Marxist – have tended either to ignore the role of *workers* in making the economic geography of capitalism or have frequently conceived of them in a passive manner. Although during the past two decades economic geographers have generated a considerable literature which seeks to understand how *capital* attempts to make the geography of capitalism in particular ways to facilitate accumulation and the reproduction of capitalist social relations, there has been much less work which examines and attempts to theorize explicitly how *workers* actively shape economic landscapes and uneven development. Labor's role in making the economic geography of capitalism has been rendered largely invisible by the analyses both of traditional main-stream neoclassical economic geographers and also, ironically, by many Marxists, for both approaches primarily present economic geographies devoid of workers as active geographical agents. In their explanations of the dynamics of the capitalist space-economy both neoclassical and Marxist approaches in geography have conceived of workers primarily from the viewpoint of how capital (in the form of transnational corporations, the firm, etc.) and, to a lesser degree, the state make investment decisions based on differences between workers located in particular places. Whereas for mainstream scholars it is the relative importance of various factors of production and consumption which determines the location of economic activity or perhaps the structure of the firm which is significant, for Marxists it is capital which acts, capital which produces landscapes in its continual search for profit. In both such views capitalists are theorized as capable of actively making economic geographies through their investment decisions whereas workers are seen rather passively either as inert 'factors' in the calculus of location or, following Harvey (1982, pp. 380–1, original emphasis), as little more than '*variable capital*, an aspect of capital itself'.

Traditionally, explanations of the genesis of economic geographies have primarily relied on understanding the decision-making processes of managers and capitalists. Although workers' activities are sometimes seen as a 'modifying' force that needs to be factored (and I use this terminology quite deliberately) into the locational equation, such an approach essentially places capital center stage both empirically and theoretically as the focus of research, whilst banishing workers to the fringes of the discipline. In essence, it tells the story of the making of the geography of capitalism through the eyes of capital(ists). In this paper, in contrast, I suggest that much insight on how the economic geography of capitalism is produced can be gained by greater analysis of the social and spatial practices of workers. Thus, whilst not rejecting the insights into the production of the geography of capitalism that can be gained by understanding the actions of capital, I would argue that at the same time it is important to recognize that workers, too, are *active* geographical agents whose activities can shape economic landscapes in ways that differ significantly from those of capital. Hence, understanding how workers actively shape economic space is important if we are to conceptualize how the geography of capitalism is made.

I wish to assert that workers have a vested interest in attempting to make space in certain ways. Workers' abilities to produce and manipulate geographic space in particular ways is a potent form of social power. Recognizing this fact raises important questions about the theoretical status of spatial relations in workers' own self-reproduction and the issue of workers' 'spatial praxis'. Whilst capital's efforts to create landscapes in particular ways have been theorized as an integral part of its self-reproduction and survival, in this paper I argue that workers, too, seek to make space in particular ways to ensure their own self-reproduction and, ultimately, survival – even if this is self-reproduction and survival *as workers in a capitalist society*. The economic geography of capitalism does not simplv evolve around workers who themselves are disconnected from the process. They are active participants in its very creation.

GEOGRAPHIES OF LABOR

Neoclassical Location Theory and Labor

Neoclassical economic geography is fundamentally about how *firms* make locational decisions *– it is firms' behavior and investment decisions which are both the activities to be explained but also the activities which define economic geographies.* For sure, workers are not always totally ignored. However, the point is that location theory (which for so long has been taken to represent the entire field of economic geography) does not conceive of them as sentient geographical actors for whom the making of the economic landscape in ways which further their own economic and social agendas is integral to their ability to reproduce themselves as workers on a daily and generational basis.

The neoclassical approach has at least two important consequences, then, when considering the conceptual marginalization of the geographic power of working class people in the literature. First, clearly, the point of view presented is that of capital. Thus, as Massey (1973, p. 34) has argued, for traditional location theory 'profit is the criterion, wages are simply labour costs'. Second, it presents an economic geography devoid of workers, both as individuals and as members of social groups. There are no workers in neoclassical explanations of the production of economic landscapes, merely crude abstractions in which labor is reduced to the categories of wages, skill levels, location, gender, union membership and the like, the relative importance of which is weighed by firms in their locational decision-making.

Marxist Economic Geography and Labor

The influx of Marxist thought in the 1970s transformed the way economic geography had traditionally been done. In particular, it sparked a welter of theorizing about the connections between the uneven development of space and the broad forces of capitalist accumulation [see, for example, Harvey, 1982; Lefebvre, 1991; Massey, 1995; Smith, 1990; Soja, 1989]. These works are truly pathbreaking. Yet, they are also somewhat problematic in the way in which they conceive of and/or marginalize the roles of workers in shaping the economic geography of capitalism. Certainly, all argue for the importance of class struggle in capitalist society. However, in terms of their actual analyses of the making of economic geographies and their practice of economic geography, these approaches pay rather scant attention to how workers' activities can directly and significantly shape the geography of capitalism and focus, instead, primarily on the geographical structure of capital and how capital structures landscapes through its own activities (such as the pursuit of profit).

For example, in *The Limits to Capital*, Harvey (1982, p. 380) has tended to conceive of workers' roles as shapers of the economic landscape only in rather limited terms of how the migration of labor affects the accumulation process. Elsewhere, in viewing class struggle in terms of 'the resistance which the working class offers to the violence which the capitalist form of accumulation inevitably inflicts upon it' Harvey's (1978, p. 124, emphasis added) epistemological priority allows labor to resist capital but apparently not to take the initiative in class struggle. Likewise, his comments that 'capital represents itself in the form of a physical landscape *created in its own image*' and that '*capital* builds a physical landscape appropriate to its own condition at a particular moment in time' (p. 124, emphasis added) highlight the extent to which he sees the geography of capitalism as largely the product of capital itself. Such statements leave little theoretical room to acknowledge that workers struggle (often successfully) to shape the economic geography of capitalism in ways which they themselves view as advantageous as part of their own practices of self-reproduction.

In short, in such a view workers are not theorized as being present at the making of the economic geography of capitalism but, instead, are seen to struggle and live within the contours of an economic and social geography created by and for capital. While capital can fashion the geography of capitalism to suit its own needs, there is little sense that workers may also do the same.

In what follows, then, I present an argument in which I attempt to build on these highly influential works so as to expand our understanding of the production of capitalism's geography. In addition to theorizing how capital produces space as part of its spatial fix, it is important to more fully theorize labor as attempting to make its own spatial fixes and to show how these, too, shape the geography of capitalism.

TOWARDS A LABOR GEOGRAPHY

The production of space in particular ways is not only important for capital's ability to survive by enabling accumulation and the reproduction of capital itself (*pace* Lefebvre and Harvey), but it is also crucial for workers' abilities to survive and reproduce themselves; just as capital does not exist in an aspatial world, neither does labor. The process of labor's self-reproduction must take place in particular geographical locations. Given this fact, it becomes clear that workers are likely to want to shape the economic landscape in ways that facilitate this self-reproduction. Struggles over the location of work, new or continued investment (public or private), access to housing and transport, all can play significant roles in allowing working class people to reproduce themselves on a daily and generational basis. Recognizing that workers may see their own self-reproduction as integrally tied to ensuring the economic landscape is made in certain ways and not in others (as a landscape of employment rather than of unemployment, for instance), allows them to be incorporated into analyses of the location of economic activities in a theoretically much more active manner than traditionally has been the case.

For example, whereas workers' involvement in local boosterist campaigns is often portrayed simply as a bad dose of false consciousness in which they help to sustain local capitals, in fact most workers would probably see retaining and/or attracting investment to their particular localities as integral to their ability to sustain their own livelihoods. Consequently, they may participate vigorously in such campaigns, not as cultural or class dupes but as active economic and geographical agents. Likewise, just as capital may find itself constrained by the structure of landscapes created at previous periods of accumulation, so may workers find that the landscapes which facilitated their social and biological reproduction in earlier times are no longer appropriate for doing so. Thus, as the structure of the family changes or as workers may have to commute greater distances to work, they may struggle for new urban designs which include centers for communally-run childcare, differently configured homes to accommodate new living arrangements (single-parenthood, for instance), new highways, and so forth – a point recognized by a number of feminist writers who have examined how women directly shape economic restructuring and the form of the built environment through their activities. All of these activities are significant for shaping the uneven development of capitalism in ways not controlled by capital itself.

Suggesting that workers have a vested interest in making the geography of capitalism in some ways and not in others allows us to say four interrelated things theoretically. First, it suggests that even if workers' struggles are less than revolutionary and even if they are still bound within the confines of a capitalist economic system, the production of the geography of capitalism is not always the prerogative of capital. Understanding only how capital is structured and operates is not sufficient to understand the making of the geography of capitalism. For sure, this does not mean that labor is free to construct landscapes as it pleases, for its agency is restricted just as is capital's – by history, by geography, by structures which it cannot control, and by the actions of its opponents. But, it does mean that a more active conception of labor and workers' geographical agency must be incorporated into explanations of the making of economic landscapes. Capital is not the only actor actively shaping the geography of capitalism *or even, in some places and times, the most significant one*, and labor is not simply a 'factor' of location in the sense in which it is so often conceived.

Second, such a conceptualization allows us to begin thinking about how the social actions of workers relate to their desire to implement in the physical landscape their own spatial visions of a geography of capitalism which is enabling of their own self-reproduction and social survival. Following Harvey's (1982) argument about how capital seeks to make a 'spatial fix' appropriate to its condition and needs at particular times in particular locations, it is also necessary to see workers' activities in terms of their desire to create particular spatial fixes appropriate to their own condition and needs at particular times in particular locations – 'labor's spatial fix'. Likewise, any examination of how workers seek to impose their spatial visions on the landscape must recognize that 'labor' is not an undifferentiated category and that different and competing groups of workers may in fact have vested interests in generating quite different spatial fixes – whereas group A may seek to keep employment in community/country A, group B may try to encourage capital flight to community/country B.

Third, conceiving of how labor might seek particular spatial fixes at particular historical junctures – fixes which are sometimes coincident with, but frequently different from, those favored by capital – allows a much less mechanistic and more deeply political theorization of the contested nature of the production of space under capitalism for, ultimately, it is the conflicts over whose spatial fix (capitalists' or workers') is actually set in the landscape that are at the heart of the dynamism of the geography of capitalism. This means that understanding processes of class formation and inter- and intra-class relations is fundamentally a geographical project. Workers often succeed in constructing landscapes in certain ways which augment their own social power and undercut that of capital. Even when they are defeated in this goal, the very fact of their social and geographical existence and struggle means they shape the process of producing space in ways not fully controlled by capital.

Fourth, I would argue that working class people, too, play their part in the generation of [spatial] scales and, hence, in making the unevenly developed geography of capitalism. Workers' struggles are frequently about constructing the very geographical scales at which capitalism itself operates. For example, whereas the extent of the urban scale is often defined in terms of Travel to Work Areas, workers' choices regarding where to live and their ability to win wage increases and shortened work hours (thereby allowing them to travel greater distances to work) clearly play an active role in determining the size and functional integrity of such TTWA's and, hence, of the urban scale. Similarly, unionized workers' abilities to create regional or even national contracts as a means to equalize conditions across space not only represent the creation of real geographical scales of bargaining but can also dramatically impact patterns of economic development and the location of work by preventing employers from whipsawing plants in different localities against each other. In turn, this directly affects the geographic development of these industries as it may prevent capital from leaving one region and migrating to another with lower wages or less restrictive work rules. In this sense, workers can be seen to be playing active roles in the creation of the economic geographies of entire industries.

Workers and the Production of the Spaces and Scales of Capitalism

Here I want to give a (very) brief outline of instances of workers playing significant roles in making the geography of capitalism, both in co-operation with, and in opposition to, certain segments of capital.

Workers and globalization

Typically, the globalization of economic and political relations has been presented in the geographic literature as the project of capital. For capital 'going global' is as much a geographical project as it is a social one. However, the history (and geography) of international labor migrations and labor internationalism suggest that this is only part of the story of globalization and that, in fact, working class people have also played very active roles in this process. The intervention of the US labor movement in Latin America during the past century or so is especially illustrative of the active role played by workers in the process of globalization.

The US labor movement has a long history of operating internationally. It has been particularly active in attempting to implement its own '"workers' Monroe Doctrine"' in Latin America, both as part of a deeply-held conviction about the US's 'civilizing' mission in this part of the world but also as a means to bring the region under the economic influence of the United States. With US manufacturers until recently largely located within the confines of the United States, many US workers have historically seen their own ability to enjoy relatively high living standards as integrally tied to US manufacturers' success in carving out new markets in the region. Such success would, workers believed, stimulate production and employment in the United States. Indeed, many workers have shared the view of John L. Lewis, president of the United Mine Workers of America, who, at a 1939 Labor Day address, argued that 'Central and South America are capable of absorbing all of our excess and surplus commodities' (quoted in Scott, 1978, p. 201). For Lewis, expansion into Latin America was one way of creating a spatial fix which would maintain US workers' livelihoods by exporting crises of underconsumption abroad.

Working to maintain access to Latin American markets has been a key element in the US labor movement's efforts to create an international spatial fix for much of this century. Consequently, many US unions have actively participated in destroying militant anti-US and anti-capitalist trade unions and political organizations in Latin America and encouraging the growth of more US-friendly bodies as a way to open up the region's markets to US capital. Developing an international spatial fix to problems of underconsumption in the US has been an important part of US labor's ability to ensure its continued self-reproduction, even if it has been at the expense of limiting the ability of workers elsewhere to do likewise. In seeking to create such a spatial fix not only has the US labor movement served as an agent of the globalization of economic and political relations, but it has also played an important part in the continued underdevelopment of the region (by helping to destroy/undermine indigenous manufacturing which threatened US competitiveness and market shares, for example) and, hence, in making the unevenly developed geography of capitalism in the Western hemisphere.

In this sense, the practice of international labor solidarity can be regarded as an effort by particular groups of workers to develop spatial fixes and to organize social relations between workers in different countries in such a way as to shape the manner in which the global space-economy is made. Building networks of solidarity is precisely about overcoming geographical and social barriers to cooperation between workers which, in the process, affect how the economic geography of capitalism evolves. At least since the nineteenth century international labor organizations have worked to develop transnational links between workers. Without question, this has not always been either a smooth or a successful project, and labor internationalism has frequently been held hostage to various nationalistic and/or ideological rivalries. Nevertheless, these activities have had very real impacts on the geography of the global economy through, for example, restricting the ability of corporations to play workers in different places against each other and shaping the investment opportunities available to corporations.

Workers and the making of one industry's economic geography

Prior to the 1950s, dockers and employers in the US East Coast longshore industry bargained for their labor contracts on a port-by-port basis. The introduction of containerization in the industry, however, unleashed powerful geographical forces which served as a catalyst to transform the political and economic geography of the industry. For dockers the growing disintegration of regional hinterlands brought about by faster waterfront handling and overland transportation of cargo raised the possibility that in some ports wages might be undercut and strikes broken as shippers were increasingly able to use other, more distant, ports to serve traditional markets. This was particularly worrisome to New York dockers who had the highest wages in the industry and who were the first to face the threat of job loss, the New York–Puerto Rico trade having been the first sea-route to be containerized. To address this problem the dockers' union (the International Longshoremen's Association) adopted a specifically geographical strategy aimed at producing a new spatial fix in the industry, one which has dramatically changed its economic geography. Two elements formed the basis of this fix.

The first element of the spatial fix pursued by the union and its members involved implementing a series of work preservation rules designed to retain certain types of cargo handling work at the piers. Whereas traditionally all cargo had been handled loose at the waterfront, with the introduction of the new technology now only the containers themselves had to be handled at the piers and the much more labor intensive work of packing and unpacking their contents (i.e. the actual freight) could be done at cheaper locations inland.

After several lengthy strikes, dockers forced steamship operators throughout the East Coast industry to agree to prohibit the conduct of certain types of container work at inland warehouses, to close various off-pier warehouses where this work was now being done, and to transfer such work to the waterfront. Although they faced substantial opposition from steamship operators, trucking companies, and even the Teamsters union which represented workers employed in such off-pier warehouses, by consciously manipulating the geography of work in the industry in this manner dockers were able to sustain pier jobs and thus their own livelihoods. In effect, by imposing their own spatial fix on the industry dockers were able to retain at the waterfront some of the work that otherwise would have migrated inland. Consciously reshaping the geography of employment in the industry was a key element in dockers' post-containerization strategy to ensure their own livelihoods were maintained. Their ability to make space in a way that benefited them was significant in shaping the evolution of the industry's economic geography during this period.

The second part of the spatial fix adopted concerned the union's ability to construct a new geographic scale of bargaining in the industry. Rank and file dockers and their leaders quickly realized that any work preservation agreements implemented solely in New York were doomed to failure because steamship operators could easily avoid the agreements' provisions by shipping instead through other ports such as Philadelphia, Boston, or even some of the southern ports, and transporting cargo overland by truck or rail to serve the New York market. Consequently, they determined that any viable strategy would have to be implemented in all ports from Maine to Texas where the union represented dockers. Likewise, many southern dockers increasingly favored a coastwide contract which adopted the provisions negotiated in New York, since this would bring them higher wages and better working conditions. Beginning in the mid-1950s the dockers' union fought to replace the system of local bargaining with one which was coastwide in scope. In effect, they attempted to reconstruct the very geographic scale at which negotiations were carried out and agreements implemented. Employers, on the other hand, opposed such an expansion of the scale of bargaining for fear it would reduce their flexibility and force them to match the higher wages paid in New York. For twenty years the union fought to impose a new geography of bargaining on the industry by developing a master contract that would equalize wages and many work conditions throughout East Coast ports. Certainly, this was not an easy process and many times the dockers' strikes were defeated. Nevertheless, through their struggles waterfront workers were eventually able to implement their own spatial and scalar fix on the industry. This they did in a number of distinct (geographical) stages.

The union's first success came in 1957 when it forced North Atlantic employers to adopt a regional master contract covering the ports from Maine to Virginia. In addition, although they were not legally bound to adopt the North Atlantic master contract, operators in the South Atlantic and Gulf ports were also increasingly pressured by dockers there to in fact do so, since for the union this would result in a system of pattern bargaining which stretched throughout the East Coast. Indeed, the union's success in imposing such a master contract and pattern bargaining system on the industry forced the employers themselves to restructure their own organization and in 1970 to form a multi-port bargaining association that covered the North Atlantic ports. Dockers' success in forcing the North Atlantic employers to develop a new bargaining association that would negotiate for the entire region was a significant achievement in this regard. Likewise, the formation of similar multiport employer bargaining associations in the South Atlantic and Gulf regions at this time were further evidence of the union's ability to force the employers to restructure their own organization so as to be able to deal with the union's efforts to remake the geography of bargaining in the industry.

[Following this success] dockers continued to push for a single legally enforceable coastwide agreement that would protect them from the consequences of containerization. This they finally gained in 1977 when they pressured employers into accepting in 34 ports from Maine to Texas a Job Security Program which was linked to a coastwide master contract. For the first time in their history, dockers from Maine to Texas were covered by a single agreement that was legally enforceable and that would ensure all were protected equally. The JSP represented the crowning moment in the union's campaign to remake the geography of bargaining in the industry from a system of locally negotiated and implemented contracts to one in which agreements were made at the scale of the entire coast.

[These examples are] important because they force us to consider seriously how workers and their organizations struggle to impose particular spatial fixes on the economic landscape and how these struggles in turn shape the geography of capitalism. For sure, US capital has had a historical interest in expanding into Latin America, but opportunities for investment abroad by US corporations were created, in part, by the very actions of workers and trade union officials. Workers' activities were important in structuring the choices open to capital and need to be accounted for in any understanding of the geography of underdevelopment in the region and the creation of a globalized economy.

Likewise, although the development of a coastwide agreement in the East Coast longshore industry may in fact have benefited some high cost operators (such as those in New York), it is important to bear in mind that the impetus for creating an industrywide agreement came from dockers in the face of ardent employer opposition. Understanding the evolution of the economic geography of the industry requires understanding how dockers successfully imposed their own geographical vision on the employers. Seeking to explain the industry's economic geography by focusing only on how capital seeks to develop spatial fixes as part of its strategy for survival and treating labor as somewhat secondary to the whole process of actively structuring the industrial landscape does not provide either a complete or a satisfactory explanation.

Linda McDowell

'Life Without Father and Ford: The New Gender Order of Post-Fordism'

***Transactions of the Institute of British Geographers* (1991)**

INTRODUCTION

In this paper I want to address the ways in which gender divisions have been included (or not) in the variants of post-Fordist theory, addressing in particular the links between economic change and the restructuring of welfare provision. Thus the focus is on women's work in two spheres: their increasing participation in wage labour and their work in the home and in the community – the work of reproduction.

Two of the most influential variants of post-Fordist theory will be examined – the regulation school, and the flexible specialization approach. Both of these bodies of work rely on similar explicit or implicit assumptions about women's secondary position in the labour market, neglect questions about the gendering of skills and ignore questions about the changing value of so-called masculine attributes in the labour market. These are questions to which a developing feminist literature on the gender composition of waged work is directing our attention. It is becoming clear that women's labour and the conditions under which they enter the labour market as bearers of specifically 'feminine' attributes is a central element of current restructuring (Jenson, 1989; McDowell, 1991; Massey, 1984; Murray, 1987; Pollert, 1988; Walby, 1989). However, I want to further argue that most of this work, reliant as it is on the notion of a patriarchal-capitalism or on the concept of a dual system (of patriarchy and capitalism) in which women's interests are theorized as being in opposition to those of men and of capital, is itself an inadequate reading of contemporary patterns of restructuring. I shall suggest that a new gender order is emerging in Britain with profound implications for the politics of economic change. In the shift from the so-called Fordist mass production regime to the new flexibility of the post-Fordist era, it seems that gender is being used to divide women's and men's interests in the labour market in such a way that *both* sexes – at least among the majority of the population – are losing out. This marks a break from the Fordist period when it is more plausible to argue that both capital and men in general benefited from the gender division of labour. In this new period, the benefits to capital of women's particular marginal and segmented position in the labour market that developed throughout the post-war period as women's participation rates rose significantly, are enhanced. In this sense there are marked continuities in women's position in both Fordist and post-Fordist times. However, what is different is that the old compact between male workers, industrial capital and the institutions of welfare Keynesianism has broken down. The new order is based on a deepening contradiction between economic and social restructuring, between the spheres of production and reproduction in both of which women's work plays an increasingly central part. While not denying that wage labour may bring (limited) advantages to many women, especially the possibility of their increased economic independence, I shall also argue that feminist critics who believe that current economic changes are beneficial for women are, like many of the post-Fordists, overly optimistic about the consequences of economic restructuring. Long-held beliefs that women's entry into waged labour has emancipatory potential may have to be re-evaluated, at least until current labour market conditions are challenged.

FORDISM, POST-FORDISM AND THE GENDER DIVISION OF LABOUR

Three variants of post-Fordist economic restructuring theory may be distinguished: the French regulationist school, the US flexible specialization school and the British flexible firm approach. Although each of the three approaches has similarities in their emphasis on mass production and mass consumption in the Fordist era and on flexibility in both production methods and the use of labour in post-Fordism, they differ in the extent to which broader social relations outside the firm, the labour market and the economy are part of their analysis. Whereas the US and British variants of flexible specialization tend to focus primarily on the firm and the internal structure of labour markets – the French regulationist school also includes greater consideration of the role of the state in the regulation of the economy and in the conditions of reproduction of labour power. Thus it is this approach that has the most to say about the nature of gender relations and the role of women in the two periods. Hence, whereas all three schools make a gesture towards women's growing significance as 'marginal' workers in the economy, only the regulation school connects changes in the spheres of production and reproduction, changes in the labour market with changes in household and family forms, although even here the latter changes are seen as a consequence of economic change and the capitalist imperative, rather than mutually constituted changes.

The Regulationist School

The French school, based on the work by Aglietta (1979, 1982) and developed by, among others, Lipietz (1986, 1987), distinguishes a number of different regimes of accumulation which are characterized by a particular labour process and by different degrees of state intervention into the regulation of the economy and the reproduction of labour power, family life and consumption (Jessop's modes of regulation and socialization). Fordism, the characteristic regime of 'intensive accumulation', is characterized by mass production, reduced working hours, relatively high wages (at least for the labour aristocracy), mass consumption based on the 'family' wage of the male breadwinner and the commodification of social life. Mass housing, the car and other standardized consumption goods are the characteristic commodities of this regime. Fordism was an era of economic growth in which high aggregate spending on consumption depended on rising real incomes, underpinned by increased public expenditure. In Britain, as in other advanced industrial economies, women were drawn in increasing numbers into tight labour markets, in part to meet the costs of the mass consumption life style. Their entry was facilitated by the provision of a range of social welfare services that, to a degree, mitigated women's family responsibilities. These included state income support, care for the elderly and other dependants and limited childcare provision. These services themselves created significant employment opportunities for women as they relied on a predominantly female labour force, and the expansion of educational and training opportunities provided women with skills and training that also improved their labour market position.

Women's entry into wage labour in this period was not, however, on equal terms with men. Even in the newly expanding public sector, women were concentrated at the bottom of the occupational hierarchy, trapped in the ghettoes of 'female' jobs where caring and servicing were seen as desirable but poorly rewarded attributes. In the manufacturing sector women also were concentrated in less skilled and low paid jobs. In Britain, in particular, part-time employment with less security and fewer occupational rights and benefits was a key strategy in expanding female employment. Thus traditional ideas about gender roles combined with labour market regulation created a labour force that was highly segmented by gender (Beechey, 1987; Dex, 1985; Rubery, 1988, Walby, 1986).

Aglietta, however, pays little attention to the role of gender in the segmentation of labour markets in the Fordist regime. Rather, he sees women as virtually irrelevant to Fordism as he argues that, despite their growing participation in the labour market, women's primary role is in the sphere of privatized consumption in the home. Thus their labour market participation is secondary, determined by their role in the nuclear family. Their wages are a supplement to the household wage and as a group, under Fordism, women are characterized as a reserve army of labour, drawn into and expelled from the labour market according to the requirements of the capitalist production process. The notion that women are a reserve is central not only to the regulationists' analysis of Fordism but is transferred wholesale into the characterization of the post-Fordist regime. And, as will be shown later, the same assumption pervades other post-Fordist analyses.

Post-Fordism, or rather neo-Fordism in Aglietta's work, is a response to the crisis of mass production and consumption in the Fordist regime. This (re)emerging flexibility is regarded by Aglietta as a form of deskilling as well as multi-skilling and neo-Fordism also sees the further extension of Taylorism in, for example, low grade clerical and service work and its intensification as in sweated manufacturing and assembly work, in the home and in workshops – all areas of 'women's work' *par excellence*.

Aglietta, however, does not pursue questions of the gendering of skills but rather places greater emphasis on the significance of the growth of a 'core' of skilled workers and an expanding service class whose rising incomes generate increased demand for a wide range of differentiated goods and services. The mode of social regulation associated with these developments is typically summarized by the withdrawal of the state from the collective provision of goods and services. State concern with the social reproduction of labour power is reduced as the core workers increasingly rely on specialist and privatized forms of provision in the market. The growing number of workers excluded from the benefits of post-Fordist restructuring – identified by Aglietta as being composed of politically marginal groups such as immigrants, ethnic minorities and rural–urban migrants – have to compensate as best they may for the reduction or withdrawal of state provision.

Despite outlining the main features of change, the regulationist school ignores the significance of gender relations in post-Fordist restructuring. Partly because most of the work in this tradition has focused on the manufacturing sector and the formal workplace of the factory, rather than on the service sector (although **Christopherson**'s work, 1989, (see Part Three, Section 3.2) is an exception) or on sweated and informal work, it has underplayed the centrality of women's labour. Thus, although in one sense there are marked continuities in women's labour market position – in their continuing and deepening segregation in 'female' occupations – in another sense post-Fordism is also witnessing the unmaking of the old Fordist gender order. As I shall demonstrate that old order, based on a stable working class, on the nuclear family supported by a male breadwinner and by women's domestic labour underpinned by Keynesian economic and welfare policies that ensured the reproduction of the working class, is passing from view.

Flexible Specialization and the Second Industrial Divide

The work of Piore and Sabel (1984) is a more limited and more optimistic version of the transition from Fordism to post-Fordism. Piore and Sabel paint a picture of new cooperative workplace relations in which the old monotony of the production line is replaced by multi-skilled, highly motivated workers equipped with technical knowhow producing goods for rapidly changing markets, based on multiple differentiated tastes.

In their discussion of the forms of institutional regulation needed to maintain innovation and flexibility, Piore and Sabel emphasize the importance of traditions of skills in locally based labour markets and the significance of training in the production of the skills required by small flexible firms, 'solar' firms (those in a subcontracting relationship with a 'core' firm) and the locally-based combinations of 'workshop factories'. Although Piore, Sabel and other theorists in the flexible specialization tradition emphasize labour market segmentation, they ignore the key role that gender divisions play in the (re)construction of valued skills in a post-Fordist restructuring. Nowhere in their book is there the recognition that skill is a socially constructed concept and hence that it is gendered; jobs themselves are gendered. They are created as masculine and feminine and their skill content is continually redrawn to assert the inferiority of women and of women's supposedly natural attributes – what Jenson (1989) refers to as the 'talents' of women as opposed to the 'skills' of the men. The prevailing societal definition of femininity is based on the idea that familiarity with machinery is somehow unfeminine or de-sexing for women. This ideology pervades the design and construction of machinery itself – the 'average' worker for whom it is designed is male – the allocation of tasks by management and, often, women's own identity and social relations at work. Women are themselves frequently reluctant to acquire technical skills which are seen as unfeminine. For all these reasons the new skilled workers on whom Piore and Sabel pin their version of the future are male.

Restructuring the labour process to privilege skilled work and workers thus recreates the gender segmentation of earlier methods of industrial production and to the extent that women (and ethnic minorities) enter the analysis at all they are there in their familiar, and familial, guise of marginal workers, participating in the secondary labour market in Fordist and post-Fordist times alike.

CAPITALISM, PATRIARCHY AND THE GENDER DIVISION OF LABOUR

Given these criticisms of the neglect of gender relations in the major approaches to recent economic change, it is now important to address the ways in which analysts influenced by feminism have portrayed the gender order of post-Fordism. In Britain the Lancaster group of sociologists have argued that patriarchal social relations must be an essential part of the analysis of economic change (Murgatroyd *et al*, 1985; Bagguley *et al*, 1990). Their conception of patriarchy is that of a system of social relations, related to but theoretically and analytically separate from capitalist relations, under which 'men benefit, directly or indirectly and to a greater and lesser extent, from the subordination of women' (Bagguley, 1990, p. 33). Thus their notion is a particular variant of socialist-feminist theory that has become known as the dual systems approach.

The capitalist-patriarchy model relies on the theoretical centrality of domestic labour – it is women's domestic labour that is appropriated by men and is also theorized as essential to the capitalist mode of production.

Men are the breadwinners of this system, gaining at work from the exclusion of women from well-paid 'male' jobs and bringing home a family wage to support the domestic labour that enables them to appear in the office or on the factory floor each day. The analogy between this argument and the regulationists' view of the mode of social regulation under Fordism is clear.

'The work performed by the woman may range from cooking and cleaning for the husband and caring for their children. Women as housewives perform this work for husbands. In these relations of production the housewife is engaged in labour for her husband who expropriates it. She is not rewarded with money for this labour, merely her maintenance (sometimes). The product of a wife's labour is labour power: that of herself, her husband and her children. The husband is able to appropriate the wife's labour power because he has possession of the labour power which she has produced. He is able to sell this labour power as if it were his own' (Walby, 1989, p. 221).

My argument with this approach is not that it has never been an adequate representation of gender relations (it is perhaps most appropriate for the classic years of Fordism – the 1950s) but that it no longer fits the current circumstances. The world of a stable working class and the nuclear family has melted into air. High divorce rates, increasing variety in household and family forms, women's entry into waged labour all challenge this classic view of patriarchal relations. It is now increasingly difficult to demonstrate that either men or capital need domestic labour.

As women enter the labour market in growing numbers it would seem inevitable that the overall amount of domestic labour performed in an economy must decline. The maintenance of the capitalist production system does not seem to have been harmed by this withdrawal.

Where the capitalist-patriarchy model also falls down is in its assumption of the unitary interests of men who, it is argued, benefit from the forms of closure and exclusion that restrict women to particular, and subordinate, positions in the labour market. While it is undeniable that the new gender order of post-Fordist times has deepened the subordination of many women, trapping them in the increasingly casualized, part-time and temporary peripheral labour market, it has also opened up opportunities for some women to join the core occupations and so increased class divisions *between* women. But what it has also succeeded in doing is turning upside down the gender divisions between large numbers of men and women. Increasing numbers of men are employed in the peripheral labour market too, on terms and conditions that traditionally were regarded as 'female'. Their life time attachment to the labour market, their 'family' wage, their conditions of employment and their skill differentials are all being eroded, so that as Phillips and Taylor (1980, p. 65) presciently suggested over a decade ago, perhaps 'we are all becoming "women workers" now' regardless of biological sex.

THE IMPACT OF RESTRUCTURING ON GENDER DIVISION: SOME EMPIRICAL EVIDENCE FROM BRITAIN

Changes in the Labour Market

It was once axiomatic among socialist-feminists that women's labour market participation was a precondition of liberation, bringing in its wake greater economic independence and a diminution of men's power over women, both at the general level and in terms of individual personal relations. It seems as if recent trends may prove this axiom wildly optimistic. Women's rising participation in the British labour market, under current economic conditions and in association with reduced state intervention in the arena of welfare, appears to be increasing their overall workload and deepening the oppression of many working class women rather than expanding opportunities for their independence. However, for other women the recent era of economic change has been one in which they have seen considerable gains.

Women's labour market participation in Britain has risen steadily throughout the entire post-war period, accelerating in particular throughout the 1970s and 1980s (Table 1). However, the growth of part-time work for women has been a particularly marked feature of post-Fordist restructuring in Britain. As the increasing feminization of the British labour market has been achieved through the use of part-time labour, the 2.0 million increase in the total number of women workers since 1971 exaggerates the opening up of opportunities for women in the tertiary sector. Rather what the 1970s and 1980s restructuring has achieved is

Table 1 Trends in men and women's employment 1971–1988 (Great Britain)

	Employees in employment (000s)					
	Men	Women	Women as % of total	% Women full-timers	% Women part-timers	% Share of part-timers of all women
1971	13 424	8224	38.0	25.3	12.6	33.5
1976	13 097	8951	40.4	24.3	16.3	39.6
1981	12 278	9108	42.6	24.7	17.8	41.9
1986	11 643	9462	44.3	25.2	19.6	42.4
1988	11 978	10096	45.7	26.2	19.5	42.8
1990	11 937	10309	46.3	26.3	20.0	43.3

Source: *Department of Employment Gazette*, various years.

the *sharing out* of employment between larger numbers of women. This feature of the 'flexible' use of women's labour through part-time employment contracts is a particular feature of the feminization of the British labour market that is not found to the same extent in the rest of Western Europe. It is partly a consequence of the social insurance system in Britain in which both employer and employee contributions are less than for full-time workers. This brings with it severely restricted entitlement to a range of social benefits such as unemployment and sick pay as well as poorer provision of work-related entitlements such as holidays and security of employment.

Women working on a part-time basis have little prospect of achieving economic independence as they are particularly poorly paid, not only in comparison with men but in comparison with other women. While the average hourly earnings of women employed full-time have remained at approximately three-quarters of those of men in full-time employment from the mid seventies, women working part-time earned only 75 per cent of the average rate for full-time women workers in 1989, a decline from 81 per cent at the beginning of the 1980s (Department of Employment, 1989). These figures are a sober indication that many of the 'new' jobs created in the 1980s match the old jobs in neither the wage levels nor the total hours.

As married women and in particular women with children have been the majority of new entrants to the labour market throughout the 1980s, it is often assumed that the expansion of part-time employment reflects women's preferences. But a range of evidence makes it increasingly clear that part-time workers themselves find the price too high. Part-time, 'flexible' work has not been created in response to 'demand' on the part of workers, whether men or women. Rather many have had part-time or temporary jobs imposed on them or have taken them for want of alternatives while continuing to seek full-time and stable work (Hakim, 1987b). As Evans (1990) has argued in a recent survey of labour market trends in Britain 'one should be wary of endorsing the view that women have a "taste" for part-time jobs, which are demonstrably exploitative jobs. It may be that the "taste" for part-time jobs is actually that of employers' (p. 53). Employers regard the use of part-time labour as an extremely desirable, if not *the* most important, element of a flexible workforce. The creation of part-time jobs has become a means of redefining the status of employees and creating new divisions across categories by separating men and women.

It is clear that women's exploited position in the labour market is not seen as an issue to be tackled. Rather the attributes associated with their gender – and particularly their continuing responsibility for domestic labour and childcare – are seen as immutable. Women continue to be constructed as marginal labour with particular characteristics that mean that their attachment to the labour market is temporary or flexible.

Women themselves, however, are demonstrating both their desire for and the economic necessity of a more permanent rather than peripheral or temporary place in the labour market, not only through their stated preferences for stable employment but also by their behaviour. For example, there has been a fall in women's voluntary turnover rates. Increasing numbers of women are acquiring educational qualifications and other labour market credentials which will give them access to more secure employment. In addition, through their rise in union membership rates and their active involvement in campaigns against the privatization of services and casualization of employment, women in Britain are demonstrating their determination to resist their continuing and deepening exploitation.

The overall impact of economic restructuring on gender relations, however, is neither straightforward nor uncontentious. The gains and losses have not been lined up straightforwardly on a gender basis, or at least not in the expected way. The pattern of gains and losses in the earnings distribution for full-time workers shows that between 1980 and 1989 all women workers across the distribution gained relative to men. But the [figures] also show that the 1980s has been a period of widening income differentials between workers. The highly paid increased their share of total income whereas the share of the low paid has fallen. Throughout the eighties relative wage rates fell for the weaker segments of the labour market – for young workers and manual workers. Male manual workers were particularly adversely affected as large numbers lost secure employment in manufacturing industries. What these changes have resulted in is a new pattern of widening differentials between workers of *the same* sex. In the 1980s the income distribution for women workers came to more closely resemble that for men than it had in previous decades as growing numbers of women with educational, vocational and professional qualifications entered the high paying occupations in the core.

Thus the 1980s have seen a widening of class divisions and a narrowing of gender divisions in the labour market. This has interesting implications for the political strategies adopted by low income workers. It is becoming clear to representatives of certain sections of low paid male workers that their interests increasingly coincide with those of women workers in the same position. Rather than men benefiting from the gender segregation of the labour market, they are losing from restructuring strategies that are defining increasing numbers of new jobs as jobs for 'women'.

Changes in Social Relations: Restructuring 'Domestic' Life?

The post-Fordist years in Britain have been marked not only by significant economic change but also by widespread changes in social relations and in 'family life'. In examining these changes, I shall suggest that, like industrial capital, the state is also less interested in the social reproduction of working class families in general and the old ideal male worker in particular. This is reflected in reduced state expenditure on social welfare provision and the declining real value of many benefits, but unemployment payments in particular. The institutions of the welfare state, that were assumed to reinforce the particular patriarchal family form that characterized Fordism, are also being restructured.

[Yet a contradiction] is evident at the general level – between a restructured economy that increasingly depends on women's labour and a restructured welfare sector that makes the same demands. The post-Fordist theorists who have argued that the new economic regime is associated with new forms of social regulation similarly fail to make clear that unlike the old Fordist compact between capital, men and the state in which the mode of accumulation and the institutions of social regulation were in (relative) harmony, the new order's reliance on women's labour in both spheres makes it inherently less stable.

Reductions in state spending on social policies have the effect of forcing many [women] back into traditional family forms for economic support, health care, care of the elderly and children and so on. However, this 'traditional' family of conservative social policy is at odds with contemporary reality. Many women are no longer available, at least on a full-time basis to undertake the caring labour that keeps 'the family' and the welfare state running in tandem. Women increasingly have less time to do this work but the state is failing to assist them. For example, rising labour market participation rates have been particularly marked among women with children and yet state-provided nursery and afterschool provision in this country remains amongst the worst in Western Europe. Most of these women, not surprisingly, are employed part-time.

Demographic and social statistics seem to indicate that, at first sight, a 'family' crisis might be underway, with women rejecting conventional living arrangements for other forms. Briefly stated women in Britain are delaying marriage and childbirth, having fewer children and increasingly outside legal matrimony and are spending longer periods living without men, both through rising longevity and rising divorce rates. Remarriage, however, remains popular but these marriages themselves are more likely to breakdown than first marriages. One of the consequences of these changes in marriage patterns is that rising numbers of women are bringing up children alone, either permanently or for periods of time. Almost all single parents are women (9 out of 10) and most of these women are single or divorced.

[Yet] it might be argued that the expansion of low wage part-time work for women over the 1980s in combination with the decline of well-paid work for working class men has actually reduced working class women's prospect of economic independence and, particularly in tight local housing markets, has made 'coupledom' almost an economic necessity.

Inequalities Between Households

Despite right wing beliefs in the 'trickle-down' theory of social progress, the gains in prosperity achieved by the most affluent over the decade were certainly not felt by the poorest and inequalities between households, as well as between individuals, increased during the 1980s. Labour market, taxation and benefit changes and welfare restructuring together rewarded the rich and penalized the poor. The opening up of class divisions between individual women was paralleled by a growing polarization between households. Of the total *disposable* income in the UK economy, the top one fifth of households received 42.2 per cent in 1986, compared with 38.1 per cent in 1976, while the bottom fifth received 5.9 per cent compared with 7 per cent a decade earlier.

Women's labour market participation has been an important part of this inequality. Despite small numbers of cross-class marriages, highly educated and well-paid women are most likely to be married to men of the same social status. Thus despite rising numbers of dual income households – a rise from 55 to 67 per cent of all married couples between 1976 and 1987 – household income differentials were not reduced. Rather there was increasing polarization between the majority of families, whether dependent on a single or a dual wage packet and the professional, dual career, double 'pay cheque' family. These latter households are the ones that have gained throughout the crises, recession, inflation and expansionary periods that have accompanied economic restructuring.

Thus during the eighties, for the many working class women propelled into the labour market by economic necessity, as well as a desire for greater independence, two incomes were essential to maintain their previous standard of living. This means that these families are now doing three jobs for the price of one previously: two in the paid labour force and one unpaid at home – the labour of household work and child rearing – if it is accepted that previously the male 'family wage' reflected some contribution towards the unpaid domestic labour of female partners. Although as indicated earlier, the number of hours devoted to domestic tasks by women who work for wages has declined, this decline has been insufficient to compensate for the increased hours in the labour market and on 'caring' tasks to compensate for cuts in state provision.

Thus the overall increase in the number of hours devoted to work, whether paid or unpaid, means that leisure and the general quality of life are severely reduced, especially for those women and their households who cannot afford to purchase in the market commodities and services that were previously provided at home. Among more affluent dual income households the tasks previously being done by the now-employed married women increasingly are being purchased: out of the home in the form of fast foods or day care, for example, or within the home by the expansion of a range of quasi-domestic service jobs. This commodification of domestic labour has itself created a rise in extremely low paid service sector employment, also usually for women in some of the most exploited and marginal forms of employment, typically on a casual basis (Lowe and Gregson, 1989). It is only for the small, but expanding, minority of women who have gained access to the core labour market and who are relieved from the double burden of routine, boring 'women's' work in the home and in the labour market, that economic restructuring and the feminization of the labour market is a liberating experience, giving them a basis from which to challenge male domination in the home and the labour market.

For working class women, unable to purchase even the low paid labour power of other women, the quality of life has deteriorated. The net effect of women's entry into the labour market has not been a redistribution of the total amount of waged and unwaged work done in Britain. Despite men's declining labour market participation rates, it is women who are doing two of the three jobs now common in an increasing number of households. A recent survey of social attitudes in Britain, published in *Social Trends* (1989) revealed that in 72 per cent of all households where both partners were in full-time work, women continued to do most of the domestic work. It was concluded that the 'acceptance of a woman's right to work outside the home does not (yet) appear to have translated itself into a sense of egalitarianism in the allocation of tasks, either actual or prescribed, within the home'.

Thus for the majority of British women, it seems hard to concur with Hartmann's (1986) argument that recent social and economic changes have contributed towards a diminution of the oppressive structures of patriarchy with the possible exception of women in the most advantageous labour market positions. However, it is also difficult to completely agree with Walby's assessment of the impact of flexible specialization on the gender division of labour that 'old forms of patriarchy are replaced by new' (1989, p. 140). In the present era, it seems as if the interests of working class men and women are drawing closer together as both sexes are adversely affected by the reconstruction of large areas of work as 'feminine'. In this latest round in the continuous struggle over the control over women's labour, the majority of women *and* men are losing. Capital is the beneficiary.

CONCLUSIONS

A great deal of work remains to be done by those convinced by the arguments that gender divisions are a crucial, but neglected, element of contemporary analyses of the post-Fordist era. Further, greater attention to the interconnections between changes in the economy and changes in what is clumsily labelled the sphere of reproduction (in the family, the community and the welfare state) is required if gender divisions become a central element of analysis.

It is also important to examine, challenge and assess these arguments through comparative case study analysis at a variety of spatial scales. Significant regional differences are apparent in the composition of local populations, in the ways in which they are divided into households and in the social characteristics of workers and potential workers who are divided by age, by race and by ethnic origins and by their social and cultural experiences.

Regions, classes and genders are being drawn into and expelled from the new order, increasingly at a world scale as multi- and transnational capital is less and less impeded by the institutional regulatory framework of nation states. Thus Third World people are at the same time reassembled as the global workforce of multinational capital in the 'world' cities of the west and exploited *in situ* in their own countries, as the workers, sectors and regions of an earlier round of accumulation are rejected. In all these changes the 'flexible' and 'marginal' labour of women plays a central part, albeit spatially and socially differentiated. Women of colour, for example, are constructed as a permanent casual labour force doing high tech work for the multinationals under peripheral Fordism and similar work or sweated labour in the new territorial production complexes in the post-Fordist west.

It is perhaps not too much to claim that the feminization of the labour market is amongst the most far-reaching of the changes of the last two decades. Geographers interested in economic and social restructuring must place the new gender order of post-Fordism on their research agenda.

Section 5.4 New(?) Social Inequalities

Doreen Massey

'Geography, Gender and High Technology'

***The South East Programme Occasional Paper Series*, Faculty of Social Sciences, Open University (1993)**

The empirical focus of this project is on high-technology scientists/engineers, predominantly involved in research, in Cambridge, England. It is a subject matter which, in most discourses, resonates with positive symbolism. 'High technology' is for instance seen by all the major political parties as one of the avenues for the escape of the British economy, in the long term, out of recession and into economic health. The term 'scientists/researchers' bears with it all the cargo of positive feelings around the new 'knowledge-based' industries; such people are highly skilled, they do interesting and challenging work in a clean environment and often small, more personal, companies. And the association of these terms with 'Cambridge' conjures up the geographical heartland of much of the UK's new growth, the Cambridge Phenomenon: the opposite of the decline of the north and west with hardly a smoke-stack or a blue collar in sight, and a healthy and environmentally attractive (in terms of the dominant discourse) local area.

It all sounds wonderful. And indeed the tale which I am going to tell is not a tale of woe. But it *is* a tale full of ambiguity and contradiction, and not a little sexism.

Previous work concerned with high-technology companies on science parks highlighted that 95% of qualified scientists and engineers in the firms on the parks were men, that they often worked very long and irregular hours, and yet that these were jobs which were held up as wonderful jobs which their holders clearly enjoyed. One of the epitomising statements made more than once in interview was, for me, 'the boundary between work and play disappears'.

A further motivation for the project derived from the need for feminists within geography to look at men and masculinity and not only at women and femininity/women's lives. Economic geographers have often noted, and tried to explain, why employers are employing women (perhaps especially indeed in some high-tech sectors, such as electronics), but much more rarely have they looked at why they are employing men. The implicit background to this stance is that it is more natural that men should be employed and thus less in need of explanation. It is a low-key example of men not being considered as gendered, of their being the unmarked category.

Moreover, there was the issue of regional variation in gender relations. The regions of the 'north' and 'west', of coal mining and trade unions, of shipbuilding and engineering have long been dubbed as particularly patriarchal in their systems of gender relations and gender definitions. And there is no doubt about the heavy sexism of, for instance, mining unions in the Welsh valleys or life in the Yorkshire coalfield. But that in itself is no reason to assume by the implicit counterposition of omission that other regions – and in particular the regions of the new, the dynamic, the high-tech – are necessarily any 'better'. The gender relations in the frontier regions of the new growth warrant investigation in their own right.

EMERGING CONCLUSIONS AND ISSUES

The Fact of Flexibility

The long and irregular hours worked by high-technology scientists were dramatically and overwhelmingly confirmed. In virtually all the companies interviewed, scientific workers were either working very long and frequently unpredictable hours, or were under considerable pressure to do so. The forms which this takes vary widely: people work through evenings, at weekends, they are called away on sudden demands in another part of the country or abroad, they are on 24-hour call, they bring work home.

The Reasons for this Flexibility

The nature of competition in these high-technology sectors

Briefly, work in this part of the economy is often done on a contract basis, contracts having to be tendered for, and the length of contract being highly variable. In this competition, design, quality and timing are more important than cost, or cost alone. Service, and relations with the client, figure largely. 'Putting the customer first' is a maxim frequently heard, and patently necessary to competitive success. It is the opposite of the Fordist approach of producing things cheaply and putting them on the market to find a buyer. And it is, of course, precisely the economic strategy being argued for by those who say we need, in order to climb into long-term economic growth, to adopt a (post-Fordist) strategy of high-quality, flexible production.

What must be recognised, however, is that such forms of production have implications for the nature of work. Designing tenders, or doing development work, does not bring in revenue and so is frequently done as additional hours, maybe in the evenings or at weekends. Speed is often an important element in gaining a competitive edge, so times are under-estimated, deadlines are too tight, there is certainly no margin to cope with the unexpected (yet predictable – for this is work in development and research) hitch. The days/weeks/months before delivery can, and frequently do, become a nightmare of tension, pressure and long hours, to get the thing out on time. The follow-up service may be as important competitively as the production of the original product, so people are constantly on call: sleep and weekends may be interrupted. 'Putting the customer first' is an excellent maxim for individual-company competitiveness, but it can exact a high toll from employees. And if a company is small, with only a few employees – precisely the archetype of go-ahead entrepreneurialism in many a rhetoric of high-tech growth – the problem is worse; for as markets expand geographically and export orders are won, the demand on the handful of high-tech scientists for both spatial and temporal 'flexibility' is enormous.

This is a deep ambiguity. The implication is that these long and unpredictable hours, and the often high-pressure lifestyle that goes with them, are necessary for 'high technology' to be a growth sector in the first place. The very way in which competition in the sector is constructed generates these demands.

The nature of the labour market for these high-tech employees

In these labour markets employees are individuals, and prized individuals at that. This is the polar opposite of the conceptualisation of labour-in-general in the infinitely replaceable, indistinguishable, 'unskilled' worker. These scientific workers in high tech are sought after as individuals, and for the individual skills, knowledge and experience which they possess. The working environment is often in small cooperative teams, sometimes small firms. Yet these very characteristics can in themselves produce pressure. The fact of these being knowledge-based jobs brings pressure on the individual who wants to be successful to keep up with what's in the professional journals, to read around the subject, to go to conferences, constantly to 'network' and to cultivate and maintain the right contacts. Working in a small team means you can't let your team-mates down – if there's a job to be finished you stay late to finish it, even if you had meant, or promised, to go home or to go out. In the small firm under pressure it is difficult to leave before the MD does even if you're only twiddling your fingers and could really have left half an hour ago . . . you have to demonstrate your commitment.

But the issue is more complicated still

For some of the pressure for these long hours, and for this tremendous commitment to paid work, comes from the scientific employees themselves. And this is so in ways quite other than those which derive from the simple fact of ambition or of individual career competitiveness. What was clear over and over again in interviews is that these employees really *love* their work. They are *into* their computers. They are absorbed by the challenges of design. They love the puzzle-solving character of it all. As many of their partners told us, these people do not just work long hours, they wake up in the night thinking about it and stumble off to their computers in the next room.

What is being said is, first: these people love their work; they really enjoy it. And that is great; it would be good if everyone could do so. Second: it throws another light on the reflection that 'the boundary between work and play disappears'. It is not just that this formulation omits consideration of domestic labour but that work itself has many of the characteristics of play. Third, however: it does mean that these employees probably work longer hours than they strictly need to, longer even than the high-pressure context requires. Both the employees themselves, when the issue is raised, and their partners, testify to this. And it is a characteristic which can be drawn on and played-to by companies in this part of the economy – what they need above all is people who are addicted to their work.

The construction of a masculinity

It is moreover arguable that this characteristic of workaholism in this particular kind of work-context, and which is celebrated by employer and employee alike, is integral to a certain type of masculinity. Or, to put it more precisely, that these are attributes socially designated as masculine (and another element of the previous research which was amply confirmed by this project is the dominance – over 90% – of men in these occupations).

There are many dimensions of masculinities, socially constructed and varying quite considerably between times, places, sectors of the economy. What is classically held up for criticism in the older regions is a conception of masculinity in terms of heroic brute strength and of the male role as that of financial provider/breadwinner. From George Orwell's lingering celebration of miners' bodies to Bea Campbell's critique of this view, this has been agreed to be the dominant characterisation of masculinity in these parts of the country/economy. In the dualistic formulations in which masculinity and femininity are typically constructed through counterposition, the feminine in this framework is caring and nurturing, the female role that of homemaker. It is distinction based on a supposed differentiation perhaps above all around physicality, the body: men as strong, protective; women as childbearers.

Yet in high-tech Cambridge what is emerging is no less patriarchal or sexist; it is merely a division constructed along different axes. This view revolves around ideas of these men as the boy into his games, as the male who is happier with machines than with people, or as the male who is best at (and maybe confined to) rationality and logic. This is a bundle of characteristics, of course, which reflects another long-held source of the constitution of masculinity and femininity (and a very different one from that which characterises the coalfields) in which masculinity is associated with the rational, the logical, the scientific: with Reason.

It may be that one significant aspect of the form of high-technology growth which took place around Cambridge in the 1980s is an expansion of this particular form of masculinity.

One question this raises, of course, is what then of femininity? This particular variant of the forms of construction of masculinity is typically established in opposition to a feminine which is emotional, caring, the soft side: Emotion to masculinity's Reason. However, the actual women who might be supposed to be the bearers of this femininity, in particular the partners of these scientists/researchers in these heterosexual households, are in most cases extremely highly trained academically and have interests in life which clearly overlap with those which are supposedly masculine. One way in which the potential conflicts/negotiations in this area are being explored in this project is through investigation of the work of 'reproduction'.

The work of 'reproduction'

There is, then, a combination of factors: from market pressures and the requirements of competition on the one hand to the characteristics of a certain kind of masculinity on the other. And these factors have together resulted in a job-design which requires such long and unpredictable hours that the holders of these jobs cannot do the work of reproduction and of caring for other people. Numerous confirmations were offered, for instance, of the fact that 'a single mother could not do this job'. (One immediate implication is that the construction of these jobs, through the interaction of the interests/characteristics of both employers and employees, is clearly exclusionary. They are ruled out for certain social groups, not through the skills or experience which they demand, but through the social form of their organisation.) Indeed, ideally, the construction of these jobs requires that the people who fill them have someone *else* to look after them.

So how do the female partners in these households negotiate these demands and tensions? How *does* the work of reproduction get done?

First, for a handful of the women there is no apparent tension at all. A number of them quite explicitly and self-consciously take advantage of the situation. They like the fact of having partners prepared to work all hours (and earn a reasonable salary), and they stay at home, tend the house and garden, maybe have children, maybe do charity work, bake bread.

Second, many of the women have tried to build careers of their own. They adopted a variety of strategies for negotiating the potential tension between this and the demands of their partners' working lives. Sometimes the woman's career is interrupted when she has children and the need to take care of them is added to that of looking after the man. Only in one case so far has the possibility of the man stopping his career in order to do the childcare been raised as an issue between the partners.

In many cases, however, the female partner, whether or not there were children, had decided to carry on with her paid employment. In these cases a whole range of further issues was raised. Where both partners were in paid employment, in almost every case the women did well over fifty per cent of the homemaking. In part this is because they make different kinds of *calculations*. If the school/nursery/childminder telephones work to say that a child is ill all the women say they simply have to leave (and it is usually the woman who is called). The men, in general, do not claim to have the same response. No deep essentialism is being evoked here; this is simply the actual difference in prioritisation which we found. Or again, a number of interviewees, both men and women, said that the women's standards of housekeeping were higher than their partners. There were a number of statements of the type 'I really don't think he'd bother if the place was dusty but I prefer to have the house reasonable.' Now, this may be simply that the women don't like living in a tip. It may be a calculation on the men's part that if they leave it long enough someone else will do it. It may be that such women are 'overworking' at housework just as the men put in extra hours at their work, the one getting paid for their additional interest, the other not. It may also be, of course, and there is more evidence for this, that the externally-interpreted responsibility for the house looking nice is imposed on the woman: if it looks awful when someone (especially the parents) come round it's her they blame, and she who feels responsible.

Now, the men are as we have seen spending very long hours at their paid work: this is where this investigation began. Very few of the women spend as much time at their paid employment. This in itself is not an 'independent' fact, but more often the result of calculation or negotiation. ('You couldn't have two people in a couple both working like that'; 'Well, someone has to pick the children up . . .') However, once established, this regime could be said to provide the men with a convincing rationale for doing less of the housework. And in some cases this is the argument which is produced, by both female and male partners. But in other cases the men either themselves express, or are reported to experience, feelings of guilt about the situation. These are men who are now embedded in a particular social stratum – the professional/technical middle class. On the whole, although not without exception, they are not unaware of some of the debates raised by feminists. They know they *ought* to be doing an equal share. Moreover this guilt is compounded (again not in all cases but in a striking number) by the fact that the long hours they spend at work and which are the reason for their underperformance at home are *enjoyable*. This is not a case of being trapped doing overtime on the line at a car-assembly plant. A number of the men expressed the desire for things to be different, and a number have managed to change things. But when the work call comes it is hard for the majority of them to resist.

One result of this complex combination of factors, though, is that the women are held back in their own careers. This happens in many different ways; what will be pointed to here are just one or two of those ways and ones which are particularly geographical.

Most basically, this prioritisation of the man's paid employment over the woman's frequently means that the decision to be in this part of the country at all is primarily a function of the man's work. This of course is overwhelmingly the case only when an explicit decision has been made to move to or stay in Cambridge. But Cambridge is a high-status place in the geography of high technology in the UK, and its attraction to people elsewhere is widely documented. One result, however, is that these women are operating in labour markets which have not been chosen in relation to their own employment.

Further, the spatial range of job choice and of possible travel-to-work for the woman may be restricted by the decision to adapt to the needs of the man's job. ('Someone has to pick up the children.') And so it is that one person's much vaunted flexibility becomes someone else's time–space constraint.

For the same reason, once in a job – and this tends to apply particularly to professional employment including the women themselves also being scientists/researchers – the women tend often to cut themselves out of those aspects of work which involve travelling, especially beyond the local area. They go to fewer conferences, they do less networking. They don't take on the international contracts. And so forth.

Finally, there is of course another way of negotiating the difficulties posed by these men's jobs for the performance of domestic labour. That is to pay for it: to buy in domestic services, to eat out, to employ a nanny. When it does occur what it implies is the further social extension of the domestic labour required to reproduce these scientists/researchers. It means another form of economic inequality structured into the growth of the area. At its strongest it would point to a labour market in part polarised between high earners (in this case largely male) on the one hand and those (mainly female) who service them on the other. Nor is the inequality only economic. In the case of nannies, for instance, the requirements for 'adaptability' and 'flexibility', for putting paid work first, also get passed on. When the high earners in the central sectors of the economy need to work late, nanny is telephoned and asked to stay late. The requirements for time-flexibility in high technology reach well beyond the portals of those high-status places of work.

SOME REFLECTIONS

(i) This form of growth, this element in the economic boom of the South East of England in the 1980s has required for its competitive success, and for the success of those employed in it, a form of work-life which by definition – in its very constitution – excludes some groups from central participation within it, and generates inequalities.

(ii) Moreover, it is imbued with a particular form of masculinity which it both trades on and further encourages. The complexities of the social organisation of this part of the economy articulate together particular forms of capitalist competition and particular forms of patriarchal relation.

(iii) The argument implies, however, that the issue is not in itself the lack of female participation in these scientist/researcher occupations but the structuring of the jobs themselves. Moreover it may be that a restructuring of these jobs – were it to be possible – would relate not only to opening them up to groups at present excluded but to wider arguments put forward by both feminists and trade unionists for a shorter working day/week.

(iv) Even the few elements which have been examined here indicate that what is being constructed in this part of the country is a version of masculinity and of gender relations very different in kind from those, say, of the coalfields, but utterly patriarchal nonetheless.

Roger Waldinger and Michael Lapp

'Back to the Sweatshop or Ahead to the Informal Sector?'

***International Journal of Urban and Regional Research* (1993)**

The concept of the 'informal sector', first introduced in the early 1970s in studies of Africa, has recently gained currency as a tool for understanding the changes under way in the advanced industrial societies. Originally, the concept was used to describe the variety of third world business enterprises characterized by their small scale, ease of entry, labour intensiveness and the evasion of government regulation. The first wave of studies identified the informal sector as a leftover from precapitalist modes of production; subsequent work portrayed the informal sector as an increasingly integral aspect of industrializing third world economies. But the most recent evidence suggests that the informal sector is a first world phenomenon as well. Increasingly, social scientists draw our attention to the growing proportion of persons working on their own account, the shift toward smaller firm size, the expanding scope of economic activities whose existence is concealed from the state, the revival of homework and the burgeoning of sweatshops. They conclude from this disparate set of phenomena that the informal sector is alive, well and growing in the postindustrial West.

This paper takes a sceptical look at this new version of the informal sector idea by examining a case that is critical for the informal sector claims – the 'sweatshop' phenomenon in New York's garment industry. That the garment industry has gone 'back to the sweatshop' is a crucial piece of evidence in the entire informal sector story.

If the sweatshop was emblematic of the conditions imposed on urban workers during the second industrial revolution, its demise is attributed both to the new regime of industrial regulation that arose with the New Deal and to the consolidation of an economy based on high-wage, standardized production. Hence, the sweatshop's return can be taken as evidence both of the collapse of the social contract that underlay the New Deal regime, and of a shift from Fordism to an industrial era in which flexible, small-scale production techniques prevail (Harvey, 1989). The presence of third world immigrants labouring in an underground economy at a time when highly educated white workers should be enjoying the vanities of the FIRE [fire, insurance, real estate] sector points to the unequal distributional consequences of the emerging economic arrangements. Finally, the growth of sweatshops explains the mystery of why third world immigrants should have come to cities at a time when low-level jobs of the type that immigrants have traditionally secured were apparently disappearing.

This article offers a different perspective. The central contention is that the sweatshop phenomenon, as conventionally interpreted, simply isn't so. Once the phenomenon is subjected to careful scrutiny, there is little evidence to support the view that massive and growing numbers of immigrants are employed in a 'rising' unregulated, informal sector.

SWEATSHOPS, IMMIGRANTS AND URBAN CHANGE

'There is a close association between areas of high immigrant concentration', note Portes and Castells, 'and those in which the US informal sector seems most vigorous' (1989, p. 23).

Proponents of 'informalization' contend that this close association stems, in part, from changes in the urban manufacturing complex. The growth of the so-called 'downgraded manufacturing sector' is one such shift, claims Sassen, involving 'the social reorganization of the work process, notably the expansion of sweatshops and industrial homework' (Sassen, 1988, p. 145). Smith [1988] echoes this argument where he maintains that 'new patterns of inequality have emerged in cities experiencing rapid economic growth' because 'low paid service workers are taking their place alongside a growing number of poorly paid industrial workers from the Third World who account for another major US central city employment growth sector in the past decade – the new immigrant sweatshop' (p. 200).

EVIDENCE REVIEWED

In our review of the literature we have found three propositions that support the contention that a large and revived sweatshop sector has emerged:

(1) A substantial segment of production has shifted underground;
(2) a substantial shadow labour force is available for these underground production facilities;
(3) a substantial portion of production has been further dispersed to a large and growing force of homeworkers.

Proposition 1: Underground Production Facilities

This proposition is central to the argument about a sweatshop revival. But how are we to know that production has gone underground? One indicator would be the ratio of production to non-production workers. The bulk of production workers in the garment industry are employed in contracting shops, specialized production facilities that make up garments to the specifications set by manufacturers or jobbers. While sweatshop contractors may be operating underground, their production is delivered to manufacturers or jobbers that all operate in a formal economy. The highly publicized case of Norma Kamali, the well-known designer-label firm that was found to use homeworkers, is just the most prominent example of the pervasive linkages between the formal and informal garment industry uncovered in the various labour standards enforcement campaigns. If there is a large sweatshop sector, the ratio of production to non-production workers in the industry should be low; if the sweatshop sector is small, the ratio should be high. Furthermore, if the informal sector has expanded, as the literature claims, then the ratio of production to non-production workers should have declined.

Time-series data for the period of purported sweatshop growth indicate no marked shift in the production worker ratio for New York City. [In addition] one indicator of a steadily growing underground labour force would be a sharp decline in the ratio of wages for production workers to total value added by manufacturing. The ratio of production workers' wages to value added by manufacturing has indeed declined. But the decline is constant, with no sudden shift in the late 1970s when the number of sweatshops and illegal immigrants purportedly burgeoned. There is no difference in the trendline between New York and those other 48 states where there are few immigrant garment workers suggesting that the downward shift is caused by productivity changes that are broadly shared throughout the industry and *not* by an increase in sweatshops That this indicator indicates no distinctive New York effect is a particularly strong finding against the informalization hypothesis.

Proposition 2: A Shadow Labour Force

A common theme in writings on the underground economy is the availability of a shadow labour force recruited for informal types of employment. The shadow labour force may consist of women, youth, or immigrants and ethnic minorities, who are under-represented in the recorded or formal labour market, but none the less appear to be engaged in the production of goods and services.

OECD (1986) notes that employment-to-population rates and hours of work are indirect indicators of the presence of such a shadow labour force. In Europe, for example, comparatively low employment–population ratios and low recorded hours of work in the Mediterranean basin countries suggest a sizeable informal sector.

Journalistic and academic accounts uniformly depict the sweatshop labour force [as] an immigrant labour force. If this is indeed the case, the above reasoning would suggest the following hypotheses: immigrants would experience below average labour participation rates; they would also be under-represented in the garment industry overall; and those immigrants employed in the garment industry would be expected to report lower than average hours of work.

Data comparing the ten largest groups in the garment industry, as of 1980, to native whites, shows that the opposite is true. Immigrants are in fact greatly over-represented in the garment industry: Chinese, for example, are over-represented by a factor of almost 7; Dominicans by a factor of almost 5. Low work-hours are not a characteristic of immigrant workers either: immigrants highly over-represented in the industry worked almost as many or more hours than native white production workers. As to labour force participation, under-representation is mainly a phenomenon of old, not new, immigrant groups; relative to native whites, Dominican women in 1980 comprised the only new immigrant group that was under-represented in the labour force, but over represented in the garment industry.

Cross-checking official statistics against other types of data casts further doubt on recent estimates of a shadow immigrant labour force. Morrison Wong (1983), Sassen (1988), and Kwong (1987) all indict New York's Chinatown as a concentration of sweatshops. While Chinatown's garment contractors may include many firms that cheat on hours and wage laws, they are clearly not underground. Data from the New York State Labor Department's 'Covered Employment Series', which come from employers' unemployment insurance reports and can be disaggregated to the zip code level, show that women's outerwear employment in Chinatown rose from 8095 in 1969 to 15,567 in 1988, a gain of 92%. During the same period, women's outerwear employment in the rest of Manhattan fell by almost 55%. These official data are entirely consistent with administrative data from the garment workers' union, which trace the explosion of Chinese-owned firms, from 8 in 1960 to 485 in 1985. Further indication of the above-ground status of the Chinatown garment industry comes from a count of the more than 400 garment firms listed in the *New York Chinatown Business Guide and Directory* (1984), available for purchase in any Chinatown bookstore.

But journalistic and scholarly accounts of the sweatshop do not simply contend that immigrants furnish the needed exploitable labour force: the key, rather, is the presence of illegal immigrants, who are supposedly so vulnerable and desperate for employment that they will accept any job, no matter how bad the conditions. Such logic has added plausibility to the claim by some researchers that the apparel industry is actually the nation's largest employer of illegal immigrants.

But support for these contentions is weak. We now know that earlier guesstimates wildly inflated the size of the undocumented population and that the 1980 Census of

Population succeeded in enumerating the great majority of illegal immigrants then resident in the country. Similarly, a decade and a half of research on illegal aliens has shown that their economic, demographic and human capital characteristics differ little from those of legal immigrants of similar ethnic backgrounds. According to a recent US Department of Labor report, 'in many instances, illegal status does not lead to significantly lower earnings, nor does it appear to impede mobility substantially' (US Department of Labor, 1989, p. 158).

Proposition 3: From Factory to Home

Inseparable from claims that sweatshops are proliferating is the contention that homework, once almost extinct, has experienced a massive rebirth. These arguments are subject to the cross-checks already developed above. If the number of homeworkers burgeoned during the 1970s to the 10,000 level, as Sassen maintains, then we should have found a much greater decline in the various ratios discussed in the previous section than we actually observed.

More direct evidence on homework comes from answers to the place of work question in the Census of Population, to which 'home' was a possible answer. Tabulations for women workers in the five largest immigrant-receiving Standard Metropolitan Statistical Areas (SMSAs) – New York, Chicago, Los Angeles and San Francisco, all of which had sizeable garment industries – provide results that run contrary to the conventional wisdom. In none of these metropolitan areas did home-sewers figure prominently among workers in the apparel industry. Nor is homeworking a phenomenon distinctive to new immigrants. In every metropolis white immigrants ranked at the top in percent employed at home.

But does the same pattern hold once one takes other factors, beside immigrant status, into account? To answer this question, we used logistic regression to estimate the probability of working at home for all employed workers in New York City in 1980. The equations for men produce no support at all for the contention of heavy immigrant employment in home-based industries. The coefficients for the equation for all employed women run contrary to claims of a burgeoning population of immigrant homeworkers with a particular concentration in apparel. The probability of employment at home is *negatively* related to employment in apparel and to immigrant status.

As Silver concludes, 'despite changes in the economy, growth in immigrant labor, and new technologies, homework is not significantly on the rise in the United States' (1989, p. 112).

RE-LOCATING IMMIGRANT INFORMAL ACTIVITY

If there is little evidence of a substantial underground garment industry, what are the implications for theories of 'informalization' in the United States?

Sweatshops, Immigration and Urban Economic Change

As noted earlier, the apparent revival of sweatshops offered scholars a clue to unravelling the puzzling coincidence of large-scale immigration and the rapid post-industrial transformation of the immigrant-receiving cities. In essence, informalization arguments explain the demand for immigrant labour in light of the emergence of new production forms, of which a major instance is the 'sweatshop'.

But such claims do not only clash with the findings of this paper; they are inconsistent with what we know about the broader economy. The available evidence does not indicate that changes in the organization of manufacturing have yielded a growing underground production sector in the United States. Rather than being a centre of goods production, the underground economy mainly involves the purchases of final goods by consumers. IRS audits indicate that the construction, retail trade and service industries accounted for more than 80% of the understatement of business receipts and profits.

Since production forms have remained stable, we suggest an alternative explanation of the puzzling relationship between immigration and urban change: the critical shift has been not on the demand but on the supply side. In New York, compositional changes – resulting from disproportionate declines in the local white population – created vacancies for immigrants at the bottom of the job ladder in industries like garments. Further openings for immigrants emerged because native-born workers dropped out of the effective labour supply in reaction to declining relative wages and working conditions. Thus, the basic structure of New York's apparel industry did not change; rather, old positions and functions were vacated and in this way entry-level opportunities for immigrant workers were created.

A similar process of ethnic succession created opportunities for immigrant entrepreneurs. High death rates among established firms owned by white ethnics and low start-up rates have provided replacement opportunities for Chinese, Dominican and Korean contractors. This pattern of replacement labour and entrepreneurship holds true more generally for the immigrant-receiving economies of New York and Los Angeles.

The Sweatshop in the Class Structure: Distributional Aspects of Informality

Exploitation is almost a synonym for sweatshops. In the social science literature, Sassen-Koob's (1984) concept of *downgraded labour* highlights the vulnerability of sweatshop workers and the collapse of working standards associated with the growth of sweatshops (see also Portes and Castells, 1989). By contrast, we will argue that the distribution of opportunities for informal income generation closely parallels the distribution of opportunities for income generation of any type.

Much of the literature on the underground economy emphasizes the incentives for workers or employers to escape state regulation. As Carson (1984) noted, incentives are a necessary but not sufficient condition for underground income generation: one must also have opportunities to evade or circumvent regulations. But not all opportunities for underground earnings are equally remunerative. For example, the first wave of research on the underground economy in the United States identified an 'irregular economy' in black ghetto communities where workers engaged in pseudo-entrepreneurial activities from which they were barred in the regular economy. Since these transactions were confined to a ghetto clientele of severely depressed incomes, the irregular economy offered little chance for surplus generation and amounted to exchanging one another's wash.

While economic marginalization confines black ghetto-dwellers to communities poor in informal resources, persons higher in the class structure than immigrants appear to enjoy even greater opportunities for participation in informal economic activities. As Marxists would predict, ownership increases both access to informal income-generating activities and the potential for hiding income from the state. Thus in the United States, it is income from rental property that is reported to the Internal Revenue Service (IRS) at the lowest rates of all. The next worst offenders, as shown by European and US data, are the self-employed. Business ownership, it turns out, confers an aspect of autonomy not fully appreciated by sociologists: namely, greater opportunity to conceal one's income. To get a feel for what these opportunities involve, consider the findings from a 1989 New York State investigation of lawyers. Their research discovered that 10% of law partners, but only 0.5% of law firm employees, failed to file state income tax returns during one of the previous three years! [Thus] a variety of studies has found that both participation in informal economic activities and tax evasion are *positively* associated with socio-economic status.

The Social Construction of a Social Problem

If there is as little to the sweatshop phenomenon as we have maintained, how then to account for its emergence as a social problem, in turn making the 'informal sector' an object of legitimate study? One clue is that the problems of the sweatshop and of undocumented immigration have been formulated in strikingly similar ways. From the start, wildly inflated estimates of the undocumented population were a major feature of the illegal immigration debate. That so many illegal immigrants were flowing into the country inexorably led to the conclusion that the problem was essentially one of social control. On the one hand, the massive illegal inflow was eroding control over movement across borders – a basic aspect of sovereignty; on the other hand, the illegals were creating or threatening to create 'an underclass outside the law' (Keely, 1982, p. 42).

Linked to undocumented immigration, the sweatshop was thus ready-made for identification as a pseudo-problem of social control. But if the sweatshop revival is a pseudo-problem, not so the conditions under which too many immigrant garment workers labour. Listen to the voice of a seasoned observer of the needle trades:

> These newcomers tend to settle in ghettoes – by choice or necessity – and to create pockets of cheap labor. Light industry entrepreneurs (and New York City manufacture is predominantly light industry) tend to set up shop in the midst of or in walking distance of these newcomers' communities. They reach out for unskilled, often female labor – offering a special inducement in the form of proximity, late arrival, early departure, and minimal penalties for days off when family life becomes demanding. In exchange, these women work for sub-standard wages.

That the date of the observation is not 1992 but 1961 offers a clue as to what the dilemma really is. After all, the existence of sweatshops presents quite a different moral and political issue from the constant depression of wages and working conditions that none the less hover just above the legal minimum. Illegal, underground 'sweatshops' are a scandal and scandals help sell the news. But the real sweatshop story is the scandal of legally low wages that do not provide an adequate standard of living. The tragedy is that low wages are an old and bitter story. After all, who wants to confront reality when the price of maintaining an industry employing several hundreds of thousands are conditions and wages that we all abhor?

Steven Pinch

'Social Polarization: A Comparison of Evidence from Britain and the United States'[1]

***Environment and Planning A* (1993)**

INTRODUCTION

One of the most complex and controversial issues in the social sciences is the extent of polarization in the social structure of cities in the advanced economies. There is certainly no shortage of data on the subject [and,] in addition to such statistical data, there is the highly visible juxtaposition of affluence and destitution on the streets of major cities such as London, New York and Los Angeles. However, a closer inspection of the polarization debate makes it difficult to disagree with Pahl's assertion that 'General statements of apparent clarity and simplicity are made which cover great complexity and confusions' (1988, p. 259).

The issue of polarization is complex for a number of reasons. To begin with, there is a very wide range of processes that are postulated to be leading towards polarization, including: deindustrialization and increasing levels of unemployment; the bifurcation of rapidly expanding service sectors into high-paid and low-paid jobs; the increase in temporary and part-time work; changes in family structure such as the increase in lone-parent, lone-elderly, and dual-income families; the continuing economic and social marginalization of ethnic minorities; and the 'residualization' of state welfare services. Given the multiplicity of such influences, it is often difficult to isolate their separate effects, especially as some influences may be mutually reinforcing whereas other influences may work in opposite directions.

In addition, there are three important methodological issues that confound the issue of polarization. First, what is the most appropriate unit of analysis to measure polarization? For example, should the unit be individuals or households? Second, what types of variables should be used to measure polarization: income, wealth, consumer goods, or issues such as job 'quality' and job satisfaction (both inside and outside the home)? Third, what is meant by polarization: an increase in low-paying jobs at the bottom of the social hierarchy; increases in high-paying and low-paying jobs; or a decrease in middle-income jobs?

INTERNATIONAL PERSPECTIVES ON SOCIAL POLARIZATION

First, such international comparisons can provide a crucial 'testing ground' for social

---

[1] Pinch, S. (1993) 'Social polarization in Britain and in the United States', *Environment and Planning A* **25**, 6: 779–96. Reprinted by permission of Pion Limited, London.

theory. Such international comparisons can therefore reveal the diversity of human experience and show that what seems inevitable in one nation does not necessarily have to be the case in another.

A second advantage of international comparisons is that they can help to avoid ethnocentrism and restricted conceptualizations. The potential for cross-fertilization of ideas between nations is again well illustrated by the polarization debate, for there has been a tendency for two main types of polarization thesis to develop on different sides of the Atlantic. In the first thesis, most extensively developed within the United States, it is argued that, because of the changing nature of occupations, we are witnessing the decline of middle-income groups and the expansion of high-paying and low-paying jobs at each end of the occupational hierarchy (Bluestone and Harrison, 1982; Harrison and Bluestone, 1988; Stanbach, 1979). These occupational shifts are usually related in complex ways to broader shifts in the nature of the economy and, in particular, to the decline of manufacturing and the growth of service employment.

If this polarization thesis is correct, then it has important implications for households. If the new jobs that are being created by the service economy pay low wages, then more members of the household may be forced to find employment and/or household members may be forced to take up more than one job in order to maintain what they regard as an adequate standard of living.

The second main type of polarization thesis has emerged from the work of Pahl in Britain (Pahl, 1984; 1988). Central to Pahl's approach is a focus upon households and all types of work undertaken within them, both paid work in the formal economy and informal work in and around the household, such as cleaning, decorating, and home improvements (termed 'self-provisioning'). Pahl argues that the majority of households in Britain have been able to mitigate their deteriorating position in the labour market by informal work and self-provisioning. However, because self-provisioning requires income, skills, and contacts that are primarily gained from the world of formal work, it is mainly those households in which all the members are unemployed that will experience the greatest disadvantage. According to this approach, rather than a disappearing middle class, social polarization takes the form of an unemployed underclass. Pahl argues that polarization in Britain is forming an 'onion-shaped' social structure, with a growing proportion of deprived households at the bottom. This is in contrast to the 'hour glass' structure that has been postulated for the United States, following the growth of both the affluent and the poor.

SIMILARITIES BETWEEN SOCIAL POLARIZATION TRENDS IN BRITAIN AND THE UNITED STATES

In the 1980s the imposition of neoliberal policies both in Britain and in the United States has led to a remarkable convergence of certain trends leading to polarization. These trends may be divided into individual and household influences, in line with the two basic approaches to social polarization outlined above.

Polarization at the Individual Level

Many studies of the United States have highlighted an association between low incomes and the expansion of service sectors (for instance, Humphries, 1988; Sassen-Koob, 1984; Stanbach and Noyelle, 1982). Similarly, in the Southampton [UK] survey, there were a number of important associations between job characteristics and industrial sector. Low incomes were concentrated in the expanding service sectors, and especially in retailing, hotels, and catering (Pinch and Storey, 1992). The public sector and financial services also tended to have bifurcated earnings patterns, with high proportions of low-wage and high-wage workers. The declining manufacturing sectors had the highest proportions of workers on middle-range incomes.

A common theme in much of the work on changing labour markets in the United States is that polarization exists, not only in relation to incomes, but also in relation to working conditions and fringe benefits. In keeping with such hypotheses, the service sectors in the Southampton study also provided the least fringe benefits, with workers in the distributive sectors most disadvantaged in this respect. The public sector also scored badly in relation to occupational benefits, although it performed better in relation to pensions. The banking and financial services sector had a polarized pattern, reflecting the bifurcation of incomes within this sector. Manufacturing sectors, in contrast, tended to provide a wide range of fringe benefits (Pinch and Storey, 1992).

Further cross-tabulations revealed that gender was deeply implicated in the process of industrial restructuring. Thus, the rapidly growing service sectors had the highest proportions of female workers. On average, women were paid less than men. The lowest wages were received by part-time workers, the majority of whom were women. With regard to fringe benefits, there were greater differences between full-time and part-time workers than between men and women. Part-time workers were also more likely to feel that they had limited chances for promotion. The low incomes, poor fringe benefits, and limited promotion prospects within the distribution sector was therefore largely a result of the large proportion of part-time workers in these sectors (Pinch and Storey, 1992).

Although these results are based upon one point in time, it is possible to infer from them that a process of polarization of occupations is underway in the Southampton economy which parallels the experience of the United States. The new jobs that are being generated by the smaller establishments in the service sectors tend to pay lower wages, provide fewer fringe benefits, and have poorer promotion prospects. In contrast, the relatively well-paying jobs with good fringe benefits provided by the larger manufacturing employers are in decline.

Polarization at the Household Level

The life chances of individuals are also influenced by the way that economic restructuring affects other members of the households to which these individuals belong. Given increasing unemployment or declining real wages, households may adopt collective strategies to cope with changing conditions in the labour market. In particular, other members of the household may be drawn into the labour market to help maintain living standards. The growth of such multiple-income households is central to Pahl's polarization thesis. He argues that there is a growing divide between those households in which there are multiple incomes, and those in which all the adult members are prevented from participating in the formal economy (Pahl, 1984).

Pahl's thesis suggests that there has been an increase in the proportion of non-earning households, and this was confirmed by the British General Household Survey (GHS) which showed an increase in such households from 25% in 1973 to 34% in 1982.

Pahl's polarization thesis also predicts an increase in the proportion of multiple-income households. However, the GHS showed that, between 1973 and 1982, there was a decline in the percentage of households in Britain with two or more incomes (the figures for these years were 39% and 34%, respectively). Dale and Bamford (1989) conclude that the increase in wives and female partners in the work force has not therefore compensated for the overall decline in employment levels. Harrison and Bluestone (1988) also show that there has been no overall increase in the proportion of families with two or more wage earners in the United States.

In the Southampton survey, however, the proportion of multiple-income households was 40%, a figure higher than that in the GHS. These data therefore suggest a considerable increase in multiple-income households in the 1980s, at least in this locality. Changing household structures had a clear impact upon perceptions of changing living standards in the last five years (Pinch and Storey, 1991). Households with multiple incomes were more likely to feel that their standard of living had got better in the last five years. These findings suggested that part-time work was playing an important part in maintaining the living standards of dual-headed families with children. These data also accord with information from Family Expenditure Surveys in Britain which show that women's earnings have become an increasingly important part of household incomes (Walker, 1988).

These findings also have broad similarities to data on the changing fortunes of families in the United States. According to Bradbury (1986), traditional families (composed of a single head aged between 34–44 years, a nonworking spouse, and two dependent children) fell behind, whereas childless two-career families pulled ahead (see also Bradbury, 1990).

DIFFERENCES BETWEEN BRITISH AND US POLARIZATION PATTERNS

Race, Income, and Lone Parents

Many of the differences in social patterns in Britain and the United States are well documented and widely acknowledged. For example, issues of race have been much more prominent in debates about social polarization in the United States – where it has principally been manifest in the 'underclass' debate – than in Britain. The significance of race for processes of social polarization should not be underestimated in the British context (see Ginsberg, 1992) but it has, so far, not assumed the same importance as in the United States.

It has also long been acknowledged that there is a wider distribution of incomes in the United States than in Europe. International comparisons of income distributions are notoriously difficult, but recent evidence suggests that the poverty rate in 1985 in the United States was 50% higher than that in Britain and double that for every continental European country (McFate, 1991). These income patterns reflect the more progressive tax regimes and higher minimum standards of income support provided by European welfare systems.

Another sphere in which US developments have been acknowledged as exceptional is that of single parenthood. Single parents are one of the most rapidly growing types of household in the United States, and by 1987 female householder families were 19.7% of all families with children. This has led to an extensive debate centred around the so-called 'feminization of poverty'. However, this is another sphere in which differences between Britain and the United States seem to have narrowed in recent years. For example, in 1990, 19% of British families with dependent children were lone parents.

Part-time Work

In this instance, this phenomenon appears to be of greater importance in Britain, for no fewer than one in four of all workers in Britain are part-timers (Walsh, 1989), whereas about one in six is in part-time employment in the United States (**Christopherson**, 1989) (see Part Three, Section 3.2). Another difference is that a higher proportion of part-timers in Britain are women. In addition, part-time jobs in the United States are less likely to be designated as low in skill.

A number of labour-market supply and demand factors can be used to account for this difference in part-time working, and these highlight some of the key differences in processes affecting social polarization.

The relatively high levels of satisfaction displayed by part-time women workers in Britain – despite their inferior pay, occupational benefits, and promotion prospects – may be explained to some degree by the domestic commitments of this group. Part-time workers tend to be women with young children, and this form of work enables them to combine paid work in the formal economy with domestic work in the home. However, the growth of part-time work is not simply, or even mostly, supply led, for there are particular benefits that British employers can derive from employing part-time workers. In addition to the possibility of using part-time workers in a flexible manner for particular times of high labour demand, either in the day or week, and also to replace temporary workers, British employers can save costs by using part-time workers. For example, below certain income thresholds employers do not have to pay National Insurance contributions for part-time workers. In addition, part-time workers may be exempt from occupational benefits such as maternity leave, sick pay, and pensions; and there are less rigorous stipulations regarding redundancy compensation payments.

The interaction of these supply and demand factors in Britain is interpreted by the British government as evidence of a harmonious match between the preferences of employers and employees. However, critics of this interpretation point to a number of features of the US situation which cast doubt upon this interpretation. It is argued that the adherence of British women to part-time work is a pragmatic compromise in the face of inadequate child-care facilities. They point to the situation in the United States in which women can receive partial reimbursement for the costs of child care through federal and state tax systems, a concession that has not been available in Britain in the 1980s.

Of considerable importance in this context is the fact that, in the United States, more advanced equal opportunities legislation, combined with higher levels of education amongst women, have enabled many women to obtain higher-paying full-time jobs which permit the procurement of child care. As Ginsberg (1992) notes, despite considerable variation in the labour-market conditions experienced by women in the United States, employers in that country are, on average, less likely to consider women as disposable workers than in the United Kingdom. The higher proportion of men in part-time work in the United States may also be explained by the lack of a comprehensive social security system for the unemployed.

Multiple-Job Holding

It is difficult to obtain reliable information on the extent of multiple-job holding or 'moonlighting' as much of this is likely to be within the hidden or informal economy. Although the extent of the informal economy may have been exaggerated in the past, it seems likely that survey responses on the issue of multiple-job holding understate its true extent. About 8% of full-time workers in Britain report having one or more other jobs compared with a figure of 15% for part-timers. However, as discussed above, part-time working in Britain is largely taken up by women with domestic commitments, and these second jobs are therefore often related to the domestic sphere such as mail-order agents, child minding, and baby-sitting. In the United States, **Christopherson** (1989) notes that, although the proportion of workers claiming they hold two jobs has remained the same since the 1960s, the increase in the labour force means that there must have been a considerable increase in multiple-job holding. The limited evidence available suggests that men have a regular job in the formal economy whereas their second job may be in the informal economy; in contrast, women are more likely to have two part-time jobs in the informal economy. Another contrast is that, whereas 'moonlighting' by men is more likely to be cyclical, second-job holding by women has shown a linear increase over time.

CONCLUSIONS

After surveying the tangled literature on polarization and the associated debates on poverty and the underclass, it is difficult to avoid Dale and Bamford's conclusion that 'whilst the concept of polarization makes an important and exciting research vehicle, it is in practice complex and difficult to put into operation with confidence' (1989, p. 498).

To grapple with this diversity it has become common to contrast two different types of economy: first, the deregulated, tax-cutting, minimum welfare state, antiunion economies, epitomized by the United States; and, second, the 'integrated', social economies of Scandinavia and continental Europe. In Esping-Anderson's (1990) analysis of the extent to which welfare benefits are 'decommodified' (that is, enable individuals to survive outside the labour market) Britain is lumped together with a group of liberal 'Anglo Saxon' economies, including the United States. However, Britain also has a strong tradition of direct state provision of welfare services, and lacks the tradition of voluntarism to be found in the United States. What does seem clear is that the imposition of neoliberal policies in Britain in the 1980s has led to a decline in trade union power and the erosion of workers' rights in a manner which has moved the country closer towards the US model – generally recognized to be the most flexible labour market of all the industrialized nations.

The above assertion is certainly supported by the evidence from a sample survey of the economically active in the Southampton city-region, which suggests a number of broad parallels with recent experience in the United States. When examined in further detail, however, a number of complex differences are apparent in British and US labour-market trends, and these have different implications for polarization. One of the most striking differences to emerge from the Southampton survey relates to part-time work. Not only are multiple-income families in Britain more likely to have one partner in part-time work, but these jobs are far more likely to be undertaken by married women with young children. Without the proliferation of part-time working in Britain it seems likely that there would be greater inequality of incomes at the household level resulting from a higher incidence both of single-income households and of dual-income families with both workers in full-time work.

But the most important difference between Britain and the United States (a difference which is not revealed by a relatively buoyant labour market such as that in Southampton) is that the US economy has had a higher rate of service-job generation than in Britain and (until recently) a correspondingly lower rate of unemployment. This high [British] rate of unemployment has had a crucial impact upon social polarization, with no less than two thirds of the poor made up of the unemployed (McFate, 1991). It is therefore not altogether surprising that in Britain the polarization debate has been dominated by the issue of unemployment. In the United States, in contrast, the issue of job quality has been much more in evidence.

Two important conclusions may be drawn from these comparisons. First, the complex web of interrelationships between labour supply and demand factors suggest that there is no one universal pattern from which individual nations deviate. Although there are global forces of competition, their impact is highly specific to particular countries. Thus, the demand for labour flexibility is met in very different ways in different countries. In Britain there is a particular set of legal and institutional factors that encourage part-time working. The advantages are less apparent in the United States, where the full-time work force is already relatively flexible by international standards. In the more regulated labour markets of France and Italy, however, trade union opposition to part-time working, and greater state regulation of wages, has led to an increased incidence of temporary working and an increased use of the informal economy.

Second, it is impossible to understand social polarization without considering both the sphere of production and the sphere of social reproduction. The high incidence of part-time work in Britain is closely bound up with the absence of a comprehensive system of child care; conversely, the higher rates of full-time working by women in Scandinavia and much of continental Europe is related to the more extensive systems of child care to be found in these countries. The relatively high levels of full-time work by women in the United States must also be related to the lower levels of welfare benefit in that country, the higher incidence of divorce and lone-parent families, and the more advanced equal opportunities legislation.

Will Hutton

'The 30–30–40 Society'

***Regional Studies* (1995)**

THE CHARACTERISTICS OF A 30–30–40 SOCIETY

In the case of the United Kingdom the 30–30–40 split is revealed by data from the Labour Force Survey which enables the identification of the following groups:

1. Those who are unemployed or non-employed with subsistence incomes (up to 50% of median income). This section of the population – the disadvantaged – includes 24.8% of the population of working age. Included in this latter group are the 1 in 4 of adult men who are either unemployed (15% of men of working age) or non-employed (10%) due to their decision to participate in government schemes, give up the quest for work (and perhaps to register themselves as permanently sick) or retire early. In the UK today the average age of retirement for men is falling towards 55 years.
2. Those with insecure jobs. This group – the marginalized and insecure – also accounts for 30% of the population, and is the fastest growing. Out of every 10 new jobs, seven are casualized, temporary, fixed contract and/or part-time. Jobs of this kind can be characterized as non-tenured. (This category excludes part-time tenured jobs defined as part-time jobs held for more than five years. Secure part-time jobs give women with family commitments the choice of combining child care with work.)
3. Those – the privileged – who have full-time tenured jobs, are self-employed workers or are part-time tenured workers. In 1975, 55% of adults were in full-time tenured jobs. Today this figure stands at 35%. Current trends are likely to accentuate this decline in the share of jobs that are full-time (40-hour per week) and that provide full holiday and pension rights.

THE CAUSES OF THE 30–30–40 SPLIT

Why has this 30–30–40 split emerged? As I shall argue, there are six main causes of which three are international and three are domestic.

The three international factors are technology, trade and international financial deregulation. Many experts attempt to downplay the dislocation caused by changes in technology and trade. An increasing body of evidence shows, however, that modern technologies allow a lot more to be done with a lot fewer people. Trade has also had an important, and will have an increasingly important, role in explaining the collapse in the demand for unskilled male labour in advanced countries. In the United States the wages of the lowest paid 20% have fallen by 20% in real terms in the last 25 years due to increased competition from products such as toys, consumer electronics and textiles made in Mexico and other newly industrializing countries. This decline in the demand for unskilled male labour is not confined to the United States and Europe but has also occurred in Japan. At the same time international financial deregulation has had a major impact on economic policy throughout the world. What international financial deregulation has done is raise the cost of capital. International financial markets are very volatile due to the existence of large scale movements into and out of different currencies. What is more, via the development of derivative markets the effects of shocks in one country are transmitted very quickly to others. This instability makes it more difficult for industrial companies to invest: industrialists will be very cautious about investing in plant and equipment if their markets are subject to large peak to trough falls in demand and sales. These peak to trough differences are greatest in the United Kingdom and United States. British Steel, for example, is at present operating at 95% capacity utilization; investment in new plant is, however, too risky due to the volatility of the economic environment and the likelihood of a sharp cyclical decline in the demand for steel.

The three national factors are labour market deregulation, the tax and benefit system, and the structure of the financial system. Of these factors the latter is of particular importance: the character of the City of London and its preferences with regard to different sectors, areas and time horizons are fundamental to an understanding of the structure of the British economy.

The key characteristics of the UK financial market are threefold. First, it is very centralized due not just to the need for liquidity but due to the existence of a unitary state and the centralisation of political power in Whitehall. Second, there is an excessive commitment to liquidity. However this concern with the degree of liquidity of the financial system, which is the abiding preoccupation of the Bank of England, raises the cost of capital. Third, there is a disengagement from, and an unwillingness to make, commitments to investment in fixed assets. At present Britain is second only to Spain in the cost of capital; the required rate of return is 20% for two-thirds of firms, and the average payback period two to three years. In the rest of the OECD returns of 12–15% suffice and the average payback period is five years. In the UK a premium is placed on shareholder rewards and a high cost of capital. These concerns manifest themselves in dividend pay-outs which, as a share of profits, are four times larger than in Japan and two times larger than in the United States. The City of London argues that dividends are a return for risk. What is not explained however is why expected returns are two times larger than in the US.

A significant cause of these differentials is the fact that markets do not accurately value companies' future returns rationally. Shareholders place a heavy emphasis on rewards in the first few years of a project, and undervalue future rewards. Due to such irrational time-discounting individual shareholders may switch from companies that provide a higher long-run rate of return to ones that provide higher short-term returns. As a result the share value of the former will decline, and the company with the superior investment programme will have to issue more shares to raise a given amount of capital. In these circumstances its cost of capital will have risen.

THE DETERMINANTS OF EXCLUSION

Companies that suffer from the undervaluation of deferred profits, the destabilizing role of changes in shareholder values and the related predominance of short-term debt often displace the consequent financial risks onto their employees. The adoption of hire and fire policies and the moves by, for example, British retailers to zero hour contracts under which their staff receive no holiday or sickness pay, have no guaranteed hours and receive no guaranteed wage are examples of this transference of financial risks.

Tendencies towards greater insecurity, marginalization and exclusion that stem from the functioning of labour markets and the underlying dynamics of the financial system are reinforced by other factors. Examples include the operation of the tax and benefit system, the marketization of health care and education, the ghettoization of recipients of income support in housing association dwellings, the rolling back of local authority expenditure on central city shopping areas and the activities of private security guards in shopping malls who will move pensioners on lest they deter shoppers from spending money, reduce sales and restrict the growth of asset and shareholder values.

An important example of these further sources of exclusion lies in the area of pension provision. With the erosion of the system of state pensions and of the intergenerational contract that lay behind it, increasing numbers of people will need a personal pension to top up their state pension. The way in which private pensions work is that individuals accumulate capital through payments into a regular savings fund. The savings are invested in the stock market. At retirement this accumulated fund is used to purchase an annuity. The size of the pension depends on the level of savings, the size of the fund, the performance of the stock market, and the competence with which the assets were managed. At present annuity rates stand at some 9%. A pension of £9,000 per year would therefore require a fund of £100,000. Creation of a fund of this size by the year 2020 would require, at current rates of return, an investment of £2,080 per year. Yet two-thirds of people in Britain have incomes that are less than the average wage of £19,500 per year. After tax and National Insurance the average wage earner receives £14,400. Out of this sum an annual investment of £2,020 is impossible. The reality is, therefore, that in the absence of significant state intervention average pensions will be far less than £9,000 per year and that many pensioners will depend on means-tested income support even setting aside many of the practical difficulties to do with the operation of private pension schemes and the evident dangers that pensions will be missold.

CONCLUSION: THE BRITISH MODE OF GOVERNANCE

The erosion of state pensions and the development of private pensions is indicative of the systematic weakness of British institutions and British public traditions. Indeed there is a convergence between the doctrines of shareholder sovereignty, the executive discretion of Government and the rentier culture of the City and of Government. Free markets do not produce the best institutions and the best outcomes. The distribution of information between market actors is asymmetrical and none is privy to perfect knowledge about the future. In these circumstances markets will lead to virtuous/vicious circles and an unequal distribution of income.

Markets must therefore be carefully sustained by social and public action. The reform of market society depends however on political reform. At a national level it will require the construction of intermediate public institutions, a decentralization of power and constitutional change, while at a wider scale Europe will have to broker a new agreement involving the construction of a social market Europe and strict regulation and controls over the financial logic of the Anglo-Saxon world.

Section 5.5 Beyond the Welfare State?

Christopher Pierson

'Continuity and Discontinuity in the Emergence of the "Post-Fordist" Welfare State'

from ***Towards a Post-Fordist Welfare State?* (1994)**

INTRODUCTION

Much of the discourse of 'Fordism', 'neo-Fordism' and 'flexibilisation' still bears the imprint of its origins in the evaluation of changes in the nature of the labour process. However, it has long been recognised that Fordism and its transformation involve far more than simple modifications in the organisation of production. Theorists of the regulation school, for example, have concerned themselves with the overall structure of 'regimes of accumulation' or 'modes of regulation' and, more particularly, with the 'modes of societalisation' or patterns of mass integration and social cohesion within which specific strategies for economic growth may be pursued. In the latter context, increasing attention has come to be focused upon changes in the nature of the welfare state in the transition from Fordism to post-Fordism. At its simplest, the argument is that the Keynesian Welfare State, with its characteristic commitment to full employment, macro-economic demand management and a growth-funded expansion of public welfare, was appropriate to the broadly Fordist regime of accumulation that dominated from the end of the Second World War until the late 1960s. However, the crisis of Fordism which developed in the early 1970s was at the same time (and, for some commentators, above all) a crisis of the Keynesian Welfare State. The present transition towards a new regime of accumulation built upon 'flexibility' will bring with it a transformation in the nature of the welfare state and whilst few have argued that the welfare state is going to 'disappear', there is widespread support for the view that under a new regime the state's provision of welfare will be quite different (Hirsch, 1991; Offe, 1987). According to Albertsen, what we are witnessing is a 'fundamental restructuring of state interventionism' in which 'Keynesian policies aiming at full employment through national regulation of general social demand' are increasingly giving way to 'austerity policies aiming at international competitiveness in wage levels and directed primarily against the public-sector service class and the lower strata of the working class' (Albertsen, 1988, p. 349).

CLASSICAL FORDISM AND THE KEYNESIAN WELFARE

In essence, the 'Golden Age' of Fordism is seen to have covered the quarter century between the end of the Second World War and the turn of the 1970s. Very broadly, this was a period of unprecedented and sustained economic growth, based upon the dominance of mass production and mass consumption (especially of consumer goods) and massified, semi-skilled labour.

Domestically, the new order was secured around:

1. Keynesian economic policies to sustain demand, to secure full employment and to promote economic growth;
2. the development of a more or less 'institutional' welfare state to deal with the dysfunctions arising from the market economy, 'to establish a minimum wage, to generalise mass consumption norms, and to coordinate the capital and consumer goods sectors' (Jessop, 1988b: 5); and
3. broad-based agreement between left and right, and between capital and labour, over these basic social institutions (a managed market economy and a welfare state) and the accommodation of their (legitimately) competing interests through elite-level negotiation.

Thus, the welfare state under Fordism was shaped by both the accumulation needs of capital (including mass consumption as an important component in the valorisation of capital) and the defensive strength of the organised working class. It provided not only the class basis for mobilisation behind the welfare state, but also the corporate basis – in the rise of both organised labour and capital – and the institutional basis with the rise of the interventionist state.

'THE CRISIS OF FORDISM'

For many commentators, 'the crisis of Fordism', increasingly apparent from the early 1970s, was itself a product of the cumulative rigidities built into the post-war Fordist settlement. Those very same Fordist arrangements which had secured the stability that made renewed capital accumulation possible in the period after 1945 had now grown 'sclerotic' and become a fetter upon continued economic growth.

The welfare state was seen to be deeply implicated in this self-precipitating 'crisis of Fordism'. First, there was the burden of funding a constantly expanding welfare budget. Social expenditure grew rapidly in the post-war period, rising across the OECD countries from 12.3 per cent of GDP in 1960 to 21.9 per cent in 1975. Increasingly, this expenditure (especially in the case of social security and pension payments) was regarded not as an investment in 'social capital' or the meeting of a social obligation, but as an 'unproductive' cost, which diverted resources away from the (shrinking) productive sectors of the economy. Rapidly rising levels of social expenditure were seen not as a way of generating 'human capital' or sustaining demand, but as an economic disincentive to both capital and labour. High marginal taxation rates, bureaucratic regulation of business and the growth of public sector employment were seen to be 'squeezing out' productive private investment. Meanwhile, the commitment to full employment and to a rising 'social wage' strengthened the defensive power of the organised working class, driving up wage costs beyond corresponding rises in productivity, hampering the process of 'structural adjustment' and consolidating the veto powers of organised labour. As economic growth faltered, the costs of the entitlement programmes of the welfare state grew, while the revenues out of which these could be funded declined, generating the much discussed 'fiscal crisis' of the mid-1970s. At the same time, the institutions of corporatist intermediation which had been established to reconcile the interests of state, capital and labour became increasingly an obstruction to economic reorganisation. In Harvey's account, 'big labour, big capital, and big government [were locked] into what increasingly appeared as a dysfunctional embrace of such narrowly defined vested interests as to undermine rather than secure capital accumulation' (Harvey,

1989a, p. 42). Thus, the institutions of the Fordist welfare state, which had once secured the grounds for capital accumulation by sustaining effective demand and managing the relations between capital and labour, had under new circumstances become a barrier to further economic growth. Governments' attempts to meet the crisis of Fordism with traditional Fordist solutions simply intensified their difficulties and, throughout the developed industrialised world, the politico-economic crisis of Fordism manifested itself in the historically unprecedented form of 'stagflation'.

'POST-FORDISM' AND THE WELFARE STATE

This crisis of Fordism and of its corresponding welfare state form is seen to have prompted a process of social and political restructuring in the quest to establish a new basis for sustained capitalist economic growth. According to Jessop, we can see this 'crisis of the welfare state as an opportunity for capital forcibly to re-impose the unity of economic and social policy in the interests of renewed accumulation' (Jessop, 1991a, p. 90).

[These restructurings] are seen to have profound consequences for the nature of the state, and especially the welfare state, under post-Fordism. There is, in fact, very limited support for the classical neo-liberal view that the problems of the welfare state can be effectively resolved by simply transferring functions from the (unproductive) state to the (value-generating) market. Indeed, there is a widespread recognition that, far from seeing a withdrawal of the state from intervention in the organisation and reproduction of labour power, the role of the state in training and the movement in and out of paid work may actually increase under post-Fordism. But such state intervention is likely to be increasingly concentrated upon the sphere of production, with even welfare provision being geared less towards the needs of clients than towards improving the international competitiveness of export-oriented industry.

Two changes in the general transition towards post-Fordism are of especial importance in recasting welfare state policy. First, there are the ways in which 'flexibilisation' of the *international* political economy has undermined the pursuit of Keynesian policies at a *national* level. Thus, the deregulation of international markets and financial institutions has tended to weaken the capacities of the interventionist state, to render all economies more 'open' and to make national capital and more especially national labour movements much more subject to the terms and conditions of international competition. In as much as the Fordist welfare state truly was a Keynesian welfare state, those changes in the international economy which have precipitated a decline of Keynesianism may be seen to have had a very material effect on the welfare state. The prospects for sustaining long-term corporatist arrangements within particular nation states (including the institutionalisation of a 'social wage') seem even less promising in a deregulated international economy.

A second challenge to the bases of the traditional welfare state comes from changes in the labour process and the organisation of employment associated with flexibilisation under post-Fordist imperatives. Changes in patterns of employment and corresponding class formation bring with them modification in both the patterns of dependency and the patterns of political support within a post-Fordist welfare state. It has been argued for some time that, partly as a result of the growth of the welfare state itself, the advanced industrial societies in which Fordism was most effectively entrenched have seen the emergence of a new line of political cleavage drawn between those dependent for their consumption respectively upon the public and the private sectors (Dunleavy, 1980; Saunders, 1986). The post-Fordist epoch is one in which the political power base of the public sector is increasingly out-powered and out-voted by the interests of the private sector. There is a consequent erosion of the basis of political support upon which the Fordist welfare state was built. For other commentators, it seems likely that division in the workforce between 'core' and 'periphery' will accelerate the transition from a 'one nation' welfare state, built around the 'objective of providing a high and rising standard of benefit . . . for all citizens as of right', towards a 'two nations' or 'Americanised' welfare state, in which there is 'a self-financed bonus for the privileged and stigmatising, disciplinary charity for the disprivileged' (Hoggett, 1994). At worst, it may lead to a wholesale residualisation of state welfare, as the securely employed middle classes and the skilled 'core' of the working class defect from public welfare, leaving the state to provide residual welfare services for an excluded minority at the least possible cost to a majority who are now sponsors but not users of these public services.

BRITAIN: TOWARDS A POST-FORDIST WELFARE STATE?

Of all the strategic responses to this 'crisis of Fordism', it is probably the prescriptions of the neo-liberals that represent the most fundamental challenge to the existing welfare state and nowhere has this neo-liberalism been more enthusiastically embraced over the past decade or so than in the UK. Certainly, the policy agenda of the Conservatives after 1979 contained many of the core elements identified with a neo-liberal strategy for transition to a post-Fordist welfare state. In terms of the 'mixed economy', there was a commitment to return publicly-owned industries to the private sector, to limit government interventions in the day-to-day management of relations between employers and employees and to 'redress' the balance of power between capital and labour. There was a commitment to sustained or enhanced economic growth, but this was to be achieved by abandoning Keynesian economics and the commitment to full employment in favour of monetarism and supply-side reforms. On the welfare state itself, there was to be a drive to cut costs by concentrating resources upon those in greatest need, to restrain the bureaucratic interventions of the 'nanny state' in the day-to-day life of its citizens, a greater role for voluntary welfare institutions and encouragement to individuals to make provision for their own and their families' welfare through the private sector. Overall, social policy was to be made much more explicitly subservient to the interests of the economy and wherever possible subject to the rigours of market-based competition.

Given an uninterrupted tenure of office, the Conservatives had considerable success in pursuing their social policy agenda during the 1980s. Unemployment was allowed to reach unheard of levels (officially in excess of three million) and a string of major corporations and utilities were returned to the private sector. Mass unemployment and changes to the laws on employment and trade union rights were used to reimpose labour market disciplines. There was a major (and popular) drive to sell off public sector housing which saw (direct) public expenditure on housing cut by a half in real terms between 1979 and 1983. Under the first two Thatcher administrations, there were increases in some NHS charges, the 'contracting out' of some ancillary services in education and the health service, some curtailing of benefit rights, a less generous basis for the upgrading of retirement pensions and tax relief to encourage the private provision of pensions and health care. The 1986 Social Security Act was quite explicit in its concern with restraining costs, 'meeting genuine need' (which implied the more effective 'targeting' of available resources) and encouraging greater labour-force participation through the time-served principle of 'less eligibility'. The election of a third Thatcher administration has been described as ushering in 'the most decisive break in British social policy since the period between 1944 and 1948' in which the modern British welfare state was created (Glennerster *et al*, 1991, p. 389; see also Le Grand, 1990). The unifying themes of these reforms of the late 1980s were the government's determination to divide welfare purchasers from welfare providers, the encouragement of private provision, the wish to relocate welfare 'within the community' and (wherever possible) the marginalisation of local government. This has led both to a devolution of day-to-day authority into the hands of local managers and a concentration of formal powers in the hands of central government. At the same time, there has been an expansion of 'private' – though often publicly subsidised – welfare for those who can afford it. There has thus been an enhancement of personal pension rights and private health care for those in secure and well-paid full-time employment.

BRITAIN: FROM 'FLAWED "FORDISM" TO FLAWED POST-FORDISM'?

How appropriate is it to understand this blend of continuity and change in British social policy as constituting the transition towards a post-Fordist form of the welfare state? There has been a transformation of the organisation of public education and health, for example, and it seems clear that the division between welfare purchasers and welfare providers is unlikely to be reversed under any foreseeable political regime. Certainly, the government has sought to introduce some of the working practices associated with post-Fordism – subcontracting, short-term contracts, casualisation, exclusion of trade unions and collective bargaining – into employment within the public sector. It has also sought to introduce within the public sector arrangements which mimic the allocational task of the market and taken steps to deregulate and diversify both the public and private provision of welfare services. However, it is perhaps more appropriate to think of these welfare reforms as a domestic response to characteristically 'post-Fordist' changes in the global political economy rather than as the inauguration of a distinctively post-Fordist welfare state.

First, there are many elements identified with the 'post-Fordist' welfare state which have long been generic features of British welfare arrangements. Those who wish to draw a sharper contrast with the past have generally given too benign an account of the decommodified or citizenship welfare state that is said to have preceded the recent epoch of reform. Elements of residualisation, of two-tier provision, of state reliance upon unpaid (female) labour within the family, of prejudice against particular minorities and of legislation subordinating welfare needs to the imperatives of labour markets are as old as the welfare state itself. Second, accounts of transformation rely upon an account of 'the crisis of the Fordist welfare-state' which is rather poorly borne out by the empirical evidence. Whilst the welfare state has faced and will face enormous problems, many of them related to limited resources, it is a mistake to understand these problems and their potential resolution within the 'crisis' logic of the 1970s and 1980s.

Third, it remains quite unclear that the reforms associated with the post-Fordist welfare state can be generically understood as 'promoting flexibility'. Certainly, some reforms may have made wages more (downwardly) flexible and the reform of pensions, for example, will mean increased choice for some. But the exercise of greater consumer choice may reduce the options not only for public welfare providers but also for other consumers. For example, in the selling off of public sector housing, enhanced choice for some has effectively restricted the choice of others. While popular sentiment may still see the welfare state as an expression of society's benevolent attitude to the less privileged, and the new right see it as a producer-dominated constraint upon the productive economy, neither of these amount to an adequate explanation of its (near universal) emergence in market-based societies. The welfare state exists in part to provide those public goods which markets cannot deliver or where market provision will have perverse outcomes. As Barr has observed, the welfare state has a 'major efficiency role' and, in a context of market failures, 'we need a welfare state for efficiency reasons, and would continue to do so even if all distributional problems had been solved' (Barr, 1987, p. 42).

Fourth, we need to recognise that welfare policy, like policy in any other area, is driven by a political as well as an economic imperative. Thus the limits of government policy may be set by a broadly post-Fordist international political economy, but within these limits government policy will be determined by a range of other considerations which may not always optimise, or indeed seek to optimise, flexibility.

Guy Standing

'Alternative Routes to Labor Flexibility'

from ***Pathways to Industrialization and Regional Development* (1992)**

INTRODUCTION

Regular wage labor is neither the present nor the future for a growing proportion of the population of industrialized countries. Almost perversely, this realization has coincided with a loss of a radical, progressive vision of the Good Society. In the past decade or so, we have seen the limit of Gramsci's brilliant insight, Fordism, which no longer seems the predominant form of the labor process. However, most of us are still searching for an alternative paradigm of the present and an alternative avenue to utopia. Without some vision, however modest, we might as well stay in our gardens or backyards. And perhaps that is a clue, for access to a garden is not just a residue of Thomas More's *Utopia*, but a part of many modern visions, encapsulating the possibility of combining different forms of paid and own account work. In the reality of Europe today, we live in disturbing times, when inequalities are worsening and when the Galbraithian strictures on American society of the 1960s are being writ large in European cities such as Rome, London, and Amsterdam, with extraordinary private affluence (of a few) coexisting with public squalor (of many).

This chapter is an attempt to reflect on two routes that different societies might follow to varying degrees in the 1990s. The first will be called *subordinated flexibility*, representing a continuation of the dominant trends of the 1980s; the second will be called *cooperative flexibility*, although one might call it integrative or social flexibility, in that it represents the vital positive value of labor security and collective regulations. The hope is that we can point to a more attractive set of options than implied by the dominant trends in most of Western Europe and North America.

SOME CONCEPTUAL PRELIMINARIES

It has been argued that since the 1970s the postwar social consensus has been disrupted by the erosion of seven forms of labor security or 'labor rights', each of which was strengthened in the preceding thirty years or so. Those are:

1. income security;
2. labor market security;
3. employment security;
4. work security;
5. job, or occupational, security;
6. labor process security; and
7. labor reproduction security.

These are defined elsewhere, but it is strategically important to distinguish between what have been called *meta* rights and *instrumental* rights. The former are those that are long-term goals reflecting fundamental values, the attainment of which is not realistic in the short term but which should he pursued steadily and consistently; instrumental rights are those that are necessary or helpful in the pursuit of real human values and aspirations. All social and economic policies should be judged by whether or not they enhance the prospect of attaining meta rights.

In this chapter it will be taken as axiomatic that the meta rights of a Good Society would include the following:

1. generalized income security for all, consistent with the economic level of development;
2. declining inequality, not the reverse;
3. labor process security, involving the strengthening of a sense of community, solidarity and active participation, through economic democracy; and
4. the right to occupation, and the right to work.

Labor process security is taken to mean that workers and particularly their representative organizations have the capacity to determine, or the possibility of influencing, the development of the labor process, defined in terms of working conditions, work structures, skill acquisition and reformulation, and so on. If, for instance, management has almost full control over labor relations, and the workers little or none, then there is definitionally labor process insecurity. One could never attain 'full' labor process security. Nevertheless, labor process security is a meta right.

Another distinction that must be preserved is between job and employment security. In the 1980s many groups lost both forms of security, some lost only one, and some conceded on one to gain on the other.

One must also distinguish between an *occupation* and a *job*. An occupation involves a career of learning and the mastery, or possession, of the mysteries of a craft or profession. There is a sense of continuity, a progression, and above all an acquisition of status, control, and autonomy. In intention, the concept of occupation refers to a positive idea of work, as creative activity, the combination of intellectual and manual activities – conception and execution – in the context of 'skill' refinement. One is tempted to believe that for many workers in the latter part of the twentieth century the development of the technical and social divisions of labor, under the guise of 'numerical' and 'functional' labor flexibility, has debased the notion of occupation.

By contrast, a 'job' is a much humbler word, conveying an activity, a limited and limiting piece of work, a narrow set of tasks. A job is what one does now, an occupation is what one is.

SUBORDINATED FLEXIBILITY: LEGACY OF THE 1980S

An essential part of the supply-side, libertarian agenda of the 1980s was that individualistic labor regulations should displace collective protective regulations.

Much of this agenda has been realized, to some extent. The individualization of employment contracts has extended way down the pay and occupational status scales and such contracts seem to have become more comprehensive as their practice has spread. While this trend is well advanced internationally, there has also been a steady erosion of collective and protective regulations. *Implicit deregulation* has been occurring by:

(a) nonimplementation of protective regulations;
(b) inadequate resources and personnel devoted to the task of policing existing regulations;
(c) erosion of the capacity to resist among those denied their rights to protection;
(d) a growing and cultivated sense of ambiguity among potential beneficiaries about the validity of such rights; and
(e) an increasing loss of entitlements to protection by virtue of their labor status.

Explicit deregulation has come about from the repeal of protective laws. Examples of explicit deregulation include the emaciation of the Wages Councils and the abolition of restrictions on night work for teenagers in the United Kingdom and legislation in the Federal Republic of Germany permitting unions and employers to derogate from working time legislation.

Although there has been both implicit and explicit erosion of protective mechanisms and collective institutions, it would be misleading to characterize this as pure deregulation. Proindividualistic regulations have been displacing procollective regulations. For proponents of what we should call a path to *subordinated flexibility*, the 1980s might be seen as a transitional era, between one based on collective regulation to one of 'contractualization'.

Four outcomes of this flexibilization deserve to be stressed – labor force fragmentation, the erosion and restructuring of economic entitlements (including welfare), the increasing need to combat the social and economic exclusion of a minority, and the continuing search for new means of raising productivity, given the partial demise or deficiencies of Taylorist methods in the more flexible labor process in the 1980s and anticipated for the 1990s.

Labor Force Fragmentation

In the context of more flexible labor processes, one can identify seven strata that deviate from the presumed norm of regular, full-time wage and salaried employment. At the top of the heap is an *elite* who have become 'capitalist employees', through profit sharing, the acquisition of subsidized shares, and also by virtue of a growing range of fringe benefits, some of which have a full value way in excess of any taxable monetized value. This elite have income security, work security, labor market security, employment security, job security, and labor process security, to the extent that they want it. It is hardly comforting that perhaps the main drawback for those involved is that their lifestyle is generally intense and fraught with stress, which ultimately threatens any individual's hold on positions of status and control.

Below this elite category one can detect what might be called frenetic proficians. A notable phenomenon of the past decade has been the spectacular growth of nominally independent 'consultants' and self-employed specialists. Key characteristics of this group are their relative youth, their frenetic work schedule and their self-satisfaction. Their expansion has been a feature of 'flexible specialization'. They have little labor security, but they have taken advantage of enterprise flexibility and the advantage to firms of using flexible specialists for short-term purposes.

A key aspect of the growth of these first two strata is that they tend to be beyond the welfare state and other regulatory institutions, increasingly having access to privatized benefits and neither contributing to nor gaining entitlement to social security benefits. This elitist detachment leads them to give political support to the associated transformation of the welfare state, from what Richard Titmus called the institutional redistributive model to the selective, residual model.

Those two strata represent the upper echelons of 'popular capitalism'. Below them are those who seem increasingly oriented to serving the interests of the upper strata, *national bureaucrats*, who retain the form of labor security obtained in the expansive times of the 1960s.

The fourth stratum down is hard to label, but might be called the *capitalist worker* stratum, consisting of those randomly fortunate wage-earning workers who, through share payments or access to successful profit-sharing schemes, accumulate savings that allow them to set up a full-time or part-time business or to live off dividends. The group may be only a tiny fraction numerically, but is ideologically rather important. They have gained in terms of income security and many have less need for employment or labor market security and so tend to be less inclined to oppose implicit or explicit erosion of those forms of labor security.

Fifth down the labor process ladder is the old *proletarian* stratum, made up of unionized, mostly male workers in regular wage employment. This was the Beveridge and Bismarkian norm for the national insurance social security system. For well-known reasons, they have been declining numerically and have had their labor process security eroded in various ways. They have lost visibly and noisily – there has been a tendency for this fifth stratum to lose in terms of income security, job security, work security, employment security, labor market security, and, most crucially, labor reproduction and labor process security.

Sixth is what might best be called the *flexiworker* stratum, which has grown enormously in many countries in recent years, encompassing many 'nonregular' forms of wage and other relatively low-income quasi-wage labor.

The category is particularly heterogeneous but what the various groups have in common is an absence of any form of labor security, most of all labor process security. Euphemisms are almost amusing. We have seen the emergence of 'permanent temporaries', 'self-employed employees', and 'in-house outworkers'. Some firms maintain workers on temporary contracts for many years; some employment agencies put workers on permanent contracts as temporaries, guaranteeing them a retainer and employment status but not any particular job.

Flexiworkers are likely to be in and out of jobs, whether full-time or part-time, but rarely in them long enough to earn entitlement to occupational welfare, social security, or even privatized insurance benefits, let alone long enough to develop the confidence to join or form unions.

This is subordinated flexibility, which can only be a threat to occupation by virtue of the incessant insecurity and casualization at the point of production. There should also be concern about the skills that such a trend emphasizes and deemphasizes. Skills that will be fostered include mobility, those underdeveloped will include understanding of production processes. This is not flexible specialization but specialized flexibility.

The scope for employment insecurity and income insecurity in this form of 'contractual fragmentation' is enormous, even though the diversity provides potential flexibility for employers and workers that might be advantageous for either or both. But perhaps the principal characteristic of flexiworkers is that they commonly lack entitlement to both enterprise/occupational welfare and national insurance-based state welfare. Some groups may have access to some benefits, some to others. But the dominant picture is one of exclusion and inadequacy of entitlement. Ironically, of course, they have a greater need for such benefits because they lack all six forms of labor security. This means that they are always threatened by the fear of floating into the seventh stratum.

This fragment, which also mushroomed in the 1980s, has attracted various epithets, the most controversial being 'underclass'. [O]ne might best describe it as the *detached stratum*, since the defining characteristic is a detachment from regular economic activity, often involving long-term or chronic recurrent unemployment, and an equally chronic need for state transfers from outside national insurance schemes.

The 'Fordist system' may have broken down, but if so it has done so not just because mass production based on regular wage employment has shrunk but because the regulatory framework has become dysfunctional, given that more and more of the population of advanced capitalist countries have become detached from productive society, that is, detached as workers.

Consumer capitalism depends on workers consuming mass-produced products, but increasingly people's identities are tied only to consumption, not to production. This applies to most of the labor fragments, most of which are not easily organizable because productively they have no collective identity. It has almost reached the stage when one cannot envisage collective *class* action any more, only sectional action, on for instance ethnic, age, or gender issues.

The labor regulatory framework built up in the era of welfare capitalism is also becoming dysfunctional, because more of the population are detached from productive society. They may be completely detached, as in the case of the long-term unemployed and many of those in 'labor market schemes'. Or they may be behaviorally detached, as with most flexiworkers, who have little continuity of employment and thus little access to nonwage components of working-class income, such as earnings-related benefits that were built up as incentives to continuity of employment, as in the Federal Republic of Germany. Historically, the social insurance welfare system has had a regulatory function, with earnings-related mechanisms expected to increase productivity, through incentive and worker commitment effects and the firms return to investment in training.

A final labor process trend closely associated with the growth of subordinated flexibility [is] the growth of what might be called 'trainingitis', which derives from the erosion of skill in the traditional senses of that term. The more occupations are split into jobs, the more labor statuses are flexible, and the larger the sixth and seventh labor force strata grow, the greater the need for *job* training and labor market training. In many European countries the role of the state in this has grown enormously, and the script is that a larger proportion of the population will have work 'careers' consisting of flexible combinations of short-term jobs preceded and succeeded by training and retraining, a pattern potentially leading to a whirlwind of jobs interspersed with training.

In sum, if the libertarian, supply-side path to flexibility persists, the following scenario seems the most likely:

1. a contractualization of the labor process, with proindividualistic regulations constraining collective action;
2. welfare pluralism, with the state as fall-back 'safety net' provider, with privatized benefits for the upper strata and with voluntary private services left to fill the gaps left by an incomplete insurance system.
3. privatization of social policy as well as of economic spheres;
4. workfare replacing means-tested and universal transfer payments for those deemed to be 'employable';
5. more policing of welfare 'scroungers';
6. a steadily growing police presence in civil society to control the losers in an aggressively competitive economic environment;
7. a neo-corporatist state based on an overt employer–government alliance to replace tripartism, with trade unions shrinking and shackled by legislation and their own fragility in the context of flexible labor markets and fragmented productive systems.

Whether or not these are exaggerated as stated, they are sufficiently present as trends to suggest that a search for alternative paths would be reasonable.

TOWARDS COOPERATIVE FLEXIBILITY

There is a nucleus of another route to labor process flexibility, which builds on the corporatist traditions kept alive in the Nordic region, but which combats a critical limitation of the Nordic models. An essential element of any viable alternative must be the avoidance of labor fragmentation and income insecurity. Yet the danger is that critics of subordinated flexibility and supply side trends of the 1980s will continue to give primacy to labor market and employment security. Neither of these is really a meta right, though both should be seen as instrumental to the promotion of other labor rights.

Full employment is always possible. They managed it fairly well in slave societies, which may not be the best recommendation. However, it is not a means of removing labor fragmentation, nor is it any more a reliable means of reversing the inequalities and erosion of labor process security. The vision must surely be a social structure in which labor process security and income security are guarded and enhanced as meta rights.

Here it might be useful merely to outline the types of reform that may be shaping an alternative path to labor flexibility.

First and foremost, unless organizations can be revived to represent the collective voice of the vulnerable segments of society, notably those in the sixth and seventh strata, the necessary impetus to sustainable nonsubordinated flexibility will not emerge or will be dissipated. That is why we need to be concerned with alternative regroupings of unions, and in particular the possibility of communal unions evolving in place of industrial or craft unions. Recently, a variant of this has also been called 'associational unionism', since it associates individualized workers who could benefit from collective representation. If workers are 'post-capitalist labor' in the sense of not being in stable proletarianized relationships, they will be uncommitted to industrial unionism, just as flexiworkers can scarcely be expected to be committed to craft unionism, besides being hard to organize or to retain. But communal unionism will also not flourish if the organization merely represents an agency for job placements, advisory services, personal loans, and a source of social security for their members, even though all those functions are desirable. If that is all unions become, the state or private commercial firms will always try to turn them into individualistic entities. Only if they are constantly concerned with the primary problem of the era, redistribution, will they develop a pivotal role. There are signs that moves to redefine communal solidarity are growing in significance, and that the promotion of economic democracy, as well as industrial and political democracy, will be high on the agenda in the 1990s. Without economic democracy, one can see no alternative to the type of subordinated flexibility sketched earlier.

More advanced in practical terms is discussion and application of experimental policies to promote flexible lifestyles that build on the fragmenting tendencies in the labor process. Many of those are double-edged, in that they could be either converted into instruments of intensifying subordinated flexibility or integrated with other policies that together promote a more active, egalitarian flexibility.

If there are more part-time employment slots, and if there is a perceived need to have quicker responses to economic restructuring – perhaps as a result of more rapid and pervasive technological innovation – then it makes little sense to hope to buck the trend. It makes much more sense to facilitate flexible work patterns on terms desired by workers. That is one reason for foreseeing an era of social experimentation. Haltingly, one sees the nucleus of what could be called a *social dividend* route to flexibility, called that because ultimately it is based on redistributing the economic surplus through ways other than wage income and welfare. This route would also give precedence to the right to work over the right to employment, bearing in mind that a right to do something can only exist if there is a matching right not to do it.

One could argue that a social dividend approach is crystalizing in the various experimental policies and institutional developments taking place in various parts of Europe. One thinks of sabbatical year and 'time bank' debates in Sweden and Finland, solidarity contracts in Belgium, partial retirement schemes, career break and parental leave arrangements, the *revenu minimum d'insertion* in France, wage earner funds, renewed interest in profit sharing, the renewed growth of cooperatives in Italy and elsewhere (including the former Soviet Union, in a big way), industrial districts in the Federal Republic of Germany and in other countries, and so on. Experimentation is the general principle of the moment. Partial retirement schemes and the removal of arbitrary retirement age notions are widely regarded as cautious steps in the general direction of lifetime flexibility, especially in the context of the ageing of European societies, though such schemes can be and have been easily turned into sources of inequity and discrimination unless developed in the appropriate institutional context.

One also notices more constructive discussion of unconditional income transfers, or citizenship income grants. That would decouple labor market status and behavior from income security and facilitate flexible combinations of productive and reproductive activities, helping in the process to legalize the shadow economy and encourage the growth of the 'informal' economy. As long as governments lack the courage to promote the genuine right to income security, that is, unconditional, universal, and individual, the 'flexible specialization' potential will be restricted, and the potential pursuit of occupation stultified. Basic security from deprivation as a citizenship right will be a necessary condition. It is not the only one, for to create an environment of cooperative flexibility will require labour process security, to prevent the vulnerable from being systematically marginalized, and thus being a threat to the working community, and to combat the coercive potential of contract law replacing collective regulation. There must be regulation to provide the basic safeguards against the structural inequalities that market mechanisms are bound to produce and intensify. That is why work security and payment system security (including minimum wage protection) will remain essential components of any path to an active flexibility society. But, ultimately, the institutional structure that promotes labor process security will be far more significant for that than any number of regulations. New forms of union, new forms of collective agency and new meanings of solidarity will need to emerge.

CONCLUDING POINTS

In sum, an alternative path to flexibility will have to be based on the promotion of the meta rights of labor process and income security, and will have to evolve through institutional mechanisms geared to create three forms of democracy – political democracy, industrial democracy (i.e. through codetermination to ensure work security, protective regulations, etc.), and economic democracy (i.e. through institutional mechanisms to redistribute economic surplus equitably, involving the collectivization of profits and citizenship income dividends). This would effectively reverse the traditional social democratic and socialist agenda, since rather than nationalize the means of production, it would privatize the management and ownership functions while socializing the surplus. The form of this alternative framework is still very far from clear, but the contours are beginning to take shape.

COPYRIGHT INFORMATION

PART TWO THE ECONOMY IN TRANSITION: GLOBALIZATION AND BEYOND?

2.2 The Globalization Debate and the Transformation of Capitalism

Hirst, P. and Thompson, G. (1996) 'Globalization – A Necessary Myth?', Chapter 1 in *Globalization in Question: The International Economy and the Possibilities of Governance*, Polity, Cambridge: 1–16. Reproduced by permission of Blackwell Publishers, Oxford.

Amin, A. (1997) 'Placing Globalization', *Theory, Culture and Society*, **14**, 2: 123–137. Reproduced by permission of Sage Publications, London.

Harvey, D (1989) 'Theorizing the Transition . . . Flexible Accumulation – Solid Transformation or Temporary Fix?', Chapters 10 and 11 in *The Condition of Postmodernity*, Blackwell, Oxford: 173–197. Reproduced by permission of Blackwell Publishers, Oxford.

Lash, S. and Urry, J. (1994) 'After Organized Capitalism', Chapter 1 of *Economies of Signs and Space*, Sage, London: 1–17. Reproduced by permission of Sage Publications, London.

2.3 Geographies of Global Integration

Warf, B. (1995) 'Telecommunications and the Changing Geographies of Knowledge Transmission in the Late 20th Century', *Urban Studies*, **32**: 529–555. Reproduced by permission of Carfax Publishing Limited, PO Box 25, Abingdon, Oxfordshire, OX14 3UE, UK.

Martin, R. (1994) 'Stateless Monies, Global Financial Integration and National Economic Autonomy: The End of Geography?', in Corbridge, S., Martin, R. and Thrift, N. (eds) *Money Power and Space*, Blackwell, Oxford: 253–278. Reproduced by permission of Blackwell Publishers, Oxford.

Fagan, R. and Le Heron, R. (1994) 'Reinterpreting the Geography of Accumulation: The Global Shift and Local Restructuring', *Environment and Planning D: Society and Space*, **12**: 265–285. Reproduced by permission of Pion Limited, London.

2.4 The Rise of New Capitalisms

Brohman, J. (1996) 'Postwar Development in the Asian NICs: Does the Neoliberal Model Fit Reality?', *Economic Geography*, **72**: 107–130. Reproduced by permission of Economic Geography, Clark University.

Daly, M.T. (1994) 'The Road to the Twenty-First Century: The Myths and Miracles of Asian Manufacturing', Chapter 8 in Corbridge, S., Martin, R. and Thrift, N. (eds) *Money Power and*

Space, Blackwell, Oxford: 165–188. Reproduced by permission of Blackwell Publishers, Oxford.

Stenning, A. and Bradshaw, M. (1998) 'Globalization and Transformation: The Changing Geography of the Post-Socialist World'. Commissioned for inclusion in this collection.

Mitchell, K. (1995) 'Flexible Circulation in the Pacific Rim: Capitalisms in Cultural Context', *Economic Geography*, **71**: 364–382. Reproduced by permission of *Economic Geography*, Clark University.

2.5 Regulating the New Capitalism

Dicken, P. (1992) 'International Production in a Volatile Regulatory Environment: The Influence of National Regulatory Policies on the Spatial Strategies of Transnational Corporations', *Geoforum*, **23**, 3: 303–316. Reproduced with permission from Elsevier Science.

Tickell, A. and Peck, J. (1995) 'Social Regulation after Fordism: Regulation Theory, Neo-Liberalism and The Global–Local Nexus', *Economy and Society*, **24**: 357–386. Reproduced by permission of Routledge, London.

PART THREE SPACES OF PRODUCTION

3.2 Reworking the Division of Labour

Sayer, A. and Walker, R. (1992) *The New Social Economy: Reworking the Division of Labour*, Blackwell, Oxford, 'Information and Substance in Products': 62–64, 'The Persistence of Industrial Goods' 65–66, 'The Extended Division of Labor': 68–69. Reproduced by permission of Blackwell Publishers, Oxford.

Lakha, S. (1994) 'The New International Division of Labour and the Indian Computer Software Industry', *Modern Asian Studies*, **28**, 2: 381–408. Reproduced by permission of Cambridge University Press.

Daniels, P.W. (1995) 'Services in a Shrinking World', *Geography*, **80**, 2: 97–110. Reproduced by permission of the Geographical Association.

Christopherson, S. (1989) 'Flexibility in the US Service Economy and the Emerging Spatial Division of Labour', *Transactions of the Institute of British Geographers*, **14**: 131–143. Reproduced by permission of the Royal Geographical Society with the Institute of British Geographers.

3.3 Rethinking the Spatial Mosaic

Amin, A. and Thrift, N. (1992) 'Neo-Marshallian Nodes in Global Networks', *International Journal of Urban and Regional Research*, **16**: 571–587. Reproduced by permission of Blackwell Publishers, Oxford.

Allen, J. (1992) 'Services and the UK Space Economy: Regionalization and Economic Dislocation', *Transactions of the Institute of British Geographers*, **17**: 292–305. Reproduced by permission of the Royal Geographical Society with the Institute of British Geographers.

Markusen, A. (1996) 'Sticky Places in Slippery Space: A Typology of Industrial Districts', *Economic Geography*, **72**, 3: 293–313. Reproduced by permission of *Economic Geography*, Clark University.

Gertler, M.S. (1995) '"Being There": Proximity, Organization, and Culture in the Development and Adoption of Advanced Manufacturing Technologies', *Economic Geography*, **71**: 1–26. Reproduced by permission of *Economic Geography*, Clark University.

Storper, M. (1995) 'The Resurgence of Regional Economies, Ten Years Later: The Region as a Nexus of Untraded Interdependencies', *European Urban and Regional Studies*, **2** (3): 191–221. Reproduced by permission of Sage Publications, London.

3.4 New Spaces of Production

Fujita, K. and Hill, R.C. (1995) 'Global Toyotaism and Local Development', *International Journal of Urban and Regional Research*, **19**: 7–22. Reproduced by permission of Blackwell Publishers, Oxford.

Florida, R. (1996) 'Regional Creative Destruction: Production Organization, Globalization, and the Economic Transformation of the Midwest', *Economic Geography*, **72**, 3: 314–334. Reproduced by permission of *Economic Geography*, Clark University.

Henry, N., Pinch, S. and Russell, S. (1996) 'In Pole Position? Untraded Interdependencies, New Industrial Spaces and the British Motor Sport Industry', *Area*, **28**, 1: 25–36. Reproduced by permission of the Royal Geographical Society with the Institute of British Geographers.

Bryson, J., Wood, P. and Keeble, D. (1993) 'Business Networks, Small Firm Flexibility and Regional Development in UK Business Services', *Entrepreneurship and Regional Development*, **5**: 265–277. Reproduced by permission of Taylor & Francis.

Selya, R.M. (1994) 'Taiwan as a service economy', *Geoforum*, **25**: 305–322. Reproduced with permission from Elsevier Science.

Sassen, S. (1994) 'Place and Production in the Global Economy', in *Cities in a World Economy*, Pine Forge, London: 1–7. Reprinted by permission of Pine Forge Press.

Roberts, S.M (1995) 'Small Place, Big Money: The Cayman Islands and the International Financial System', *Economic Geography*, **71**, 3: 237–256. Reproduced by permission of *Economic Geography*, Clark University.

PART FOUR SPACES OF CONSUMPTION

4.2 Capitalizing on Consumption

Bocock, R. (1992) 'Consumption and Lifestyles', in Bocock, R. and Thompson, K. (eds) *Social and Cultural Forms of Modernity*, Polity Press in association with the Open University: 120–154. Reproduced by permission of Blackwell Publishers, Oxford.

Harvey, D. (1989) 'Time–Space Compression and the Postmodern Condition', *The Condition of Postmodernity*, Blackwell, Oxford: 285–289. Reproduced by permission of Blackwell Publishers, Oxford.

Zukin, S. (1991) 'Real Cultural Capital', in *Landscapes of Power: From Detroit to Disney World*, University of California Press: 259–267. Reproduced by permission of the Regents of the University of California.

Cook, I. (1994) 'New Fruits and Vanity: The Role of Symbolic Production in the Global Food Economy', in Busch, L. (ed) *From Columbus to Conagra: The Global Station of Agriculture and Food Order*, University of Kansas Press, Kansas: 232–248. Reproduced by permission of University of Kansas Press.

4.3 Landscapes of Consumption

Knox, P.L. (1991) 'The Restless Urban Landscape: Economic and Sociocultural Change and the Transformation of Metropolitan Washington, DC', *Annals Association of American Geographers*, **81**: 181–209. Reproduced by permission of Blackwell Publishers, Malden, MA, USA.

Wrigley, N. and Lowe, M. (1995) 'New Landscapes of Consumption', from *Retailing, Consumption and Capital: Towards the New Economic Geography of Retailing*, Longman, London: 24–28. Reproduced by permission of Addison Wesley Longman Ltd.

Goss, J. (1993) 'The "Magic of the Mall": An Analysis of Form, Function, and Meaning in the Contemporary Retail Built Environment', *Annals of the Association of American Geographers*, **83**, 1: 18–47. Reproduced by permission of Blackwell Publishers, Malden, MA, USA.

4.4 Pick 'N' Mix: The Commodification of Identity

Clammer, J. (1992) 'Aesthetics of the Self: Shopping and Social Being in Contemporary Urban Japan', in Shields, R. (ed) *Lifestyle Shopping: The Subject of Consumption*, Routledge, London: 195–215. Reproduced by permission.

Thrift, N. and Leyshon, A. (1992) 'In the Wake of Money: The City of London and the Accumulation of Value', in Budd, L. and Whimster, S. (eds) *Global Finance & Urban Living: A Study of Metropolitan Change*, Routledge: London, 282–311. Reproduced by permission.

PART FIVE WORK, EMPLOYMENT AND SOCIETY

5.2 The Transformation of Work

Pollert, A. (1988) 'Dismantling Flexibility', *Capital and Class*, **34**: 42–75. Reproduced by permission of the author.

Beynon, H. (1992) 'The End of the Industrial Worker?', in Abercrombie, N. and Warde, A. (eds) *Social Change in Britain*, Polity Press, Cambridge, 167–183. Reproduced by permission of Blackwell Publishers, Oxford.

Allen, J. and du Gay, P. (1994) 'Industry and the Rest: The Economic Identity of Services', *Work, Employment and Society*, **8**, 2: 255–271. © BSA Publications, reproduced with permission of Cambridge University Press.

5.3 Rethinking Labour

Pearson, R. (1988) 'Female Workers in the First and Third Worlds: The Greening of Women's Labour', in Pahl, R. (ed) *On Work: Historical, Comparative and Theoretical Approaches*, Blackwell, Oxford, 449–466 and K. Purcell (1986) The Changing Experience of Employment, Macmillan. Reproduced by permission of Macmillan Ltd.

Pugliese, E. (1993) 'Restructuring of the Labour Market and the Role of Third World Migrations in Europe', *Environment and Planning D: Society and Space*, **11**, 5: 513–522. Reproduced by permission of Pion Limited, London.

Herod, A. (1997) 'From a Geography of Labor to a Labor of Geography: Labor's Spatial Fix and the Geography of Capitalism', *Antipode*, **29**, 1: 1–31. Reproduced by permission of Blackwell Publishers, Malden, MA, USA.

McDowell, L. (1991) 'Life Without Father and Ford: The New Gender Order of Post-Fordism', *Transactions of the Institute of British Geographers,* **16**: 400–419. Reproduced by permission of the Royal Geographical Society with the Institute of British Geographers.

5.4 New(?) Social Inequalities

Massey, D. (1993) 'Geography, Gender and High Technology', *South East Programme Occasional Paper Series No. 7,* Open University, Milton Keynes. This paper was first delivered as the J.B. Willans Lecture delivered at the University of Wales, Aberystwyth, in 1993. Reproduced by permission of the author and the University of Wales, Aberystwyth.

Waldinger, R. and Lapp, M. (1993) 'Back to the Sweatshop or Ahead to the Informal Sector?', *International Journal of Urban and Regional Research,* **17**, 1: 6–29. Reproduced by permission of Blackwell Publishers, Oxford.

Pinch, S. (1993) 'Social Polarization in Britain and in the United States', *Environment and Planning A,* **25**, 6: 779–796. Reproduced by permission of Pion Limited, London.

Hutton, W. (1995) 'The 30–30–40 Society', *Regional Studies,* **29** 8: 719–721. Reproduced by permission of Carfax Publishing Limited, PO Box 25, Abingdon, Oxfordshire, OX14 3UE, UK.

5.5 Beyond the Welfare State?

Pierson, C. (1994) 'Continuity and Discontinuity in the Emergence of the "Post-Fordist" Welfare State', in Burrows, R. and Loader, S. (eds) *Towards a Post-Fordist Welfare State?,* Routledge, London, 95–113. Reproduced by permission.

Standing, G. (1992) 'Alternative Routes to Labor Flexibility', in Storper, M. and Scott, A.J. (eds) *Pathways to Industrialization and Regional Development*, Routledge, London: 255–275. Reproduced by permission.

REFERENCES

Addison T and Demery L (1988) Wages and labour conditions in East Asia: A review of case-study evidence, *Development Policy Review* **6**: 371–93.

Adelman C C (ed) (1988) *International Regulation: New Rules in a Changing World Order*, San Francisco: Institute for Contemporary Studies.

Adorno T and Horkheimer M (1969) *The Dialectic of Enlightenment*, New York: Continuum.

Aglietta M (1979) *A Theory of Capitalist Regulation: The US Experience*, London: New Left Books.

Aglietta M (1982) World capitalism in the eighties, *New Left Review* **136**: 25–36.

Aglietta M (1985) The creation of international liquidity, in L Tsoukalis (ed) *The Political Economy of International Money*, London: Sage and Royal Institute of International Affairs, pp. 171–202.

Akwule R (1992) *Global Telecommunications: The Technology, Administration, and Policies*, Boston: Focal Press.

Albertsen N (1988) Postmodernism, post-Fordism and critical social theory, *Environment and Planning D: Society and Space* **6**: 339–65.

Albrow M (1997) Travelling beyond local cultures: Socioscapes in a global city, in J Eade (ed) *Living the Global City*, London: Routledge.

Aldrich H E and Zimmer C R (1986) Entrepreneurship through social networks, in D Sexton and R Smilor (eds) *The Art and Science of Entrepreneurship*, Cambridge: Ballinger.

Allen J (1988a) Fragmented firms, disorganised labour? in J Allen and D Massey (eds) *Restructuring Britain: The Economy in Question*, London: Sage, pp. 185–227.

Allen J (1988b) Service industries: Uneven development and uneven knowledge, *Area* **20**: 15–22.

Allen J (1988c) The geographies of service, in D Massey and J Allen (eds) *Uneven Development: Cities and Regions in Transition*, London: Hodder and Stoughton, pp. 124–41.

Allen J and du Gay P (1994) Industry and the rest: the economic identity of services, *Work, Employment and Society* **8**: 255–71.

Allen J and Massey D (eds) (1988) *Restructuring Britain: The Economy in Question*, London: Sage.

Altvater E (1992) Fordist and post-Fordist international division of labor and monetary regimes, in M Storper and A Scott (1992), pp. 21–45.

Altvater E (1993) *The Future of the Market*, London: Verso.

Amadeo E and Banuri T (1991) Policy, governance, and the management of conflict, in T Banuri (ed) *Economic Liberalization: No Panacea. The Experiences of Latin America and Asia*, New York: Oxford University Press.

America's Heartland (1994) The Midwest's role in the global economy, *Business Week*, 11 July: 116–24.

Amin A (1989) Flexible specialisation and small firms in Italy: Myths and Realities, in F Pyke, G Becattini and W Sengenberger (eds) *Industrial Districts and Inter-Firm Cooperation in Italy*, Geneva: ILO, pp. 185–219.

Amin A (1991) Small firms in Italy: Myths and realities, in A Pollert (ed) *Farewell to Flexibility?* Oxford: Blackwell.

Amin A and Dietrich M (1991) From hierarchy to hierarchy: The dynamics of contemporary corporate restructuring in Europe, in A Amin and M Dietrich (eds), *Towards a New Europe?* Aldershot: Edward Elgar.

Amin A and Graham S (1997) The ordinary city, *Transactions of the Institute of British Geographers* **22**, 4: 411–29.

Amin A and Thrift N (1992) Neo-Marshallian nodes in global networks, *International Journal of Urban and Regional Research* **16**: 571–87.

Amirahmadi H (1989) Development paradigms at a crossroad and the South Korean experience, *Journal of Contemporary Asia* **19**: 167–85.

Amsden A (1989) *Asia's Next Giant*, New York: Oxford University Press.

Andreff W (1984) The international concentration of capital, *Capital and Class* **22**: 59–80.

Appadurai A (1990) Disjuncture and difference in the global cultural economy, in M Featherstone (ed) *Global Culture: Nationalism, Globalization and Modernity*, London: Sage.

Appelbaum R and Henderson J (1992) *States and Development in the Asian Pacific Rim*, Newbury Park California: Sage.

Ariff M and Hill H (1986) *Export-Oriented Industrialization: The ASEAN Experience*, Boston: Allen and Unwin.

Arthur W B (1988) Urban systems and historical path dependence, in J Ausubel and R Herman (eds), *Cities and Their Vital Systems*, Washingdon D.C.: National Academy of Engineering, pp. 85–97.

Arthur W B (1989) Competing technologies, increasing returns and lock-in by historical events, *The Economic Journal* **99**: 116–31.

Arthur W B (1990) Silicon Valley locational clusters: When do increasing returns imply monopoly? *Mathematical Social Sciences* **19**: 235–51.

Ascher K (1987) *The Politics of Privatization: Contracting Out Public Services*, London: Macmillan.

Athukorala P (1989) Export performance of 'new exporting countries': How valid is the optimism? *Development and Change* **20**: 89–120.

Atkinson J (1984) Manpower strategies for flexible organisations, *Personnel Management*, August: 29–31.

Aucamp J (1978) How to locate and select a marketing research consultancy, in A Rawnsley (ed) *Manual of Industrial Marketing Research*, Chichester: Wiley.

Aydalot P (1986) *Milieux Innovateurs en Europe*, Paris: GREMI.

Aydalot P and Keeble D (eds) (1988) *High Technology Industries and Innovative Environments: The European Experience*, London: Routledge.

Bachelard G (1964) *The Poetics of Space*, Boston: Beacon Press.

Baerwald T (1978) The emergence of a new 'downtown', *Geographical Review* **68**: 308–18.

Bagguley P, Mark-Lawson J, Shapiro D, Urry J, Walby S and Warde A (1990) *Restructuring: Place, Class and Gender*, London: Sage.

Bagnasco A (1977) *Tre Italie*, Bologna: Il Mulino.

Bakhtin M M (1984) *Rabelais and His World*, Bloomington: Indiana University Press.

Balassa B (1981) *Structural Adjustment Policies in Developing Economies*, World Bank Working Paper 464, Washington D.C.: World Bank.

Balassa B (1988) The lessons of East Asian development: An overview, *Economic Development and Cultural Change* **36**: 273–90.

Balassa B (1991) *Economic Policies in the Pacific Area Developing Countries*, London: Macmillan.

Banuri T (1991) Introduction in T Banuri (ed) *Economic Liberalization: No Panacea. The Experiences of Latin America and Asia*, New York: Oxford University Press, pp. 1–28.

Barr A and York P (1987) Work: Just the job, *The Observer*, 8 Nov.

Barr N (1987) *The Economics of the Welfare State*, London: Weidenfeld and Nicolson.

Barrett Brown M (1988) Away with all the great arches: Anderson's history of British capitalism, *New Left Review* **167**: 22–51.

Barthes R (1973) *Mythologies*, London: Paladin.

Bartlett C and Ghoshal S (1987) Managing across borders: New strategic requirements, *Sloan Mangement Review* **7**, Summer: 7–17.

Bartlett C and Ghoshal S (1989) *Managing Across Borders*, Boston: Harvard Business School Press.

Barton C (1983) Trust and credit: Some observations regarding business strategies of overseas Chinese traders in South Vietnam, in L Lim and P Gosling (eds) *The Chinese in Southeast Asia*, Singapore: Maruzen Asia, pp. 46–63.

Baudrillard J (1981) *For a Critique of the Political Economy of the Sign*, St Louis: Telos Press.

Becattini G (1990) The Marshallian industrial district as a socio-economic notion, in F Pyke, G Becattini and W Sengenberger (eds) *Industrial Districts and Inter-Firm Co-operation in Italy*, Geneva: International Institute for Labour Studies, pp. 37–51.

Beddington N (1982) *Design for Shopping Centres*, London: Butterworth Scientific.

Beechey V (1987) *Unequal Work*, London: Verso.

Begg I G and Cameron G C (1988) High technology location and the urban areas of Great Britain, *Urban Studies* **25**: 366–79.

Bell D (1973) *The Coming of Post-Industrial Society*, New York: Basic Books.

Bello W and Rosenfeld S (1990a) *Dragons in Distress: Asia's Miracle Economies in Crisis*, San Francisco: Institute for Food and Development Policy.

Bello W and Rosenfeld S (1990b) High speed industrialization and environmental devastation in Taiwan, *Ecologist* **20**: 125–32.

Benjamin W (1973) *Charles Baudelaire: A Lyric Poet in the Era of High Capitalism*, London: New Left Books.

Beresford D (1991) Adoring South African farmers name a fruit after Thatcher, bearer of unworthy praise, *The Guardian*, 17 May.

Beynon H (1992) The end of the industrial worker? in N Abercrombie and A Warde (eds) *Social Change in Britain*, Cambridge: Polity Press, pp. 167–83.

Bhagwati J (1986) Rethinking trade strategy, in J Lewis and V Kallab (eds) *Development Strategies Reconsidered*, New Brunswick: Transaction Books, pp. 91–194.

Bienefeld M (1992) Financial deregulation: Disarming the nation state, *Studies in Political Economy* **37**: 31–58.

Birch D and Jain S M (1989) *America's Future Industrial Needs*, Washington: National Association of Industrial and Office Parks.

Birley S (1985) The role of networks in the entrepreneurial process, *Journal of Business Venturing* **1**: 107–17.

Bluestone B and Harrison B (1982) *The Deindustrialization of America*, New York: Basic Books.

Boehning R W (1984) *Studies in International Labor Migrations*, London: Macmillan.

Bond R (1989) Feeling safe again, *Shopping Center World*, November: 181–4.

Boorstin D J (1961) *The Image: A Guide to Pseudo-Events in America*, New York: Harper Colophon.

Bourdieu P (1984) *Distinction: A Social Critique of the Judgement of Taste*, translated by R Nice, London: Routledge. Also published in 1987.

Bourdieu P (1987) What makes a social class? On the theoretical and practical existence of groups, *Berkeley Journal of Sociology* **3**: 1–17.

Boyer C (1987) The return of the aesthetic to city planning: Future theory as a departure from the past, paper presented at Rutgers University Center for Urban Policy Research Conference on Planning Theory in the 1990s, Washington, D.C.

Bradbury K L (1986) The shrinking middle class, *New England Economic Review*, September/October: 42–55.

Bradbury K L (1990) The changing fortunes of American families in the 1980s, *New England Economic Review* (July/September): 25–40.

Bradford C (1987) Trade and structural change: NICs and next tier NICs as transitional economies, *World Development* **15**: 299–316.

Bradshaw M (1990) New regional geography, foreign area studies and Perestroika, *Area* **22**, 4: 315–22.

Bradshaw M (ed) (1991) *The Soviet Union: A New Regional Geography?* London: Belhaven.

Bradshaw M (ed) (1997a) *Geography and Transition in the Post-Soviet Republics*, Chichester: Wiley.

Bradshaw M (1997b) Introduction: Transition and geographical change, in M Bradshaw (ed) *Geography and Transition in Post-Soviet Republics*, Chichester: Wiley, pp. 3–7.

Bradshaw M (1997c) The geography of foreign investment in Russia, 1993–1995, *Tijdschrift voor Economische en Sociale Geografie* **8**, 1: 77–84.

Bradshaw M and Hanson P (1998) Understanding regional patterns of economic change in Russia: An introduction, *Communist Economies and Economic Transformation* **10**, 3: 285–304.

Bradshaw M and Lynn N (1994) After the Soviet Union: The post-Soviet states in the world system, *Professional Geographer* **46**, 4: 439–49.

Bradshaw M, Stenning A and Sutherland D (1998) Economic restructuring and regional change in Russia, in J Pickles and A Smith (eds) *Theorising Transition*, London: Routledge, pp. 147–71.

Breheny M J and McQuaid R (1988) *The Development of High Technology Industries: An International Survey*, London: Routledge.

Brenner N (1996) The global city/territorial state nexus, mimeo, Department of Political Science, University of Chicago.

Bressand A, Distler C and Nicolaidis K (1989) Networks at the heart of the service economy, in A Bressand and A Nicolaidis (eds) *Strategic Trends in Services: An Inquiry into the Global Service Economy*, New York: Harper & Row, pp. 17–33.

British Tourist Agency (BTA) (1989) *Digest of tourist statistics 13*, London: British Tourist Authority.

Britton S (1990) Role of services in production, *Progress in Human Geography* **14**: 529–46.

Browett J (1985) The newly industrializing countries and radical theories of development, *World Development* **13**: 789–803.

Brusco S (1986) Small firms and industrial districts: The experience of Italy, in D Keeble and E Wever (eds) *New Firms and Regional Development in Europe*, London: Croom Helm, pp. 184–202.

Bryan R (1987) The state and the internationalisation of capital: An approach to analysis, *Journal of Contemporary Asia* **17**: 253–75.

Bryson J R and Daniels P W (eds) (1998) *Service Industries in the Global Economy*, Cheltenham: Edward Elgar.

Buchner B (1988) Social control and the diffusion of modern telecommunications technologies: A cross-national study, *American Sociological Review* **53**: 446–53.

Budd L and Whimster S (eds) (1992) *Global Finance and Urban Living: A Study of Metropolitan Change*, London: Routledge.

Buderi R (1992) Global innovation: Who's in the lead? *Business Week*, 3 August: 68–73.

Bull A, Pitt M and Szarka J (1991) Small firms and industrial districts, structural explanations of small firm viability in three countries, *Entrepreneurship and Regional Development* **3**: 83–99.

Burawoy M (1992) The end of Sovietology and the renaissance of modernization theory, *Contemporary Sociology* **21**: 774–85.

Burgess J (1990) The production and consumption of environmental meanings in the mass media: A research agenda for the 1990s, *Transactions Institute of British Geographers* NS **15**: 139–61.

Burmeister L (1990) State industrialization and agricultural policy in Korea, *Development and Change* **21**: 194–223.

Butchart R L (1987) A new UK definition of the high technology industries, *Economic Trends* **400**: 82–8.

Buttel F H, Larson O and Gillespie G W (1990) *The Sociology of Agriculture*, Westport: Greenwood Press.

Camagni R (ed) (1991) *Innovation Networks: Spatial Perspectives*, London: Belhaven Press.

Cantwell J (1995) The globalisation of technology: What remains of the product cycle? *Cambridge Journal of Economics* **19**: 155–74.

Carson C (1984) The underground economy: An introduction (Part 1), *Survey of Current Business* **84**, 5: 21–37.

Casey B (1987) The extent and nature of temporary work in Great Britain, *Policy Studies*, 8 July.

Castells M (1989a) High technology and the new international division of labour, *Labour and Society* **14**: 23.

Castells M (1989b) *The Informational City: Information Technology, Economic Restructuring and the Urban-Regional Process*, Oxford: Blackwell.

Castles S and Kosack G (1973) *Immigrant Workers and the Class Structure in Western Europe*, London: Oxford University Press.

Cawson A (1985) Varieties of corporatism: The importance of the meso-level of interest intermediation, in A Cawson (ed), *Organised Interests and the State*, London: Sage.

Cayman Islands Government (1993) *1992 Compendium of Statistics*, Grand Cayman: Economics and Statistics Office.

Cerny P G (1991) The limits of deregulation: Transnational interpenetration and policy change, *European Journal of Political Research* **19**: 173–96.

Cerny P G (1995) Globalisation and the changing logic of collective action, *International Organization* **34**, 2: 595–625.

Champion A G and Townsend A R (1990) *Contemporary Britain: A Geographical Perspective*, London: Edward Arnold.

Chang H J (1993) The political economy of industrial policy in Korea, *Cambridge Journal of Economics* **17**: 131–57.

Chapman M and McCullough B (eds) (1992) *Local Communities and Japanese Transplants*, Conference proceedings, Bloomington: Illinois Wesleyan University.

Chinitz B (1960) Contrasts in agglomeration: New York and Pittsburgh, *American Economic Association, Papers and Proceedings* **40**: 279–89.

Christopherson S (1989) Flexibility in the US service economy and the emerging spatial division of labour, *Transactions of the Institute of British Geographers*, NS **14**: 131–43.

Christopherson S (1993) Market rules and territorial outcomes: The case of the United States, *International Journal of Urban and Regional Research* **17**: 274–88.

Clapp R A J (1988) Representing reciprocity, reproducing domination: Ideology and the labor process in Latin American contract farming, *Journal of Peasant Studies* **16**, 1: 5–39.

Clark G (1986) The crisis of the midwest auto industry, in A Scott and M Storper (eds) *Production, Work and Territory*, Hemel Hempstead: Allen & Unwin.

Clark G and Kim W B (1993) Commentary: Industrial restructuring and regional adjustment in Asian NIEs, *Environment and Planning A* **25**: 1–4.

Cliff T (1974) *State Capitalism in Russia*, London: Pluto.

Cloar J A (1990) *Centralized Retail Management: New Strategies for Downtown*, Washington: Urban Land Institute.

Coakley J and Harris L (1983) *The City of Capital: London's Role as a Financial Centre*, Oxford: Basil Blackwell.

Coffey W and Bailly A (1991) Producer services and flexible production: An exploratory analysis, *Growth and Change* **22**: 95–117.

Cohen S S and Zysman J (1987) *Manufacturing Matters: The Myth of the Post-Industrial Economy*, New York: Basic Books.

Commonwealth Secretariat (1990) *Fruit and Tropical Products*, London.

Connor P (1989) 'Silent salesmen' at work inside and out, *Chain Store Age Executive*, November: 191–5.

Cooke K and Lehrer D (1993) The Internet: The whole world is talking, *The Nation* **257**: 60–3.

Cooke P (1989) *Localities: The Changing Face of Urban Britain*, London: Unwin Hyman.

Cooke P and Morgan K (1991) *The Network Paradigm: New Departures in Corporate and Regional Development*, Regional Industrial Research Report No. 8, Cardiff: University of Wales.

Cooke P, Moulaert F, Swyngedouw E, Weinstein O and Wells P (1992) *Towards Global*

Localization: The Computing and Telecommunications Industries in Britain and France, London: UCL Press.

Corbridge S (1984) Crisis, what crisis? Monetarism, Brandt II and the geopolitics of debt, *Political Geography Quarterly* **3**: 331–45.

Corbridge S (1988) The asymmetry of interdependence: The United States and the geopolitics of international financial relations, *Studies in Comparative International Development* **23**: 3–29.

Council of Great Lakes Governors (1994) *North America's High-Performance Heartland*, Chicago: Council of Great Lakes Governors.

Country Life/Knight, Frank and Rutley (1987) *Buying a Country House*, London: Country Life.

Cowling K and Sugden R (1987) Market exchange and the concept of a transnational corporation, *British Review of Economic Issues* **9**: 57–68.

Crandall R (1993) *Manufacturing on the Move*, Washington, D.C.: Brookings Institution.

Crawford M (1992) The world in a shopping mall, in M Sorkin (ed) *Variations on a Theme Park: The New American City and the End of Public Space*, New York: Hill and Wang, pp. 3–30.

Creigh S, Roberts C, Gorman A and Sawyer P (1986) Self employment in Britain. Results from the Labour Force Surveys, *Employment Gazette*, June.

Crewe L and Lowe M S (1995) Gap on the map? Towards a geography of consumption and identity, *Environment and Planning A* **27**, 12: 1877–98.

Crook S, Pakulski J and Waters M (1992) *Postmodernization: Change in Advanced Society*, London: Sage.

Curran J, Jarvis R, Blackburn R and Black S (1991) Small firms and networks: constructs, methodological strategies and some preliminary findings, paper presented at the UKEMRA conference, Blackpool.

Cusumano M (1985) *The Japanese Automobile Industry*, Cambridge: Harvard University Press.

Cybriwsky R (1991) *Tokyo: The Changing Profile of an Urban Giant*, Boston: Hall and Co.

Dale A and Bamford C (1989) Social polarization in Britain 1973–82. Evidence from the General Household Survey: A comment on Pahl's hypothesis, *International Journal of Urban and Regional Research* **13**: 481–500.

Daly M T (1993) No economy is an island, in G Rees, G Rodley and F Stilwell (eds) *Beyond the Market: Alternatives to Economic Rationalism*, Sydney: Pluto Press, pp. 72–90.

Daniels P W (1982) *Service Industries: Growth and Location*, Cambridge: Cambridge University Press.

Daniels P W (1983) Service industries: Supporting role or centre stage? *Area* **15**: 301–9.

Daniels P W (1985) *Service Industries: A Geographical Appraisal*, London: Methuen.

Daniels P W (1991) A world of services? *Geoforum* **22**: 359–76.

Daniels P W (1993) *Service Industries in the World Economy*, Cambridge: Blackwell.

Daniels P W (1995) Internationalization of advertising services in a changing regulatory environment, *The Service Industries Journal* **15**: 276–94.

Davis H and Scase R (1985) *Western Capitalism and State Socialism: An Introduction*, Oxford: Blackwell.

Davis M (1993) Who killed LA? A political autopsy, *New Left Review* **197**: 3–28.

Davis T C (1991) Theatrical antecedents of the mall that ate downtown, *Journal of Popular Culture* **24**, 4: 1–15.

De Carmoy H (1990) *Global Banking Strategy*, Oxford: Blackwell.

de Cecco M (1987) *Money and Innovation*, Oxford: Blackwell.

De Certeau M (1984) *The Practice of Everyday Life*, Berkeley: University of California Press.

De Certeau M (1985) Practices of space, in M Blonsky (ed) *On Signs*, Cambridge: Blackwell, pp. 122–45.

Department of Employment (1989) *New Earnings Survey*, London: HMSO.

Dex S (1985) *The Sexual Division of Work*, Brighton: Wheatsheaf.

Dicken P (1990) Seducing foreign investors – the

competitive bidding strategies of local and regional agencies in the United Kingdom, in M Hebbert and J C Hansen (eds) *Unfamiliar Territory: The Reshaping of European Geography*, Aldershot: Averbury, Chapter 12.

Dicken P (1992) *Global Shift: The Internationalization of Economic Activity*, 2nd edition, London: Paul Chapman.

Dicken P (1993) The changing organization of the global economy, in R Johnston (ed) *The Challenge for Geography – A Changing World: A Changing Discipline*, Oxford: Blackwell, pp. 31–53.

Dietz J (1992) Overcoming under-development: What has been learned from East Asian and Latin American Experiences?, *Journal of Economic Issues* **26**: 373–83.

Diniz C C and Borges Santos F (1995) Manaus: A satellite platform in the Amazon region, Working Paper, Brazil: CEDEPLAR, Universidad Federal de Minas Gerais.

Diniz C C and Razavi M (1994) Emergence of new industrial districts in Brazil: Sao Jose dos Campos and Campinas cases, Working Paper, Brazil: CEDEPLAR, Universidad Federal de Minas Gerais.

Doner R (1991) *Driving a Bargain: Automobile Industrialization and Japanese Firms in Southeast Asia*, Berkeley: University of California Press.

Dosi G and Orsenigo L (1985) Order and change: An exploration of markets, institutions and technology in industrial dynamics, SPRU Discussion Paper no. 22, Brighton.

Dosi G, Pavitt K and Soete L (1990) *The Economics of Technical Change and International Trade*, New York: NYU Press.

Douglass M (1993) Social, political and spatial dimensions of Korean industrial transformation, *Journal of Contemporary Asia* **23**: 149–72.

Dowd K (1988) *Private Money*, London: Institute of Economic Affairs.

Drucker P (1993) *Post-Capitalist Society*, Oxford: Butterworth-Heinemann.

Duncan S (1989) Uneven development and the difference that space makes, *Geoforum* **20**: 131–9.

Dunford M (1990) Theories of regulation, *Environment and Planning D: Society and Space* **8**: 297–322.

Dunford M and Kafkalas G (1992) The global-local interplay, corporate geographies and spatial development strategies in Europe, in M Dunford and G Kafkalas (eds) *Cities and Regions in the New Europe: The Global-Local Interplay and Spatial Development Strategies*, London: Belhaven, pp. 3–38.

Dunleavy P (1980) The political implications of sectoral cleavages and the growth of state employment, *Political Studies* **28**: 364–84 and 527–49.

Dunning G H and Norman G (1987) The location choice of offices of international companies, *Environment and Planning A* **19**: 613–31.

Eade J (ed) (1997) *Living the Global City*, London: Routledge.

Eco U (1986) Function and sign: The semiotics of architecture, in M Gottdiener and A Ph Lagopoulos (eds) *The City and the Sign*, New York: Columbia University Press, pp. 55–86.

Economist, The (1992) A survey of financial centres, *The Economist*, 27 June.

Economist, The (1993) A survey of international banking, *The Economist*, 10 April.

Elfring T (1988) *Service Employment in Advanced Economies: A Comparative Analysis of its Implications for Economic Growth*, Aldershot: Avebury.

Elger T (1987) Flexible futures? New technology and the contemporary transformation of work, *Work, Employment and Society* **1**, 4: December.

Elger T (1990) Technical innovation and work reorganization in British manufacturing in the 1980s: Continuity, intensification or transformation, *Work, Employment and Society* **4**, 2, Special Supplement: 67–101.

Emmerij L (1987) *Development Policies and the Crisis of the 1980s*, Paris: Development Centre for the OECD.

Emmott B (1992) *Japan's Global Reach*, London: Century.

Entertainment anchors: New mall headliners (1989) *Chain Store Age Executive* **54**: 63–5.

Epstein G and Gintis H (1992) International capital markets and the limit of national economic policy, in T Banuri and J B Schor (eds) *Financial Openness and National Autonomy*, Oxford: Clarenden Press, pp. 166–97.

Erickson R (1983) The evolution of the suburban space economy, *Urban Geography* **4**: 95–121.

Ernst D (1985) Automation and the worldwide restructuring of the electronics industry: Strategic implications for developing countries, *World Development* **13**, 3: 342–3.

Esping-Anderson G (1990) *The Three Worlds of Welfare Capitalism*, Cambridge: Polity Press.

Euromonitor (1993) *European Advertising, Marketing and Media Data 1992*, London: Euromonitor.

Evans L (1990) The 'demographic dip': A golden opportunity for women in the labour market? *National Westminster Bank Quarterly Review*, February: 48–69.

Evans P B and Tigre P B (1989) Going beyond clones in Brazil and Korea: A comparative analysis of NIC strategies in the computer industry, *World Development* **17**, 11: 1751–2.

Fagan R H (1986) Australia's BHP Ltd: An emerging transnational resources corporation, *Raw Materials Report* **4**: 46–56.

Fagan R H (1989) Social relations and spatial structures in global capitalism: The case of the Australian steel industry, *Environment and Planning A* **21**: 671–3.

Fagan R H (1990) Elders IXL Ltd: Finance capital and the geography of corporate restructuring, *Environment and Planning A* **22**: 647–66.

Fagan R H and Bryan R (1991) Australia and the changing global economy: Background to social inequality in the 1980s, in J O'Leary and R Sharp (eds) *Social Inequality in Australia: The Social Justice Collective*, Melbourne: Heinemann, pp. 7–31.

Fagan R H and Rich D C (1991) Industrial restructuring in the Australian food industry: Corporate strategy and the global economy, in P Wilde and R Hayter (eds) *Industrial Transformation and Challenge in Australia and Canada*, Ottawa: Carleton University Press, pp. 175–94.

Featherstone M (1987) Lifestyle and consumer culture, *Theory, Culture and Society* **4**, 1: 55–70.

Febre R (1986) Contract work in the recession, in K Purcell, S Wood, A Waton, A and S Allen (eds) *The Changing Experience of Employment*, Basingstoke: Macmillan.

Fielding A J (1992) Migration and social mobility: South east England as an escalator region, *Regional Studies* **26**: 1–15.

Fine B and Harris L (1985) *The Peculiarities of the British Economy*, London: Lawrence and Wishart.

Fiske J, Hodge R and Turner G (1987) *Myths of Oz: Readings in Australian Popular Culture*, Boston: Unwin Hyman.

Fitting a shopping center to downtown (1991) *Urban Land*, July: 28–29.

Fleron F and Hoffman E (1993) Communist studies and political science: Cold war and peaceful coexistence, in E Fleron and E Hoffman (eds) *Post-Communist Studies and Political Science: Methodology and Empirical Theory in Sovietology*, Boulder: Westview Press, pp. 3–23.

Florida R (1991) The new industrial revolution, *Futures* **23**: 559–76.

Florida R and Jenkins D (1996) *Transfer and Adoption of Organizational Innovation: Japanese Transplants in the United States*, Pittsburgh: Carnegie Mellon University, Center for Economic Development.

Florida R and Kenney M (1990) Silicon Valley and Route 128 won't save us, *California Management Review* **33**: 68–88.

Florida R and Kenney M (1991) The Japanese transplanted, production organization and regional development, *Journal of the American Planning Association* **21**: 21–38.

Florida R and Kenney M (1992) Restructuring in place: Japanese investment, production organization, and the geography of steel, *Economic Geography* **68**: 146–73.

Fonseca J W (1976) The semi-urban Landscape, *Landscape* **21**: 23–5.

Forsythe J (1993) The sponsor's choice, *Newsweek*, 17 May: 14.

Forum (1988) Landscape architecture **78**: 66–75.

Foucault M (1970) *The Order of Things: Archaeology of the Human Sciences*, London: Tavistock.

Fraser W H (1981) *The Coming of the Mass Market 1850–1914*, London: Macmillan.

Freeman V (1992) British flair stays in pole position, *The Times*, 31 March: 7.

Frieden B J and Sagalyn L B (1989) *Downtown, Inc: How America Rebuilds Cities*, Cambridge: MIT Press.

Friedmann J (1986) The world city hypothesis, *Development and Change* **17**: 69–83.

Friedmann J (1991) New wines, new bottles: The regulation of capital on a world scale, *Studies in Political Economy* **36**: 9–42.

Friedmann J (1995) Where we stand: A decade of world city research, in P Knox and P Taylor (eds) *World Cities in a World System*, Cambridge: Cambridge University Press.

Friedmann J and Wolff G (1982) World city formation: An agenda for research and action, *International Journal of Urban Region Research* **6**: 309–44.

Froebel F, Heinrichs J and Kreye O (1980) *The New International Division of Labour*, Cambridge: Cambridge University Press.

Frow J (1991) Tourism and the semiotics of nostalgia, *October* **57**: 123–51.

Fujita K and Hill R C (1989) Global production and regional 'hollowing out' in Japan, *Comparative Urban and Community Research* **2**: 200–28.

Fujita K and Hill R C (1993) Toyota city: Industrial organization and the local state in Japan, in K Fujita and R C Hill (eds) *Japanese Cities in the World Economy*, Philadelphia: Temple University Press.

Fuller L and Smith V (1991) Consumers' reports: Management by customers in a new economy, *Work, Employment and Society* **5**, 1: 1–16.

Gaddy C and Ickes B (1998) Russia's virtual economy, *Foreign Affairs* **77**, 5: 53–67.

Garrahan P and Stewart P (1992) *The Nissan Enigma: Flexibility at Work in a Local Economy*, New York: Mansell.

Garreau J (1991) *Edgecity: Life on the New Frontier*, New York: Doubleday.

General Agreement on Tariffs and Trade (GATT) (1985–6 and 1990–1) *International Trade* (annual), Geneva: GATT.

General Agreement on Tariffs and Trade (GATT) (1992) *The Uruguay Round*, Geneva: GATT.

Gereffi G and Wyman D (eds) (1990) *Manufacturing Miracles: Paths of Industrialization in Latin America and East Asia*, Princeton: Princeton University Press.

Gerlach M (1989) Keiretsu organization in the Japanese economy: Analysis and trade implications, in C Johnson, L Tyson and J Zysman (eds) *Politics and Productivity: How Japan's Development Strategy Works*, New York: Ballinger.

Gershuny J I (1978) *After Industrial Society?* London and Basingstoke: Macmillan.

Gershuny J I and Miles I D (1982) *The Service Economy: The Transformation of Employment in Industrial Societies*, London: Frances Pinter.

Gershuny J I and Miles I D (1983) *The New Service Economy*, New York: Praeger.

Gertler M S (1988) The limits to flexibility: Comments on the post-Fordist vision of production and its geography, *Transactions of the Institute of British Geographers* **13**, 4: 419–32.

Gertler M S (1992) Flexibility revisited: Districts, nation-states, and the forces of production, *Transactions of the Institute of British Geographers* NS **17**: 259–78.

Gertler M S (1993) Implementing advanced manufacturing technologies in mature industrial regions: Towards a social model of technology production, *Regional Studies* **27**: 665–80.

Gibson K D (1990) Australian coal in the global context: A paradox of efficiency and crisis, *Environment and Planning A* **22**: 629–46.

Gibson K D (1991) Company towns and class processes: A study of coal towns in Central

Queensland, *Environment and Planning D: Society and Space* **9**: 285–308.

Giddens A (1981) *A Contemporary Critique of Historical Materialism*, London: Macmillan.

Giddens A (1985) Time, space and regionalization, in D Gregory and J Urry (eds) *Social Relations and Spatial Structures*, New York: St. Martin's, pp. 265–95.

Giddens A (1990) *The Consequences of Modernity*, Cambridge: Polity.

Gill S and Law D (1988) *Global Political Economy: Perspectives, Problems and Policies*, Hemel Hemstead: Harvester Wheatsheaf.

Gillespie A and Williams H (1988) Telecommunications and the reconstruction of comparative advantage, *Environment and Planning A* **20**: 1311–21.

Gillette H (1985) The evolution of the planned shopping center in suburb and city, *Journal of the American Planning Association* **51**, 4: 449–60.

Gilpin R (1987) *The Political Economy of International Relations*, Princeton: Princeton University Press.

Ginsberg N (1992) *Divisions of Welfare*, London: Sage.

Girdner E J (1987) Economic liberalisation in India: The new electronics policy, *Asian Survey* **XXVII**, 11: 1188–204.

Glasmeier A (1988) Factors governing the development of high tech industry agglomerations: A tale of three cities, *Regional Studies* **22**: 287–301.

Glennerster H, Power A and Travers T (1991) A new era for social policy: A new enlightenment or a new Leviathan? *Journal of Social Policy* **20**: 389–414.

GOI (1986) *Policy on Computer Software Export, Software Development and Training*, New Delhi: Department of Electronics, Government of India.

Goldstein N (1984) The women left behind: Technical change and restructuring in the electronics industry in Scotland, paper presented at Workshop on Women and Multinationals, University of East Anglia.

Goldthorpe J *et al* (1968–9) *The Affluent Worker in the Class Structure*, 3 vols, Cambridge: Cambridge University Press.

Golob E, Gray M, Markusen A and Park S O (1995) Valley of the heart's delight: Silicon Valley reconsidered. Working Paper, Project on Regional and Industrial Economics, Rutgers University, presented at the Regional Science Association annual meetings, Niagara Falls, Canada, November, 1994.

Gordon D M (1988) The global economy: New edifice or crumbling foundations? *New Left Review* **168**: 24–64.

Gordon I (1995) The impact of economic change on minorities and migrants in Western Europe, in K McFate, R Lawson and W J Wilson (eds) *Poverty, Inequality and the Future of Social Policy. Western States in the New World Order*, New York: Russell Sage Foundation.

Gordon R (1989) Beyond entrepreneurialism and hierarchy: The changing social and spatial organization of innovation, paper presented at the Third International Workshop on Innovation, Technological Change and Spatial Impacts, Selwyn College, Cambridge, UK, 3–5 September.

Gordon R (1990) Systèmes de production, réseaux industrial, et régions: les transformations dans l'organisation sociale et spatiale de l'innovation, *Révue d'Economie Industrielle* **51**: 304–39.

Gordon R (1992) Global networks and the innovation process in high technology SMEs: The case of Silicon Valley, in P Maillat and J C Perrin (eds) *Entreprises Innovatrices et Reseaux Locaux*, Paris: ERESA-Economica.

Gorzelak G (1996) *The Regional Dimensions of Transformation in Central Europe*, London: Kingsley.

Goss J D (1992) Modernity and postmodernity in the retail built environment, in F Gayle and K Anderson (eds) *Ways of Seeing the World*, London: Unwin Hyman.

Gottdiener M (1986) Recapturing the center: A semiotic analysis of shopping malls, in M Gottdiener and A Lagopoulos (eds) *The City and the Sign*, New York: Columbia University Press, pp. 288–302.

Gottdiener M and Gephart G (1991) The multinucleated metropolitan region: A comparative analysis, in R Kling, S Olin and M Poster (eds) *Postsuburban California*, Berkeley: University of California Press, pp. 31–54.

Gottman J (1983) *The Coming of the Transactional City*, College Park: University of Maryland Institute for Urban Studies.

Grabher G and Stark D (1997) *Restructuring Networks in Post-Socialism: Legacies, Linkages and Localities*, Oxford: Oxford University Press.

Graham E and Krugman P (1991) *Foreign Direct Investment in the United States*, Washington: Institute for International Economics.

Graham J (1993) Firm and state strategy in a multipolar world: The changing geography of machine tool production and trade, in H Noponen, J Graham and A Markusen (eds) *Trading Industries, Trading Regions*, New York: Guilford Press, pp. 140–74.

Graham J *et al* (1988) Restructuring in US manufacturing: The decline of monopoly capitalism, *Annals of the Association of American Geographers* **78**: 473–90.

Granovetter M (1985) Economic action and social structure: The problem of embeddedness. *American Journal of Sociology* **91**, 3: 481–510.

Gray M, Golob E and Markusen A (1996) Big firms, long arms, wide shoulders: the 'hub-and-spoke' industrial district in the Seattle region, *Regional Studies* **30**: 651–66.

Great Britain, Foreign and Commonwealth Office (1973) *Cayman Islands*, London: HMSO.

Green M (1990) *Mergers and Acquisitions: Geographical and Spatial Perspectives*, London: Routledge.

Grieco J M (1984) *Between Dependency and Autonomy, India's Experience with the International Computer Industry*, Berkeley and Los Angeles: University of California Press, pp. 84–93.

Griffiths J (1990) Search for quality backing, *Financial Times*, 26 January: 5.

Gros D and Steinherr A (1995) *Winds of Change: Economic Transition in Central and Eastern Europe*, London: Longman.

Gruen V (1973) *Centers for the Urban Environment: Survival of the Cities*, New York: Van Nostrand Reinhold.

Gruen V and Smith L (1960) *Shopping Towns USA: The Planning of Shopping Centers*, New York: Van Nostrand Reinhold.

Gulati S (1988) Capital flight: Causes, consequences and cures, *Journal of International Affairs* **24**, 1: 165–85.

Haggard S (1990) *Pathways from the Periphery: The Politics of Growth in the Newly Industrializing Countries*, Ithaca: Cornell University Press.

Hakim C (1987a) Homeworking in Britain, *Employment Gazette*, February, **95**, 2: 92–104.

Hakim C (1987b) Trends in the flexible workforce, *Employment Gazette* **95**, 11: 549–61.

Halal W (1986) *The New Capitalism*, New York: Wiley.

Hall P (1988) *Cities of Tomorrow*, New York: Blackwell.

Hall P, Breheny M, McQuaid R and Hart D (1987) *Western Sunrise: The Genesis and Growth of Britain's Major High Tech Corridor*, London: Allen & Unwin.

Hall S (1980) Encoding/decoding, in S Hall *et al* (eds) *Culture, Media, Language*, London: Croom Helm, pp. 128–39.

Hamilton F (1967) Models of industrial location, in R Chorley and P Haggett (eds) *Models in Geography*, London: Methuen, pp. 381–6.

Hamilton G (1991) The organizational foundations of Western and Chinese commerce: A historical and comparative analysis, in G Hamilton (ed) *Business Networks and Economic Development in East and Southeast Asia*, Hong Kong: Centre of Asian Studies, pp. 48–65.

Hannerz U (1990) Cosmopolitans and locals in world culture, *Theory, Culture and Society* **7**: 237–51.

Hannerz U (1996) *Transnational Connections*, London: Routledge.

Harding E U (1989) After IBM's exit an

industry arose, *Software Magazine*, International Edition, November: 49.

Harris J M (1991) Global institutions and ecological crisis, *World Development* **19**: 111–22.

Harris L (1988) Alternative perspectives in the financial system, in L Harris, J Coakley, M Crossdale and T Evans (eds) *New Perspectives in the Financial System*, London: Croom Helm, pp. 7–35.

Harris N (1987) *The End of the Third World: Newly Industrializing Countries and the Decline of an Ideology*, London: Penguin.

Harrison B (1990) Industrial districts old wine in new bottles? Working Paper 90-35, School of Urban and Public Affairs, Carnegie Mellon University, Pittsburgh.

Harrison B (1992) Industrial districts: Old wine in new bottles? *Regional Studies* **26**: 469–83.

Harrison B (1994) *Lean and Mean: The Changing Landscape of Corporate Power in the Age of Flexibility*, New York: Basic Books.

Harrison B and Bluestone B (1988) *The Great U-Turn*, New York: Basic Books.

Hart J F (1983) The bypass strip as an ideal landscape, *Geographical Review* **73**: 218–23.

Hart-Landsberg M (1993) *The Rush to Development: Economic Change and Political Struggle in South Korea*, New York: Monthly Review Press.

Hartmann H (1986) Changes in women's economic and family roles in post-world war II United States, in L Benaria and C Stimpson (eds) *Women, Households and the Economy*, Rutgers: Rutgers University Press, pp. 33–64.

Harvey D (1978) The urban process under capitalism: A framework for analysis, *International Journal of Urban and Regional Research* **2**: 101–31.

Harvey D (1982) *The Limits to Capital*, Oxford: Blackwell.

Harvey D (1985) The geopolitics of capitalism, in D Gregory and J Urry (eds) *Social Relations and Spatial Structures*, London: Macmillan, pp. 128–63.

Harvey D (1988) The geographical and geopolitical consequences of the transition from Fordism to flexible accumulation, in G Sternleib and J Hughes (eds) *America's New Market Geography: Nation, Region and Metropolis*, New Brunswick: Rutgers University Press, pp. 101–34.

Harvey D (1989) *The Condition of Postmodernity: An Enquiry into the Origins of Cultural Change*, Oxford: Blackwell.

Harvey D (1990) Between space and time: reflections on the geographical imagination, *Annals of the Association of American Geographers* **80**: 418–34.

Hatch R (1988) *Flexible Manufacturing Networks*, Washington: Corporation for Enterprise Development.

Haug W F (1986) *Critique of Commodity Aesthetics*, Minneapolis: University of Minnesota.

Hay C (1994) Crisis, what crisis? Re-stating the problem of regulation, *Lancaster Working Paper in Political Economy* 47, Department of Sociology, University of Lancaster.

Hayek F A (1976) *Denationalization of Money*, London: Institute of Economic Affairs.

Hazel D (1992) Crime in the malls: A new and growing concern, *Chain Store Age Executive*, February: 27–9.

Held D (1995) *Democracy and the Global Order*, Cambridge: Polity Press.

Henderson J W (1986) The new international division of labour and American semiconductor production in Southeast Asia, in C J Dixon, D Drakakis-Smith and H D Watts (eds) *Multinational Corporations and the Third World*, London: Croom Helm, pp. 91–117.

Henry A (1994) Britain keeps revving past the chequered flag, *The Guardian*, 9 September: 37.

Henry N (1992) The new industrial spaces: Locational logic of a new production era? *International Journal of Urban and Regional Research* **16**: 375–96.

Hepworth M (1986) The geography of technological change in the information economy, *Regional Studies* **20**: 407–24.

Hepworth M (1990) *Geography of the Information Economy*, London: Guildford Press.

Hewison R (1987) *The Heritage Industry*, London: Methuen.

Higgot R *et al* (1985) Theories of development and underdevelopment: Implications for the study of Southeast Asia, in R Higgot and R Robison (eds) *Southeast Asia: Essays in the Political Economy of Structural Change*, London: Routledge and Kegan Paul, pp. 45–6.

Hill R C and Fujita K (1993) Global interdependence and urban restructuring in Japan, in K Fujita and R C Hill (eds) *Japanese Cities in the World Economy*, Philadelphia: Temple University Press.

Hill R C and Lee Y J (1994) Japanese multinationals and East Asian development: The case of the automobile industry, in L Sklair (ed) *Capitalism and Development*, London: Routledge.

Hill R C, Indergaard M and Fujita K (1989) Flat Rock home of Mazda: The social impact of a Japanese company on an American community, in P Arnesen (ed) *The Auto Industry Ahead: Who's Driving*, Ann Arbor: University of Michigan Center for Japanese Studies.

Hirsch J (1991) From the Fordist to the post-Fordist state, in B Jessop, H Kastendiek, K Nielson and O Pederson (eds) *The Politics of Flexibility*, London: Edward Elgar.

Hirst P and Thompson G (1992) The problem of globalization: International economic relations, national economic management, and the formation of trading blocs, *Economy and Society* **21**, 4: 357–96.

Hirst P and Thompson G (1996a) Globalisation: Ten frequently asked questions and some surprising answers, *Soundings* **4**: 47–66.

Hirst P and Thompson G (1996b) *Globalization in Question*, Cambridge: Polity Press.

Hirst P and Zeitlin J (1991) Flexible specialisation vs. post-Fordism: Theory, evidence and policy implications, *Economy and Society* **20**, 1: 1–56.

Hodgson G (1988) *Economics and Institutions: A Manifesto for a Modern Institutional Economics*, Cambridge: Polity Press.

Hodgson G (1993a) *Economics and Evolution: Bringing Life Back into Economics*, Cambridge: Polity Press.

Hodgson G (ed) (1993b) (ed) *The Economics of Institutions*, Aldershot: Edward Elgar.

Hoggett P (1994) The politics of the modernisation of the UK welfare state, in R Burrows and B Loader (eds) *Towards a Post-Fordist Welfare State?* London: Routledge.

Holland S (1987) *The Global Economy*, London: Weidenfeld and Nicolson.

Holmstrom B (1985) The provision of services in a market economy, in R Inman (ed) *Managing the Service Economy: Prospects and Problems*, Cambridge: Cambridge University Press, pp. 183–213.

Hoogvelt A (1990) Extended review: Rethinking development theory, *Sociological Review* **38**: 352–61.

Hopkins J S P (1990) West Edmonton Mall: Landscape of myths and elsewhereness, *Canadian Geographer* **34**, 1: 2–17.

Hounshell D (1984) *From the American System to Mass Production, 1800–1932*, Baltimore: Johns Hopkins University Press.

Howells J and Green A (1988) *Technological Innovation, Structural Change and Location in UK Services*, Aldershot: Avebury.

Howes C and Markusen A (1993) Trade, industry and economic development, in H Noponen, J Graham and A Markusen (eds) *Trading Industries, Trading Regions*, New York: Guilford Press, pp. 1–44.

Howkins A (1986) The discovery of rural England, in R Cols and P Dodd (eds) *Englishness, Politics and Culture 1880–1920*, London: Croom Helm, pp. 62–88.

Hudson R (1989) Labour-market changes and new forms of work in old industrial regions: Maybe flexibility for some but not flexible accumulation, *Environment and Planning D: Society and Space* **7**: 5–30.

Huffman F (1989) Mall Street, USA, *Entrepreneur*, August: 95–9.

Hughes A and Singh A (1991) The world economic slowdown and the Asian and Latin American economies: A comparative analysis of economic structure, policy, and performance, in T Banuri (ed) *Economic Liberalization: No Panacea. The Experiences of Latin*

America and Asia, New York: Oxford University Press, pp. 57–98.

Hughes H (ed) (1988) *Achieving Industrialization in East Asia*, Cambridge: Cambridge University Press.

Humphries J (1988) Women's employment in restructuring America: The changing experience of women in three recessions, in J Rubery (ed) *Women and Recession*, Andover: Routledge, Chapman and Hall, pp. 13–47.

Hutton W (1995) *The State We're In*, London: Jonathan Cape.

Huws U (1993) *Teleworking in Britain*, Research Series No. 18, Sheffield: Employment Department.

Hymer S (1971) The multinational corporation and the law of uneven development, in J W Bhagwati (ed) *Economics and World Order*, New York: Macmillan.

IDS (1986) Flexibility at Work, *IDS Study* 360, April, London: Income Data Services.

Jackson P (1989) *Maps of Meaning*, London: Unwin Hyman.

Jackson P (1994) Consumption and identity: A theoretical agenda and some preliminary findings, paper presented at the Annual Conference of the Institute of British Geographers, Nottingham.

Jacobs J (1988) *The Mall: An Attempted Escape from Everyday Life*, Prospects Heights: Waveland Press.

Jager A (1986) Class definition and the aesthetics of gentrification: Victoriana in Melbourne, in N Smith and P Williams (eds) *Gentrification of the City*, Boston: Allen & Unwin, pp. 78–91.

Jahn J (1986) Some remarks on the notion of labour society: Reply to Sina Aho, *Acta Sociologica* **29**: 61–8.

James B G (1989) *Trojan Horse: The Ultimate Japanese Challenge to Western Industry*, London: Mercury.

Jeffrey D and Hubbard N J (1988) Foreign tourism, the hotel industry and regional economic performance, *Regional Studies* **22**: 319–29.

Jenkins C and Sherman B (1979) *The Collapse of Work*, London: Eyre Methuen.

Jenkins R (1985) Internationalization of capital and the semi-industrialized countries: The case of the motor industry, *Review of Radical Political Economics* **17**: 59–81.

Jenson J (1989) The talents of women, the skills of men: Flexible specialisation and women, in S Wood (ed) *The Transformation of Work?* London: Unwin Hyman, pp. 141–55.

Jessop B (1983) Accumulation strategies, state forms and hegemonic projects, *Kapitalistate* **10–11**: 89–112.

Jessop B (1988a) Post-Fordism, state theory and class struggle: More than a reply to Werner Bonefeld, *Capital and Class* **34**: 147–68.

Jessop B (1988b) *Conservative Regimes and the Transition to Post-Fordism: The Cases of Britain and West Germany*, Essex Papers in Politics and Government, Colchester: University of Essex.

Jessop B (1989) *Thatcherism: The British Road to Post-Fordism*, Essex Papers in Politics and Government 68, Colchester: University of Essex.

Jessop B (1991a) The welfare state in the transition from Fordism to post-Fordism, in B Jessop, H Kastendiek, K Nielson and O Pedersen (eds) *The Politics of Flexibility*, London: Edward Elgar.

Jessop B (1991b) Thatcherism and flexibility: The white heat of a post-Fordist revolution, in B Jessop, H Kastendiek, K Nielsen and O Pedersen (eds) *The Politics of Flexibility*, London: Edward Elgar.

Jessop B (1992a) Fordism and post-Fordism: A critical reformulation, in Storper and Scott (eds) (1992), pp. 46–69.

Jessop B (1992b) Post-Fordism and flexible specialisation: Incommensurable, contradictory, complementary, or just plain different perspectives? in H Ernste and V Meier (eds) *Regional Development and Contemporary Industrial Response: Extending Flexible Specialisation*, London: Belhaven Press, pp. 25–44.

Jessop B (1992c) Regulation and politics: The integral economy and the integral state, mimeo, University of Lancaster (available from the author at Department of Sociology,

Bailrigg, Lancaster University: Lancaster, LAI 4YL).

Jessop B (1994a) Post-Fordism and the state, in A Amin (ed) *Post-Fordism: A Reader*, Oxford: Blackwell, pp. 251–79.

Jessop B (1994b) The transition to post-Fordism and the Schumpeterian workfare state, in R Burrows and B Loader (eds) *Towards a Post-Fordist Welfare State*, London: Routledge, pp. 13–37.

Jessop B (1995) The future of the national state: Erosion or reorganisation? mimeo, Department of Sociology, Lancaster University.

Jessop B, Bonnett K, Bromley S and Ling T (1988) *Thatcherism: A Tale of Two Nations*, Cambridge: Polity Press.

JETRO (1991) *White Paper on Foreign Direct Investment*, Tokyo: Japan External Trade Organization.

Joffe A (1990) Fordism and post-Fordism in Hungary, *South African Sociological Review* **2**, 2: 67–88.

Johnson C (1982) *MITI and the Japanese Economic Miracle: The Growth of Industry Policy, 1925–1975*, Stanford: Stanford University Press.

Johnson R (1986) The story so far, and further transformations? in D Punter (ed) *Introduction to Contemporary Cultural Studies*, Harlow: Longman.

Jones N P and North J (1991) Japanese motor industry transplants: The West European dimension, *Economic Geography* **67**: 105–23.

Joseph K J (1989) Growth performance of Indian electronics under liberalisation, *Economic and Political Weekly* **XXIV**, 33: 1915.

Julius D (1990) *Global Companies and Public Policy: The Growing Challenge of Foreign Direct Investment*, London: Frances Pinter.

Kao C (1991) 'Personal trust' in the large businesses in Taiwan: A traditional foundation for contemporary economic activities, in G Hamilton (ed) *Business Networks and Economic Development in East and Southeast* Asia, Hong Kong: Centre of Asian Studies, pp. 66–76.

Kato H (1993) Local auto output in US set to exceed exports from Japan, *Nikkei Weekly*, 6 Dec.

Kearney R C (1990) Mauritius and the NIC model redux, or how many cases make a model? *Journal of Developing Areas* **24**: 195–216.

Keeble D (1988) High-technology industry and local environments in the United Kingdom, in P Aydalot and D Keeble (eds) *High Technology Industry and Innovative Environments: The European Experience*, London: Routledge.

Keeble D (1989) High-technology industry and regional development in Britain: The case of the Cambridge phenomenon, *Environment and Planning C: Government and Policy* **7**: 153–72.

Keeble D (1990) Small firms, new firms and uneven regional development in the United Kingdom, *Area* **22**: 234–45.

Keeble D, Bryson J R and Wood P (1991) Small firms, business service growth and regional development in the United Kingdom: Some empirical findings, *Regional Studies* **25**: 439–57.

Keely C (1982) Immigration and the American future, in L Liebman (ed), *Ethnic Relations in America*, New York: Prentice Hall.

Kennedy P (1987) *The Rise and Fall of the Great Powers*, New York: Random House.

Kenney M and Florida R (1993) *Beyond Mass Production*, New York: Oxford University Press.

Keohane R O (1984) *After Hegemony: Cooperation and Discord in the World Political Economy*, Princeton: Princeton University Press.

Kester W C (1991) *Japanese Takeovers*, Boston: Harvard Business School Press.

Kim W B (1993) Industrial restructuring and the dynamics of city-state adjustments, *Environment and Planning A* **25**: 27–46.

King A D (1990) *Global Cities: Post Imperialism and the Internationalization of London*, London: Routledge.

Kirn T J (1987) Growth and change in the service sector of the United States: A spatial perspective, *Annals of the Association of American Geographers* **77**: 353–72.

Knox P (1991) The restless urban landscape: Economic and sociocultural change and the transformation of metropolitan Washingtion, DC, *Annals of the Association of American Geographers* **81**: 181–209.

Kobrin S J (1987) Testing the bargaining hypothesis in the manufacturing sector in developing countries, *International Organization* **41**: 609–38.

Kodama E (1989) How research investment decisions are made in Japanese industry, in D Evered and S Harnett (eds), *The Evaluation of Scientific Research*, Chichester: Wiley, pp. 201–14.

Kohli A (1989) Politics of economic liberalisation in India, *World Development* **17**, 3: 305–28.

Kojima K (1977) *Japan and a New World Economic Order*, Boulder: Westview Press.

Konig W (1975) *Towards an Evaluation of International Subcontracting Activities in Developing Countries*, Report on Maquiladoras in Mexico, Mexico City: UNECLA.

Kotz M (1990) A comparative analysis of the theory of regulation and the social structure of accumulation theory, *Science and Society* **54**: 5–28.

Kowinski W S (1985) *The Malling of America: An Inside Look at the Great Consumer Paradise*, New York: William Morrow.

KPMG Peat Marwick (1988) *Captive Insurance in the Cayman Islands*, Grand Cayman: Peat Marwick.

Krueger A (1986) Changing perspectives on development economies and World Bank research, *Development Policy Review* **4**: 195–210.

Kumar K (1978) *Prophecy and Progress*, Harmondsworth: Penguin Books.

Kuo S W Y, Ranis G and Fei J C H (1981) *The Taiwan Success Story: Rapid Growth With Improved Distribution in the Republic of China, 1952–1979*, Boulder: Westview.

Kwong P (1987) *The New Chinatown*, New York: Hill and Wang.

Laclau E (1990) *New Reflections on the Revolution of our Times*, London: Verso.

Lal D (1983) *The Poverty of Development Economics*, Cambridge: Harvard University Press.

Landry C and Bianchini E (1995) *The Creative City*, London: Demos/Comedia.

Lane D (1978) *The Socialist Industrial State: Towards a Political Sociology of State Socialism*, London: Allen & Unwin.

Langdale J (1985) Electronic funds transfer and the internationalisation of the banking and finance industry, *Geoforum* **16**: 1–13.

Langdale J (1989) The geography of international business telecommunications: The role of leased networks, *Annals of the Association of American Geographers* **79**: 501–22.

Lash S and Urry J (1987) *The End of Organized Capitalism*, Cambridge: Polity.

Lash S and Urry J (1994) *Economies of Signs and Spaces*, London: Sage.

Le Grand J (1990) The state of welfare, in J Jills (ed) *The State of Welfare: The Welfare State in Britain Since 1974*, Oxford: Clarendon Press.

Le Heron R B (1988) The internationalisation of New Zealand forestry companies and the social reappraisal of New Zealand's exotic forest resource, *Environment and Planning A* **20**: 489–515.

Le Heron R B (1990) Goodman Fielder Wattie Ltd: Internationalisation and performance, in M de Smit and E Wever (eds) *The Corporate Firm in the Changing Global Economy*, Andover: Routledge, Chapman & Hall, pp. 100–19.

Le Heron R B, Roche M and Anderson G (1989) Reglobalisation of New Zealand's food and fibre system: Organisational dimensions, *Journal of Rural Studies* **5**: 395–404.

Lecesse M (1988) Brave old world, *Landscape Architecture* **78**: 56–75.

Lefebvre H (1971) *Everyday Life in the Modern World*, New York: Harper & Row.

Lefebvre H (1991) *The Production of Space*, Trans. by D. Nicholson-Smith, Oxford: Blackwell.

Leidner R (1991) Serving hamburgers and

selling insurance: Gender, work, and identity in interactive service jobs, *Gender and Society* **5**: 154–77.

Leo P and Philippe J (1991) Networked producer services: Local markets and global development, in P W Daniels (ed) *Services and Metropolitan Development: International Perspectives*, London: Routledge.

Levine J (1990) Lessons from Tysons Corner, *Forbes* April 30: 186–7.

Levitt T (1983) Globalization of markets, *Hansard Business Review* **61**: 92–102.

Lewis P F (1983) The galactic metropolis, in H Platt and G Macinko (eds) *Beyond the Urban Fringe*, Minneapolis: University of Minnesota Press, pp. 23–49.

Leyshon A and Thrift N (1992) Liberalisation and consolidation: The single European market and the remaking of European financial capital, *Environment and Planning A* **24**: 49–81.

Leyshon A and Tickell A (1994) Money order? Discursive construction of Bretton Woods and the making and breaking of regulatory space, *Environment and Planning A* **26**: 1861–90.

Leyshon A, Thrift N and Tommey C (1989) The rise of the British provincial financial centre, *Progress in Planning* **31**: 151–229.

Li Ka-shing (1991) The best and the richest, *HongKong Inc.*, January: 22–9.

Lilwall C and Allcock K (1988) Value at a glance, in Country Life/Knight, Frank and Rutley, *Buying a Country House. A County Guide to Value*, London: Country Life.

Lim L (1978) *Women Workers in Multinational Corporations in Developing Countries: The Case of the Electronics Industry in Malaysia and Singapore*, Women's Studies Program Occasional Paper No. 9, University of Michigan.

Lin C (1988) East Asia and Latin America as contrasting models, *Economic Development and Cultural Change* **36**: 153–97.

Lin J (1989) Beyond neoclassical shibboleths: A political-economic analysis of Taiwanese economic development, *Dialectical Anthropology* **14**: 283–300.

Lipietz A (1984) Imperialism or the beast of the apocalypse? *Capital and Class* **22**: 81–110.

Lipietz A (1986) New tendencies in the international division of labour: Regimes of accumulation and modes of regulation, in A J Scott and M Storper (eds) *Production, Work, Territory: The Geographical Anatomy of Industrial Capitalism*, London: Allen & Unwin, pp. 16–40.

Lipietz A (1987) *Mirages and Miracles: The Crises of Global Fordism*, London: Verso.

Lipietz A (1992a) The regulation approach and capitalist crisis: An alternative compromise for the 1990s, in M Dunford and G Kafkalas (eds) *Cities and Regions in the New Europe: The Global-Local Interplay and Spatial Development Strategies*, London: Belhaven Press, pp. 309–34.

Lipietz A (1992b) *Towards a New Economic Order: Postfordism, Ecology and Democracy*, Oxford: Oxford University Press.

Lipsig-Mumme C (1983) The renaissance of homeworking in developed economies, *Relations Industrielles* **38**: 545–67.

List F [1844] (1916) *The National System of Political Economy*, Trans. by S. Lloyd, London: Longman Press.

Liu K (1990) *Orthodoxy in Imperial China*, Berkeley: University of California Press.

Lloyd J (1988) Death of the honourable Englishman, *GQ*, December: 152–5.

Lohr S (1988) The growth of the global office, *New York Times*, 18 October.

Longmore A (1994) Football puts its skill on profit, *The Times*, 19 December: 27.

Lovering J (1990) Fordism's unknown successor: A comment on Scott's theory of flexible accumulation and the re-emergence of regional economies, *International Journal of Urban and Regional Research* **14**: 159–74.

Lovering J (1991) The changing geography of the military industry in Britain, *Regional Studies* **25**: 279–93.

Lowe M and Gregson N (1989) Nannies, cooks, cleaners and au pairs: New issues for feminist geography, *Area* **21**: 415–17.

LRD (1986) Flexibility Examined, *Bargaining*

Report, London: Labour Research Department.

Luger M and Goldstein H (1990) *Technology in the Garden*, Chapel Hill: University of North Carolina Press.

Lundvall B A (1988) Innovation as an interactive process: From user-producer interaction to the national system of innovation, in C Dosi, C Freeman, R Nelson, C Silverberg and L Soete (eds) *Technical Change and Economic Theory*, London: Frances Pinter, pp. 349–69.

MacEwan A and Tabb W K (eds) (1989a) *Instability and Change in the World Economy*, New York: Monthly Review Press.

MacEwan A and Tabb W K (1989b) The economy in crisis: National power and international instability, *Socialist Review* **19**: 67–91.

Maitland B (1990) *The New Architecture of the Retail Mall*, New York: Van Nostrand Reinhold.

Malecki E J (1984) Technology and regional development: A survey, *APA Journal* **50**, 3: 262–6.

Malecki E J (1987) Comments on Scott's 'High tech industry and territorial development: The rise of the Orange County complex, 1955–1984', *Urban Geography* **8**: 77–81.

Mann S (1987) *Local Merchants and the Chinese Bueaucracy, 1750–1950*, Stanford: Stanford University Press.

Marceau J (1989) *A Family Business? The Making of an International Business Elite*, Cambridge: Cambridge University Press.

Marcus G E (1986) Contemporary problems of ethnography in the modern world system, in J Clifford and G E Marcus (eds) *Writing Culture. The Poetics and Politics of Ethnography*, Los Angeles and Berkeley: University of California Press, pp. 165–93.

Marcus G E and Fischer M J (1986) *Anthropology as Cultural Critique: An Experimental Moment in the Human Sciences*, Chicago: University of Chicago Press.

Marcuse H (1964) *One-dimensional Man*, Boston: Beacon Press.

Marden P (1992) 'Real' regulation reconsidered, *Environment and Planning A* **24**: 751–67.

Marginson P, Edwards P, Martin R, Pucell, J and Sisson K (1988) *Beyond the Workplace*, Oxford: Blackwell.

Margolis J (1995) Wheels of fortune, *Sunday Times*, Style, 1 January: 6.

Marketing Strategies for Industry (1988) *Fruit UK: Marketing Database*, Surrey: Marketing Strategies for Industry.

Markusen A (1985) *Profit Cycles, Oligopoly and Regional Development*, Cambridge: MIT Press.

Markusen A (1991) The military-industrial divide, *Environment and Planning D: Society and Space* **9**: 391–416.

Markusen A (1994) Studying regions by studying firms, *The Professional Geographer* **46**: 477–90.

Markusen A, Hall P and Glasmeier A (1986) *High Tech America: The What, How, Where and Why of the Sunrise Industries*, Boston: Allen & Unwin.

Markusen A, Hall P, Campbell S and Deitrick, S (1991) *The Rise of the Gunbelt*, New York: Oxford University Press.

Marshall A (1920) *Principles of Economics*, London: Macmillan.

Marshall J N *et al* (1988) *Services and Uneven Development*, Oxford: Oxford University Press.

Marshall J N, Damesick N and Wood P (1987) Understanding the location and the role of producer services in the United Kingdom, *Environment and Planning A* **19**: 575–93.

Marshall M (1987) *Long Waves of Regional Development*, London: Macmillan.

Martin J (1985) If baseball can't save cities any more, what can a festival market do? *Journal of Cultural Geography* **3**: 33–46.

Martin P (1990) Trolley fodder, *Sunday Times Magazine*, 4 November: 24–36.

Martin R (1993) Economic theory and human geography, in D Gregory, R Martin and G E Smith (eds) *Human Geography: Society, Space and Social Science*, London: Macmillan, pp. 21–53.

Masai Y (1989) Greater Tokyo as a global city, in R Knight and G Gappert (eds) *Cities in a Global Society*, New Park: Sage.

Massey D (1973) Towards a critique of industrial location theory, *Antipode* **5**: 33–9.

Massey D (1984) *Spatial Divisions of Labour: Social Structures and the Geography of Production*, London: Macmillan.

Massey D (1988) Uneven development: Social change and spatial divisions of labour, in D Massey and J Allen (eds) *Uneven Redevelopment: Cities and Regions in Transition*, London: Hodder and Stoughton, pp. 250–76.

Massey D (1993) Power-geometry and a progressive sense of place, in J Bird *et al* (eds) *Mapping the Futures: Local Cultures, Global Change*, London: Routledge.

Massey D (1994) *Space, Place and Gender*, Cambridge: Polity.

Massey D (1995) *Spatial Divisions of Labour: Social Structures and the Geography of Production*, second edition, London: Macmillan.

Massey D and Henry N (1992) *Something New, Something Old: A Sketch of the Cambridge Economy*, South East Occasional Paper Series No. 2, Milton Keynes: Open University.

Massey D and Meegan R (1982) *The Anatomy of Job Loss*, Andover: Methuen.

Maswood S J (1989) *Japan and Protection*, New York: Routledge.

Mayer M (1992) The shifting local political system in European cities, in M Dunford and G Kafkalas (eds) *Cities and Regions in the New Europe: The Global-Local Interplay and Spatial Development Strategies*, London: Belhaven, pp. 255–78.

Mayer M (1994) Post-Fordist city politics, in A Amin (ed) *Post-Fordism: A Reader*, Oxford: Blackwell, pp. 316–39.

McCloud J (1989) Fun and games is serious business, *Shopping Center World*, July: 28–35.

McCloud J (1991) Today's high-tech amenities can increase owners' profits, *Shopping Center World*, July: 25–8.

McDonald K (1985) The commercial strip, *Landscape Architecture* **28**, 2: 12–19.

McDowell L (1991) Restructuring production and re-production: Some theoretical and empirical issues relating to gender or women in Britain, in M Gottdiener and C Pickvance (eds) *Urban Life in Transition*, London: Sage.

McDowell L and Court G (1994) Missing subjects: Gender, power and sexuality in merchant banking, *Economic Geography* **70**, 3: 229–51.

McFate K (1991) *Poverty, Inequality and the Crisis of Social Policy*, Washington: Joint Centre for Political and Economic Studies.

McGrath M E and Hoole R W (1992) Manufacturing's new economies of scale, *Harvard Business Review* **70**: 94–102.

McGrew A G and Lewis P (1992) *Global Politics*, Cambridge: Polity.

McKinsey Global Institute (1993) *Manufacturing Productivity*, Washington: McKinsey Global Institute.

McMichael P and Myhre D (1991) Global regulation versus the nation-state, *Capital and Class* **43**: 83–106.

Medlick S (1985) *Paying Guests*, London: Confederation of British Industry.

Meyer D (1989) Midwest industrialization and the American manufacturing belt in the nineteenth century, *Journal of Economic History* **49**: 921–37.

Millward N and Stevens M (1986) *British Workplace Industrial Relations, 1980–1984*, The DE/ESRC/PSI/ACAS Surveys, Aldershot: Gower.

Mingione E and Redclift N (1986) *Beyond Employment*, Oxford: Blackwell.

Mitter S (1986) Industrial restructuring and manufacturing homework, *Capital and Class* **27**: 37–80.

Mohan J (1988a) Spatial aspects of health-care employment in Britain 1: Aggregate trends, *Environment and Planning A* **20**: 7–23.

Mohan J (1988b) Restructuring, privatization and the geography of health-care provision in England, 1983–87, *Transactions of the Institute of British Geographers*, NS **13**: 449–65.

Mollenkopf J and Castells M (eds) (1992) *Dual City: Restructuring New York*, New York: Russell Sage Foundation.

Montagu-Pollock M (1991) All the right connections, *Asian Business* **27**, 1: 20–4.

Moran M (1991) *The Politics of the Financial Services Revolution*, London: Macmillan.

Morgan K and Sayer A (1988) *Microcircuits of Capital: Sunrise Industry and Uneven Development*, Cambridge: Polity.

Morita A (1992) A critical moment for Japanese management, *Economic Eye* **13**: 4–9.

Morris J (ed) (1991) *Japan and the Global Economy*, London: Routledge.

Morris-Suzuki T (1991) Reshaping the international division of labor: Japanese manufacturing investment in southeast Asia, in J Morris (ed) *Japan and the Global Economy*, London: Routledge.

Moss M (1987a) Telecommunications, world cities and urban policy, *Urban Studies* **24**: 534–46.

Moss M (1987b) Telecommunications and international financial centres, in J Brotchie, P Hall and P Newton (eds) *The Spatial Impact of Technological Change*, London: Croom Helm.

Moss M and Dunau A (1986) Offices, information technology, and locational trends, in J Black, K Roark and L Schwartz (eds) *The Changing Office Workplace,* Washington: Urban Land Institute, pp. 171–82.

Moulaert F, Swyngedouw E A and Wilson P (1988) Spatial responses to Fordist and post-Fordist accumulation and regulation, *Papers of the Regional Science Association* **64**: 11–23.

Mukhi V and Chellam R (1988) Software: An emerging business, *Business India*, 22 Supplement, August–4 September: 131.

Murgatroyd L, Savage M, Shapiro R, Urry J, Walby S and Warde A (1985) *Localities, Class and Gender*, London: Pion.

Murray F (1987) Flexible specialisation in the 'Third Italy', *Capital and Class* **33**: 84–95.

Murray R (1992) Flexible specialisation and development strategy: The relevance for Eastern Europe, in H Ernste and V Meier (eds) *Regional Development and Contemporary Industrial Response: Extending Flexible Specialisation*, London: Belhaven, pp. 197–218.

Nalven C and Tate P (1989) Taking the offshore option, *Datamation* **35**, 3: 72–6.

Nash T (1989) India's lead in offshore software, *ELSOFTEX Newsletter*, Electronics and Computer Software Export Promotion Council **2**, 1: 4.

Nau H (1990) *The Myth of America's Decline*, Oxford: Oxford University Press.

Nederveen Pieterse J (1997) Futures of capitalism: Going global, *Development and Change* **28**, 2: 376–82.

NEDO (1986) *Changing Working Patterns*, Report prepared by the Institute of Manpower Studies for the National Economic Development Office in association with the Department of Employment, London: NEDO.

Nelson K (1986) Labor demand, labor supply and the suburbanization of low-wage office work, in A Scott and M Storper (eds) *Production, Work, Territory*, Boston: Allen & Unwin.

Nelson R and Winter S (1982) *An Evolutionary Theory of Economic Change*, Cambridge: Harvard University Press.

Nicholls A (1993) Is bigger better? *Media and Marketing Europe*, July: 30–1.

Nicol L (1985) Communications technology: Economic and spatial impacts, in M Castells (ed) *High Technology, Space, and Society*, Beverly Hills: Sage, pp. 191–209.

North D (1981) *Structure and Change in Economic History*, New York: Norton.

Obayashi M (1993) Kanagawa: Japan's brain center, in K Fujita and R C Hill (eds) *Japanese Cities in the World Economy*, Philadelphia: Temple University Press.

O'Brien R (1992) *Global Financial Integration: The End of Geography*, London: Frances Pinter.

O'Brien S and Ford R (1988) Can we say at last goodbye to social class? *Journal of the Market Research Society* **30**, 3: 289–332.

Offe C (1987) Democracy against the Welfare State? *Political Theory* **15**: 501–37.

Ogle G (1990) *South Korea: Dissent Within the Economic Miracle*, London: Zed.

Ohmae K (1985) *Triad Power: The Coming Shape of Global Competition*, New York: Macmillan.

Ohmae K (1989) *The Borderless World*, London: Fontana.

Ohmae K (1990) *The Borderless World: Power and Strategy in an Interdependent Economy*, New York: Harper Business.

Ohmae K (1993) The rise of the region state, *Foreign Affairs*, Spring: 78–87.

Ohmae K (1995a) Putting global logic first, *Harvard Business Review*, January/February: 119–25.

Ohmae K (1995b) *The End of the Nation State*, New York: Free Press.

Oldenburg R (1989) *The Great Good Life*, New York: Paragon House.

Olson M (1982) *The Rise and Decline of Nations*, New York: Yale University Press.

Oman C and Wignaraja G (1991) *The Postwar Evolution of Development Thinking*, London: Macmillan.

Ong A (1984) Global industries and Malay peasants in peninsular Malaysia, in J Nash and M P Fernandez-Kelly (eds) *Women, Men and the International Division of Labour*, Albany: SUNY.

Onis Z (1991) The logic of the developmental state, *Comparative Politics* **24**: 109–26.

Organization for Economic Cooperation and Development (1986) *Flexibility in the Labour Market: A Technical Report*, Paris: OECD.

Organization for Economic Cooperation and Development (1992) *Statistics on International Transactions in Services*, Paris: OECD.

Organization for Economic Cooperation and Development (1994) *The Performance of Foreign Affiliates in the OECD Countries*, Paris: OECD.

Orwell G (1957) Down the mine, in *Inside the Whale and other Essays*, Harmondsworth: Penguin.

Osborne T (1988) Revolutionizing the retail landscape, *Marketing Communications*, October: 17–25.

Ostry S (1990) *Governments and Corporations in a Shrinking World*, New York: Council on Foreign Relations Press.

Page B and Walker R (1991) From settlement to Fordism: The agro-industrial revolution in the American Midwest, *Economic Geography* **67**: 281–305.

Pahl R E (1984) *Divisions of Labour*, Oxford: Blackwell.

Pahl R E (1988) Some remarks on informal work, social polarization and the social structure, *International Journal of Urban and Regional Research* **12**: 247–67.

Palan R and Abbott J (1996) *State Strategies in the Global Political Economy*, London: Frances Pinter.

Pallot J and Shaw D (1981) *Planning in the Soviet Union*, London: Croom Helm.

Park K (1993) Women and development: The case of South Korea, *Comparative Politics* **25**: 127–45.

Park S O and Markusen A (1994) Generalizing new industrial districts: A theoretical agenda and an application from a nonWestern economy, *Environment and Planning A* **27**: 81–104.

Parkin F (1982) Max Weber, Horwood: Chichester.

Parsonage J (1992) Southeast Asia's 'growth triangle': A subregional response to global transformation, *International Journal of Urban and Regional Research* **16**: 307–20.

Pawlak E J *et al* (1985) A view of the mall, *Social Service Review*, June: 305–17.

Pecchioli R (1983) *The Internationalization of Banking: The Policy Issues*, Paris: OECD.

Peck J (1992a) Labor and agglomeration: control and flexibility in local labour markets, *Economic Geography* **68**: 325–47.

Peck J (1992b) 'Invisible threads': Homeworking, labour market relations, and industrial restructuring in the Australian clothing trade, *Environment and Planning D: Society and Space* **10**: 671–89.

Peck J and Jones M (1995) Training and enterprise councils: Schumpeterian workfare state, or what? *Environment and Planning A* **27**, 9: 1361–96.

Peck J and Tickell A (1992) Local modes of social regulation: Regulation theory, Thatcherism and uneven development, *Geoforum* **23**: 347–64.

Peck J and Tickell A (1994) Searching for a new

institutional fix: The *after*-Fordist crisis, in A Amin (ed) *Post-Fordism: A Reader*, Oxford: Blackwell, pp. 280–314.

Peet R (1991) *Global Capitalism*, London: Routledge.

Petras J and Hui P (1991) State and development in Korea and Taiwan, *Studies in Political Economy* **34**: 179–98.

Phillips A and Taylor B (1980) Sex and skill: Notes towards a feminist economics, *Feminist Review* **6**: 79–88 also in Feminist Review (ed) (1986) *Waged Work: A Reader*, London: Virago.

Picciotto S (1991) The internationalization of the state, *Capital and Class* **43**: 43–64.

Pickles J and Smith A (eds) (1998) *Theorising Transition: The Political Economy of Post-Communist Transformations*, London: Routledge.

Pilkington H (1998) *Migration, Displacement and Identity in Post-Soviet Russia*, London: Routledge.

Pinch S and Storey A (1991) Social polarization in a buoyant labour market – the case of Southampton: A response to Pahl, Dale and Bamford, *International Journal of Urban and Regional Research* **15**: 453–60.

Pinch S and Storey A (1992) Labour-market dualism: Evidence from a survey of households in the Southampton city-region, *Environment and Planning A* **24**: 571–89.

Piore M and Sabel C (1984) *The Second Industrial Divide: Possibilities for Prosperity*, New York: Basic Books.

Pollert A (1988) Dismantling flexibility, *Capital and Class* **34**: 42–75.

Pooley S (1991) The state rules, OK? The continuing political economy of nation-states, *Capital and Class* **43**: 65–82.

Porter M E (1990) *The Competitive Advantage of Nations*, London: Macmillan.

Portes A and Castells M (1989) World underneath: The origins, dynamics and effects of the informal economy, in A Portes, M Castells and L Benton (eds) *The Informal Economy: Studies in Advanced and Less Developed Countries*, Baltimore: Johns Hopkins University Press.

Poster M (1990) *The Mode of Information: Poststructuralism and Social Context*, Chicago: University of Chicago Press.

Pred A (1977) *City Systems in Advanced Economies*, London: Hutchinson.

Preston L E and Windsor D (1992) *The Rules of the Game in the Global Economy*, Boston: Kluwer Academic.

Price D G and Blair A M (1989) *The Changing Geography of the Service Sector*, New York: Belhaven.

Pringle R (1992) Financial markets versus governments, in T Banuri and J Schor (eds) *Financial Openness and National Autonomy*, Oxford: Clarenden Press, pp. 89–109.

Pryke M (1991) An international city going 'global': Spatial change in the City of London, *Environment and Planning D: Society and Space* **9**: 197–222.

Pudup M B (1992) Industrialization after (de)industrialization: A review essay, *Urban Geography* **13**: 187–200.

Radice H (1984) The national economy: A Keynesian myth?, *Capital and Class* **22**: 111–40.

Raj M (1990) India's Computer Network Revolution, *Asia Technology*, March: 22–4.

Rajan A and Fryatt J (1988) *Create or Abdicate: The City's Human Resource Choice for the 90s*, London: Witherby.

Rasiah R (1988) The semiconductor industry in Penang: Implications for the new international division of labour, *Journal of Contemporary Asia* **18**, 1: 24–46.

Rathbun R D (1990) *Shopping Centers and Malls* 3, New York: Retail Reporting Corporation.

Redding G (1990) *The Spirit of Chinese Capitalism*, New York: Walter de Gruyter.

Regional Studies (1998) *Special Issue: Globalization, Institutions, Foreign Investment and the Reintegration of East and Central Europe and the Former Soviet Union with the World Economy* **32**, 7.

Regulska J (1997) Decentralisation or (re)centralisation: Struggle for political power in Poland, *Environment and Planning C: Government and Policy* **20**, 3: 643–80.

Reich R (1983) *The Next American Frontier*, New York: Times Books.

Reich R (1991) *The Work of Nations: Preparing Ourselves for 21st-Century Capitalism*, London: Simon and Schuster.

Relph E (1987) *The Modern Urban Landscape*, Baltimore: Johns Hopkins University Press.

Republic of China (1975, 1991) *Statistical Yearbook of the Republic of China*, Annual, Taipei: Directorate-General of Budget, Accounting and Statistics, Executive Yuan.

Retail Uses (1991) *Urban Land*, March: 22–6.

Reynolds M (1990) Food courts: Tasty! *Stores*, August: 52–4.

Richards G (1990) Atmosphere key to mall design, *Shopping Center World*, August: 23–9.

Ricks R B (1991) Shopping center rules misapplied to older adults, *Shopping Center World*, May: 52–6.

Riddle D I (1986) *Service-Led Growth: The Role of the Service Sector in World Development*, New York: Praeger.

Riedel J (1988) Economic development in East Asia: Doing what comes naturally? in H Hughes (ed) *Achieving Industrialization in East Asia*, Cambridge: Cambridge University Press, pp. 1–38.

Rimmer P (1993) Reshaping western Pacific rim cities: Exporting Japanese planning ideas, in K Fujita and R C Hill (eds) *Japanese Cities in the World Economy*, Philadelphia: Temple University Press.

Roberts S M (1994) Fictitious capital, fictitious spaces? The geography of offshore financial flows, in S Corbridge, R Martin and N Thrift (eds) *Money, Power and Space*, Oxford: Blackwell, pp. 91–115.

Robertson R (1992) *Globalization: Social Theory and Global Culture*, London: Sage.

Rodan G (1987) The rise and fall of Singapore's 'second industrial revolution', in R Robison, K Hewison and R Higgot (eds) *Southeast Asia in the 1980s: The Politics of Economic Crisis*, Sydney: Allen & Unwin, pp. 158–75.

Romer P (1990) Endogenous technological change, *Journal of Political Economy* **98**, 5: 71–101.

Rosenberg N (1976) *Perspectives on Technology*, Cambridge: Cambridge University Press.

Rosenberg N (1982a) *Inside the Black Box*, Cambridge: Cambridge University Press.

Rosenberg N (1982b) Technological progress and economic growth, in N Rosenberg and L Jorberg (eds) *Technical Change, Employment and Investment*, Lund: Department of Economic History, University of Lund, pp. 7–27.

Rouse J W (1962) Must shopping centers be inhuman? *Architectural Forum*, June: 105–7.

Rowe P G (1991) *Making a Middle Landscape*, Cambridge: MIT Press.

Rubery J (ed) (1988) *Women and Recession*, London: Routledge.

Russell S (1994) In Pole Position: A Study of the British Motor Sport Industry, Unpublished dissertation, University of Southampton.

Russo M (1985) Technical change and the industrial district: The role of inter-firm relations in the growth and transformation of ceramic tile production in Italy, *Research Policy* **14**: 329–43.

Sabel C F (1989) Flexible specialisation and the re-emergence of regional economies, in P Hirst and J Zeitlin (eds) *Reversing Industrial Decline? Industrial Structure and Policies in Britain and Her Competitors*, Oxford: Berg.

Sabel C F, Herrigel G, Kazis R and Deeg R (1987) How to keep mature industries innovative, *Technology Review*, April: 27–35.

Sachs J (1985) External debt and macroeconomic performance in Latin America and East Asia, *Brookings Papers on Economic Activity* **2**: 523–64.

Sadler D (1992) *The Global Region: Production, State Policies and Uneven Development*, Oxford: Pergamon.

Salih K, Young M L and Rasiah R (1988) The changing face of the electronics industry in the periphery: The case of Malaysia, *International Journal of Urban and Regional Research* **12**, 3: 375–403.

Salt J (1989) A comparative overview of international trends and types, *International Migration Review* **3**: 431–56.

Samuel R (1989) Introduction: Exciting to be English, in R Samuel (ed) *Patriotism: The*

Making and Unmaking of British National Identity, Volume 1. History and Politics, London: Routledge, pp. xviii–lxvii.

Sassen S (1988) *The Mobility of Capital and Labor*, New York: Cambridge University Press.

Sassen S (1991) *The Global City: New York, London, Tokyo*, Princeton: Princeton University Press.

Sassen S (1993) *Migration Systems*, Working Paper, New York: Russell Sage Foundation.

Sassen S (1994) *Cities in a World Economy*, London: Pine Forge.

Sassen-Koob S (1984) The new labor demand in world cities, in M P Smith (ed) *Cities in Transformation*, Newbury Park: Sage, pp. 139–71.

Saunders P (1986) *Social Theory and the Urban Question*, second edition, London: Hutchinson.

Sawicki D S (1989) The festival marketplace as public policy: Guidelines for future policy decisions, *American Planning Association Journal*, Summer: 347–61.

Saxenian A (1985) The genesis of Silicon Valley, in P Hall and A Markusen (eds) *Silicon Landscapes*, Boston: Allen & Unwin, pp. 20–34.

Saxenian A (1994) *Regional Advantage: Culture and Competition in Silicon Valley*, Cambridge: Harvard University Press.

Sayer A (1985) Industry and space: A sympathetic critique of radical research, *Environment and Planning D: Society and Space* **3**: 3–29.

Sayer A (1989) Post-Fordism in question, *International Journal of Urban and Regional Research* **13**, 4: 666–95.

Sayer A and Walker R (1992) *The New Social Economy: Reworking the Division of Labour*, Oxford: Blackwell.

Scanlon R (1989) New York City as global capital in the 1980s, in R Knight and G Gappert (eds) *Cities in a Global Society*, Newbury Park: Sage.

Schiller H (1993) The information highway: Public way or private road? *The Nation* **257**: 64–5.

Schoenberger E (1988) From Fordism to flexible accumulation: Technology, competitive strategies and international location, *Environment and Planning D: Society and Space* **6**: 245–62.

Schor J (1990) Financial openness and national autonomy, Discussion Paper 1523, Harvard University Institute of Economic Research, Cambridge.

Schott J J (1991) Trading blocs and the world trading system, *The World Economy* **14**, 1: 1–18.

Schumpeter J (1942) *Capitalism, Socialism, and Democracy*, New York: Harper & Row.

Schwanke D (1990) *Development Trends 1989*, Washington: Urban Land Institute.

Schwartz M and Romo F (1993) *The Rise and Fall of Detroit: How the Automobile Industry Destroyed its Capacity to Compete*, SUNY: Department of Sociology, Stony Brook.

Scott A (1986) High tech industry and territorial development: The rise of the Orange County complex, 1955–1984, *Urban Geography* **7**: 3–45.

Scott A J (1988a) Flexible production systems and regional development, *International Journal of Urban and Regional Research* **11**: 171–85.

Scott A J (1988b) *New Industrial Spaces: Flexible Production, Organisation and Regional Development in North America and Western Europe*, London: Pion.

Scott A J (1991) Flexible production systems: analytical tasks and theoretical horizons – a reply to Lovering, *International Journal of Urban and Regional Research* **15**: 130–4.

Scott A J and Paul A (1990) Collective order and economic coordination in industrial agglomerations: The technopoles of Southern California, *Environment and Planning C: Government and Policy* **8**: 179–93.

Scott A J and Storper M (1987) High technology industry and regional development: A theoretical critique and reconstruction, *International Social Science Journal* **112**: 215–32.

Scott J (1978) *Yankee Unions, Go Home! How the AFL Helped the U.S. Build an Empire in Latin America*, Vancouver: New Star Books.

Scott N K (1989) *Shopping Centre Design*, London: Van Nostrand Reinhold.

Segal, Quince and Wicksteed (1985) *The Cambridge Phenomenon: The Growth of High Technology Industry in a University Town*, Cambridge: SQW.

Selya R M (1975) Trading under duress: The case of Taiwan, *Asian Profile* **3**: 441–6.

Selya R M (1982) Contrasting measures of changing industrial distribution in Taiwan, *Asian Geographer* **1**: 35–50.

Selya R M (1993) Economic restructuring and spatial changes in manufacturing in Taiwan, 1971–1986, *Geoforum* **24**: 115–26.

Selya R M (1994) *Taipei*, London: Wiley.

Shaw S L and Williams J F (1991) Role of transportation in Taiwan's regional development, *Transportation Quarterly* **45**: 271–96.

Shefter M (1993) *Capital of the American Century: The National and International Influence of New York City*, New York: Russell Sage Foundation.

Shields R (1989) Social spatialization and the built environment: West Edmonton Mall, *Environment and Planning D: Society and Space* **7**: 147–64.

Shimogawa K (1986) Product and labour strategies in Japan, in S Tolliday and J Zeitlin (eds) *The Automobile Industry and its Workers: Between Fordism and Flexibility*, Cambridge: Polity.

Shomina E S (1992) Enterprises and the urban environment in the USSR, *International Journal of Urban and Regional Research* **16**: 222–33.

Shutt J and Whittington R (1987) Fragmentation strategies and the rise of small units, *Regional Studies* **21**: 13–23.

Sibley D (1988) Survey 13: The purification of space, *Environment and Planning D: Society and Space* **6**: 409–21.

Silin R (1972) Marketing and credit in a Hong Kong wholesale market, in W Willmott and L Crissman (eds) *Economic Organization in Chinese Society*, Stanford: Stanford University Press, pp. 327–52.

Silver H (1989) The demand for homework: Evidence from the US Census, in E Boris and C R Daniels (eds) *Homework: Historical and Contemporary Perspectives on Paid Labor at Home*, Urbana: University of Illinois Press.

Simmel G (1903) The metropolis and mental life, in D Levine (ed) (1971) *On Individuality and Social Form*, Chicago: University of Chicago Press.

Sinclair J (1987) *Images Incorporated: Advertising as Industry and Ideology*, London: Croom Helm.

Sivanandan A (1989) New circuits of imperialism, *Race and Class* **30**, 4: 4.

Smith A (1998) *Reconstructing the Regional Economy: Industrial Transformation and Regional Development in Slovakia*, Cheltenham: Edward Elgar.

Smith A D (1992) *International Financial Markets: The Performance of Britain and Its Rivals*, Cambridge: Cambridge University Press.

Smith G (1996) *The Nationalities Question in the Post-Soviet States*, London: Longman.

Smith M P (1988) *City, State and Market*, New York: Blackwell.

Smith N (1990) *Uneven Development: Nature, Capital and the Production of Space*, second edition, first published 1984, Oxford: Blackwell.

Smith N and Williams P (eds) (1986) *Gentrification of the City*, Boston: Allen & Unwin.

Social Trends (1989) London: HMSO.

Soja E W (1989) *Postmodern Geographies: The Reassertion of Space in Critical Social Theory*, London: Verso.

Solinas G (1982) Labour market segmentation and the workers' careers: The case of the Italian knitwear industry, *Cambridge Journal of Economics*, No.6.

Sommer J W (1975) Fat city and hedonopolis: The American urban future, in R Abler *et al* (eds) *Human Geography in a Shrinking World*, North Scituate: Duxbury Press, pp. 132–48.

Sowton E (1989) Faster, better and more profitable, *The Banker*, August: 23–5.

Spitz B (ed) (1983) *Tax Havens Encyclopaedia*, Issue 15, London: Butterworths.

Sridharan E (1989) World trends and India's software products, *Dalaquest*, December: 56.

Stanbach T (1979) *Understanding the Service Economy*, Baltimore: Johns Hopkins University Press.

Stanbach T (1987) *Computerization and the Transformation of Employment*, Boulder: Westview Press.

Stanbach T and Noyelle T (1982) *Cities in Transition*, Totowa: Allenhead, Osman.

Standing G (1986) *Unemployment and Labour Market Flexibility: The United Kingdom*, Geneva: International Labour Organization.

Stark D (1992) The great transformation? Social change in eastern Europe, *Contemporary Sociology* **21**, 3: 299–304.

Stenning A (1996) Post-socialist geographies and the conceptualisation of transformation, in D Turnock (ed) *Frameworks for Understanding Post-Socialist Processes*, Leicester University Geography Department, Occasional Paper 36, pp. 79–82.

Stenning A (1997) Economic restructuring and local change in the Russian Federation, in M Bradshaw (1997a), pp. 145–62.

Stenning A (1998) The Changing Politics of Local Economic Development in the Russian Federation, unpublished PhD thesis, School of Geography, University of Birmingham.

Stewart S (1984) *On Longing: Narratives of the Miniature, the Gigantic, the Souvenir, the Collection*, Baltimore: Johns Hopkins University Press.

Stoker G (1990) Regulation theory, local government and the transition from Fordism, in D S King and J Pierre (eds) *Challenges to Local Government*, London: Sage, pp. 242–64.

Stokvis J R and Cloar J A (1991) CRM: Applying shopping center techniques to downtown retailing, *Urban Land*, April: 7–11.

Storper M (1985) Technology and spatial production relations: Disequilibrium, interindustry relationships, and industrial evolution, in M Castells (ed) *High Technology, Space and Society*, Beverly Hills: Sage.

Storper M (1986) Technology and new regional growth complexes: The economics of discontinuous spatial development, in P Nijkamp (ed) *Technological Change and Employment: Urban and Regional Dimensions*, Berlin: Springer.

Storper M (1989) The transition to flexible specialisation in the US film industry: The division of labour, external economies and the crossing of industrial divides, *Cambridge Journal of Economics* **13**, 2: 273–305.

Storper M (1992) The limits to globalization: Technology districts and international trade, *Economic Geography* **68**: 60–93.

Storper M (1993) Regional 'worlds' of production: Learning and innovation in technology districts of France, Italy and the USA, *Regional Studies* **27**: 433–56.

Storper M (1995) The resurgence of regional economies, ten years later: The region as a nexus of untraded interdependencies, *European Urban and Regional Studies* **2**, 3: 191–221.

Storper M and Harrison B (1991) Flexibility, hierarchy and regional development: The changing structure of industrial production systems and their forms of governance in the 1980s, *Research Policy* **20**: 407–22.

Storper M and Scott A J (1989) The geographical foundations and social regulations of flexible production complexes, in J Wolch and M Dear (eds) *The Power of Geography*, London: Unwin Hyman, pp. 21–40.

Storper M and Walker R (1989) *The Capitalist Imperative: Territory, Technology, and Industrial Growth*, Oxford: Blackwell.

Stowsky J (1987) *The Weakest Link: Semiconductor Production Equipment, Linkages, and the Limits to International Trade*, Working Paper 27, Berkeley Roundtable on the International Economy, University of California at Berkeley.

Strange S (1971) *Sterling and British Policy*, Oxford: Oxford University Press.

Strange S (1986) *Casino Capitalism*, Oxford: Blackwell.

Strange S (1988) *States and Markets*, London: Frances Pinter.

Suchman D R (1990) Housing and community development, in D Schwanke (ed) *Development Trends*, Washington: Urban Land Institute, pp. 28–37.

Sutherland D and Hanson P (1996) Structural change in the economies of Russia's regions, *Europe-Asia Studies* **48**, 3: 367–92.

Swyngedouw E (1986) *The Socio-Spatial Implications of Innovations in Industrial Organisations*, Working Paper 20, Johns Hopkins European Center for Regional Planning and Research.

Swyngedouw E A (1989) The heart of the place: The resurrection of locality in an age of hyperspace, *Geografiska Annaler* **71**: 31–42.

Swyngedouw E A (1992) The Mammon quest: 'Glocalisation', interspatial competition and the monetary order: The construction of new scales, in M Dunford and G Kafkalas (eds) *Cities and Regions in the New Europe*, London: Belhaven, pp. 39–67.

Teaford I C (1990) *The Rough Road to Renaissance: Urban Revitalization in America, 1940–1985*, Baltimore: Johns Hopkins University Press.

Thomas R J (1985) *Citizenship, Gender and Work: Social Organization of Industrial Agriculture*, Los Angeles and Berkeley: University of California Press.

Thompson E P (1963) Time, work discipline, and industrial capitalism, *Past and Present*, No. 38: 56–97.

Thompson W (1962) Locational differences in inventive effort and their determinants, in R Nelson (ed) *The Rate and Direction of Inventive Activity*, Princeton: Princeton University Press, pp. 253–71.

Thrift N (1986) The geography of world economic disorder, in R J Johnston and P J Taylor (eds) *A World in Crisis? Geographical Perspectives*, Oxford: Blackwell, pp. 12–76.

Thrift N (1987) The fixers: The urban geography of international commercial capital, in J Henderson and M Castells (eds) *Global Restructuring and Territorial Development*, London: Sage, pp. 203–33.

Thrift N (1989a) Images of social change, in C Hamnett, L McDowell and P Sarre (eds) *The Changing Social Structure*, London: Sage, pp. 12–42.

Thrift N (1989b) New times and new spaces? The Perils of transition models, *Environment and Planning D: Society and Space* **7**: 127–9.

Thrift N (1990) The perils of the international finance system, *Environment and Planning A* **22**: 1135–40.

Thrift N (1994) On the social and cultural determinants of international financial centres: The case of the City of London, in S Corbridge, R Martin and N Thrift (eds) *Money, Space and Power*, Oxford: Blackwell.

Thrift N (1996) 'Not a Straight Line but a Curve', or, Cities are not Mirrors of Modernity, mimeo, Department of Geography, University of Bristol.

Thrift N and Leyshon A (1992) In the wake of money: The City of London and the accumulation of value, in L Budd and S Whimster (eds) *Global Finance and Urban Living: A Study of Metropolitan Change*, London: Routledge, pp. 282–311.

Tickell A and Peck J A (1992) Accumulation, regulation and the geographies of post-Fordism: Missing links in regulationist research, *Progress in Human Geography* **16**: 190–218.

Toffler A (1970) *Future Shock*, London: Bodley Head.

Townsend A R (1992) The attraction of urban areas, *Tourism Recreation Research* **17**, 2: 24–32.

Treyvish A, Pandit K and Bond A (1993a) Macrostructural employment shifts and urbanisation in the former USSR: An international perspective, *Post-Soviet Geography* **34**, 3: 157–71.

Treyvish A, Pandit K and Bond A (1993b) Post-industrial transformation in the European regions of the former USSR, 1959–1985, *Post-Soviet Geography* **34**, 10: 613–30.

Trigilia C (1990) Work and politics in the Third Italy's industrial districts, in F Pyke, G Becattini and W Sengenberger (eds) *Industrial Districts and Inter-Firm Co-operation in Italy*, Geneva: International Institute for Labour Studies, pp. 160–84.

Tuan V F (1988) The city as a moral universe, *Geographical Review* **78**, 3: 316–24.

Turner V (1982) *From Ritual to Theater*, New York: Performing Arts Publications.

UNCTAD (1994) *World Investment Report*, Geneva: United Nations Centre for Trade and Development.

United Nations (1991) *The Triad in Foreign Direct Investment*, New York: United Nations Centre on Transnational Corporations.

United States Department of Labor (1989) *The Effects of Immigration on the US Economy and Labor Market*, Washington: GPO.

Urry J (1987) Some social and spatial aspects of services, *Environment and Planning D, Society and Space* **5**: 5–26.

Urry J (1990a) Work, production and social relations, *Work, Employment and Society* **4**, 2: 271–80.

Urry J (1990b) *The Tourist Gaze: Leisure and Travel in Contemporary Societies*, London: Sage.

Veblen T (1953) *The Theory of the Leisure Class*, New York: Mentor Books.

Vernon R (1966) International investment and international trade in the product cycle, *Quarterly Journal of Economics* **80**: 190–207.

Versluysen E (1981) *The Political Economy of International Finance*, New York: St. Martin's Press.

Vogel E (1986) Pax Nipponica? *Foreign Affairs* **64**: 752–67.

Vogel E (1992) *The Four Little Dragons: The Spread of Industrialization in East Asia*, Cambridge: Harvard University Press.

von Hippel E (1987) Cooperation between rivals: Informal know-how trading, *Research Policy* **16**: 291–302.

Wachtel H M (1986) *The Money Mandarins: The Making of a Supranational Economic Order*, New York: Pantheon Books.

Wachtel H M (1987) Currency without a country: The global funny money game, *The Nation*, 26 December, **245**: 784–90.

Wacker T J (1989) A macroview of private banking: Marketing challenges and choices, in D B Zenoff (ed) *Marketing Financial Services*, Cambridge: Ballinger, pp. 71–81.

Wade R (1990) *Governing the Market: Economic Theory and the Role of Government in East Asian Industrialization*, Princeton: Princeton University Press.

Wade R (1992) East Asia's economic success: Conflicting perspectives, partial insights, shaky evidence, *World Politics* **44**: 270–320.

Wade R (1993) Managing trade: Taiwan and South Korea as challenges to economics and political science, *Comparative Politics* **25**: 147–67.

Walby S (1986) *Patriarchy at Work*, Cambridge: Polity.

Walby S (1989) Flexibility and the sexual division of labour, in S Wood (ed) *The Transformation of Work?* London: Unwin Hyman, pp. 127–40.

Walker J (1988) Women, the state and the family in Britain: Thatcher economics and the experience of women, in J Rubery (ed) *Women and Recession*, London: Routledge, pp. 218–50.

Wallace I (1990) *The Global Economic System*, London: Unwin-Hyman.

Wallerstein I (1991) *Geopolitics and Geoculture*, Cambridge: Cambridge University Press.

Walsh T (1989) Part-time employment and labour market policies, *National Westminster Bank*, May: 43–55.

Walter A (1991) *World Power and World Money: The Role of Hegemony and International Monetary Order*, Brighton: Wheatsheaf.

Walter I (1988) *Global Competition in Financial Services: Market Structure, Protection, and Trade Liberalization*, Cambridge: Ballinger.

Walter I (1989) *Secret Money*, London: Unwin Hyman.

Warde A (1986) Industrial restructuring, local politics and the reproduction of labour power: Some theoretical considerations, *Environment and Planning D, Society and Space* **6**: 75–96.

Warf B (1989) Telecommunications and the globalization of financial services, *Professional Geographer* **41**: 257–71.

Warf B (1991) The internationalization of New York services, in P Daniels (ed) *Services and Metropolitan Development: International Perspectives*, London: Routledge, pp. 245–64.

Waterson M (1992) International advertising expenditure statistics, *International Journal of Advertising* **11**: 14–67.

Watts M J (1989) The agrarian question in Africa: Debating the crisis, *Progress in Human Geography* **13**, 1: 1–41.

Watts M J (1992) Living under contract: Work, production, politics and the manufacture of discontent in a peasant society, in A Pred and M Watts (eds) *Reworking Modernity: Capitalism and Symbolic Discontent*, New Brunswick: Rutgers University Press, pp. 65–105.

Webster J (1986) Word processing and the secretarial division of labour, in K Purcell *et al* (eds) *The Changing Experience of Employment*, London: Macmillan.

Weinstein A (1977) Foreign investments by service firms: The case of multinational service agencies, *Journal of International Business Studies*, July/September.

White G (ed) (1988) *Developmental States in East Asia*, New York: St Martin's Press.

Williams R (1973) *The Country and the City*, London: Chatto & Windus.

Womack J, Jones D and Roos D (1990) *The Machine that Changed the World*, New York: Rawson Associates.

Wong G (1991) Business groups in a dynamic environment: Hong Kong 1976–1986, in G Hamilton (ed) *Business Networks and Economic Development in East and Southeast Asia*, Hong Kong: Centre of Asian Studies, pp. 126–54.

Wong M (1983) Chinese sweatshops in the US: A look at the garment industry, in I H Simpson and R L Simpson (eds) *Research in the Sociology of Work*, Volume 11, Greenwich: JAI Press.

Wong S (1991) Business groups in a dynamic environment: Hong Kong 1976–1986, in G Hamilton (ed) *Business Networks and Economic Development in East and Southeast Asia*, Hong Kong: Center of Asian Studies, pp. 126–54.

Woo J E (1991) *Race to the Swift*, New York: Columbia University Press.

Wood J S (1988) Suburbanization of center city, *Geographical Review* **78**: 325–29.

Wood P A (1991a) Conceptualising the role of services in economic change, *Area* **23**: 66–72.

Wood P A (1991b) Flexible accumulation and the rise of business services, *Transactions of the Institute of British Geographers* **16**: 160–72.

Wood P A, Bryson J R and Keeble D (1991) *Regional Patterns of Small Firm Development in the Business Services: Preliminary Evidence from the UK*, Working Paper 14, Small Business Research Centre, Cambridge University.

World Bank (1982, 1983, 1985, 1987, and 1992) *World Development Report* (annual), New York: Oxford University Press/World Bank.

Wright P (1985) *On Living in an Old Country*, London: Verso.

Yannopoulos G N (1983) The growth of transnational banking, in M Casson (ed) *The Growth of International Business*, London: Allen & Unwin, pp. 236–57.

Zukin S (1982) *Loft Living: Culture and Capital in Urban Change*, Baltimore: Johns Hopkins University Press.

Zukin S (1990) Socio-spatial prototypes of a new organization of consumption: The role of real cultural capital, *Sociology* **24**, 1: 37–56.

Zukin S (1991) *Landscapes of Power: From Detroit to Disney World*, Berkeley: University of California Press.

Zysman J (1996) The myth of a 'global economy': Enduring national foundations and emerging regional realities, *New Political Economy* **1**, 2: 157–84.

INDEX

Index compiled by Geoffrey C. Jones